A History of Probability and Statistics and Their Applications before 1750

A History of Probability and Statistics and Their Applications before 1750

ANDERS HALD

Formerly Professor of Statistics
University of Copenhagen
Copenhagen, Denmark

WILEY

A Wiley-Interscience Publication

JOHN WILEY & SONS

New York • Chichester • Brisbane • Toronto • Singapore

Copyright © 1990 by John Wiley & Sons, Inc.

Library of Congress Cataloging in Publication Data:

Hald, Anders, 1913–
    A history of probability and statistics and their applications
    before 1750/Anders Hald.
        p. cm.—(Wiley series in probability and mathematical
    statistics. Probability and mathematical statistics, ISSN
    0271–6232)
        Bibliography: p.
        Includes index.
        ISBN 0-471-50230-8
        1. Probabilities—History—17th century. 2. Probabilities—
    History—18th century. 3. Mathematical statistics—History—
    17th century. 4. Mathematical statistics—History—18th
    century.
    I. Title. II. Series.   89–33269
    QA273.A4H35   1989   CIP
    519.2'09—dc20

Printed in the United States of America

10 9 8 7 6 5 4 3 2

Preface

Until recently a book on the history of statistics in the 19th century was badly needed. When I retired six years ago, I decided to write such a book, feeling that I had a good background in my statistical education in the 1930s, when the curriculum in statistics was influenced mainly by the writings of Laplace, Gauss, and Karl Pearson. Studying the original works of these authors I found no difficulty in understanding Gauss and Pearson, but I soon encountered difficulties with Laplace. The reason is of course that Gauss and Pearson are truly 19th century figures, whereas Laplace has his roots in the 18th century.

I then turned to the classical authors and worked my way back to Cardano through de Moivre, Montmort, Nicholas and James Bernoulli, Huygens, Fermat, and Pascal. Comparing my notes with Todhunter's *History*, I found to my surprise that his exposition of the topics in probability theory that I found most important was incomplete, and I therefore decided to write my own account.

The present book, covering the period before 1750, is an introduction to the one I had in mind. It describes the contemporaneous development and interaction of three topics: probability theory and games of chance; statistics in astronomy and demography; and life insurance mathematics.

Besides the story of the life and works of the great natural philosophers who contributed to the development of probability theory and statistics, I have told the story of important problems and methods, in this way exhibiting the gradual advance of solving these problems. I hope to have achieved a better balance than had been achieved before in evaluating the contributions of the various authors; in particular, I have stressed the importance of the works of John Graunt, Montmort, and Nicholas Bernoulli.

The contents of the book depend heavily on research carried out by many authors during the past 40 years. I have drawn freely on these sources and acknowledged my debt in the references. The manuscript was written during the years 1985–1987, so works published in 1986 and 1987 are not fully integrated in the text. Some important books and papers from 1988 are briefly mentioned.

With hesitation, I have also included some background material on the history of mathematics and the natural and social sciences because I have always felt that my students needed such knowledge. I realize of course that my qualifications for doing so are rather poor since I am no historian of science. These sections and also the biographies are based on secondary sources.

The plan of the book is described in Section 1.2.

I am grateful to Richard Gill for advice on my English in Chapters 2 and 3, to Steffen L. Lauritzen for translating some Russian papers, and to Olaf Schmidt for a discussion of Chapter 10. In particular, I want to thank Søren Johansen for discussions on the problem of the duration of play.

I am grateful to two anonymous reviewers from the publisher for valuable comments on the manuscript and for advice resulting in considerable reduction of the background material. I thank the copy editor for improving my English and transforming it into American.

I thank the Institute of Mathematical Statistics, University of Copenhagen, for placing working facilities at my disposal.

I thank the Almqvist & Wiksell Periodical Company for permission to use material in my paper published in *Scandinavian Actuarial Journal*, 1987; the International Statistical Institute for permission to use material from three papers of mine published in *International Statistical Review*, 1983, 1984, and 1986; and Springer-Verlag for permission to use material from my paper published in *Archive for History of Exact Sciences*, 1988.

I am grateful to the Department of Statistics, Harvard University, for permission to quote from Bing Sung's *Translations from James Bernoulli*, Technical Report No. 2, 1966, and to Thomas Drucker for permission to quote from his (unpublished) translation of Nicholas Bernoulli's *De Usu Artis Conjectandi in Jure*.

My first book on statistics, written fifty years ago, was dedicated to G. K., so is this one.

ANDERS HALD

September 1988

Contents

A History of Probability and
Statistics and Their
Applications before 1750

CHAPTER 1

The Book and Its Relation to Other Works

1.1 PRINCIPLES OF EXPOSITION

This book contains an exposition of the history of probability theory and statistics and their applications before 1750 together with some background material. A history should of course give an account of the time and place of important events and their interpretations. However, opinions differ greatly on where to put the main emphasis of interpretation.

We have attempted to cover three aspects of the history: problems, methods, and persons. We describe probabilistic and statistical problems and their social and scientific background; we discuss the mathematical methods of solution and the statistical methods of analysis; and we include the background and general scientific contributions of the persons involved, not only their contributions to probability and statistics.

Since history consists of facts and their interpretation, history continually changes because new facts are found in letters, archives, and books, and new interpretations are offered in the light of deeper understanding based, in this case, on the latest developments in probability theory, statistics, and the history of science.

In the 17th and 18th centuries many problems were formulated as challenge problems, and answers were given without proofs. Some books on probability were written for the educated public and therefore contained statements without proofs. In such cases we have tried to follow the author's hints and construct a proof which we believe represents the author's intentions.

The material has been ordered more according to problems and methods than according to persons in an attempt to treat the achievements of the various authors as contributions to a general framework.

A leading principle of the exposition of probability theory and life insurance mathematics has been to rewrite the classics in uniform modern terminology and notation. It is clear that this principle may be criticized for distorting the facts. Many authors prefer to recount the old proofs with the original notation to convey the flavor of the past to the reader. There are two essential steps in modernization that we have made here. The first is to use a single letter, p, say, to denote a probability instead of the ratio of the number of favorable cases to the total number of cases, $a/(a + b)$, say, where a and b are positive integers. This change of notation conceals the fact that nearly all the probabilities discussed were constrained to rational fractions. The advantage of this notation was noted by de Moivre (1738, p. 29) who writes, "Before I make an end of this Introduction, it will not be improper to shew how some operations may often be contracted by barely introducing one single Letter, instead of two or three, to denote the Probability of the happening of one Event" and, further (on p. 30), that "innumerable cases of the same nature, belonging to any number of Events, may be solved without any manner of trouble to the imagination, by the mere force of a proper Notation." However, de Moivre did not rewrite the *Doctrine of Chances* with the new notation; he used it only in his *Annuities upon Lives* (1725 and later editions). We have followed the advice of de Moivre and rewritten the proofs in the new notation, feeling confident that the reader will keep in mind that most probabilities were defined as proper rational fractions, a fact which is nearly always obvious from the context.

The second great simplification of the proofs is obtained by the introduction of subscripts. In analyzing some complicated games of chance, for example, Waldegrave's problem, Nicholas Bernoulli and de Moivre had to use the whole alphabet divided into several sections to denote probabilities and expectations of the players corresponding to various states of the game. De Moivre achieved some simplification by using superscripts in a few cases. In many problems they gave the solution for two, three, and four players only and concluded that "the continuation of this rule is manifest," in this way avoiding a general proof which would have been rather unintelligible. Using modern notation with subscripts, it is easy to rewrite such proofs in much shorter form without invalidating the idea of the proof; in fact, we believe that our readers will get a clear idea of the proof because they are accustomed to this symbolism, just as readers in the past understood the original form of the proof because they were educated in that notational tradition.

Comparison of proofs and results in a uniform notation makes evaluating the contributions of various authors easier and minimizes the danger of attributing too much to an individual author. Furthermore, the importance of the results to the following period and to today becomes evident.

The same principle of exposition cannot be used for statistics, because statistics before 1750 was nonmathematical. We shall therefore illustrate the development of statistical methods by typical examples, giving both the original data and their analysis at the time and adding some comments from a modern point of view.

The book is written in textbook style, since our main purpose is to give an account of the most important results in the classical literature. Like most histories of mathematics and science, our exposition concentrates on results which have proved to be of lasting importance.

The persons who laid the foundation of probability theory and statistics were natural philosophers having a broader background and outlook than scientists today. The word "scientist" was coined about the middle of the 19th century, reflecting an ongoing specialization and professionalization. Nevertheless, we shall often use the words "mathematician" and "scientist" to stress certain characteristics of the persons involved.

To convey the flavor of classical works, we shall present quotations of programs from the prefaces of books, the formulation of important problems, and some heated disputes of priority.

We shall point out priorities, but the reader should be aware of the uncertainty involved by taking note of Stigler's Law of Eponymy, (Stigler, 1980), which in its simplest form states that, "No scientific discovery is named after its original inventor."

The driving force behind the development of probability theory and statistics was pressure from society to obtain solutions to important problems for practical use, as well as competition among mathematicians. When a problem is first formulated and its solution indicated, perhaps only by a numerical example, the problem begins a life of its own within the mathematical community; this leads to improved proofs and generalizations of the problem, and we shall see many examples of this phenomenon.

Finally, it should be noted that any history is necessarily subjective, since the weight and interpretation of the events selected depend on the author's interests.

For the serious student of the history of probability theory and statistics, we can only recommend that he or she follow the advice given by de Moivre (1738, p. 235), discussing the works of James and Nicholas Bernoulli on the binomial distribution: "Now the Method which they have followed has been briefly described in my *Miscellanea Analytica*, which the Reader may consult if he pleases, unless they rather chuse, which perhaps would be the best, to consult what they themselves have writ upon that Subject."

1.2 PLAN OF THE BOOK

A fuller title of the book would be *A history of probability theory and statistics and their applications to games of chance, astronomy, demography, and life insurance before 1750, with some comments on later developments.* The topics treated may be grouped into five categories:

Background in mathematics, natural philosophy, and social conditions
Biographies
Probability theory and games of chance
Statistics in astronomy and demography
Life insurance mathematics

Probability theory before 1750 was inspired mainly by games of chance. Dicing, card games, and lotteries, public and private, were important social and economic activities then as today. It is no wonder that intellectual curiosity and economic interests led to mathematical investigations of these activities at a time when the mathematization of science was going on. We shall distinguish three periods.

The period of the foundation of probability theory from 1654 to 1665 begins with the correspondence of Pascal and Fermat on the problem of points, continues with Huygens' treatise on *Reckoning at Games of Chance,* and ends with Pascal's treatise on the *Arithmetical Triangle* and its applications. The correspondence was not published until much later. In his treatise, Pascal solves the problem of points by recursion and finds a division rule, depending on the tail probability of the symmetric binomial. In their correspondence, he and Fermat had solved the same problem also by combinatorial methods. Huygens uses recursion to solve the problem numerically. He also considers an example with a possibly infinite number of games, which he solves by means of two linear equations between the conditional expectations of the two players. All three of them solved the problem of the Gambler's Ruin without publishing their method of solution.

After a period of stagnation of nearly 50 years, there followed a decade with astounding activity and progress from 1708 to 1718 in which the elementary and fragmentary results of Pascal, Fermat, and Huygens were developed into a coherent theory of probability. The period begins with Montmort's *Essay d'Analyse sur les Jeux de Hazard,* continues with de Moivre's *De Mensura Sortis,* Nicholas Bernoulli's letters to Montmort, James Bernoulli's *Ars Conjectandi,* Nicolaas Struyck's *Reckoning of Chances in Games,* and ends with de Moivre's *Doctrine of Chances.* Hence, by 1718 four comprehensive textbooks were available. We shall mention the most important results obtained. They discussed elementary rules of probability calculus, conditional probabilities and expectations, combinatorics, algorithms and recursion formulae, the method of inclusion and exclusion, and examples of using infinite series and limiting processes. They derived the binomial and negative binomial distributions, the hypergeometric distribution, the multivariate version of these distributions, the occupancy distribution, the distribution of the sum of any number of uniformly distributed variables, the Poisson approximation to the binomial, the law of large numbers for the binomial, and an approximation to the tail of the binomial. They solved the problem of points for a game of bowls and for the game of tennis, Waldegrave's problem, the problem of coincidences, and the problem of duration of play, and found the minimax solution for the strategic game Her.

The third period, from 1718 to 1738, was a period of consolidation and steady progress in which de Moivre derived the normal approximation to the binomial distribution, developed a theory of recurring series, improved his solution of the problem of the duration of play, and wrote the second edition of the *Doctrine of Chances*, which became the most important textbook before the publication of Laplace's *Théorie Analytique des Probabilités* in 1812.

We shall discuss these books in detail. We have, however, singled out the most important problems for separate treatment to show how they were solved by joint effort, often in competition among several authors.

Many problems were taken up by the following generation of mathematicians and given solutions that have survived until today. We shall comment on these later developments, usually ending with Laplace's solutions.

The successful development of probability theory did not immediately lead to a theory of statistics. A history of statistical methods before 1750 must therefore build on typical examples of data analysis; we have concentrated here on examples from astronomy and demography.

Astronomers had been aware of the importance of both systematic and random errors since antiquity and tried to minimize the influence of such errors in their planning of observations and data analysis. We shall discuss some data by Tycho Brahe from the end of the 16th century as an example. The mathematization of science in the beginning of the 17th century naturally led many scientists to determine not only the mathematical form of natural laws but also the values of the parameters by fitting equations to data. They inserted the best sets of observations in the equations, as many as the number of parameters, solved for the parameters, calculated the expected values, and studied the deviations between observed and calculated values. Prominent examples are Kepler's three laws on planetary motion derived from his physical theories and data collected by Copernicus and Tycho Brahe. Kepler's data were used by Newton to check his axiomatic theory. Galileo used several sets of observations on the new star of 1572 to compare two hypotheses on the position of the star. We shall also see how Newton used an interpolation polynomial to find the tangent to the orbit of a comet.

A paragon for descriptive statistical analysis of demographic data was provided by Graunt's *Natural and Political Observations made upon the Bills of Mortality* in 1662. Graunt's critical appraisal of the rather unreliable data, his study of mortality by cause of death, his estimation of the same quantity by several different methods, his demonstration of the stability of statistical ratios, and his life table set up new standards for statistical reasoning. Graunt's work led to three different types of investigations: political arithmetic; testing the stability of statistical ratios; and calculation of expectations of life and survivorship probabilities.

Petty also employed Graunt's method of analysis, although less critical, to economic data and coined the term "political arithmetic" for analyses of data of political importance. Similar methods were used by natural philosophers and theologians to analyze masses of data on human and animal populations. The many regular patterns observed were taken as proof of the existence of a supreme being and His "original design." We shall remark only slightly on this line of thought.

It is surprising that probabilists at the time recognized the importance of Graunt's work and without hesitation used their theory on games of chance to describe demographic phenomena. They wrote about the chance of a male birth and the chance of dying at a certain age.

Graunt gave a detailed description and analysis of the yearly variation of the sex ratio at birth in London and Romsey and suggested that similar investigations should be carried out in other places. Arbuthnott used some of Graunt's data extended to his own time to give a statistical proof, based on the symmetric binomial, for the existence of divine providence, a proof that was further strengthened by 'sGravesande. Nicholas Bernoulli compared the observed distribution of the yearly number of male births with a skew binomial distribution, the parameter being estimated from the data, and discussed the probability of the observed number of outliers. His investigation is the first attempt to fit a binomial to data and to test the goodness of fit. Some years later, Daniel Bernoulli used the normal approximation to the binomial in his analysis of deviations between observed and expected values of the number of male births to decide between two hypothetical values of the sex ratio.

Huygens used Graunt's life table to calculate the median and the average remaining lifetime for a person of any given age. He also showed how to calculate survivorship probabilities and joint-life expectations. His results were, however, not published, but similar results were published without proof by James Bernoulli and later proved by Nicholas Bernoulli.

The usefulness of probability theory was convincingly demonstrated by application to problems of life insurance. In the 16th and 17th centuries, states and cities sold life annuities to their citizens to raise money for public purposes. The yearly benefit of an annuity was fixed as a percentage of the capital invested, often as twice the prevailing rate of interest and independent of the nominee's age. In a report from 1671, de Witt showed how to calculate the value of an annuity by means of a piecewise linear life table combined with the age of the nominee and the rate of interest. De Witt's life table was hypothetical, although he referred to some investigations of the mortality of annuitants. In 1693 Halley constructed a life table from observations of the yearly number of deaths in Breslau, calculated the first table of values of annuities as a function of the nominee's age, and explained how to calculate joint-life annuities.

After these ingenious beginnings one would have expected rapid development of both mathematical and practical results in view of the fact that many economic contracts in everyday life depended on life contingencies, but nothing happened for about 30 years. The breakthrough came in 1725 with de Moivre's *Annuities upon Lives*, greatly simplifying both the mathematics and the calculations involved; however, as shown by Simpson, de Moivre went too far in his simplifications. Simpson therefore constructed his own life table for the population of London, and by recursion he calculated tables of values of single- and joint-life annuities for various rates of interest. In the strong competition between de Moivre and Simpson, a comprehensive theory of life annuities was created, and the necessary tables for practical applications provided.

In some chapters in this book we have supplemented the text with problems for the reader, mostly taken from the classical literature.

Although we have not included every classical paper, or every paper commenting on the classical literature, we believe that we have covered the most important ones. However, for various reasons two important results before 1750 have been omitted. The first is Cotes's rule (1722) for estimating a true value by a weighted mean, when observations are of unequal accuracy (see Stigler, 1986, p. 16); the second is Daniel Bernoulli's results (1738) on the theory of moral expectation, the utility of money, and the Petersburg problem (see Todhunter, 1865, Jorland, 1987, and Dutka, 1988).

We shall discuss our reasons for stopping our history at 1750.

By 1750 probability theory had been recognized as a mathematical discipline with a firm foundation and its own problems and methods as described by de Moivre in the *Doctrine of Chances*. A new development began with the introduction of inverse probability by Bayes (1764) and Laplace (1774b).

By 1750 statistics had still not become a mathematical discipline; a mathematical theory of errors and of estimation emerged in the 1750s, as described by Stigler (1986).

Also about 1750, the first phase of the development of a theory of life insurance had been completed. In the 1760s life insurance offices arose so that new and more accurate mortality observations became available. A theory of life assurances was developed, and new ways of calculating and tabulating the fundamental functions were invented.

The reader should note that formulae are numbered with a single number *within sections*. When referring to a formula in *another chapter* the decimal notation is used, (20.5.25) say, denoting formula (25) in §5 of Chap. 20. *Within a chapter* the chapter number is omitted so that only section and formula numbers are given.

1.3   A COMPARISON WITH TODHUNTER'S BOOK

The unquestioned authority on the early history of probability theory is Isaac Todhunter (1820–1884) whose masterpiece, *A History of the Mathematical Theory of Probability from the Time of Pascal to that of Laplace*, was published in 1865. Kendall (1963) has written a short biography of Todhunter in which he gives a precise characterization of his work: "The *History of the Mathematical Theory of Probability* is distinguished by three things. It is a work of scrupulous scholarship; Todhunter himself contributed nothing to the theory of probability except this account of it; and it is just about as dull as any book on probability could be."

We consider Todhunter's *History* an invaluable handbook giving a chronological review of the classical literature grouped according to authors. For the period before 1750, however, we shall argue that Todhunter's account of important topics is incomplete, that he has overlooked the significance of important contributions, and that the trend in the historical development is lost by his organization of the material.

In the many references to Todhunter's *History* in the following we shall omit the year of publication (1865) and give page references only.

As a mathematician Todhunter concentrates on the mathematical theory of probability and disregards the general background, the lives of the persons involved, and the application of their theories to statistics and life insurance.

Todhunter's book is ordered chronologically according to authors; each important author is allotted a separate chapter in which his works are reviewed page by page and commented upon. This method makes it easy for the reader to locate the contributions of each author but difficult to follow the advances made by various authors to the solution of a given problem. We have avoided this dilemma by reviewing the works of each author and referring the detailed treatment of the most important topics to separate chapters that show the historical development for each topic.

More important, however, are the different weights given to many topics by Todhunter and by us. It is not surprising that the significance of a theorem or method differs when viewed from the perspectives of 1865 and today. Todhunter meticulously reports proofs of many results which are without interest today; conversely, he omits proofs of results of great importance. We shall give some examples.

Today one of the most important and interesting topics is the development from James Bernoulli's law of large numbers for the binomial distribution through Nicholas Bernoulli's improved version of James's theorem and his approximation to the binomial tail probability to de Moivre's normal approximation. These three results are treated by Todhunter in less than two pages (pp. 72, 131, 192). He states Bernoulli's theorem without giving his proof; he has overlooked the significance of Nicholas' contribution and gives neither theorem nor proof; he states de Moivre's result for $p = 1/2$ only and indicates the proof by the remark, "Thus by the aid of Stirling's Theorem the value of Bernoulli's Theorem is largely increased." Todhunter has completely overlooked de Moivre's long struggle with this problem, the importance of de Moivre's proof as a model for Laplace's proof, and de Moivre's statement of the theorem for any value of p. Instead of giving the historical development of the method of proof, he gives Laplace's proof (pp. 548–552) because, as he says, previous demonstrations are now superseded by that. This is of course a very peculiar argument for a historian.

It is a common misunderstanding, perhaps due to Todhunter's incomplete account, that de Moivre gave the normal approximation only for the symmetric binomial.

The deficiency of Todhunter's method is most conspicuous in his analysis of the correspondence between Montmort and Nicholas Bernoulli, published in the second edition of Montmort's *Essay* (1713). These closely intertwined letters contain formulations of new problems, usually as a challenge to the recipient; theorems without proofs, sometimes with hints for solution; replies to previous questions; a running commentary on progress with the solution of various problems; and remarks on the contributions of other authors. A single letter often treats five to ten different topics. It is of course impossible to get the gist of these letters in a page-by-page review; rather, it is necessary to give an overview of the contents grouped by subject matter. Todhunter therefore does not realize the importance of Nicholas Bernoulli's work; perhaps he was also under the influence of de Moivre who in the later editions of the *Doctrine* tried to conceal the importance of Bernoulli's results to his own work.

The most difficult topic in probability theory before 1750 was the problem of the duration of play. It was formulated by Montmort in 1708; the first explicit solution was given by Nicholas Bernoulli in 1713. Two solutions were given by de Moivre in 1718, and these were worked out in more detail in 1730 and 1738. Todhunter gives up analyzing this important development, instead he uses Laplace's solution from 1812 to prove de Moivre's theorems. Furthermore, he does not comment on Laplace's solution from 1776 by solving a partial difference equation because this method "since [has] been superseded by that of Generating Functions" (Todhunter, p. 475).

The same procedure is used by Todhunter in his discussion of Waldegrave's problem, the probability of winning a circular tournament, which was solved incompletely by Montmort and de Moivre. A general solution was given by Nicholas Bernoulli, but Todhunter gives only Laplace's proof without noting that Bernoulli's is just as simple.

Todhunter's discussion of the strategic game Her is rather incomplete. He has overlooked the fact that Montmort gives the general form of the player's expectation under randomized strategies and that Waldegrave solves the problem numerically arriving at what today is called the minimax solution. Misled by Todhunter's account, Fisher (1934) solved the "old enigma of card play" by randomization and reached the same solution as Waldegrave did 221 years before.

Todhunter gives unsatisfactory accounts of James Bernoulli's and Montmort's probabilistic discussion of the game of tennis, of the problem of points in a game of bowls, of Montmort's discussion of the occupancy problem, of Simpson's solution of the theory of runs, and of several other problems mentioned in the following chapters.

Kendall's characterization that "it is just as dull as any book on probability could be" applies equally well to several sections of the present book. Detailed proofs of elementary theorems illustrating the historical development are necessarily dull for us, even if they were exciting for them. Pascal, Fermat, Huygens, Hudde, James Bernoulli, Montmort, Nicholas Bernoulli, de Moivre, and Struyck were all intensely interested in solving the problem of the Gambler's Ruin, which today is considered elementary. For statisticians who find examples of games of chance rather dull, it must be a consolation to know that dicing and card playing have their equivalents in sampling from infinite and finite populations, respectively.

1.4 WORKS OF REFERENCE

Gouraud's *Histoire* (1848, 148pp.) gives a nonmathematical and rather uncritical exposition of probability theory and insurance mathematics beginning with Pascal and Fermat and ending with Poisson and Quetelet. It contains many references and was therefore useful for Todhunter when he wrote his *History* (1865, 624pp.).

The first two chapters of Czuber's *Entwicklung der Wahrscheinlichkeitstheorie* (1899, 279pp.) covers nearly the same period as the present book but in less detail. Czuber indicates some methods of proof without giving complete proofs.

The books by Edwards (1987, 174pp.), *Pascal's Arithmetical Triangle*, and Hacking (1975, 209pp.), *The Emergence of Probability*, may be read as an introduction to the present one; they give a more detailed treatment of certain aspects of the history up to the time of Newton and Leibniz.

David (1962, 275pp.) gives a popular history of probability and statistics from antiquity through the time of de Moivre, stressing basic ideas and providing background material for the lives of the great probabilists.

Jordan's book (1972, 619pp.) contains a mathematical account of classical probability theory organized according to topics, with some references to the historical development.

The first 81 pages of Maistrov's book (1974, 281pp.) gives a sketch of the history of probability theory before 1750.

Daston's *Classical Probability in the Enlightenment* (1988, 423pp.) gives a comprehensive, nonmathematical study of the basic ideas in classical probability theory in their relation to games of chance, insurance, jurisprudence, economics, associationist psychology, religion, induction, and the moral sciences, with references to a wealth of background material. Daston's discussion of the history of probabilistic ideas is an excellent complement to our discussion of mathematical techniques and results.

Turning to books on the history of statistics, we mention first Karl Pearson's *The History of Statistics in the 17th and 18th Centuries*, Lectures given at University College, London, 1921–1933, edited by E. S. Pearson (1978, 744pp.). This is a fascinating book with an unusual freshness that conveys Pearson's enthusiasm and last-minute endeavors in preparing his lectures. It describes "the changing background of intellectual, scientific and religious thought," and gives lively biographies with digressions into the fields of mathematics and history of science. Pearson does not discuss statistics in the natural sciences but is mainly concerned with political arithmetic, demography, and the use of statistics for theological purposes. Pearson does not conceal his strong opinions on the subjects treated and the persons involved, which occasionally lead to biased evaluations.

Stigler's *The History of Statistics* (1986, 410pp.) is the first comprehensive history of statistics from 1750 to 1900; it also contains a discussion of Bernoulli's law of large numbers and de Moivre's normal approximation to the binomial.

Westergaard's *Contributions to the History of Statistics* (1932, 280pp.) gives the history of political arithmetic, population statistics, economic statistics, and official statistics before 1900, as well as a short survey of statistical theory. It is a nonmathematical, well-balanced, and scholarly work, with valuable references to the vast literature on descriptive and official statistics.

John's *Geschichte der Statistik* (1884, 376pp.) contains a description of the development of German political science, at that time called statistics, and of political arithmetic and population statistics before 1835.

Meitzen's *Geschichte, Theorie und Technik der Statistik* (1886, 240pp.) discusses the history of official statistics with the main emphasis on its development in Germany.

Following the pioneering work by M. G. Kendall and F. N. David in the 1950s and 1960s, there has been growing interest in the history of probability and statistics, and a great number of papers have been published; the most important, relating to the period before 1750, are listed in the References at the end of this book. Several important papers have been reprinted in *Studies in the History of Statistics and Probability*, Vol. 1 edited by E. S. Pearson and M. G. Kendall (1970) and Vol. 2 edited by M. G. Kendall and R. L. Plackett (1977). A *Bibliography of Statistical Literature Pre-1940* has been compiled by Kendall and Doig (1968).

A comprehensive account of the development of life insurance and its social, economic, and political background before 1914, with some remarks on mathematical results has been given by Braun in *Geschichte der Lebensversicherung und der Lebensversicherungstechnik* (1925, 433pp.).

For the biographies we have of course used the *Dictionary of Scientific Biography*, edited by C. C. Gillispie (1970–1980) and the individual biographies available.

As reference books for the history of mathematics we have used Cantor (1880–1908) and Kline (1972).

For long periods of time there existed a considerable backlog of publications of the Academies at London, Paris, Turin, etc., so that papers were read some years before they were published. Referring to such papers we have used the *date of publication*; in the list of references, however, we have usually added the date of communication to the Academy.

CHAPTER 2

A Sketch of the Background in Mathematics and Natural Philosophy

2.1 INTRODUCTION

The first mathematical analyses of games of chance were undertaken by Italian mathematicians in the 16th century. The main results, which remained unpublished for nearly a century, were obtained by Cardano about 1565.

It was almost 100 years after Cardano before probability theory was taken up again, this time in France by Pascal and Fermat (1654). Their work was continued by Huygens (1657) in the Netherlands. He wrote the first published treatise on probability theory and its application to games of chance.

About the same time a statistical analysis of data on the population of London was carried out by Graunt (1662). He did not have any knowledge of probability theory.

The first contributions to life insurance mathematics were made by de Witt (1671) in the Netherlands and by Halley (1694) in England. They combined Huygens' probability theory with Graunt's life table.

Error theory and the fitting of equations to data were developed in astronomy and navigation. Outstanding contributions are due to the Danish astronomer Tycho Brahe in the late 16th century, the German astronomer and mathematician Kepler in the beginning of the 17th century, and the Italian natural philosopher Galileo.

The problems taken up were of great current interest scientifically, socially, and economically. Their solutions depended on the mathematical background and sometimes required the development of new mathematical tools.

All these activities were well under way just before the Newtonian revolution, which was of decisive importance both mathematically and philosophically to further development.

For historical background we shall sketch the principal progress in mathematics and natural philosophy of importance for our subject before 1650. However, these fields only constitute a small part of the cultural background at the time. Most natural philosophers had a very broad education, worked in many different areas, and entertained ideas which today would be called superstitious. Belief in astrology, alchemy, and magic was widespread. Cardano and Tycho Brahe are outstanding examples of the versatile men of the Renaissance. Besides being a great mathematician, physician, and scientist, Cardano believed in and practiced divination, occultism, and healing by magic. The astronomer Tycho Brahe worked also in astrology and alchemy and produced many medicaments, for example, an elixir against the then common and dangerous epidemic diseases. Both men also made important technical inventions and thus bear witness to the close relationship between science and technology.

The purpose of the present chapter is to refresh the reader's memory on some of the salient historical facts before 1650. It is, however, not possible to point to a simple causal explanation of the development of probability and statistics in terms of these facts, but the record should make it easier for the reader to review and to grasp the spirit of the time.

The exposition is necessarily very brief; it is also biased in the sense that it concentrates on the most conspicuous events in the development of mathematics and natural philosophy, and it emphasizes those events that are of particular interest for the history of probability and statistics.

2.2   ON MATHEMATICS BEFORE 1650

Classical Greek mathematics had been nearly forgotten in Western Europe in the early Middle Ages. The Crusades and increasing trade and travel in the Mediterranean countries from about 1100 brought the Europeans into contact with the Arabs and the Byzantines who had preserved the Greek works. During the later Middle Ages and the Renaissance, the classical works were translated, commented upon by European mathematicians, and put to good use in connection with many practical applications, such as navigation, surveying, architecture, and commercial arithmetic. A survey of the existing mathematical knowledge with a view to applications was given by Luca Pacioli (c. 1445–c. 1517) in 1494.

In the 16th century considerable progress was made in arithmetic, algebra, and trigonometry. Zero was accepted as a number, and negative and irrational numbers came gradually into use. Complex numbers occurred in the solution of quadratic equations, but they were considered "useless." The decimal system of notation was introduced for fractions, replacing the ratio of two integers.

Two prominent Italian mathematicians, Niccolò Tartaglia (c. 1499–1557) and Girolamo Cardano (1501–1576), wrote textbooks containing new results in arithmetic and algebra. For example, they gave methods for the solution of equations of the third and fourth degrees, and Cardano noted that the number of roots equaled the degree of the equation.

The French mathematician François Vieta (1540–1603) published several works on plane and spherical trigonometry in which he systematized and extended the formulae for right and oblique plane triangles and for spherical right triangles. He also found many trigonometric identities, for example, the important expression for $\sin nx$ in terms of $\sin x$. Vieta's trigonometric research was mainly inspired by problems in astronomy and surveying; however, he also showed how to use trigonometric formulae for the solution of certain algebraic equations.

Progress in algebra was hampered by the tradition that geometry was the only real mathematics, and algebraic results had therefore to be given a geometrical interpretation. For example, algebraic equations had to be written in homogenous form of at most the third degree. Vieta realized, however, that algebra could be used to prove geometrical results and to handle quantities whether or not they could be given a geometrical interpretation. Thus algebra gradually became a separate mathematical discipline independent of geometry.

The increasing use of mathematics in practice resulted in the computation and publication of many tables, particularly tables of trigonometric functions.

Texts on arithmetic and algebra in the Renaissance were written in a verbal style with abbreviations for special words: for example p for plus, m for minus and R for square root. According to Kline (1972, p. 260), the expression $(5 + \sqrt{-15})(5 - \sqrt{-15}) = 25 - (-15) = 40$ was written by Cardano as

$$5p: Rm: 15$$

$$5m: Rm: 15$$

$$25m: m: 15 \, qd \, est \, 40.$$

Gradually, symbols were introduced for the unknowns and exponents for powers. A decisive step was taken by Vieta, who used letters systematically also as coefficients in algebraic equations. The sign $=$ for equality was proposed about the middle of the 16th century but was not universally accepted. Descartes used $\propto$ as a stylized ae (from *aequalis*), and this was still used by Bernoulli in his *Ars Conjectandi* in 1705. The symbol ∞ for infinity was introduced by Wallis in 1655. The letter π was introduced in the beginning of the 18th century but de Moivre still used c (derived from circumference) as late as 1756 in his *Doctrine of Chances*.

Essential steps in the free use of letters and special signs for mathematical symbols were first taken by Descartes, Newton, and Leibniz.

The most important advance in arithmetic in the 17th century was the invention of logarithms. The German mathematician Michael Stifel (1486–1567) considered in 1544 the correspondence between terms of an arithmetic and a geometric series and stated the "four laws of exponents," but he did not take the decisive step of introducing logarithms. This was done by John Napier (1550–1617), Laird of Merchiston in Scotland, a prominent politician and defender of the Protestant faith. He published two books on logarithms, the *Descriptio* (1614) and the *Constructio* (1619), the first giving the definitions and working rules of logarithms and a seven-figure table of logarithmic sines and tangents, the second containing theory and proofs. Napier considered the synchronized motion of two points, each moving on a straight line, the one with constant velocity, and the other with a decreasing velocity proportional to the distance remaining to a fixed point, the initial velocity being the same. In modern notation his model may be written as

$$dx/dt = r, \qquad x(0) = 0,$$
$$dy/dt = -y, \qquad y(0) = r,$$

with the solution $x(t) = rt$, and

$$y(t) = re^{-t} = re^{-x/r}, \qquad t \geqslant 0.$$

It follows that the Naperian logarithm, $\log_N y = x$, is a linear function of the natural (or hyperbolic) logarithm

$$\log_N y = r \log_e \frac{r}{y}.$$

Napier constructed his table of logarithms by means of a detailed tabulation of the function $g(x) = r(1 - \epsilon)^x$, $x = 0, 1, \ldots$, for $r = 10^7$ and for small positive values of ϵ. He carried out these calculations personally during a period of nearly 20 years.

Since $\log_N r = 0$, and

$$\log_N(10y) = \log_N y - 23{,}025{,}842.34,$$

Napier realized that his definition of logarithms was unpractical, and in cooperation with Henry Briggs (1561–1630), professor of mathematics first in London and later in Oxford, he proposed the system of common (or Briggsian) logarithms with base 10. From 1615 Briggs devoted the main part of his time to the construction of logarithmic tables. In 1617 he published the first table of common logarithms of the natural numbers from 1 to 1000. This was followed by his *Arithmetica Logarithmetica* (1624) containing the logarithms of the natural numbers from 1 to 20,000 and from 90,000 to 100,000 to 14 decimal places, with an introduction on the construction of the table and examples of arithmetical and geometrical applications. Posthumously occurred his *Trigonometria Brittanica* (1623) containing sines, tangents, and their logarithms to 14 decimal places. Many other tables were published about the same time so that 20 years after Napier's book, a wealth of logarithmic tables was available, and for the next 300 years logarithmic tables were the most important tools for computational work. In *Napier Tercentenary Memorial Volume* (1915) Glaisher writes,

> By his invention Napier introduced a new function into mathematics, and in his manner of conceiving a logarithm he applied a new principle; but even these striking anticipations of the mathematics of the future seem almost insignificant by comparison with the invention itself, which was to influence so profoundly the whole method of calculation and confer immense benefits upon science and the world.

For more details on the history of logarithms, we refer to Naux (1966, 1971) and Goldstine (1977).

The great progress in physics and astronomy in the beginning of the 17th century by Galileo and Kepler had a profound influence on the direction of mathematics. By fitting mathematical equations to data, they obtained simple descriptions of physical phenomena, and they thus demonstrated the usefulness of mathematics in science and technology. All the great contributions to mathematics in the following centuries came from men who were as much scientists as mathematicians.

As natural tools for his work in astronomy Johannes Kepler (1571–1630) worked on interpolation; logarithms; tabulation of trigonometric functions and logarithms; the mathematics of conics, for example, the gradual change of one conic into another by a change of the parameters; the length of curves;

and the areas and volumes limited by curves and surfaces. He calculated such areas and volumes as the sum of a large number of small sections.

In the 1630s and 1640s many mathematicians worked on the area (integration) problem. Cavalieri (1598–1647) invented a method of "indivisibles," a geometrical method for finding areas and volumes by means of an infinite number of equidistant parallel linesegments and areas, respectively. Other mathematicians, such as Fermat, Roberval, Pascal, and Wallis, solved concrete problems either by Cavalieri's method or by approximating the area under a curve by the sum of the areas of suitably chosen rectangles with bases of the same length, letting the number of rectangles increase indefinitely and keeping only the main term of the sum. Using the latter method, Fermat, for example, worked out the integral of x^n over a finite interval for all rational n except -1. A general method of integration (and differentiation) had, however, to wait for the works of Newton and Leibniz in the latter part of the century.

Practical problems in optics, perspective, and cartography led Girard Desargues (1591–1661) to use projection and section as a general method in geometry, and he thus founded modern projective geometry. He studied transformation and invariance for the purpose of deriving properties of the conics from those already proved for the circle.

The most influential natural philosopher and mathematician in the first half of the 17th century was René Descartes (1596–1650). Here we shall only mention some mathematical results contained in his *La Géométrie* (1637). Descartes continued Vieta's attempts to introduce better symbolism. For example, he introduced the rule of using the first letters of the alphabet for constants and coefficients and the last for variables and unknowns. He also continued Cardano's and Vieta's algebraic works. He asserted that the number of roots in a polynomial equation $f(x) = 0$ equals the degree of $f(x)$ and, furthermore, that $f(x)$ is divisible by $x - a$ if and only if $f(a) = 0$. He considered algebra to be an extension of logic and independent of geometry and used algebra to solve geometrical construction problems. He founded what today is called analytic or coordinate geometry by introducing (oblique) coordinate axes and defining a curve as any locus given by an algebraic equation. In particular, he studied the conics and the correspondence between their algebraic and geometrical expressions. He also began the study of curves of higher degrees. The important problem of finding the tangent of a curve was solved by a combination of geometrical and algebraic reasoning.

About the same time, and independently, Fermat solved nearly the same problems in analytical geometry, but since his works were not published (they only circulated in manuscript), he did not have the same influence as Descartes.

Frans van Schooten (1615–1660), professor of mathematics in Leiden, translated *La Géométrie* into Latin, which was still the international scientific language. He also added his own commentary and taught Cartesian geometry to his students. Because his translation was much in demand, he published a much enlarged second edition, which besides his own commentaries contained essential contributions by his students Huygens, de Witt, and Hudde, whom we shall meet later in their capacities as probabilists.

The early history of combinatorics is rather obscure (see Biggs, 1979). From numerical examples given by the Indian mathematician Bhaskara about 1150, it seems that he knew the general formulae for the number of permutations of *n* objects and the number of combinations of *r* among *n* objects. From the Hindus this knowledge spread to the Europeans through the Arabs.

The binomial expansion and the corresponding arithmetical triangle of coefficients originated also among Hindu and Arab mathematicians. The arithmetical triangle and its construction are explained by the Arab mathematician al-Tusi in 1265 but is not found in European works before the 16th century.

The relationship between the combinatorial formula and the binomial coefficient was recognized by Marin Mersenne (1588–1648) in 1636. Finally, a unified theory of the combinatorial numbers, the figurate numbers, and the binomial coefficients was developed by Pascal in 1654 and published in 1665.

The previous remark that the early history of combinatorics is rather obscure is no longer true after the publication of *Pascal's Arithmetical Triangle* by Edwards (1987). Here the history of the arithmetical triangle is traced back to Pythagorean arithmetic, Hindu combinatorics, Arabic algebra, and Chinese and Persian mathematics, and a meticulous study of the development in Europe is given, comprising contributions by Tartaglia, Cardano, Stifel, Mersenne, and Pascal and ending with the use of the binomial coefficients in the works of Wallis, Newton, Leibniz, and Bernoulli. We shall return to this book in §§4.3 and 5.2.

The above sketch of the history of mathematics before 1650 is essentially based on the book by Kline (1972), where details and references may be found.

2.3 ON NATURAL PHILOSOPHY BEFORE 1650

The 12th century saw the rise of the European university with its four faculties: theology, law, medicine, and the arts. The curriculum of the arts embraced grammar, logic, rhetoric, arithmetic, music, geometry, and astronomy. Most teachers and scholars had a clerical education, and Latin, the official language of the Church, became the universal scholarly language. University teaching and research were dominated by Christian theology and the heritage of the Greeks. The classical Greek works were read in Latin translation, often obtained by translating Arab editions. The most important works were by Aristotle in philosophy, ethics, and logic; Euclid in geometry; Ptolemy in astronomy; and Galen in medicine.

In the 12th and 13th centuries many philosophers wrote commentaries on the Scriptures and the Greek classics for the purpose of developing a general philosophy uniting these two lines of thought. The attempts to reconcile Aristotelian ideas with Christian theology resulted in a firmly established philosophy which has been called Scholasticism. Among the scholastic philosophers the most famous was Thomas Aquinas (1225–1274) whose system of thought was later authorized by the Catholic Church as the only right one. It became the dominant philosophy for about 400 years.

Aristotle's scientific results were considered authoritative, but his inductive–deductive method by which these results had been obtained was pushed into the background. Scholasticism took over Aristotelian logic with its laws of reasoning (the doctrine of syllogism) and used it to create a system of theological explanations of both natural and supernatural phenomena. It also adopted Aristotle's natural philosophy with its teleological explanations, its distinction of sublunary matter into four elements (earth, water, air, and fire), and its conception of an immutable universe of celestial bodies set into motion by God. The natural motion of terrestrial elements was supposed to be linear, whereas the motion of the heavenly bodies was circular. The immutability of the universe was in agreement with the Scriptures and with the deterministic outlook of Christian theology, which supposed that everything was created by the will of an all-powerful God.

The exegetic and speculative nature of scholasticism led to opposition, particularly among English philosophers, who advocated the importance of observation, experimentation, and induction in natural philosophy as a supplement to the revelations in the Scriptures. The most prominent spokesman of this school was William of Ockham (1285–1349), who is most known today for the maxim called "Ockham's Razor": "Entities are not to be multiplied without necessity." This philosophical principle of economy of the number of concepts used in the construction of a theory had great importance for the development of logic, mathematics, and natural philosophy.

A long period of gradual progress in wealth and knowledge was disrupted by the Black Death in 1348. In just a few years about one-third of the population of Europe died of the plague. It took about 100 years for Europe to recover from this catastrophy, which was further aggravated by the Hundred Years' War (1338–1453) between England and France.

The second half of the 15th century was a period with many inventions made by practical men, such as artisans, architects, shipbuilders, and engineers. A considerable metal-working and mining industry was developed. For example, rails were used to further transport in mines, and suction pumps were constructed for drainage. Many inventions of importance for shipping and warfare were made. Charts and instruments for navigation, such as the magnetic compass, the quadrant, and the astrolabe, were developed, and ocean-going sailing ships were built. The effectiveness of gunpowder in warfare was greatly improved by the construction of cannons and handguns. These inventions were the necessary conditions for the great voyages of exploration about 1500 and the following conquests overseas with their wide-ranging consequences for daily life in Europe.

About 1450 printing by movable type was invented, and Gutenberg set up his printing press in Mainz. The printing of illustrations from engraved metal plates also became common. The new techniques spread rapidly all over Europe. The transition from handwritten to printed books was a technological advance with revolutionary effects not only in the world of learning but in religion, politics, art, and technology as well. The next 50 years saw the printing of thousands of books both in Latin and the vernacular such that the knowledge hitherto accumulated in libraries for the few suddenly became available for the many, particularly for the laity. Printed books helped to spread the culture of the Renaissance and to further the Reformation.

The extraordinary growth of trade and industry in the Italian city-states in the 15th century created a new class of merchants and bankers who used their great wealth to support not only artists but artisans and scientists as well for the purpose of furthering the development of useful methods for the new competitive capitalist economy. The intellectual outlook gradually changed from the authoritative scholastic philosophy to a more independent way of thinking based on observations and experiments.

Freedom of thought also spread to religious matters through the Reformation in the first half of the 16th century. Protestantism in its various forms (Lutherans, Calvinists, Huguenots, Puritans, Presbyterians) spread all over Europe (except for Italy and Spain, which remained Catholic). The bible was translated from Greek into the national languages, commentaries were issued, and the laity were encouraged to read and interpret the Scriptures themselves. The struggle between Protestantism and Catholicism, mixed with strong economical and political interests, led in many countries to civil war, with increased intolerance and orthodoxy on both sides.

Among the measures taken by the Catholic Church to identify and suppress its opponents were the establishment of the Holy Office, the Inquisition in Rome in 1542, and the *Index librorum prohibitorum* (Index of prohibited books) in 1559. This list came to comprise the famous books by Copernicus, Galileo, Kepler, and Descartes. In due course, when the

Lutheran, Calvinian, and Anglican Churches acquired secular power, they also forced their dogmas on society, in particular on schools and universities. Protestantism, however, did not develop an all-embracing philosophy covering not only religious and moral but also scientific matters, and the conflict between the Reformed Churches and the ongoing scientific revolution therefore became less severe.

The works of Pythagoras and Plato were studied with renewed interest during the Renaissance, and their idea of a rational and harmonious universe describable in mathematical terms was taken over. The Church accepted the idea that an omniscient God has created the universe according to simple mathematical laws, originally unknown to man. It therefore became a praiseworthy enterprise to study and disclose these laws to obtain a better understanding of God and his "original design."

The Catholic Church, however, considered the laws of nature found by the scientists as hypotheses or practical computational devices, which were accepted as true explanations of nature only if they did not contradict the dogmas of the Church. This attitude gave rise to endless conflicts between scientists and the dogmatists of the Church, which seriously restricted the lives and modes of expression of such men as Copernicus, Cardano, Galileo, Kepler, and Descartes.

The scientific revolution in the 16th and 17th centuries, which forms the foundation of modern mathematics and science, may summarily be considered to begin with the publication of Nicholas Copernicus' *De Revolutionibus Orbium Coelestium* (On the Revolutions of the Celestial Spheres) in 1543 and reaching its culmination by Isaac Newton's *Philosophiae Naturalis Principia Mathematica* (The Mathematical Principles of Natural Philosophy) in 1687.

Among astronomers, Copernicus' book was received as a great work in mathematical astronomy comparable only to Ptolemy's *Almagest*, which had been the basis for all astronomy since about A.D. 150. However, Copernicus' heliocentric model gave no better predictions of phenomena than did Ptolemy's geocentric model, and Copernicus did not provide any empirical evidence for his hypotheses. The strength of the Copernican model was its simple and harmonious explanations of planetary motions and the natural ordering of the earth and the planets in relation to the sun. A few astronomers accepted the Copernican ideas and published books with improved versions of Copernicus' mathematics and tables, but general acceptance had to wait until the beginning of the 17th century.

Beyond the small circle of astronomers, Copernicus' main ideas of the diurnal rotation of the earth and the earth's yearly revolution around the sun just as another planet were generally rejected and ridiculed as being at variance with experience and with the Scriptures. Luther, Melanchthon, and

Calvin immediately rejected these ideas, and somewhat later the Catholic Church realized that the Copernican ideas might be destructive to the authority of the Church, and *De Revolutionibus* was therefore put on the *Index* in 1616.

To decide between the two competing systems, more accurate observations were necessary. These were provided by Tycho Brahe (1546–1601) who redetermined the positions of the planets and the stars with considerably greater accuracy than before by regular observations throughout 25 years in the last part of the 16th century.

Building upon the Copernican model and Tycho Brahe's observations, and guided by his physical ideas on the functioning of the planetary system, Kepler derived his three fundamental laws for the motion of the planets. He published his results in *Astronomia Nova* (The New Astronomy) in 1609. The greater simplicity and accuracy of Kepler's model soon became generally accepted. The final step in this development was Newton's derivation of Kepler's laws (and many other results) from the law of gravitation in 1687.

The most spectacular part of the scientific revolution is the developments in astronomy; however, it should not be forgotten that great progress was also made in physics, technology, and medicine.

The technological knowledge that had accumulated over many centuries now became exposited in books so that it became available not only to other artisans and engineers but also to scientists. Experiences in the mining industries in Germany were described by Agricola (Georg Bauer) in his *De re Metallica* (On Metallurgy) in 1556. In England books were published about techniques in agriculture, such as animal breeding, agricultural chemistry, rotation of crops, and land draining. To help navigators determine the longitude at sea, work on mechanical clocks was intensified. In the beginning of the 17th century, the telescope, the microscope, and the barometer were invented.

The English philosopher and essayist Francis Bacon (1561–1626) became known for his attacks on scholastic philosophy and on Aristotle, in particular in his *Novum Organum* (The New Instrument [of Thought]) from 1620. He stressed the necessity of systematic collection and classification of observations and experiments to get a broad basis for induction, and he warned against rash generalizations because observed correlations might be spurious. He advocated the use of the inductive–deductive method and pointed out that the validity of the deductive phase depends on a broad inductive base. His opinions were really not so much a criticism of Aristotle as of scholastic misapprehensions. He did not rate mathematics highly as a scientific tool, and his ideas became of more importance in the natural and social sciences than in physics. His ideas on the importance of tabulation and classification of facts as a basis for analysis influenced John Graunt and

William Petty in their works on vital statistics and political arithmetic. Bacon argued for the cooperation of science and technology, collaborative research, and the social utility of research, ideas which first became accepted in the latter part of the century. His opinions also influenced John Locke, the founder of empiricism in the theory of knowledge.

In 1600 William Gilbert (1544–1603) published his important book *De Magnete*, which contains his experiments and theories on magnetism. He described the attraction of magnets, explained the behavior of the magnetic compass, and generalized from the properties of small spherical magnets to the magnetic field of the earth. From this work Kepler constructed his theory on magnetic forces governing the planetary system.

The Aristotelian dogma on the immutability of the heavens was first effectively challenged by Tycho Brahe's observation of a new star in 1572 and his publication and discussion of his observations the following year. The brilliancy of the star gradually diminished until it finally disappeared 18 months after its first appearance. Tycho found that no parallax could be observed and that the position of the star was fixed in relation to neighboring stars in Cassiopeia and therefore concluded that the star belonged to the sphere of fixed stars. This conclusion was impugned by other astronomers who believed in the traditional explanation that comets and new stars were atmospheric phenomena and thus belonged to the sublunary changeable sphere.

Information on the construction of a telescope in the Netherlands in 1608 led Galileo Galilei (1564–1642) in 1609 to construct his own telescope with a thirtyfold magnifying power. He used his new instrument for astronomical observations and obtained a number of spectacular results which he published in 1610. Among his discoveries were that the moon was not a perfect sphere but a body similar to the earth with valleys and rather high mountains, that the Milky Way consisted of many faint stars, that many more stars existed than observable with the naked eye, and that Jupiter had four satellites revolving around Jupiter just as the moon revolved around the earth. In the next few years he also discovered the phases of Venus and the occurrence of sunspots. His telescopic observations thus supplemented traditional positional astronomy with information on the physical properties of the planets and the sun. These observations naturally shocked adherents of the Aristotelian and Ptolemaic doctrines, and Galileo used them as support for the Copernican system. Later he also tried to explain the tides in terms of motions of the earth.

In 1616 the Inquisition decided against the heliocentric system and placed *De Revolutionibus* on the *Index*. When Galileo in 1632 published *Dialogo sopra i due massimi sistemi del mondo* (Dialogue on the Two Chief World Systems), in which he discussed the Ptolemaic and the Copernican systems without really concealing his arguments and sympathy for the latter, he was called before the inquisition in 1633 and forced to recant his opinions. The *Dialogo* was put on the *Index*, and Galileo was sentenced to lifelong house arrest under surveillance.

Galileo was of a polemic nature and constantly at war with contemporary Aristotelian philosophers. He wrote all his books in Italian for the educated layman. He was not given to philosophical speculations but worked mainly on the solution of specific engineering and physical problems. He made a new delimitation of physical sciences by distinguishing between objective qualities of matter, such as shape, size, weight, position, and motion, and subjective qualities, such as color, taste, and smell. He restricted physics to the description of objective qualities of bodies and relations between them.

Galileo's use of physical astronomy as support for Copernicanism was, however, not his only great achievement. He realized that the Aristotelian laws of motion were at variance with the motions of the earth and that a new theory of motion had to be constructed. His greatest achievement was to establish a new method of scientific inquiry based on the method of abstraction. He realized that "natural laws" are valid only in an abstract world, and he thus laid the foundation for the axiomatization and mathematization of physics. Based on "extrapolation" of observations and experiments, he considered such abstract experiments as the motion of a perfect ball rolling on a frictionless horizontal plane, the free fall of a body in a vacuum, and the trajectory of a projectile moving without air resistance under influence of both horizontal and vertical forces. The three examples mentioned led him to a form of the law of inertia, to the theory of uniform acceleration for freely falling bodies, and to the parabolic form of the trajectory of a projectile. He also studied the motion of the pendulum and found the isochronism and the relation of period to length of the pendulum. He stressed the fundamental principle for application of the axiomatic method in science: that the validity of axioms has to be tested by comparing results of actual experiments with results deduced from the axioms. In 1636 he completed his manuscript of *Discorsi e dimostrazioni mathematiche intorno a due nuove scienze* (Discourses and Mathematical Demonstrations Concerning Two New Sciences), which was published in the Netherlands in 1638. This book contains his discussion on the strength of materials and the results of his researches for more than 30 years on kinematics.

René Descartes received his basic education at a Jesuit college in France and subsequently took a degree in law. He then spent more than ten years studying, soldiering, and traveling all over Europe before he settled in the Netherlands from 1628 to 1649 to write his great works on mathematics, physics, physiology, and philosophy. He died in Stockholm as philosopher at the court of Queen Christina.

Descartes set himself the ambitious task of constructing a new all-embracing natural philosophy based on the rich development of the natural sciences in the 16th century and the beginning of his own century. His purpose was to replace scholasticism with a mechanistic philosophy, to minimize the influence of God on cosmology and the natural sciences, and to leave the interpretation of the Scriptures to theologians.

Descartes was a great mathematician, his greatest invention being analytical geometry. He took the methods of mathematics as a model for his work in other fields, beginning with simple and evident ideas, clearly and distinctly true, like the axioms of geometry. He perceived God to be a perfect being who had created the universe consisting of matter in motion once and for all. He had a dual conception of man consisting of both matter and mind, these "substances" being independent of each other. Even though he himself carried out a great many experiments, his philosophy was essentially deductive and resulted in a purely mechanistic (kinematic) view of nature, including the human body.

Fundamental to his theory of matter and motion was the idea that the universe was filled with matter in the form of particles of three kinds: (1) fine, irregular particles able to fill out all space between other particles; (2) larger, globular particles able to move smoothly; and (3) coarser, irregular, and heavier particles. This composition of matter gave him sufficient freedom to construct *models* of many physical and physiological phenomena. Motion of one body (a collection of particles) resulted in the simultaneous motion of other bodies to avoid a vacuum. Motion was thus caused by collision or pressure. He postulated that God had created innumerable vortices of matter (particles in whirlpool motion) so that each of the heavenly bodies was at rest each in its own vortex. By means of this (peculiar) construction he kept the earth at rest and at the same time revolving around the sun by letting the vortex of the earth revolve around the vortex of the sun. From the idea that the amount of matter and motion was given once and for all, he derived the law of inertia and the law on the conservation of "quantity of motion," the latter concept being defined as magnitude (size) times speed.

Descartes also made essential contributions to optics (the law of refraction) and to physiology by elaborating on Harvey's model of the heart.

His main works are *Discours de la méthode* from 1637 and *Principia Philosophiae* from 1644.

Descartes' mechanistic philosophy unavoidably brought him into conflict with the Catholic Church, and he therefore left France and lived in the more tolerant Calvinistic Holland. However, here also he was severely attacked by the Church. After his death his books were put on the *Index*.

Weaknesses in Descartes' philosophy and discrepancies between some of his specific results and those of Kepler, Galileo, and Harvey, say, had been pointed out in Descartes' own lifetime. Nevertheless, his general ideas were accepted, dominating for more than a century, even after Newton's criticism of Descartes' cosmology. It should be remembered that the only alternative universal philosophy was scholasticism with its dependence on Aristotelian concepts and theological arguments. Descartes' *Principia* thus became very popular also outside scientific circles. This was due to his excellent style, his rational thinking with emphasis on clear ideas, his easily understandable mechanistic view of causation, and the coherence and universality of his philosophy. His work led to the discarding of many obsolete ideas from previous times. He greatly influenced Pascal, Huygens, and Newton.

Whatever religion they adhered to, the men who created the scientific revolution shared the belief in the Pythagorean idea that God had designed an orderly and rational world in accordance with simple numerical (mathematical) principles. Kepler, for example, called the book in which he derived his third law the *Harmony of the World*. The Pythagorean commitment prevailed also in the 18th century and influenced the basic outlook of statisticians and probabilists.

A typical example from statistics is to be found in the paper by John Arbuthnott (1712) with the following opening sentences:

Among innumerable Footsteps of Divine Providence to be found in the Works of Nature, there is a very remarkable one to be observed in the exact Ballance that is maintained, between the Numbers of Men and Women; for by this means it is provided, that the Species may never fail, nor perish, since every Male may have its Female, and of a proportionable Age. This Equality of Males and Females is not the Effect of Chance but Divine Providence, working for a good End.

Another example, taken from the *Doctrine of Chances* (1756, p. 252) by the probabilist Abraham de Moivre, runs as follows:

Again, as it is thus demonstrable that there are, in the constitution of things, certain Laws according to which Events happen, it is no less evident from Observation, that those Laws serve to wise, useful and beneficient purposes; to preserve the steadfast Order of the Universe, to propagate the several Species of Beings, and furnish to the sentient Kind such degrees of happiness as are suited to their State. But such Laws, as well as the original Design and Purpose of their Establishment, must all be from *without*; the *Inertia* of matter, and the nature of all created Beings, rendering it impossible that any thing should modify its own essence, or give to itself, or to anything else, an original determination or propensity. And hence, if we blind not ourselves with metaphysical dust, we shall be led, by a short and obvious way, to the acknowledgement of the great MAKER and GOVERNOUR of all; *Himself all-wise, all-powerful* and *good*.

The above sketch of the history of natural philosophy before 1650 is greatly influenced by Hall (1983), who has given a description and analysis of the scientific revolution in all its complexity.

CHAPTER 3

Early Concepts of Probability and Chance

3.1 TWO CONCEPTS OF PROBABILITY

The concept of probability is an ambiguous one. It has gradually changed content, and at present it has many meanings, in particular in the philosophical literature. For our purpose it is sufficient to distinguish between two kinds of probability, each characterized by one of several adjectives, depending on the context.

Objective, statistical, or aleatory probabilities are used for describing properties of random mechanisms or experiments, such as games of chance, and for describing chance events in populations, such as the chance of a male birth or the chance of dying at a certain age. Such probabilities are derived from symmetry considerations or estimated from relative frequencies. Based on an idealized game of chance with a finite number of equally possible outcomes the *classical* probability or chance of a compound event is defined as the ratio of the number of favorable outcomes to the total number of outcomes.

Subjective, personal, or epistemic probabilities are used for measuring the degree of belief in a proposition warranted by evidence which need not be of a statistical nature. Such probabilities refer to our imperfect knowledge and thus only indirectly to the things or events about which a statement is made. Some philosophers consider epistemic probability, which they call logical probability as a measure of the strength of a logical relation between two propositions, i.e., a weaker relation than that of logical consequence.

A clear distinction between the two kinds of probability is due to James Bernoulli (1713) (see §15.7). Most of the present book is concerned with objective probability, also called *chance* in the older literature. The title of

28

de Moivre's textbook on probability theory is the *Doctrine of Chances: or, A Method of Calculating the Probability of Events in Play* (1718). Before the Renaissance, probability was nonmathematical. It was not until the beginning of the 16th century that Italian mathematicians began to discuss the odds of various outcomes of games of chance based on the fundamental idea that the possible outcomes of a single game are equally likely. This elementary calculus of chances was assumed to be part of arithmetic and algebra, and it is discussed in Chapter 4.

In the present chapter we shall consider the development of concepts of probability before the mid-17th century. The two concepts defined above will serve as frame of reference. In recent years many papers have been written on this topic; however, opinions differ greatly about the interpretation of the classical texts, so we can do no better than give a brief survey of the most important papers, where references to the original sources may be found.

3.2   PROBABILITY IN ANTIQUITY AND THE MIDDLE AGES

Rudimentary concepts of probability, chance, and randomness occur since ancient times in connection with sortilege, fortune-telling, games of chance, philosophy, law, insurance, sampling inspection, and errors of prediction in the various sciences, for example, astronomy and medicine.

In view of the achievements of the Greeks in mathematics and science, it is surprising that they did not use the symmetry of games of chance or the stability of relative frequencies to create an axiomatic theory of probability analogous to their geometry. However, the symmetry and stability which is obvious to us may not have been noticed in ancient times because of the imperfections of the randomizers used. David (1955, 1962) has pointed out that instead of regular dice, *astragali* (heel bones of hooved animals) were normally used, and Samburski (1956) remarks that in a popular game with four astragali, a certain throw was valued higher than all the others despite the fact that other outcomes have smaller probabilities, which indicates that the Greeks had not noticed the magnitudes of the corresponding relative frequencies.

Hasover (1967) and Rabinovitch (1969; 1970a; 1973) have given examples of the uses of random mechanisms mentioned in Talmudic and rabbinic literature. Drawing of lots were used in religious ceremonies, for allocation of daily duties among priests in the temple, and for various legal purposes.

As suggested by David, the use of astragali and the drawing of lots in divination may have prevented a scientific study of the outcomes of games of chance for religious reasons. Kendall (1956) points to other reasons too, among them the absence of a notion of chance events, and concludes that "it is in the basic attitude towards the phenomenal world" that the lack of a probability theory should be sought. A similar conclusion is reached by Sambursky who holds that the philosophy of Plato and Aristotle limited the outlook of Greek scientists so that they looked for regularity only in mathematics and the heavens.

This point of view leads to a study of the philosophical works of Aristotle. According to van Brakel (1976) and Schneider (1980), Aristotle classified events into three types: (1) certain events that happen necessarily; (2) probable events that happen in most cases; and (3) unpredictable or unknowable events that happen by pure chance. Furthermore, he considered the outcomes of games of chance to belong to the third category and therefore not accessible to scientific investigation, and he did not apply the term probability to games of chance. Aristotle's concept of probability is epistemic and nonquantitative. His probable events are events that, on the given evidence, happen with a high degree of probability.

Aristotle's classification of events was adopted by Roman philosophers. Scholastic philosophers, however, had to reconcile Aristotle's philosophy with the deterministic philosophy of the Church. Events unpredictable or unknowable to man were considered to be predetermined by God. In that way chance events in a deterministic world may be characterized by subjective probabilities describing the state of information of the individual.

In discussing probability, Thomas Aquinas distinguishes among science or certain knowledge, opinion or probable knowledge, and the accidental or chance. Byrne (1968) has given a detailed discussion of the concept of probability in the writings of Aquinas. He holds that probability for Aquinas is a qualificative of an opinion or a proposition and thus a precursor of logical probability. However, Byrne also states that Aquinas measures the contingent in terms of the necessary by giving a weight to the contingent, in that way introducing a rudimentary frequency interpretation of the probable. One of Byrne's conclusions is the following: "There is a similarity (A) between Thomas's theory of probability and the contemporary logical theory of probability and (B) between Thomas's theory of contingency and the contemporary frequency theory of probability." Byrne emphasizes the continuity of the development of the concept of probability from the Middle Ages to modern times.

3.3 PROBABILITY FROM THE RENAISSANCE TO THE MID-17TH CENTURY

During the Renaissance, probability was still a nonnumerical epistemological concept, whereas chance was expressed as a ratio, and a calculus of chances became part of algebra. However, from the beginning of the 18th century the terms probability and chance have been used synonymously, which means that probability had become numerical. Hacking (1975) is concerned with the problem of how and when this transformation took place.

Hacking's seminal book *The Emergence of Probability* (1975) is the first comprehensive and coherent exposition of the development of the modern dual concept of probability. The subtitle of the book is A Philosophical Study of Early Ideas about Probability, Induction and Statistical Inference. It is not a study of the history of probability and statistics; even if Hacking also gives a fascinating account of the breakthrough of these subjects in their many manifestations around 1660, it is as stated, a philosophical study of ideas and is therefore difficult to follow and evaluate for a statistician.

Hacking bases his explanation on an investigation of three concepts: opinion, evidence, and signs. It is important to distinguish between two forms of evidence: the evidence of testimony, called external evidence, and the evidence of things, called internal evidence, a distinction formulated explicitly in Arnauld and Nicole's *La logique, ou l'art de penser* (1662).

Hacking holds that in the Renaissance a probable opinion was one "which was approved by authority, or by the testimony of respected judges," and that until the end of the Renaissance there was no concept of inductive (internal) evidence. This is a narrower definition of a probable opinion than that current in Aristotelian and scholastic philosophy, which also recognized the existence of internal evidence and its ability to confer probability on opinions.

Hacking argues that a chief concept of the low sciences, such as alchemy and medicine, was that of a sign, which came to be considered as the testimony of nature. On the one hand, a sign considered as testimony made an opinion probable; on the other, the predictive value of a sign could be measured by the frequency with which the prediction holds. "This transformation from sign into evidence is the key to the emergence of a concept of probability that is dual." However, many authors challenge this postulate and hold that both interpretations of signs have existed since ancient times.

The first examples of numerical probabilities are given by Arnauld and Nicole (1662). From then on the calculus of chances was applied also to probability. The doctrine of chances developed into a theory of probability with applications in many different fields but without much concern about the interpretation of the probability in question, the only exception being James Bernoulli.

Hacking's explanation of the emergence of the modern concept of probability has led to a great deal of discussion and a critical examination of his basic concepts. As indicated above most critics argue for the simultaneous existence of probability and chance in rudimentary forms from ancient times and for a gradual development to the modern form. This point of view, based on philosophical, theological, and linguistic arguments, has been advanced by Shafer (1976, 1978), Schneider (1976, 1980), Garber and Zabell (1979), and Bellhouse (1988). For example, Garber and Zabell write

> Cicero, Quintilian, John of Salisbury, Nicole Oresme, writers of diverse backgrounds writing in different centuries, all but Oresme well before the Renaissance. Yet the concepts of probability they use bear striking resemblances to the modern concept of probability, the concept that Hacking claims originated only in the 17th century. In each case, there is a connection between probability and rational belief on the one hand, and probability, for-the-most part truth, and frequency of occurrence on the other. And while probability is not yet fully mathematical, it has degrees, in Oresme closely identified with relative frequencies.

Hacking (1980) defends himself by writing that "what is important is not the occurrence of a few probability ideas in antique texts but a use for them, a use that spans morals, politics, economics and social affairs, and which engenders a new era of conjecturing on the one hand and a new mode of representing reality on the other." He adds that the transformation which took place about 1660 amounts to a new way of thinking—the emergence of a new (statistical) style in reasoning.

Turning from philosophy to law we note that different types of testimony and evidence have often been weighted relative to full proof and that the court's verdict depended on an "accumulation" of such weighting.

Coumet (1970), Daston (1980), and Schneider (1980b) have also discussed the theological and legal aspects of risk-tasking for the development of the concept of expectation in probability theory. The Church condemned gambling and usury as morally wrong, but it was impossible to disregard the existence of risks in commercial life, analogous to gambling risks, and these risks had to be accepted as legal. Hence, a new class of contracts, called aleatory contracts, came into existence embracing marine insurance, life contingencies, inheritance expectations, lotteries, and risky investments in business. The basis of such contracts became the specification of conditions for the equity of the parties involved, which required assessment of risks combined with the possible gains and losses. An aleatory contract thus corresponded to a fair game, that is, a game in which the participants have equal expectations.

A survey of the occurrence of probability concepts in scientific literature from Aristotle to Bernoulli, with many quotations from original works, has been given by Sheynin (1974).

CHAPTER 4

Cardano and *Liber de Ludo Aleae*, c. 1565

*The most fundamental principle of all in gambling is simply
equal conditions, e.g. of opponents, of bystanders, of money,
of situation, of the dice box, and of the die itself. To the extent
to which you depart from that equality, if it is in your
opponents favour, you are a fool, and if in your own, you are unjust.*
                                                    —*CARDANO, c. 1565.*

4.1 ON GAMES OF CHANCE

There is no evidence of a calculus of chances in the Middle Ages. Some early
instances of analyses of plays with three dice have been reported by Kendall
(1956). About 960, Bishop Wibold of Cambrai correctly enumerated the 56
different outcomes (without permutations) of playing with three dice. The
216 ways in which three dice can fall, ordered according to the number of
points scored, were listed in a Latin poem, *De Vetula*, with a section on
sports and games, presumably from the 13th century.

In the 14th century dice games, which had been popular since antiquity,
were supplemented by card games. According to Kendall (1957), no reference
to playing cards has been found in Europe before 1350. Whether the Italian
playing cards were invented independently of playing cards from the East is
unknown. By the 14th century, paper making and block printing, the technical
background for the manufacture of playing cards, had been established in
Europe. In the beginning, cards were rather expensive and mostly used by
wealthy people playing for high stakes. The Church preached against card
playing, and in some states card playing was forbidden. The use of cards

therefore spread slowly, and it was several hundred years before card games became more popular than dicing.

Lotteries had been used by the Roman emperors. In the Middle Ages and the Renaissance, lotteries became important as a means of financing Government expenditures. Private lotteries also flourished, but because of unreasonable conditions they were suppressed or declared illegal, although private lotteries were later authorized to assist charities and the fine arts. In his *Treatise of Taxes and Contributions* (1662), the English economist William Petty writes this about lotteries:

> Now in the way of Lottery men do also tax themselves in the general, though out of hope of Advantage in particular: A Lottery therefore is properly a Tax upon unfortunate self-conceited fools; ... Now because the world abounds with this kinde of fools, it is not fit that every man that will, may cheat every man that would be cheated; but it is rather ordained, that the Sovereign should have the Guardianship of these fools, or that some Favourite should beg the Sovereigns right of taking advantage of such mens folly, even as in the case of Lunaticks and Idiots.... This way of Lottery is used but for some Leavies, and rather upon privato-publick accompts, (then for maintaining Armies or Equipping Fleets), such as are Aque-Ducts, Bridges and perhaps Highways, etc.

Such realistic evaluations did not, however, restrain people from participating in lotteries then or in later times.

In view of the recreational and economic importance of gaming, it is no wonder that mathematicians sought to analyze games of chance to determine the odds of winning and thus the stakes in a fair game. The development sketched above is reflected in the mathematical literature on games of chance. The 16th and 17th century works are mainly occupied with problems of dicing, ball games, table games, and lotteries, whereas problems of card games abound in the beginning of the 18th century.

The terminology employed was naturally closely dependent on the problem studied. The fundamental concepts were the *number of chances*, from which the *odds* were derived; and, taking the stakes or the prizes into account, the *value* or *expectation* of a throw or a drawing, from which the fairness of a game could be judged. The term "probability" was not used regularly in the context of games of chance before the beginning of the 18th century, but it is of course easy today to interpret the writings in terms of probability. The arguments used were purely mathematical, without any recourse to observations of relative frequencies.

From about 1500 leading Italian mathematicians tried to solve the division problem, that is, the problem of an equitable division of the stakes in a prematurely stopped game. Since combinatorial theory did not exist at the time, the problem presented a real challenge, and they did not succeed.

Cardano took up the study of a theory of gambling and derived some elementary results about 1565. The division problem was first solved by Pascal and Fermat in 1654, and the birth of probability theory is usually associated with that date.

4.2 EARLY ATTEMPTS TO SOLVE THE PROBLEM OF POINTS

Two players, A and B, agree to play a series of fair games until one of them has won a specified number of games, s, say. For some accidental reason, the play is stopped when A has won s_1 and B s_2 games, s_1 and s_2 being smaller than s. How should the stakes be divided?

This is the formulation of the *division problem* in abstract form. The problem is also known as the *problem of points* because we may give the winner of each game a certain number of points and count the number of points instead of the number of games.

The division problem is presumably very old. It is first found in print by Pacioli (1494) for $s = 6$, $s_1 = 5$, and $s_2 = 2$. Pacioli considers it as a problem in proportion and proposes to divide the stakes as s_1 to s_2.

The duration of the play originally agreed upon is at least s games and at most $2(s - 1) + 1 = 2s - 1$ games. Without any reason, Pacioli introduces the maximum number of games and argues that the division should be as $s_1/(2s - 1)$ to $s_2/(2s - 1)$, or as mentioned above as s_1 to s_2. There is no probability theory or combinatorics involved in Pacioli's reasoning.

The next attempt to solve the problem is by Cardano (1539). He shows by example that Pacioli's proposal is ridiculous and proceeds to give a deeper analysis of the problem. We shall return to this after a discussion of some other, more primitive, proposals.

Tartaglia (1556) criticizes Pacioli and is sceptical of the possibility of finding a mathematical solution. He thinks that the problem is a juridical one. Nevertheless, he proposes that if s_1 is larger than s_2, A should have his own stake plus the fraction $(s_1 - s_2)/s$ of B's stake. Assuming that the stakes are equal, the division will be as $s + s_1 - s_2$ to $s - s_1 + s_2$.

Forestani (1603) formulates the following rule: First A and B should each get a portion of the total stake determined by the number of games they have won in relation to the maximum duration of the play, i.e., the proportions $s_1/(2s - 1)$ and $s_2/(2s - 1)$, as also proposed by Pacioli. But then Forestani adds that the remainder should be divided equally between them, because Fortune in the next play may reverse the results. Hence the division will be as $2s - 1 + s_1 - s_2$ to $2s - 1 - s_1 + s_2$. Comparison with Tartaglia's rule will show that s has been replaced by $2s - 1$.

Cardano (1539) is the first to realize that the division rule should not depend on (s, s_1, s_2) but only on the number of games each player lacks in winning, $a = s - s_1$ and $b = s - s_2$, say. He introduces a new play where A, starting from scratch, is the winner *if he wins a games before B wins b games*, and he asks what the stakes should be for the play to be fair. He then takes for a fair division rule in the stopped play the ratio of the stakes in this new play and concludes that the division should be as $b(b + 1)$ to $a(a + 1)$.

His reasons for this result are rather obscure. Considering an example for $a = 1$ and $b = 3$, he writes

> He who shall win 3 games stakes 2 crowns; how much should the other stake. I say that he should stake 12 crowns for the following reasons. If he shall win only one game it would suffice that he stakes 2 crowns; and if he shall win 2 games he should stake three times as much because by winning two games he would win 4 crowns but he has had the risk of losing the second game after having won the first and therefore he ought to have a threefold compensation. And if he shall win three games his compensation should be sixfold because the difficulty is doubled, hence he should stake 12 crowns.

It will be seen that Cardano uses an inductive argument. Setting B's stake equal to 1, A's stake becomes successively equal to 1, $1 + 2 = 3$, and $1 + 2 + 3 = 6$. Cardano then concludes that in general A's stake should be $1 + 2 + \cdots + b = b(b + 1)/2$. He does not discuss how to go from the special case $(1, b)$ to the general case (a, b), but presumably he has just used the symmetry between the players.

There are of course traces of probability arguments in Cardano's reasoning, but his arguments are unclear and do not lead to the correct division rule.

Without reference to Cardano, Peverone (1558) published a translation in Italian of Cardano's examples, and so Cardano's division rule has sometimes been ascribed to Peverone (see Kendall, 1956 and David, 1962).

The survey given above is based on the exposition by Cantor (Vol. 2, 1900) and the comprehensive discussion by Coumet (1965), who gives more details and references (see also Schneider, 1985).

4.3   CARDANO AND *LIBER DE LUDO ALEAE*

In view of the passion for gambling and the development of mathematics in Italy in the 16th century, it is not surprising that one of the prominent mathematicians, Girolamo Cardano, or Jerome Cardan (1501–1576), took up the study of a theory of gambling. Cardano also had the special qualification that he was an inveterate gambler and was, furthermore, possessed of an indomitable urge to transmit any knowledge he had to the public in writing. Cardano was an illegitimate son of a Milanese lawyer, who also lectured on mathematics. Cardano studied medicine in Padua and received his degree as Doctor of Medicine about 1526. He applied for admission to the College of Physicians in Milan but was refused, formally because of his illegitimate birth but presumably more because of his strange character. He had a sharp tongue and was known as an eccentric and a gambler. From 1526 he practiced as a country doctor, but returned to Milan in 1532 and was appointed lecturer in mathematics. In 1539 he published his first two mathematical books. His repeated requests for admission to the College of Physicians were without avail, so he fought back by publishing a book in 1536 entitled *On the bad practices of medicine in common use.* Finally, in 1539 he was elected a member of the College, and within a few years he became rector of the guild and the most prominent physician in Milan. His services were much in demand all over Europe. He also lectured on mathematics at the university in Pavia. He wrote many books on medicine, mathematics, astronomy, physics, games of chance, chess, death, the immortality of the soul, wisdom, and many other topics. In 1562 he moved to Bologna as professor of medicine. Accused of being a heretic in 1570, he was arrested, dismissed, and denied the rights to lecture publicly and to have his books printed. He continued to write and wrote his autobiography *De Vita Propria Liber* (The Book of My Life) before he died in Rome.

In 1545 Cardano published a textbook on algebra which for the first time contained a method for finding the roots of a cubic equation. This result has become known as Cardano's formula, even though he stated that the method was due to Scipione del Ferro around 1500 and, independently, to Niccolò Tartaglia in 1535. The publication gave rise to a heated dispute with Tartaglia who had disclosed his method to Cardano about 1539 for his promise to keep it secret. Cardano's name has also been connected with several mechanical inventions, namely, Cardan's suspension, the Cardan joint, the Cardan shaft.

Several books have been written about Cardano. For a statistician, the most interesting is by Ore (1953) on which the short biography given above is based.

Among Cardano's many unpublished manuscripts was *Liber de Ludo Aleae* (The Book on Games of Chance), which was published for the first time in the ten-volume edition of his works in 1663. An English translation by S. H. Gould may be found in the book by Ore (1953); a reprint of the translation was published separately in 1961. It consists of 32 short chapters. From a remark in Chap. 20 it follows that Cardano was writing that chapter in 1564; it is not known when the manuscript was completed. In Chap. 5

(Why I have dealt with gambling), he writes, "Even if gambling were altogether an evil, still, on account of the very large number of people who play, it would seem to be a natural evil. For that very reason it ought to be discussed by a medical doctor like one of the incurable diseases."

De Ludo Aleae is a treatise on the moral, practical, and theoretical aspects of gambling, written in colorful language and containing some anecdotes on Cardano's own experiences. In his autobiography Cardano admits to "an immoderate devotion to table games and dice. During many years—for more than forty years at the chess boards and twenty-five years of gambling—I have played not off and on but, as I am ashamed to say, every day." In *De Ludo Aleae* the experienced gambler gives practical advice to the reader, as indicated by the following chapter headings: On conditions of play; Who should play and when; The utility of play, and losses; The fundamental principle of gambling; The hanging dice box and dishonest dice; Conditions under which one should play; On frauds in games of this kind; On luck in play; On timidity in the throw; Do those who teach also play well; On the character of players. These chapters are very entertaining reading, in particular because the author feels obliged to recommend prudence and to preach morality at the same time recounting some of his own adventures of a very different nature.

Most of the theory in the book is given in the form of examples from which general principles are or may be inferred. In some cases Cardano arrives at the solution of a problem through trial and error, and the book contains both the false and the correct solutions. He also tackles some problems that he cannot solve and then tries to give approximate solutions. We shall only refer to the most important correct results and disregard the many unclear statements and confusing numerical examples regarding various games. A more detailed commentary has been given by Ore (1953).

The main results about dicing are found in Chap. 9–15 and 31–32. He clearly states that the six sides of a die are equally likely, if the die is honest, and introduces chance as the ratio between the number of favorable cases and the number of equally possible cases. As a gambler, he naturally often uses odds instead of chance. In Chap. 14, he defines the concept of a fair games in the following terms:

> So there is one general rule, namely, that we should consider the whole circuit [the total number of equally possible cases], and the number of those casts which represents in how many ways the favorable result can occur, and compare that number to the remainder of the circuit, and according to that proportion should the mutual wagers be laid so that one may contend on equal terms.

For pedagogical reasons, he introduces "dice" with three, four, and five equally likely sides.

The enumeration of the equally possible cases is carried out as follows. First the number of different types of outcomes are found, for example, for three dice he gives the number of triplets as 6, the number of doublets and one different face as 30, and the number of cases with all faces different as 20. Next he finds the number of permutations for each type, in the present example 1, 3 and 6, and the final result is given as $6 \times 1 + 30 \times 3 + 20 \times 6 = 216$. He carries out this analysis in detail for two and three ordinary dice and for four 4-sided dice. Based on these enumerations, he gives several examples of the chance of different types of outcomes.

By addition of a number of equally possible cases, he derives the chance of compound events. First he tabulates the distribution of the sum of the points in games by two dice and three dice, respectively. Next he finds the chance of various combinations of points for two dice. Let 1 denote the occurrence of at least one ace, let 2 denote the occurrence of at least one ace or a deuce, and so on. Counting only disjoint cases, Cardano gives the result as shown in the following table:

Favorable Cases for Combination of Points

	TWO DICE	THREE DICE
Cases for 1 point	11	91
Additional for: 2	9	61
3	7	37
4	5	19
5	3	7
6	1	1
Total:	36	216

Besides application of the addition rule in connection with the enumerations mentioned above, Cardano's most advanced result is the "multiplication rule" for finding the odds for and against the recurrence of an event every time in a given number of repetitions of a game. Let the number of equally possible cases in the game be t, and let r be the number of favorable cases, so that the odds are $r/(t - r)$. By trial and error and by analyzing games with dice having three, four, five, and six faces, he obtains the important result that in n repetitions the odds will be $r^n/(t^n - r^n)$. Setting $p = r/t$, the result becomes $p^n/(1 - p^n)$, which is the form used today. We shall encounter this problem and its solution many times in the further development of the calculus of chances. Cardano gives several examples of applications, the most complicated being the following. In a game with three dice there are 91 favorable cases of 216 equally possible cases for getting at least one ace. The odds for this result turning up every time in three games are $91^3/(216^3 - 91^3) = 753,571/9,324,125$, which is a little less than 1 to 12.

He also gives examples of the computation of the average number of points in various games with dice.

Cardano writes several chapters on card games, in particular on Primero, the medieval version of poker. He demonstrates how to find the chance of certain simple outcomes by drawing cards from a deck.

According to Edwards (1987), several current textbooks on mathematics in the 16th century contain discussions of the arithmetical triangle and the figurate numbers. Tartaglia (1556) gives the binomial coefficients up to $n = 12$ obtained by means of the addition rule and arranged in an arithmetical triangle. Without proof he also gives a rule for finding the number of ways n dice can fall exhibited as a table of figurate numbers, i.e., as the repeated sums of $(1, 1, 1, \ldots, 1)$, giving $(1, 2, 3, \ldots, 6)$; $(1, 3, 6, \ldots, 21)$; and so on. Cardano (1539) derived the formula $2^n - n - 1$ for the number of combinations of n things taken two or more at a time, and in 1570 he pointed out the correspondence between the figurate numbers and the combinatorial numbers C_k^n, the number of combinations of n things taken k at a time, which he tabulated up to $n = 11$. Furthermore, he showed that

$$C_k^n = \frac{n - k + 1}{k} C_{k-1}^n,$$

from which he derived the multiplicative form

$$C_k^n = \frac{n(n - 1) \cdots (n - k + 1)}{1 \cdot 2 \cdots k}.$$

However, Cardano does not refer to or use these results or those of Tartaglia in *De Ludo Aleae*.

It is strange that Cardano, who was a practical man as well as a mathematician, does not give any empirical data in *De Ludo Aleae*; not even a single relative frequency is recorded from his extensive experience of gambling. In this he is in accord with the great probabilists Pascal, Huygens, Bernoulli, Montmort, and de Moivre and modern writers of textbooks on probability, who consider probability theory to be a mathematical discipline based on a set of axioms.

On the other hand, Cardano's book contains many important practical observations. We have chosen the epigraph of this chapter to demonstrate the remarkable fact that Cardano clearly formulated the fundamental condition for application of probability theory in practice, which is that the observations should be taken under essentially the same conditions.

Because of the late publication of *De Ludo Aleae* (1663) it did not influence further development directly, but it seems reasonable to assume that

Cardano's results were known by the mathematical community in the late 16th century. At the time of Pascal (1654), Cardano's formula $p^n/(1 - p^n)$ was considered elementary.

4.4 GALILEO AND THE DISTRIBUTION OF THE SUM OF POINTS OF THREE DICE, c. 1620

The next paper on probability, *Sopra le Scoperte dei Dadi* (On a discovery concerning dice), is due to Galileo Galilei. An English translation by E. H. Thorne may be found in the book by David (1962). According to David, the paper is presumably written between 1613 and 1623; it was first published in Galileo's collected works in 1718.

Galileo's paper is written in reply to the following question. Playing with three dice, 9 and 10 points may each be obtained in six ways. How is this fact compatible with the experience "that long observation has made dice-players consider 10 to be more advantageous than 9"? In his answer Galileo lists all three-partitions of the numbers from 3 to 10, finds the number of permutations for each partition, and tabulates the distribution of the sum of points. In that way he proves "that the sum of points 10 can be made up by 27 different dice-throws, but the sum of points 9 by 25 only." His method and result are the same as Cardano's; he takes for granted that the solution should be obtained by enumerating all the equally possible outcomes and counting the number of favorable ones. This indicates that this method was generally accepted among mathematicians at the time. Cardano and Galileo did not go further than three dice. The problem for n dice was taken up and solved by de Moivre (1712), James Bernoulli (1713), and Montmort (1713) (see §14.3).

The incident is remarkable not only because of the recorded observation (10 occurs more often than 9) but also because somebody ("he who has ordered me to produce whatever occurs to me about such a problem") asked for and actually got an explanation in terms of a probabilistic model. Recently there has been some discussion on the problem of whether such small differences between probabilities as 27/216 and 25/216 or 27/52 and 25/52 could be detected in practice. The important point is, however, that somebody observed a difference between relative frequencies, whether significant or not, and asked for an explanation.

The reader who knows some statistical theory may solve the following problems: How many throws with three dice will be necessary to conclude with reasonable confidence that the probability of 10 points is larger than that of 9? Discuss various experimental designs to test the hypothesis that the odds are 1:1 against the alternative that they are 27:25.

CHAPTER 5

The Foundation of Probability Theory by Pascal and Fermat in 1654

Thus, joining the rigour of demonstrations in mathematics with the uncertainty of chance, and conciliating these apparently contradictory matters, it can, taking its name from both of them, with justice arrogate the stupefying name: The Mathematics of Chance (Aleae Geometria).

—PASCAL, in an address to the Académie Parisienne de Mathématiques, 1654

5.1 PASCAL AND FERMAT

The direct cause of the new contributions to probability theory was some questions on games of chance from Antoine Gombaud, Chevalier de Méré, to Blaise Pascal in 1654. Pascal communicated his solutions to Pierre de Fermat for approval, and a correspondence ensued. At that time scientific journals did not exist, so it was a widespread habit to communicate new results by letters to colleagues.

Like Cardano and Galileo, Pascal and Fermat used the basic principle of enumeration of the equally possible cases and among them the cases favorable to each player, but they now had a combinatorial theory at their disposal. They thus laid the foundation of what has later been called *combinatorial chance*. They wrote about the number of chances ("hasards") and did not use the term probability. They took *the addition and multiplication rules* for probabilities for granted. They introduced the *value* of a game as

42

the probability of winning times the total stake, the fundamental quantity which Huygens three years later called the *expectation*. Psacal also introduced *recursion or difference equations* as a new method for solving probability problems. They solved *the division problem* by means of the binomial and the negative binomial distributions, and they also discussed the problem of the Gambler's Ruin.

Pierre de Fermat (1601–1665) came of a family of wealthy merchants. He studied law at the universities of Toulouse and Orléans, and he also studied mathematics at Bordeaux, where he was much influenced by the works of Vieta. He made his living as a lawyer and a jurist and had mathematics as a lifelong hobby. In 1631 he became Counsellor of the Parlement of Toulouse and was gradually promoted to higher offices at the same place where he was to stay for the rest of his life. As a man of wide-ranging learning, not only in mathematics and the sciences but also in the humanities, he corresponded with many colleagues, those outside France as well. He communicated his many important mathematical results, often without proof or with incomplete proof, in letters or in manuscripts to his friends, in that way guarding his freedom and status as an "amateur" without obligation to publish and at the same time getting the recognition that he nevertheless desired. This attitude naturally led to priority disputes with other prominent contemporary mathematicians, such as Descartes and Wallis. Rather late in his career, 9 August 1654, he wrote to Carcavi and proposed that Carcavi and Pascal should undertake to amend, edit, and publish his treatises without disclosing the author's name; however, Pascal was not interested in that kind of work so nothing came of the proposal. An incomplete edition of Fermat's works was first published posthumously in 1679.

The basis of Fermat's mathematics was the classical Greek treatises combined with Vieta's new algebraic methods. Independently of Descartes, yet simultaneously, he laid the foundations of analytic geometry, but whereas Descartes' *Géométrie* (1637) assumed great importance, Fermat's contribution only circulated in manuscript form. Fermat discussed the polynomial equation of the second order in two variables and derived the relations between the coefficients and the curves known as conic sections. He contributed to the beginnings of the calculus by devising methods for finding the tangent of a curve and for determining maxima and minima. He gave without proof a recursion formula for finding sums of powers of integers which he used for calculating areas under polynomial curves by summing the areas of a large number of inscribed and circumscribed rectangles and deriving the common limiting value. Today the fame of Fermat rests mainly on his contributions to the theory of numbers in which he stated many theorems without proof, although indicating a general method of proof, the method of infinite descent. Many mathematicians have since labored to prove his theorems. His "last theorem," which says that no integral solutions of the equation $x^n + y^n = a^n$ are possible for $n > 2$, is still unproved. He made a lasting contribution to the sciences as well. Fermat's Principle of Least Time states that light traveling from one point to another always takes the path requiring least time. A biography of Fermat, with a discussion of his mathematical works, their background, and their importance for his contemporaries, has been written by Mahoney (1973).

Blaise Pascal (1623–1662) was an infant prodigy. His mother died when he was three and his father, a civil servant, took care of his education and brought him at an early age to the meetings of Mersenne's Academy, one of the many private gatherings of scientists, which developed into the Académie Royale des Sciences in 1666. Influenced by the work of Desargues, Pascal, 16 years old, wrote a small paper *Essai pour les coniques* (1640) in which he presented a theory of conic sections considered to be projections of a circle, giving a number of definitions and theorems without proof, and indicating his intention to continue this research. For many years he worked on a treatise on projective geometry and conic sections, but it was never published, and only the introductory section is extant.

To lighten his father's accounting work as a tax official, Pascal invented a calculating machine for addition and subtraction and took care of its construction and sale.

Inspired by works of Galileo and Torricelli disproving the Aristotelian doctrine that nature abhors a vacuum, Pascal in 1646 began a series of barometric experiments to support and extend Torricelli's results. His reports and his conclusion on the existence of the vacuum and the influence of altitude on the weight of the air caused many disputes before they were accepted.

Together with Fermat he laid the foundation of probability theory in 1654, and at the same time he wrote the important treatise on the arithmetic triangle. From 1658 to 1659 he wrote a series of papers on the properties of the cycloid in which he used summation of approximating rectangles to find the area of the segments of the cycloid, and he likewise computed the volumes of solids generated by the rotation of such segments.

In the 1640s he suffered the first attacks of the illness which intermittently disabled him for the rest of his life. After the death of his father in 1651, he lived on his inherited fortune. In 1646 the Pascal family converted to Jansenism, and late in 1654 Pascal had his "second conversion," and from that time on he was mostly concerned with religious problems. For a time he retreated to a group of Jansenists at Port Royal. Attacks of the Jesuits on the Jansenists, in particular on Antoine Arnauld, professor of theology at the Sorbonne, caused him to take part in polemic by writing his *Lettres provinciales* in 1656. In his last years he worked on a treatise on religious philosophy, an *Apology for the Christian Religion*. Fragments found after his death were published in 1670 as the *Pensées*.

Pascal was a controversial and paradoxical man, a great writer with an unusual talent for polemic. He was deeply split between his commitment to mathematics and natural science and his religious commitment. The literature on Pascal is enormous. The sketch above is mainly based on the biography by Taton (1974), where further details and references may be found.

The famous correspondence of Pascal and Fermat consists of seven letters exchanged between July and October 1654. Three letters from Pascal to Fermat were published in the 1679 edition of Fermat's works; these letters were reprinted together with four letters from Fermat to Pascal, one of them addressed to Carcavi, in the 1779 edition of Pascal's works. The whole correspondence may be found in *Oeuvres de Fermat*, Vol. 2, 1894, or in any of the many editions of *Oeuvres de Pascal*. English translations may be found in the books by Smith (1929) and David (1962).

Ore (1960) has given a vivid account of the relation between de Méré and Pascal. Short biographies of Pascal and Fermat have been given by David (1962) along with a critical analysis of their correspondence.

Besides the letters, Pascal's contributions to probability theory are given in his *Traité du triangle arithmétique, avec quelques autres petits traités sur la même matière*, printed in 1654 but only released to the public in 1665. From one of Fermat's letters it is clear that Fermat has a copy in 1654, and the *Traité* should therefore be considered an integral part of the correspondence. An English translation of the first part of the *Traité* may be found in Smith (1929).

5.2 PASCAL'S ARITHMETIC TRIANGLE AND SOME OF ITS USES

The arithmetic triangle is an array of numbers, beginning with unity, such that for each triangular set of numbers, the sum of the first two on a diagonal equals the third on the next diagonal, as shown in the following table:

Pascal's arithmetic triangle

```
1   1   1   1   1   1   1
1   2   3   4   5   6
1   3   6  10  15
1   4  10  20
1   5  15
1   6
1
```

The fundamental properties of these numbers had been known for some time before Pascal wrote his treatise. The importance of Pascal's work lies in its clear and systematic exposition, his rigorous proofs, and his application of the arithmetic triangle to the solution of many related problems concerning the figurate numbers, the number of combinations, the binomial expansion, sums of powers of equidistant integers, and the problem of points. In the following we shall give a summary of the contents of Pascal's treatise apart from his discussion of the problem of points, which is discussed in §5.3.

As indicated above, Pascal defines the numbers in the arithmetic triangle by recursion. Consider a two-way table, and let the number in the $(m + 1)$st row and the $(n + 1)$st column, t_{mn}, say, be defined by *the recursion formula*

$$t_{mn} = t_{m-1,n} + t_{m,n-1}, \quad (m, n) = 0, 1, \ldots, \quad (m, n) \neq (0, 0),$$

the boundary conditions

$$t_{m,-1} = t_{-1,n} = 0, \quad (m, n) = 1, 2, \ldots,$$

and the generator $t_{00} = 1$. Pascal gives the values of t_{mn} up to $m = n = 9$; an abridged version has been shown above. Besides the numbers Pascal also labels the cells by different letters for use in the proofs. The many latin and Greek letters make his proofs rather awkward. We have therefore introduced the notation t_{mn} as shown in the following table, and we have transcribed his proofs accordingly.

Pascal's arithmetic triangle

$m \backslash n$	0	1	2	3	4	5	6
0	t_{00}	t_{01}	t_{02}	t_{03}	t_{04}	t_{05}	t_{06}
1	t_{10}	t_{11}	t_{12}	t_{13}	t_{14}	t_{15}	
R_{24} 2	t_{20}	t_{21}	t_{22}	t_{23}	t_{24}		
3	t_{30}	t_{31}	t_{32}	t_{33}			
4	t_{40}	t_{41}	t_{42}				
5	t_{50}	t_{51}					
D_{64}—6	t_{60}						

C_{24}

Pascal points out that one may use a generator different from 1, and in several places he indicates the effects of such a change.

Pascal proves 20 propositions about t_{mn}, ending with the multiplicative form

$$t_{mn} = \frac{(m+n)(m+n-1)\cdots(m+1)}{n(n-1)\cdots 1}.$$

All his results are given in verbal form, making use of row sums, column sums, and diagonal sums, which we shall write as

$$R_{mn} = \sum_{j=0}^{n} t_{mj}, \qquad C_{mn} = \sum_{i=0}^{m} t_{in},$$

$$D_{m+n,n} = \sum_{j=0}^{n} t_{m+n-j,j}, \qquad D_{m+n} = D_{m+n,m+n},$$

as indicated in the table above. The notation corresponds closely to Pascal's verbal formulation; compared with other notations it has the advantage that symmetry properties are easily exhibited.

Most of the propositions are fairly simple, and Pascal proves them by repeated application of the recursion formula. We shall list the first 11 propositions without comment and leave the proof and interpretation to the reader:

$$t_{0n} = t_{m0} = t_{00} = 1. \tag{1}$$

Properties of row and column sums:

$$t_{mn} = R_{m-1,n} = C_{m,n-1}, \tag{2) and (3}$$

$$t_{mn} - 1 = \sum_{i=0}^{m-1} R_{i,n-1} = \sum_{j=0}^{n-1} C_{m-1,j}. \tag{4}$$

Properties of symmetry:

$$t_{mn} = t_{nm}; \qquad \{t_{mj}, \, j = 0, 1, \ldots\} = \{t_{jm}, \, j = 0, 1, \ldots\}. \tag{5) and (6}$$

Properties of diagonal sums:

$$D_{k+1} = 2D_k; \qquad D_k = 2^k, \tag{7) and (8}$$

$$\sum_{i=0}^{k} D_i = D_{k+1} - 1, \tag{9}$$

$$D_{k,n} = D_{k-1,n} + D_{k-1,n-1}, \tag{10}$$

$$t_{mm} = 2t_{m,m-1} = 2t_{m-1,m}. \tag{11}$$

Pascal points out that Proposition (12) about the *ratio*

$$\frac{t_{m+1,n-1}}{t_{mn}} = \frac{n}{m+1}, \tag{12}$$

is of particular importance. The proof is by induction. Incomplete induction had been used since Euclid, but the formulation given by Pascal is explicit and complete. He writes

> Since this proposition covers an infinity of cases, I shall give a short proof based on two lemmas. The first lemma, which is evident, says that the proposition holds for $m + n = 1$, because it is clear that the ratio is as 1 to 2. The second lemma says, that if the proposition holds for a given base [i.e. for a given value of $m + n$], then it necessarily holds also for the following base.

Combining the two lemmas, he concludes that the proposition holds for $m + n = 2$, and then by Lemma 2 for $m + n = 3$, and so on to infinity.

Pascal's proof is as follows. Setting $m + n = k$, say, the proposition may be written as

$$\frac{t_{k-r,r}}{t_{k-r-1,r+1}} = \frac{r+1}{k-r}, \qquad r = 0, 1, \ldots, k-1.$$

Considering the ratio for $k + 1$ and using the recursion formula, he obtains

$$\frac{t_{k+1-r,r}}{t_{k-r,r+1}} = \frac{t_{k-r,r} + t_{k+1-r,r-1}}{t_{k-r-1,r+1} + t_{k-r,r}}.$$

Dividing both numerator and denominator by $t_{k-r,r}$ and using the hypothesis, the ratio becomes

$$\frac{1 + r/(k - r + 1)}{1 + (k - r)/(r + 1)},$$

which equals $(r + 1)/(k + 1 - r)$, as was to be proved.

By means of (12) it is easy to prove the following properties of the ratios:

$$\frac{t_{m+1,n}}{t_{mn}} = \frac{m+n+1}{m+1}; \qquad \frac{\cdot\, t_{m,n+1}}{t_{mn}} = \frac{m+n+1}{n+1}. \tag{13 and 14}$$

Combining these results with (2) and (3), it follows that

$$\frac{R_{mn}}{t_{mn}} = \frac{m+n+1}{m+1}; \qquad \frac{R_{mn}}{R_{m+1,n-1}} = \frac{m+2}{n}, \qquad \text{(15) and (16)}$$

$$\frac{R_{mn}}{C_{mn}} = \frac{n+1}{m+1}; \qquad \frac{R_{mn}}{R_{nm}} = \frac{n+1}{m+1}. \qquad \text{(17) and (18)}$$

From (11) and (14), it follows that

$$\frac{t_{m+1,m+1}}{4t_{mn}} = \frac{2m+1}{2m+2}. \tag{19}$$

Finally, since

$$\frac{t_{mn}}{t_{m+n,0}} = \frac{t_{mn}}{t_{m+1,n-1}} \frac{t_{m+1,n-1}}{t_{m+2,n-2}} \quad \cdots \quad \frac{t_{m+n-1,1}}{t_{m+n,0}},$$

repeated applications of (12) give

$$t_{mn} = \frac{(m+n) \quad (m+n-1) \quad \cdots \quad m+1}{n(n-1) \quad \cdots \quad 1}, \tag{20}$$

which is the multiplicative form of t_{mn}.

The recursion formula may be looked upon as a partial difference equation, and Pascal has thus given one of the first examples of an explicit solution of such an equation and a discussion of the properties of the solution.

Today these results are included in textbooks on combinatorics and probability but usually starting with the definition of the binomial coefficients as

$$t_{mn} = \binom{m+n}{m} = \frac{(m+n)!}{m!n!}, \tag{21}$$

so that we start with Pascal's result (20). Obviously, the recursion formula becomes

$$\binom{m+n}{m} = \binom{m+n-1}{m-1} + \binom{m+n-1}{n-1}, \tag{22}$$

and (2) may be written as

$$\sum_{j=0}^{n} \binom{j+m-1}{m-1} = \binom{m+n}{m}. \tag{23}$$

In *Traité des ordres numériques* and *De numericis ordinibus tractatus*, Pascal applies his results to a discussion of the figurate numbers, which he defines as the repeated sums of 1. These numbers had been studied before Pascal, partly because they enumerate the number of (equidistant) points contained within certain figures (hence the name) and partly because of their simple additive structure. He arranges these numbers in a two-way table as shown below:

Pascal's Table of Figurate Numbers

	ROOT				
ORDER	1	2	3	4	5
1	1	1	1	1	1
2	1	2	3	4	5
3	1	3	6	10	15
4	1	4	10	20	35

Any number, 10, say, at row 4 and column 3 is obtained as the sum of the numbers in the row above ending at the given column $10 = 1 + 3 + 6$. This property is, however, exactly that proved as Proposition (2) above. The figurate numbers are therefore in a one-to-one correspondence to the numbers in the arithmetic triangle, which is also obvious by looking at the numbers in the diagonals, and the properties of the figurate numbers therefore follow from the results above.

Pascal shows how to solve the equation

$$t_{m-1,n} < c \leqslant t_{mn},$$

for given n and c by a simple numerical procedure. He further discusses the properties of the product $m(m+1) \cdots (m+n-1)$ and shows how the largest integral root of the equation $m^n = c$ may be found.

In *Usage du Triangle Arithmétique pour trouver les Puissances des Binômes et Apotomes*, Pascal shows by example the connection between the binomial coefficients and the numbers in the arithmetic triangle, i.e.,

$$(a+b)^k = \sum_{i=0}^{k} t_{k-i,i} a^{k-i} b^i, \qquad k = 1, 2, \ldots. \tag{24}$$

He refrains from giving a general proof, partly because it is somewhat obvious, and partly because it has been given by Hérigone and others.

In *Usage du Triangle Arithmétique pour les Combinaisons* and *Combinationes*, Pascal first defines the different combinations of k elements from a set of n elements, say, as all the different sets of k elements without regard to order. Pascal does not introduce a notation for the number of combinations; we shall use here the notation C_k^n, which later became popular.

Pascal begins by proving four lemmas on C_k^n. The first three,

$$C_k^n = 0 \quad \text{for} \quad k > n; \qquad C_n^n = 1; \qquad \text{and} \qquad C_1^n = n;$$

follow directly from the definition. Lemma 4,

$$C_k^{n+1} = C_k^n + C_{k-1}^n, \tag{25}$$

is demonstrated by various examples. Pascal's reasoning is as follows. The $n + 1$ elements may be regarded as n elements plus a specific element, A, say, and C_k^{n+1} consists of all the combinations without A and those containing A. The first equals C_k^n, and the second equals C_{k-1}^n because only $k - 1$ elements should be selected besides A.

Next follows the basic proposition on the relation between the row sum in the arithmetic triangle and the number of combinations,

$$R_{n-k,k} = C_k^{n+1}, \qquad k = 0, 1, \ldots, n.$$

The proof is by induction. Suppose that $R_{n-1-k,k} = C_k^n$. Then,

$$R_{n-k,k} = R_{n-k,k-1} + t_{n-k,k}$$
$$= R_{n-1-(k-1),k-1} + R_{n-1-k,k} \quad \text{[from (2)]}$$
$$= C_{k-1}^n + C_k^n \quad \text{(from the hypothesis)}$$
$$= C_k^{n+1} \quad \text{[from (25)]}.$$

Since the proposition obviously holds for $n = 0$, the proof is complete.

It then follows immediately that

$$C_k^{n+1} = t_{n+1-k,k} = \binom{n+1}{k}, \tag{26}$$

which establishes the relation between the number of combinations and the numbers in the arithmetic triangle.

In the *Combinationes* Pascal gives a rather trivial discussion of the properties of C_k^n derived from the properties of $t_{n-k,k}$. The most important part of the paper is the section on *Sommation des Puissances Numériques* in which Pascal solves the problem of summation of powers of the terms of an arithmetic series by recursion. Pascal points out that this result may be used to find the area under a parabolic curve of any order. Pascal's proof is as follows. Let

$$a_i = a + id, \qquad i = 0, 1, \ldots,$$

where a and d are positive integers, and set

$$s_m = \sum_{i=0}^{n-1} a_i^m, \qquad m = 0, 1, \ldots.$$

The problem is to find s_m in terms of $s_0, s_1, \ldots, s_{m-1}$. By means of the binomial theorem, Pascal finds

$$a_{i+1}^{m+1} - a_i^{m+1} = (a_i + d)^{m+1} - a_i^{m+1}$$
$$= \sum_{x=1}^{m+1} \binom{m+1}{x} a_i^{m+1-x} d^x.$$

Summation from $i = 0$ to $i = n - 1$ gives

$$a_n^{m+1} - a_0^{m+1} = \sum_{x=1}^{m+1} \binom{m+1}{x} s_{m+1-x} d^x.$$

Solving for s_m, $m \geq 1$, Pascal finds

$$(m+1)ds_m = a_n^{m+1} - a_0^{m+1} - \sum_{x=2}^{m+1} \binom{m+1}{x} s_{m+1-x} d^x.$$

For $a = d = 1$, Pascal's formula becomes a recursion formula for the sums of powers of the first n integers,

$$(m+1)s_m = (n+1)^{m+1} - 1 - \sum_{x=2}^{m+1} \binom{m+1}{x} s_{m+1-x}, \qquad m = 1, 2, \ldots. \quad (27)$$

Beginning with $s_0 = n$, it follows that s_m may be written as a polynomial in n of degree $m + 1$, the main term being $n^{m+1}/(m+1)$. It would thus have been easy to tabulate the first of these polynomials. However, Pascal does not mention this problem; it was left to J. Bernoulli (1713) to give an explicit formula for s_m (see §15.4).

The history of the formula for sums of powers of integers has been studied by Edwards (1982a) and Schneider (1983a, b). It turns out that the German mathematician Johannes Faulhaber (1580–1635) between 1614 and 1631 published the polynomial expressions for sums of powers of integers up to the 17th power. Faulhaber had found that

$$s_{2m} = s_2 f_{m-1}(s_1); \qquad s_{2m+1} = s_3 g_{m-1}(s_1),$$

where

$$s_1 = \tfrac{1}{2}[(n+1)n], \qquad s_2 = \tfrac{1}{6}[(2n+1)(n+1)n], \qquad s_3 = s_1^2,$$

and f_{m-1} and g_{m-1} denote polynomials of degree $m-1$ with undetermined coefficients, which he found from the m linear equations involving the power sums of the first m natural numbers.

Edwards' (1987) history of Pascal's arithmetic triangle was published after the completion of the present chapter. It contains many important new observations on developments before Pascal, both in the East and in Europe; we shall, however, confine ourselves to a few remarks on developments in Europe. Edwards writes, "Pascal was, as we shall see, a little forgetful about his sources. Practically everything in the *Traité* except the solution to the important 'Problem of Points' will have been known to Mersenne's circle by 1637."

Edwards devotes separate chapters to the early history of the figurate numbers, the combinatorial numbers, and the binomial numbers. The figurate and the binomial numbers are defined by their additive properties and are mainly used in pure mathematical contexts, whereas the combinatorial numbers are the outgrowth of many practical problems on the composition of letters, syllables, musical notes, ingredients of a mixture, prime numbers, and so on. Essential problems have been to establish the correspondence between the three kinds of numbers and to derive the multiplicative form from the additive property.

We shall give a chronological list of important results without mentioning the original works; references may be found in Edwards' book.

The first general results on combinatorics in Europe are due to Levi ben Gerson (1321), who lived in France. He derived the formulae $n!$, $n^{(r)} = n(n-1)\cdots(n-r+1)$, and $C_r^n = n^{(r)}/r!$ for the number of permutations of n different things, the number of permutations of n things taken r at a time, and the number of combinations of n things taken r at a time, respectively. His proofs are by complete induction, and he is now credited with being the first to have used this method correctly (see Rabinovitch, 1970b). Previously, first Bernoulli and then Pascal got the credit.

Stifel (1544, 1545) and Scheubelius (1545), both from Germany, used the figurate and binomial numbers in connection with the extraction of roots.

Tartaglia (1556) found that the number of ways n dice can fall is given by a figurate number. Cardano (1570) found the multiplicative form of the figurate numbers and pointed out the correspondence between the combinatorial and the figurate numbers.

Vieta (1591) found that the coefficients in his formula for $\sin nx$ in terms of $\sin x$ were figurate numbers.

Briggs (1633) used the figurate and binomial numbers in binomial expansions, in trigonometric calculations like Vieta, and in interpolation formulae.

In 1636 Fermat found the multiplicative form of the figurate numbers and used these numbers in his formula for the summation of power of integers; he communicated his results to Mersenne.

Mersenne (1636, 1637) extended Cardano's table of C_r^n up to $n = 36$ and discussed the three formulae given by Levi ben Gerson. He also gave the formula

$$\frac{n!}{a!\,b!\,c!\cdots}$$

for the number of permutations of n things of which a things are of one kind, b of another, and so on. Mersenne used these results in connection with the arrangement of musical notes.

No doubt Pascal knew Mersenne's books; Pascal's contributions are thus essentially his proofs and the clear demonstration of the relations between the figurate, binomial, and combinatorial numbers.

5.3 THE CORRESPONDENCE OF PASCAL AND FERMAT AND PASCAL'S TREATISE ON THE PROBLEM OF POINTS

Besides the problem of points, Pascal and Fermat also discussed two rather simple dice games, which we shall discuss first.

In his letter of 29 July 1654, Pascal writes to Fermat on the simplest of de Méré's problems:

> If one undertakes to throw a *six* with one die, the advantage of undertaking it in 4 throws is as 671 to 625. If one undertakes to throw a *double-six* with two dice,

there is a disadvantage of undertaking it in 24 throws. And nevertheless 24 is to 36 (which is the number of faces of two dice) as 4 to 6 (which is the number of faces of one die).

Pascal writes that he does not have time to give the solution and that Fermat will easily find it himself. The method for solving this problem had been given earlier by Cardano. It amounts to finding the smallest number of trials, n, say, such that

$$\frac{1-q^n}{q^n} \geq 1 \quad \text{or} \quad q^n \leq \tfrac{1}{2}.$$

For $p = 1/6$, one finds $n = 4$ and the odds $671/625$ as given by Pascal, and for $p = 1/36$, one gets $n = 25$ and the approximate odds $506/494$, whereas $n = 24$ gives $491/509$.

Obviously, de Méré's reasoning presupposes that n is proportional to 6^k, $k = 1, 2, \ldots$. Since $p = 6^{-k}$ and $(1-p)^n \cong e^{-np}$, the equation $(1-p)^n = \tfrac{1}{2}$ has the approximate solution

$$n = (\ln 2)6^k = 0.6931 \times 6^k, \qquad k = 1, 2, \ldots,$$

which gives $n = 4.16$ for $k = 1$ and $n = 24.95$ for $k = 2$. Since n must be an integer, the strict proportionality is destroyed. The approximate solution is due to de Moivre (1712), see §14.4.

In an undated letter from Fermat to Pascal, presumably the earliest one of those preserved, they discuss the value of a throw in a dice game in which the player undertakes to throw a 6 (i.e. at least one 6) in eight throws. Fermat says that after the stakes have been made, the player should have 1/6 of the total stake in compensation for not making the first throw. If, after this has been settled, they further agree that the player should not make the second throw, he should have 1/6 of the remaining stake as compensation, i.e. 5/36, of the total stake, and so on. Fermat calls this the *value* of the throw and remarks that this result is in agreement with the one communicated to him by Pascal. The value of the kth throw is thus $(1/6)(5/6)^{k-1}$, $k = 1, 2, \ldots, 8$, times the total stake, and by summation the value of the eight throws becomes $1 - (5/6)^8 = 0.767$ times the total stake, which is the compensation the player should have for giving up the play altogether. (This last result is not given by Fermat.) It will be seen that Fermat's value equals the probability of winning times the total stake. Suppose now that the player has made three throws without success. What is then the value of the fourth throw? It seems that Pascal in a previous letter, not longer extant, has given this value as 125/1296, whereas Fermat states that according to his principle, the value is

$1/6$ of the remaining stake, i.e., $1/6$ of the total stake. Fermat stresses that this principle holds no matter the number of unsuccessful throws. Presumably he means that the value of the remaining five throws should be found as for the eight original throws, i.e., as $(1/6)(5/6)^{k-1}$, $k = 1, 2, \ldots, 5$, so that the value of the five remaining throws becomes $1 - (5/6)^5 = 0.598$ times the total stake. Hence if the play is interrupted after m unsuccessful throws, the player should have $1 - (5/6)^{8-m}$, $m = 0, 1, \ldots, 7$, times the total stake in compensation for giving up the play. Expressed in modern terminology it will be seen that Fermat's reasoning is based on the geometric distribution $p(1 - p)^{k-1}$, $k = 1, 2, \ldots$, for $p = 1/6$.

Before discussing the correspondence further, we shall give an account of Pascal's *Usage du Triangle Arithmétique pour déterminer les partis qu'on doit faire entre deux joueurs qui jouent en plusieurs parties* in which he solves the division problem, first by recursion on the expectations and next by use of the arithmetic triangle. He always assumes that the two players have an equal chance of winning in a single game.

He begins by stating two obvious division principles (axioms): (1) If a player gets a certain sum, whatever the outcome of the game this sum should be allotted to him; (2) if the total stake falls to the winner with probability $1/2$ and the game is not played, the stake should be divided equally between the two players. Formulated in another way we may say that the *value* of the game or the player's *expectation* equals $t/2$ if the total stake equals t.

From these principles he proves Corollary 1: If a player gets the amount s, when he loses, and the amount $s + t$ when he wins, and the game is not played, then he should have the amount $s + t/2$. In the proof he first uses principle (1), which says that the player should first have the amount s, which belongs to him independently of the outcome of the game, and furthermore he should have $t/2$ accordingly to principle (2). In Corollary 2 he simply remarks that the same result may be obtained by adding the two amounts and dividing by 2. It will be seen that Pascal in these corollaries derives the player's expectation for the case where the two outcomes of the game correspond to different prizes, which need not be limited to the total stake and zero as in principle (2).

Like Cardano (see §4.2), Pascal explains that the division rule should depend only on the number of games each player lacks in winning. We shall let $e(a, b)$ denote A's share of the total stake, or A's expectation, if the play is interrupted when A lacks a games and B lacks b games in winning. Pascal begins by discussing six examples for a total stake equal to 8; later he changes to the stake 1, which means that the expectation equals A's probability of winning. Since Pascal keeps strictly to the division problem, he always speaks of the *fraction* of the stake going to A; he does not use the term probability or chance in this context. For convenience we shall usually set the total stake

Table 5.3.1. Pascal's recursive calculation of $e(a, b)$

	a	b	$e(a-1,b)$ $e(a,b-1)$	$e(a,b)$
(1)	0	$n > 0$		1
(2)	n	n		1/2
(3)	1	2		
(3a) A wins	0	2	1	3/4
(3b) A loses	1	1	1/2	
(4)	1	3		
(4a) A wins	0	3	1	7/8
(4b) A loses	1	2	3/4	
(5)	1	4		
(5a) A wins	0	4	1	15/16
(5b) A loses	1	3	7/8	
(6)	2	3		
(6a) A wins	1	3	7/8	11/16
(6b) A loses	2	2	1/2	

equal to unity so that $e(a, b)$ represents both A's expectation and his probability of winning; the interpretation will be clear from the context. Pascal usually sets the stake equal to a power of 2 to avoid fractions in $e(a, b)$. Table 5.3.1 gives Pascal's results divided by 8, i.e., A's probability of winning.

The results stated in the first two examples are obvious. In the third and the following examples Pascal considers the two possible outcomes of the next game: Either A wins, or he loses. For each of these cases the expectation is known from the previous calculations, and the final expectation is then found as the average of the two intermediate ones according to Corollary 2. Pascal concludes that the method may be used in general. It will be seen that he uses a conditional argument corresponding to the formula

$$E\{X\} = \{E(X|A) + E(X|\bar{A})\}/2.$$

Using modern notation, Pascal's procedure may be written as

$$e(0, n) = 1 \quad \text{and} \quad e(n, n) = \tfrac{1}{2}, \quad n = 1, 2, \ldots,$$
$$e(a, b) = \tfrac{1}{2}[e(a-1, b) + e(a, b-1)], \quad (a, b) = 1, 2, \ldots. \tag{1}$$

This is a simple example of a partial difference equation. It also provides a good illustration of the interdependence of mathematics and probability theory. To obtain an explicit expression for $e(a, b)$ the difference equation has to be solved, which Pascal did by means of his results for the arithmetic triangle. In the 18th century Pascal's recursive method became very popular and helped to solve more difficult problems, which induced by Moivre, Lagrange, and Laplace to develop general methods for the solution of difference equations, as discussed in Chap. 23.

For large values of (a, b), numerical recursion as shown in Table 5.3.1 is cumbersome. Possibly for this reason Pascal looked for a more expedient solution, which he found by comparing the recursion formula for the expectations with that for the numbers in the arithmetic triangle, noting that they differ only by a factor of $1/2$ and by the boundary conditions. He therefore proposes that

$$e(a, b) = \frac{1}{D_{a+b-1}} \sum_{j=a}^{a+b-1} t_{a+b-1-j,j}, \qquad a+b = 2, 3, \ldots, \qquad (2)$$

the numerator being equal to the last b terms of the diagonal sum and the denominator being equal to 2^{a+b-1}. Because of symmetry the last b terms equal the first b so that

$$e(a, b) = \frac{D_{a+b-1,b-1}}{D_{a+b-1}}. \qquad (3)$$

Pascal proves this formula by induction. Since it obviously holds for $a + b = 2$, he assumes that it holds for $a + b = k$ and proves that it holds for $a + b = k + 1$. To find $e(a, k + 1 - a)$, he uses the recursion formula

$$e(a, k + 1 - a) = \frac{e(a - 1, k + 1 - a) + e(a, k - a)}{2}$$

$$= \frac{D_{k-1,k-a} + D_{k-1,k-a-1}}{2D_{k-1}}$$

$$= \frac{D_{k,k-a}}{D_k},$$

which follows from (2.7) and (2.10). This concludes the proof.

It will be seen that (2) and (3) may be written as

$$e(a, b) = \sum_{i=a}^{a+b-1} \binom{a+b-1}{i} \left(\frac{1}{2}\right)^{a+b-1} = \sum_{i=0}^{b-1} \binom{a+b-1}{i} \left(\frac{1}{2}\right)^{a+b-1}. \qquad (4)$$

Pascal goes on to investigate some special cases. For $a < b$ he finds *the fraction of B's stake going to A*, assuming that both stake the same amount. Obviously, this fraction equals $2[e(a, b) - 1/2]$. By means of (3) and the properties of the triangular numbers, he finds

$$2\left[e(1,b) - \frac{1}{2}\right] = 1 - \frac{1}{D_{b-1}} = 1 - \left(\frac{1}{2}\right)^{b-1}, \tag{5}$$

$$2\left[e(b-1,b) - \frac{1}{2}\right] = \frac{t_{b-1,b-1}}{D_{2b-2}} = \binom{2b-2}{b-1}\left(\frac{1}{2}\right)^{2b-2}, \tag{6}$$

$$2\left[e(b-2,b) - \frac{1}{2}\right] = \frac{2t_{b-2,b-1}}{D_{2b-3}} = \binom{2b-3}{b-2}\left(\frac{1}{2}\right)^{2b-4}, \tag{7}$$

and he remarks that the last result is two times the last but one, or that A's gain by winning the first point is the same as the gain by winning the second. This means that he is considering *the fraction of B's stake going to A when A is winning a single game or point.* He does not, however, derive the general result

$$2[e(a,b) - e(a+1,b)] = \frac{t_{a,b-1}}{D_{a+b-1}} = \binom{a+b-1}{a}\left(\frac{1}{2}\right)^{a+b-1}, \tag{8}$$

which is easily proved by means of the recursion formula and (3). This result shows that the value of winning a point measured in terms of the opponent's stake equals the binomial probability (8).

Pascal's paper is purely mathematical, and he does not attempt any combinatorial interpretation of his results even though he mentions that the division rule may be found also by combinatorial methods. This topic is discussed in the correspondence with Fermat.

Several letters of the correspondence are missing, among them Pascal's first letter with the formulation of the problem and his solution and Fermat's answer in which he proposes to solve the problem by combinatorics. However, in Pascal's letter of 24 August 1654, he describes in detail Fermat's procedure. Fermat says that if the players lack a and b games, respectively, in winning then the play will be over after *at most* $a + b - 1$ further games. Imagining that these games are actually played, there will be 2^{a+b-1} equally possible outcomes and the cases favorable to A in relation to 2^{a+b-1} will give his fraction of the stake. For $(a,b) = (2,3)$ there will be $2^4 = 16$ outcomes, which may be listed as AAAA, AAAB, AABA, AABB, ..., BBBB, the letters indicating the winner of each game. Since there are 11 cases favorable to A, Fermat's rule gives $e(2,3) = 11/16$. In Pascal's letter of 29 July he makes the following comment on this procedure: "Your method is very safe and is the one which first came to my mind in this research; but because the trouble of the combinations is excessive I have found an abridgement and indeed another method that is much shorter and more neat, which I would like to tell you here in a few words." Pascal then proceeds to give an exposition of his recursive method by means of three examples like those we have already discussed.

By example he also proves (5) and (6). Presumably he was working on his treatise when he wrote the letter because a month later, Fermat in a letter of 29 August acknowledges receipt of a copy of the *Traité*. Pascal's letter is not as clear, systematic, and comprehensive as the *Usage de Triangle Arithmétique*, but it contains the main ideas and a complete table for the case $b = 6, a = 1, 2, \ldots, 6$ and a stake of 512. Presumably Pascal has first calculated $e(a, b)$ by recursion; he gives a table of $e(a, b) - 1/2$, and by differencing he gets a table of $e(a, b) - e(a + 1, b)$, i.e., A's share of B's stake for each game won. He does not understand certain features of this table, which indicates that he does not know the general result expressed by Equation (8).

Returning to the discussion of the combinatorial method, Pascal writes that he has communicated Fermat's solution to some colleagues in Paris and that Roberval, professor of mathematics, has objected to it because the division rule depends on games which possibly will not be played. Pascal replies that this is immaterial because the two procedures, the actual and the hypothetical play, will always give the same result for the two players, since the cases in which A wins or loses in the actual play will also be the cases in which he wins or loses in the hypothetical play. This is easily seen from the example above, since in each combination of four letters only one of the players can win, and the same winner comes out whether all four games are played or the play is stopped after the winning score.

In his reply of 25 September, Fermat approves of Pascal's explanation and adds two important remarks to convince Roberval. First he says that it does not affect the division how far the fictitious play is extended over $a + b - 1$ games. If in the example the play is extended to five instead of four games, the effect will be that each of the 16 combinations will be changed to two by adding an A or a B, and this does not change the relative number of cases in which A wins.

Although Fermat illustrates his second remark by an example with three players, we shall use it here in connection with the example above for $(a, b) = (2, 3)$. It is clear that A may win in two, three, or four games, corresponding to the arrangements AA, ABA and BAA, and ABBA, BABA, BBAA, which each contains two A's *with an A at the last place*. For two games there are 2^2 equally possible outcomes with one favorable to A; for three games there are 2^3 outcomes with two favorable ones; and for four games, 2^4 outcomes with three favorable ones. The chances for A to win may thus be found as the sum

$$e(2,3) = \frac{1}{2^2} + \frac{2}{2^3} + \frac{3}{2^4} = \frac{11}{16},$$

in agreement with the result previously found. For the case (a, b) it is clear that A may win at either the ath, the $(a + 1)$st, ..., or the $(a + b - 1)$st game. The number of different arrangements of a A's and i B's with an A at the last place equals the number of combinations of $a - 1$ A's and i B's, so that

$$e(a, b) = \sum_{i=0}^{b-1} \binom{a - 1 + i}{a - 1} \left(\frac{1}{2}\right)^{a + i}. \tag{9}$$

Fermat does not give this general result explicitly, but he writes

> And this rule is good and general for all cases, so that without recourse to fictitious games, the actual combinations in each number of games give the solution and show what I said at the outset, that the extension to a certain number of games is nothing else than the reduction of various fractions to the same denominator.

Hence, Pascal and Fermat both solved the problem of points by combinatorial methods; Pascal also solved it by recursion and showed that the explicit solution of the difference equation is identical to the combinatorial solution; finally Fermat used what is called today *a waiting-time argument* to find the solution (9), which erroneously has been called Pascal's distribution.

It is thus clear that Pascal solved the problem of points by means of *the binomial distribution* for $p = \frac{1}{2}$. Possibly there will not be general agreement about the statement above that Fermat solved the problem by means of *the negative binomial distribution* (9), as it is called today. This result is usually ascribed to Montmort (see §14.1). The reader should check that (4) and (9) give identical results.

To summarize and clarify the relations between the important concepts and results of the correspondence, we have shown the arithmetic triangle in a random walk diagram in Fig. 5.3.1.

The random walk starts at $(0,0)$. If A wins a point the path goes one step to the right, and if B wins it goes one step up, the two paths being equally likely. It follows that only the states lying on a diagonal are equally likely and that the triangular numbers (binomial coefficients, number of combinations) give the number of paths from $(0,0)$ to the state in question, as shown on the kth diagonal. Player A wins the stake if he gets a points

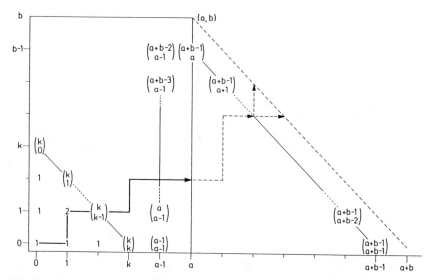

Fig. 5.3.1. Random walk diagram for the problem of points. The number of paths from $(0, 0)$ to (a, b) equals $\binom{a+b}{a}$; A wins if he gets a points before B gets b.

before player B gets b, which means that A wins if the random walk ends at the vertical line from $(a, 0)$ to $(a, b - 1)$. These states are, however, not equally likely, and the number of paths can therefore not be added to find the number of chances for A. Consider now a fictitious continuation of the play from the b winning states on the vertical until $a + b - 1$ games have been played; i.e., until the random walk ends at the $(a + b - 1)$st diagonal. It is easy to see that the paths going through the winning states and the paths ending at the $(a + b - 1)$st diagonal for the abscissas a to $a + b - 1$ are the same. The number of chances for A may therefore be found by adding the number of paths ending at this diagonal, which leads to Pascal's result (2). The dashed part of the path represents the fictitious games, and Fermat's argument about the continuation of the fictitious play has been illustrated by the $(a + b)$th diagonal. Fermat's result (9) corresponds to the number of paths ending at the vertical from $(a - 1, 0)$ to $(a - 1, b - 1)$ and continuing horizontally to the vertical through $(a, 0)$. The result (8) may also be derived immediately from the diagram by comparing the $(a + b - 1)$st and the $(a + b)$th diagonals.

The problem of *the division rule for three players*, A, B, and C, lacking 1, 2, and 2 games, respectively, is discussed in Pascal's letter of 24 August, presumably after some preliminary remarks by Fermat in a letter no longer extant. Pascal writes, "When there are only *two* players, your combinatorial method is very sure, but when there are *three* I think I can demonstrate that it is incorrect unless you proceed in some other manner which I do not understand." He then gives an analysis of the play analogous to the one he has just given for two players by listing the 27 different arrangements of three letters: AAA, AAB, AAC, ABA,..., CCC, since the play will be completed in at most three games. The problem is now how these arrangements should be allotted to the three players. First he counts all the arrangements containing one A, two B's, and two C's, respectively, which gives the distribution 19:7:7. He then says that this method is erroneous because some of the arrangements are included two times as favorable to two players, such as ABB. Next he amends this procedure by allotting $\frac{1}{2}$ to each of the two players, which leads to the distribution $16:5\frac{1}{2}:5\frac{1}{2}$. Finally, he says that this solution is also wrong, because his general method (recursion) gives the result 17:5:5. In his reply of the 25 September, Fermat points out that the *order* of the letters has to be taken into account because what happens after one of the players has won is of no importance. If that is done, the combinatorial method and the recursive method will give the same result. The fictitious games only serve to make all the arrangements equally likely.

The methods developed by Pascal and Fermat were used by John Bernoulli, Montmort, and de Moivre to solve the problem of points for players with different probabilities of winning a single game (see §14.1).

After Montmort in his *Essay* (1713) gave a short history of the development of probability theory from 1654 to 1713, many authors have commented on the correspondence of Pascal and Fermat and Pascal's *Traité*. We shall limit ourselves to mention some of the more recent contributions in chronological order: David (1962), Coumet (1970), Jordan (1972), Dupont (1975–1976 and 1979), Seneta (1979), and Edwards (1982b, 1987).

The correspondence of Pascal and Fermat was resumed in 1656 when Pascal posed Fermat a problem which today is known as the problem of the Gambler's Ruin. Through Carcavi the problem was passed on to Huygens who included it in his treatise as *the fifth problem* to be solved by the reader. Pascal, Fermat, and Huygens all solved the problem numerically without disclosing their methods (see Edwards, 1983). We shall discuss this problem in §6.3 and §14.2.

5.4   PASCAL'S WAGER

After his conversion in November 1654, Pascal for a time retreated to the religious community at Port Royal where he worked together with the theologians Pierre Nicole and Antoine Arnauld for the Jansenist cause. For use in the upper classes of the school attached to the community, Nicole and

Arnauld wrote *La logique on l'art de penser* (1662), also called the *Logique de Port Royal*. The final chapters of this book contain elements of a probability calculus and its applications to aspects of life other than games of chance and with a frequency interpretation of some of the probabilities involved. Possibly, these chapters were written under Pascal's influence. The book ends with an indication of Pascal's probability arguments for leading a pious life.

Pascal's *Pensées* contains a short section "Infini-rien: Le pari," usually called Pascal's wager, which has been the subject of much exegesis. For a statistician the methodological analysis given by Hacking (1975) in terms of decision theory is the most interesting. Briefly told, Pascal imagines two states of the world: God is or he is not; two possible actions: a pious life or a worldly one; a probability distribution over the states with *finite* probability for the existence of God, and "utilities" of the four combinations of states and actions such that the three utilities are finite, but the fourth corresponding to the pious life when God exists is *infinite* because it leads to an infinite life of infinite happiness. Hence, the expected utility of a pious life will be infinite for any admissible probability distribution and therefore larger than the expected utility of a worldly life. As pointed out by Hacking, this curious incident is interesting because Pascal's formulation, even if rather obscure, contains the basic notions of modern decision theory.

The present section is based on Hacking (1975, Chap. 8 and 9) and Heyde and Seneta (1977, §5.8).

CHAPTER 6

Huygens and *De Ratiociniis in Ludo Aleae*, 1657

Though I would like to believe, that if someone studies these things a little more closely, then he will almost certainly come to the conclusion that it is not just a game which has been treated here, but that the principles and the foundations are laid of a very nice and very deep speculation.

—*HUYGENS, 1657*

6.1 HUYGENS AND THE GENESIS OF HIS TREATISE

Christiaan Huygens (1629–1695) was born into a wealthy Dutch family whose members held high posts in the civil and diplomatic services of the Netherlands. Under the supervision of their father, Christiaan and his brothers received a comprehensive education both in the humanities and the sciences. From 1645 to 1649 he studied law and mathematics at the universities of Leiden and Breda. He was much influenced by his mathematics teacher Frans van Schooten, who introduced him to Cartesian mathematics; he also studied the newly published Cartesian philosophy intensely. During the years from 1650 to 1666, he lived mostly at home, supported economically by his father, and concentrated on research in mathematics, optics, astronomy, and physics with results that established him as the foremost scientist in Europe. He visited Paris and London several times and was elected a member of the newly established Royal Society at London in 1663. When the Académie Royale des Sciences was founded in 1666, Huygens was offered and accepted membership and moved to Paris where he stayed until 1681, supported by Louis XIV. The small group of salaried academicians

65

worked not only on theoretical problems but also carried out technological work of importance for the government. Because of illness he returned to Holland in 1681, and he did not go back to Paris, perhaps because his position as a Protestant and a foreigner was becoming less pleasant. At that time he had completed his most important scientific work. He spent his last years in The Hague.

In mathematics Huygens continued the works of Descartes, Fermat, and Pascal by solving problems derived from mechanics and physics. For many special functions he found areas and volumes corresponding to certain sections, centers of gravity, and centers and radii of curvature. He was a master, able to invent the tricks required to solve each specific problem, but he did not develop a general method.

In astronomy he continued the three lines of research begun by Galileo. He constructed better telescopes; he used these to observe new planets and stars, and he constructed a pendulum clock for more accurate measurement of time. In 1655 he improved the methods of grinding and polishing lenses and built a telescope of much greater power than previously known. By observation of Saturn he detected an unknown satellite and gave the correct description and explanation of Saturn's ring. His invention of the pendulum clock in 1656 made regular and more accurate time measurements possible, one of the consequences being that the angular distance between stars could be found easily and accurately. The clocks were also important for determining longitude at sea, but since the pendulum was not sufficiently stable at sea, he later used springs instead.

In the *Horologium oscillatorium* (The Pendulum Clock, 1673) Huygens continued the work of Galileo on the theory of the motion of bodies and demonstrated a wealth of new results. We shall only mention a few. For the mathematical pendulum he found the relation between the time of oscillation, the length of the pendulum, and the acceleration due to gravity, a formula that today (for small deviations from the vertical) is written as $T = 2\pi\sqrt{l/g}$. A curve with the property that a body oscillating on it has a time of oscillation independent of the magnitude of oscillation is called a tautochrone. Huygens proved that the cycloid is tautochronic, so that in principle he was able to construct a clock, the cycloidal pendulum, with time of oscillation independent of the amplitude. In this connection he developed the thory of evolutes and proved that the evolute of a cycloid is again a cycloid. For a compound pendulum he defined the center of oscillation and showed how to compute the length of a simple pendulum oscillating isochronously with the compound. Using the result $T = 2\pi\sqrt{l/g}$, he made a rather accurate experimental determination of the gravitational constant. His studies of circular motion led him to the fundamental formula for centrifugal force $f \propto v^2/r$.

Another main achievement was his wave theory of light, completed in 1678 but not published until 1690 as the *Traité de la lumière*. In accordance with his mechanistic philosophy, inherited from Descartes, he considered light to be the effect of waves in an ether consisting of very small particles, and he derived the laws of refraction and reflection from this hypothesis. In his works on the collision of bodies and on light, he realized that the ideas of Descartes were too vague and in many cases also at variance with experimental results. He therefore refined the Cartesian ideas by introducing more classes of elementary particles, by allowing a vacuum between particles, and by studying the laws of impact. In this way he was able to defend the mechanistic explanation of natural phenomena, for example, gravity.

Huygens carried on a vast correspondence which has been published as the first ten volumes of his collected works. Like Fermat and Newton, he was reluctant to publish his works, and some of his books were printed only many years after they had been written. It is presumably only due to the efforts of van Schooten that Huygens' treatise on probability theory was published shortly after its completion.

Huygens' collected works with many notes and commentaries have been published in 22 volumes by the Holland Society of Science. A comprehensive biography has been written Vollgraff (1950); shorter ones are due to Dijksterhuis (1953), David (1962), and Bos (1972).

In 1656 Huygens wrote a small treatise, 16 pages, with the title *Van Rekeningh in Speelen van Geluck* (On Reckoning at Games of Chance). Anyone interested in the detailed history of Huygens' work on probability theory should consult Huygens' collected works, Vol. 14, pp. 1–179 (1920), which contains a reprint of his treatise, a translation into French, nine appendices with Huygens' later work on related problems, and an introduction and many notes by Korteweg based on Huygens' correspondence. The letters by Huygens on these matters have been printed in volumes 1 and 5; precise references have been given by Korteweg.

Huygens visited Paris for the first time in the autumn of 1655. He met Roberval and Mylon, a friend of Carcavi, but did not meet Carcavi and Pascal. According to himself he heard about the probability problems discussed the year before but was not informed about the methods used and the solutions obtained. This sounds odd in view of the fact that Roberval was fully informed about the Pascal–Fermat correspondence; however, Roberval does not seem to have been much interested in probability theory. On his return to Holland Huygens solved the problems, and in April 1656 he sent van Schooten his manuscript of the treatise. At the same time he wrote to Roberval and asked for his solution of the most difficult problem (Proposition 14) in the treatise for comparison with his own.

Van Schooten proposed that Huygens' treatise would be published at the end of his *Exercitationum Mathematicarum*, which he was preparing for publication. This would require a translation of the treatise into Latin, and after some discussion van Schooten offered to do this himself.

In the meantime, since Roberval did not answer his letter, Huygens wrote to Mylon who, through Carcavi, sent the problem on to Fermat. In a letter from Fermat to Carcavi and from him to Huygens (22 June 1656), Fermat gave the solution without proof, a solution which to Huygens' satisfaction agreed with his own. Furthermore, Fermat posed five problems to Huygens, which Huygens immediately solved. He sent the solutions to Carcavi on 6 July and asked him to inform Mylon, Pascal, and Fermat to determine whether their solutions agreed with his. Huygens later used two of Fermat's problems as problems 1 and 3 at the end of his treatise. Carcavi's answer of 28 September convinced Huygens that the methods employed by Pascal and himself were in agreement. Further, this letter contained a problem posed by Pascal to Fermat, which Huygens solved and included as the fifth problem in his treatise. Carcavi's letter contained the solutions given by Fermat and Pascal without the proofs. Huygens' solution, also without proof, is stated in his letter to Carcavi on 12 October.

In March 1657 van Schooten sent Huygens the Latin version for final additions and corrections. Huygens then added Proposition 9 and five problems for the reader, among them the three by Fermat and Pascal mentioned above. In a letter to van Schooten, which was used as Huygens' preface, Huygens stressed the importance of this new topic and stated that "for some time some of the best mathematicians of France have occupied themselves with this kind of calculus so that no one should attribute to me the honour of the first invention." The *De Ratiociniis in Ludo Aleae* was published in September 1657; the Dutch version was not published until 1660.

Huygens' treatise is the first published work on probability theory; all the previously mentioned works were published after 1657. It won immediate recognition and became the standard text on probability theory for the next 50 years. This is surprising in view of the fact that Pascal's treatise, published in 1665, is more comprehensive. The reason may be that Huygens' treatise was available in Latin and easily accessible as part of the widely circulated text by van Schooten.

6.2 *DE RATIOCINIIS IN LUDO ALEAE*

The *De Ludo Aleae* is composed as a modern paper on probability theory. From an axiom on the value of a fair game Huygens derives three theorems on expectations. He uses the theorems to solve a number of problems of current interest on games of chance by recursion. Finally, he poses five problems, gives the answers to three of them, and leaves the proofs to the reader.

The three theorems and the eleven problems are usually called Huygens' 14 propositions. David (1962, pp. 116–117) has translated the propositions from the *De Ludo Aleae* into English. We shall present a translation of the Dutch version; there is no difference in meaning, merely slight differences in wording. In the Dutch version Huygens does not use the word "expectation" but writes about the value of a game.

In a short introduction Huygens makes the following (somewhat obscure) statement: "I take as fundamental for such games that the chance to gain something is worth so much that, if one had it, one could again get the same chance in a fair game, that is, a game in which nobody stands to lose." He illustrates this axiom by the following example:

Suppose that somebody has 3 shillings in his one hand and 7 in the other and that I am asked to chose between them; this is so much worth to me as if I had 5 shillings for certain. Because if I have 5 shillings I can establish a fair game in which I have an even chance of getting 3 or 7 shillings, as will be shown below.

The first three propositions are as follows:

1. "If I have equal chances of getting a or b, this is so much worth for me as $(a + b)/2$."

2. "If I have equal chances of getting a, b or c, this is so much worth for me as if I had $(a + b + c)/3$."

3. "If the number of chances of getting a is p, and the number of chances of getting b is q, assuming always that any chance occurs equally easy, then this is worth $(pa + qb)/(p + q)$ to me."

Today we consider this the definition of mathematical expectation. For Huygens, however, these propositions required proof. As mentioned above, his basic notion is a lottery with as many tickets and players as there are chances in the given game, each player having the same chance of winning and therefore staking the same amount, x, say. To make the lottery equivalent to the given game, the winner of the total stake agrees to pay certain amounts to the losers. *The value of the game or the expectation of a player is defined as the stake per participant in the equivalent lottery.*

To prove Proposition 1, Huygens introduces a lottery with two players who agree that the winner shall pay the amount a to the loser so that the winner gets $2x - a$. Setting $2x - a = b$, the lottery becomes equivalent to the given game, and $x = (a + b)/2$.

The proof of Proposition 2 is analogous. Huygens remarks that this result may be extended to any number of players so that the value of a game with a finite number of equal chances equals the arithmetic mean of the prizes.

To prove Proposition 3 about the weighted mean, Huygens reasons as follows. Consider a lottery with $p + q$ players including myself. With q of the players I make the agreement that each should pay me the amount b if he wins and receive b from me if I win, and with each of the remaining $p - 1$ players I agree that he should pay me the amount a if he wins and receive a from me if I win. Hence, I have one chance in $p + q$ of getting the amount $(p + q)x - qb - (p - 1)a$, q chances of getting b, and $p - 1$ chances of getting a. This lottery is equivalent to the original game for $(p + q)x - qb - (p - 1)a = a$, so that $x = (pa + qb)/(p + q)$.

Huygens does not state the generalization $(pa + qb + rc)/(p + q + r)$, say, explicitly, but in Proposition 13 he considers an example with three outcomes, which he solves by using Proposition 3 two times in succession, first finding the weighted average of a and b and then the weighted average of this intermediate result and c.

Huygens solves the first nine problems by the same method, the method of *recursion*. Considering a given state of the play, *he imagines one more game carried out, lists all the possible outcomes and the corresponding number of chances, and uses Proposition 1, 2, or 3 to find the (marginal) expectation.* The resulting difference equation is used recursively to find the numerical solution of the problem based on the expectation in the first (or the last) game which is known or easy to find. *Huygens only discusses the problems numerically,* he does not give the difference equation formally and does not attempt an explicit solution. It will be seen that Huygens employs the same method as Pascal. In the following exposition we shall use the notation from §5.3 to write the recursion formula corresponding to Huygens' numerical examples.

In Propositions 4–7 Huygens discusses the problem of points for two players.

Proposition 4. "Suppose that I play against another person about three games, and that I have already won two games and he one. I want to know what my proportion of the stakes should be, in case we decide not to continue the play and divide the stakes equitably between us."

This is the problem of points for $(a, b) = (1, 2)$. Propositions 5–7 correspond to $(a, b) = (1, 3), (2, 3)$, and $(2, 4)$. Using Proposition 1, Huygens solves these problems by recursion exactly like Pascal, see (5.3.1), and finds the same results as those given in Table 5.3.1.

In Proposition 8 Huygens discusses the problem of points for three players, and in Proposition 9 he generalizes to any number of players.

Proposition 8. "Let us suppose that three persons play together, and that the first lacks one game, the second one game and the third two games."

Proposition 9. "To calculate the proportion due to each of a given number of players, who each lacks a given number of games, it is necessary to find out what the player, whose proportion we want to know, would get, if he or any of the other players wins the following game. Adding these parts and dividing by the number of players give us the proportion sought."

Using Proposition 2, his solution may be written

$$e(a, b, c) = \frac{1}{3}[e(a - 1, b, c) + e(a, b - 1, c) + e(a, b, c - 1)]. \tag{1}$$

He tabulates 17 values of $e(a, b, c)$ for small values of (a, b, c). His result for $(1, 2, 2)$ agrees with Pascal's result.

In Propositions 10 and 11 Huygens considers de Méré's problem.

Proposition 10. "To find how many turns one should take to throw a six with one die."

Proposition 11. "To find how many turns one should take to throw two sixes with two dice."

Using Proposition 3, these problems are solved by recursion. Huygens first considers a game in which A wagers the amount t, say, that he will get at least one 6 in n throws. Let A's expectation be e_n. It follows that $e_1 = t/6$ and that $e_{n+1} = (1/6)t + (5/6)e_n$. This leads to $e_2 = 11t/36$, $e_3 = 91t/216$, $e_4 = 671t/1296$, and so on. Hence, for $n = 4$, the odds are $671:625$, as also found by Pascal and Fermat.

Proposition 11 is solved by the same method, with the chances 1 and 5 replaced by 1 and 35, which gives $e_1 = t/36$ and $e_2 = 71t/1296$. To speed up the recursion, Huygens goes directly from e_2 to e_4 by the formula

$$e_4 = \frac{71t + 1225e_2}{1296} = \frac{178,991t}{1679,616}.$$

He derives this formula by interpreting 71 and 1225 as the number of chances for getting a double 6 in two throws. He then states that he has used this method to find e_8, e_{16}, and e_{24} and that $n = 24$ gives odds slightly less than $1:1$, whereas $n = 25$ gives odds larger than $1:1$.

Huygens does not indicate the explicit solution of the equation

$$e_{n+1} = pt + qe_n, \qquad e_1 = pt, \qquad n = 1, 2, \ldots, \tag{2}$$

which is easily found to be $e_n = (1 - q^n)t$.

Proposition 12. "To find how many dice one should take to throw two sixes at the first throw."

This is obviously a generalization of Propositions 10 and 11. Huygens reformulates the problem as follows: To find how many throws of a die are necessary to have at least an even chance of getting two sixes. Let e_n be A's expectation when he wagers the amount t that he will get at least two 6's in n throws. In Proposition 10 it was shown that A's expectation when he wagers to get at least one 6 in two throws equals $11t/36$. Using Proposition 3, Huygens then finds

$$e_{n+1} = \frac{1}{6}\left(\frac{11t}{36}\right) + \left(\frac{5}{6}\right)e_n.$$

Beginning with $e_2 = t/36$, he gets $e_3 = 16t/216$, and so on. He writes that $n = 10$ will give odds slightly larger than 1:1.

Obviously, Huygens' recursion formula may be written as

$$e_{n+1} = p(1 - q^n)t + qe_n, \qquad n = 2, 3, \ldots. \tag{3}$$

Proposition 13. "Suppose that I play one single throw with two dice against another person on the condition, that if the outcome is 7 points I win, if the outcome is 10 points he wins, and otherwise the stakes should be divided equally between us. To find what proportion each of us should have."

As mentioned before, Huygens solves this problem by successive applications of Proposition 3. First he notes that the number of chances for the three outcomes are 6, 3, and 27, respectively. He then disregards the last outcome and finds the expectation $(6/9)t + (3/9) \times 0 = 2t/3$. The final expectation then becomes $(9/36)(2t/3) + (27/36)(t/2) = 13t/24$.

Proposition 14. "Suppose that I and another player take turns in throwing with two dice on the condition, that I win if I throw 7 points, and he wins if he throws 6 points, and I let him have the first throw. To find the ratio of my chance to his."

This is a case of the problem of points different from the previous ones in that there is no upper limit to the number of games. Since the usual recursive method is not directly applicable, *Huygens invents a new method* which was used later to solve other problems of the same nature. James Bernoulli called this method *Huygens' analytical method*. Huygens proves Propositions 4–13 numerically by recursion, but in Proposition 14 he uses two algebraic equations (analysis) for the solution.

Denoting the players by A and B, it follows that A's probability of winning a single game is 6/36 and B's 5/36. Let the total stake be t, and let A's expectation before the play begins be x so that B's expectation becomes $t - x$. Each time it is B's turn to throw, A's expectation is x, but when it is A's turn, his (conditional) expectation will have a larger value, y, say. When it is B's turn to throw, it follows from Proposition 3 that A's expectation is $(5/36) \times 0 + (31/36)y = x$. When it is A's turn to throw, we have, similarly, $(6/36)t + (30/36)x = y$. Solving for x we get $x = 31t/61$, so that the ratio of x to $t - x$ becomes 31:30.

It will be seen that the solution is obtained by means of two linear equations between the two (conditional) expectations

Huygens does not comment further on the solution. To see the connection with the method of recursion, let e_n denote A's expectation when n games have been played without anybody winning. Let the order of the players be BA BA BA..., and let the probability of winning a single game be p_1 for A and p_2 for B, $p_1 + q_1 = p_2 + q_2 = 1$. Huygens' usual way of reasoning leads to the equations

$$e_{2n} = q_2 e_{2n+1} \qquad e_{2n+1} = p_1 t + q_1 e_{2n+2}, \qquad n = 0, 1, \dots . \qquad (4)$$

Presumably, Huygens has noticed that the *periodicity* of the play and the *independence* of the infinitely many games mean that $e_{2n} = e_0$ and $e_{2n+1} = e_1$. Huygens writes that "it is obvious that each time B is going to throw the expectation of A must again be equal to x." This means that there are really only two equations,

$$e_0 = q_2 e_1 \qquad \text{and} \qquad e_1 = p_1 t + q_1 e_0, \qquad (5)$$

which have the solution $e_0 = p_1 q_2 t/(1 - q_1 q_2)$. In his comments to Huygens' treatise, James Bernoulli (1713) points out the periodicity and the independence.

Neither Huygens nor Bernoulli notices the fact that the recursion (4) leads to an infinite geometric series for e_0 with the sum given above.

Huygens' treatise ends with five problems for the reader which are discussed in §6.3.

It will be seen that Huygens, directly or indirectly, derives all his results by recursion. His method is very clear and easily understandable. Numerically it is, however, very cumbersome. It is remarkable that he does not use combinatorial methods.

There exist three comprehensive, annotated editions of Huygens' treatise: (1) the first part of *Ars Conjectandi* by James Bernoulli (1713); (2) Korteweg's edition (1920) in Huygens collected works, Vol 14; and (3) the edition by Dupont and Roero (1984).

According to Todhunter, *De Ludo Aleae* was translated into English by Arbuthnott (1692) and Browne (1714). A translation into German was done by Haussner (1899), into French by Korteweg (1920), and into Italian by Dupont and Roero (1984). Besides the papers already mentioned we refer to von der Waerden (1975), Hacking (1975), Sheynin (1977), Schneider (1980), and Holgate (1984).

6.3 HUYGENS' FIVE PROBLEMS AND HIS SOLUTIONS

The solutions given by Huygens are as usual only numerical. We shall, however, give the formulae corresponding to his reasoning.

Problem 1. A and B play against each other with two dice on the condition that A wins if he throws six points and B wins if he throws seven points. A has the first throw, B the following two, then A the following two, and so on, until one or the other wins. The question is, What is the ratio of A's chances to B's? *Answer*: 10,355 to 12,276.

This is the problem posed by Fermat in his letter to Huygens in June 1656. Obviously, it is a generalization of Proposition 14. It is solved by Huygens in his letter to Carcavi of 6 July 1656.

The order of the players is ABBA ABBA .... Let e_n be A's expectation just before game number n, given that the previous games have been unsuccessful. Because of the periodicity and the independence, it is only necessary to consider $e_1, \ldots, e_4$. Huygens' reasoning may be expressed by the four equations

$$e_4 = p_1 t + q_1 e_1, \qquad e_3 = q_2 e_4, \qquad e_2 = q_2 e_3, \qquad e_1 = p_1 t + q_1 e_2.$$

Solving for e_1, one find

$$\frac{e_1}{t} = \frac{p_1(1 + q_1 q_2^2)}{1 - q_1^2 q_2^2}.$$

For $p_1 = 5/36$ and $p_2 = 6/36$, one gets $e_1/t = 10{,}355/22{,}631$, in agreement with Huygens' answer.

Problem 2. Three players, A, B, and C, having 12 chips of which four are white and eight black, play on the condition that the first blindfolded player to draw a white chip wins, and that A draws first, B next, and then C, then A again, and so on. The question is, What are the ratios of their chances to each other?

Huygens' solution is given in a note from 1665 (see *Oeuvres*, Vol. 14, p. 96). He assumes that the drawings are with replacement, since he uses $p = 4/12 = 1/3$ as the probability of winning in each drawing. The order of the players is ABC ABC.... Let x, y, and z denote the expectations of the three players before the play begins. If A wins at the first drawing, he gets the stake, and if he loses he becomes number three in the following series of drawings with an expectation equal to z. Hence, $x = pt + qz$. If A loses, which happens with probability q, then B moves to the first place in the following series of drawings and thus gets the expectation x, so that $y = qx$. Similarly, $z = qy$. The solution of the three equations is

$$\frac{x}{t} = \frac{p}{1-q^3}, \qquad \frac{y}{t} = \frac{pq}{1-q^3}, \qquad \frac{z}{t} = \frac{pq^2}{1-q^3}.$$

with the ratios $1:q:q^2$, which become $9:6:4$ for $p = 1/3$, as found by Huygens.

Problem 3. A wagers with B that out of 40 cards, there being 10 of each color, he will draw four cards, so that he gets one of each colour. The chances of A to those of B are found to be 1000 to 8139.

This problem was posed by Fermat in his letter of June 1656; the answer without proof is given in Huygens' letter to Carcavi of 6 July. Presumably, Huygens has reasoned as shown below.

Suppose that n cards have been drawn and that they are of different colors. Let the total stake be unity, and let A's expectation be e_n. The condition for A to win the next drawing is that he draws a card among the $10(4 - n)$ cards of colors different from those already drawn. Hence, $e_n = 10(4 - n)e_{n+1}/(40 - n)$. Since $e_4 = 1$, it follows by recursion that $e_0 = (10/37)(20/38)(30/39)(40/40) = 1000/9139$, in agreement with Huygens' result.

Problem 4. As before, the players have twelve chips, four white and eight black; A wagers with B that by drawing seven chips blindfolded, he will get three white chips. The question is, What is the ratio of A's chances to B's?

Huygens' solution is given in a note from 1665 (see *Oeuvres*, Vol. 14, pp. 97–99).

Let $e(a, b)$ be A's expectation when a white and b black chips have been drawn. Among the remaining $12 - a - b$ chips there are $4 - a$ white and $8 - b$ black. Hence,

$$e(a, b) = \frac{(4 - a)e(a + 1, b) + (8 - b)e(a, b + 1)}{12 - a - b}, \qquad 0 \leqslant a \leqslant 4, \qquad 0 \leqslant a + b \leqslant 7,$$

with the boundary conditions $e(3, 4) = t$, $e(a, 7 - a) = 0$ otherwise, $e(a, 6 - a) = 0$ for $a = 0, 1$, and $e(0, 5) = 0$.

Starting from $e(3, 4) = t$, Huygens gets $e(3, 3) = 5t/6$, and so on, until after 19 iterations, he reaches $e(0, 0) = 35t/99$. He notes that this result may also be found as solution to the "complementary problem," where five chips are drawn and the composition sought is $(1, 4)$. He carries out this iteration, which only has nine steps.

Huygens also considers the modification proposed by Hudde, that the outcome should be three or more white chips, that is, $(3, 4)$ or $(4, 3)$, and by iteration he gets the result $42t/99$ (see *Oeuvres*, Vol. 14, pp. 100–101).

Problem 5. A and B each having 12 counters play with three dice on the condition that if 11 points are thrown, A gives a counter to B and if 14 points are thrown, B gives a counter to A and that he wins the play who first has all the counters. Here it is found that the number of chances of A to that of B is $244, 140, 625$ to $282, 429, 536, 481$.

This is the problem posed by Pascal to Fermat and through Carcavi to Huygens in a letter of 28 September 1656 that contains the answers given by Pascal and Fermat. Huygens' answer is in his letter to Carcavi of 12 October 1656. His proof is given in a note from 1676 (see *Oeuvres*, Vol. 14, pp. 151–155). The problem is known as the Gambler's Ruin problem.

In Pascal's formulation the players start with the score $(0, 0)$, and the winner is the one who first leads with 12 points. This is also the form that Huygens uses in his proof.

The number of chances to win a point is 15 for A and 27 for B; we shall set $p = 15/42 = 5/14$ and $q = 27/42 = 9/14$. Let $e(a, b)$ denote A's expectation when A has a points and B has b points. The problem is to find $e(0, 0)$.

Huygens begins by analyzing the simple case where the play ends when one of the players leads with two points. He lists all the possible outcomes and their probabilities in a diagram, as shown in Fig. 6.3.1.

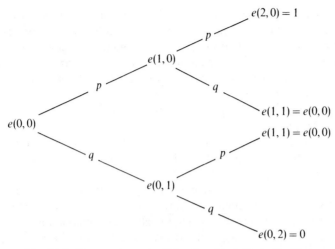

Fig. 6.3.1. Huygens' graph of the Gambler's Ruin Problem.

It follows immediately from Fig. 6.3.1 that the following equations hold:

$$e(1,0) = pe(2,0) + qe(1,1) = p + qe(0,0),$$

$$e(0,1) = pe(1,1) + qe(0,2) = pe(0,0),$$

$$e(0,0) = pe(1,0) + qe(0,1),$$

with the solution $e(0,0) = p^2/(p^2 + q^2)$.

Huygens next turns to the case where the winner has to be four points ahead. Using the result above he takes two steps at a time; that is, he goes from $(0,0)$ through $(2,0)$ to $(4,0)$, say, with probabilities proportional to p^2. This gives three equations with the solution $e(0,0) = p^4/(p^4 + q^4)$. He finally remarks that if a lead of eight points is necessary for winning, then $e(0,0) = p^8/(p^8 + q^8)$, and so on.

If a lead of three points is required to win, he goes from $(0,0)$ to $(1,0)$ with probability p and then to $(3,0)$ with probability p^2 and similarly for the other paths in the diagram, which leads to the equations

$$e(1,0) = \frac{p^2 e(3,0) + q^2 e(1,2)}{p^2 + q^2} = \frac{p^2 + q^2 e(0,1)}{p^2 + q^2},$$

$$e(0,1) = \frac{p^2 e(2,1) + q^2 e(0,3)}{p^2 + q^2} = \frac{p^2 e(1,0)}{p^2 + q^2},$$

$$e(0,0) = pe(1,0) + qe(0,1),$$

with the solution $e(0,0) = p^3/(p^3 + q^3)$. He generalizes to $p^6/(p^6 + q^6)$.
The cases considered so far only require the solution of three equations. If a lead of $n = 5$ points is needed to win, the number of equations becomes considerably larger, and Huygens remarks that a similar solution may be obtained as for $n = 3$ but that it takes a longer time. Finally, he states that in general the ratio of A's and B's expectations is $p^n:q^n$.

The answer given at the end of Problem 5 will be seen to be $5^{12}:9^{12}$.

No indications exist as to the methods used by Pascal and Fermat to solve the Gambler's Ruin problem. Edwards (1983) has suggested that Pascal solved the difference equation and that Fermat used a combinatorial argument, see Problem 7 in §6.5.

Huygens' five problems became a challenge to mathematicians at the time and their solutions, interpretations (drawings with or without replacement), and generalizations were discussed by Huygens himself and by Hudde, Spinoza, Bernoulli, Montmort, de Moivre, and Struyck, see §14.2.

6.4 OTHER CONTRIBUTIONS BY HUYGENS

After 1657, Huygens intermittently returned to probability problems, but he did not publish his results.

His most important contribution is the probabilistic interpretation of Graunt's life table, given in correspondence with his brother Lodewijk in 1669, which we shall discuss in §8.1.

In 1665 he had a lengthy correspondence with Hudde on the solution of Problems 2 and 4 and some more difficult problems. They solved these problems by recursion; a summary has been given by Korteweg (1920).

In 1679 and 1688 he solved problems on popular card games.

6.5 PROBLEMS

1. Show that the difference equation from Proposition 12,

$$e_{n+1} = p(1 - q^n) + qe_n, \qquad e_2 = p^2, \quad n = 2, 3, \ldots,$$

has the solution

$$e_n = 1 - npq^{n-1} - q^n.$$

Give an interpretation of this result. Compare with Huygens' solution.

2. *Generalization of Problem 1.* Let A win if he gets at least m successes in n trials, and let B win otherwise. Let e_{mn} be B's expectation (probability of winning) and show that

$$e_{mn} = pe_{m-1,n-1} + qe_{m,n-1}, \qquad m = 0, 1, \ldots, n-1,$$

$e_{0n} = 0$ and $e_{nn} = 1 - p^n$. Prove that the solution of this difference equation equals the sum of the first m terms of the binomial distribution.

3. From the discussion of Proposition 14, it follows that

$$e_{2n} = p_1 q_2 t + q_1 q_2 e_{2n+2}, \qquad n = 0, 1, \ldots.$$

Show that this leads to an infinite series for e_0 with a sum in agreement with Huygens' result. Give an interpretation of the terms of the series.

4. In Huygens' Problem 1, let e_n be A's expectation when n games have been played without anybody winning. Derive the recursions for $e_{4n}, \ldots, e_{4n+3}$ and find the infinite series for e_0 and its sum.

5. Modify Huygens' Problem 1 by letting A and B take turns, A having two throws and B three. Show by Huygens' method that the solution is

$$\frac{e_1}{t} = \frac{p_1(1 + q_1)}{1 - q_1^2 q_2^3} = \frac{72{,}360}{159{,}811}.$$

6. Solve Huygens' Problems 3 and 4 by the combinatorial method and compare with Huygens' method.

7. *The Gambler's Ruin Problem.* (a) Solve the Gambler's Ruin problem for $n = 5$ by Huygens' method. (b) Edwards (1983) suggests that Pascal solved the Gambler's Ruin problem by solving the difference equation $e(m) = pe(m + 1) + qe(m - 1)$, where $e(m)$ denotes A's expectation when he leads by m points, $m = -12, -11, \ldots, 12$, $e(-12) = 0$, and $e(12) = 1$. Find $e(0)$ and show that the corresponding odds are $p^{12}:q^{12}$. (c) A wins the play if it ends with the points $(12, 0), (13, 1), (14, 2), \ldots$, and, similarly, B wins for $(0, 12), (1, 13), (2, 14), \ldots$. Edwards (1983) suggests that Fermat based his reasoning on the odds for the successive symmetric pairs of these series. Show that for each pair the odds are $p^{12}:q^{12}$.

8. A and B take turns at a play in which the players for each lost game have to increase the stake by unity. The first time a player wins he gets the

whole stake. Let the probabilities of A and B of winning a single game be p_1 and p_2, respectively. Find their expectations. For a given value of p_1, find the value of p_2 that makes the expectations equal.

This problem is formulated and solved by Huygens in his correspondence with Hudde, see Korteweg (1920).

CHAPTER 7

John Graunt and the
Observations Made upon
the Bills of Mortality, 1662

Now having (I know not by what accident) engaged my
thoughts upon the Bills of Mortality, and so far succeeded
therein, as to have reduced several great confused Volumes
into a few perspicuous Tables, and abridged such Observations
as naturally flowed from them, into a few succinct Paragraphs,
without any long series of multiloquious Deductions, . . .

— *JOHN GRAUNT, 1662*

7.1 ON THE ORIGIN OF THE WORD "STATISTICS"

Surveys of people and property have been carried out from ancient times
for fiscal and military purposes. In Republican Rome the members and
property of every family were recorded quinquennially, and Augustus
extended the census to cover the whole Roman empire. With the fall of the
Roman empire the census disappeared, and regular censuses covering a whole
national state were not taken up again until the beginning of the 18th
century. In the Middle Ages and the Renaissance, surveys were carried out
intermittently, mainly for fiscal reasons, such as the Domesday Book of 1086
in England and the comprehensive survey in Florence in 1427. It is, however,
only recently that a *statistical analysis* of these data was begun after they were
transferred to electronic computers.

The word "statistics" is of Italian origin. It is derived from *stato* (state),
and a *statista* is a man who deals with affairs of the state. The original

81

meaning of statistics is thus a collection of facts of interest to a statesman.

Statistics was used in this sense in Italy in the 16th century, and from there it spread to France, Holland, and Germany, where it was taught at the universities in the 17th and 18th centuries. It embraced the political constitutions of states and mainly verbal descriptions of the important characteristics of states, such as population, economy, and geography. The use of the word "statistics" in this sense died out in the beginning of the 19th century.

A systematic collection of numerical data on population and economy began again in the Italian city-states, notably Venice and Florence, during the Renaissance. A descriptive statistical analysis of such data occurred first in 1662 when John Graunt analyzed the weekly reports on vital statistics for London, which had been published regularly since 1604. Graunt's methods of analysis, extended also to economic data, were continued by Petty, Gregory King, and others and called *Political Arithmetic*. About 1800 this line of investigation began to be called statistics. According to K. Pearson, it was J. Sinclair who first used the word "statistics" in this modern sense in his *The Statistical Account of Scotland drawn up from the communications of the ministers of the different parishes* [1791–1799, 21 volumes; see Plackett (1986)].

More information on this matter may be found in Westergaard (1932, Chap. 2), K. Pearson (1978, Chap. 1), M. G. Kendall (1960), and Cullen (1975, pp. 10–11).

7.2   GRAUNT'S DISCUSSION OF THE PLAGUE MORTALITY

The high mortality in Europe before the 19th century was due to epidemic diseases, wars, and famine, which often interacted to produce terrible results. Among the epidemic diseases the plague was the worst. After the Black Death in 1348–1350, the plague recurred frequently for nearly 400 years.

The plague is an infectious disease transmitted by fleas carried mainly by rats, gerbils, and squirrels. This was unknown at the time, and it was generally believed that contagion depended upon corrupt air. Quarantine or isolation was used as a measure to reduce contagion.

In London in the 1530s a warning system was established by requesting the parish clerks to submit weekly reports on the number of plague deaths and all other deaths in case of a beginning epidemic. These weekly bills of mortality served to tell the authorities when measures should be taken against the epidemic, and they were also used by the wealthy part of the population as an indication of when to leave the city for the fresh air of the country.

In 1538 parish registers were instituted for the purpose of recording all weddings, christenings, and burials within the Church of England. People of dissenting faiths were not recorded. Hull (1899, p. XCI) estimates that the number of burials recorded should be increased by about 15% to get a correct picture of the mortality in London for the period investigated by Graunt.

Starting in 1604, weekly bills of mortality for the parishes of London were published by the Company of Parish Clerks, and a bill for the whole year was published at the end of each year. The ages of the deceased were not recorded before 1728. The great interest in the weekly bills is shown by the fact that they were printed from 1625, and a subscription arrangement was established for the general public. The weekly bills were published regularly until 1842 when they were superseded by the publication from the Registrar General.

This large amount of data had not been analyzed statistically before John Graunt published his book in 1662.

Graunt states that the continued weekly bills for London begin the 29th of December 1603 and that there were bills published before in times of great mortality, for example, in the years 1592–1594. He gives examples of the yearly bills for 1624, a year with nearly no plague, and for 1625, a great plague year. From 1629 the burials and christenings are given separately for males and females. He gives detailed information on how London covers an increasing number of parishes during his period of investigation from 1604 to 1661. Finally, he describes how the deaths are classified according to cause of death by two "Searchers, who are ancient Matrons, sworn to their Office." They are called to the place where the deceased lies and "by view of the same, and by other enquiries," they decide upon the cause of death and report to the parish clerk. The bill for 1632 contains 63 diseases and casualties ordered alphabetically. The weekly bill was published every Thursday "and dispersed to the several Families who will pay four shillings per Annum for them." The original information in the Hall books is more detailed than in the printed bills, and Graunt has presumably used these books, since his fundamental table embraces 81 causes of death.

After having described the history and provenance of the data, Graunt, a true statistician, turns to a discussion of the trustworthiness of the data. We shall return to that problem in general in §7.3. Here we shall concentrate on Graunt's discussion on the mortality of the plague. His data are shown in the following table.

Number of Burials, Plague Deaths, and Christenings in London

	PART OF THE YEAR[a]			THE WHOLE YEAR		
YEAR	BURIALS	PLAGUE DEATHS	%	BURIALS	CHRISTENINGS	%
1592	25,886	11,503	44	26,490	4,277	16
1593	17,844	10,662	60	17,844	4,021	23
1603	37,294	30,561	82	38,244	4,784	13
1625	51,758	35,417	68	54,265	6,983	13
1636	23,359	10,400	45	23,359	9,522	41

[a] Part of the year means March or April to December, depending on the beginning of the plague; for 1593, the whole year.
Source: Graunt, 1662, Chap. 4.

Graunt does not express the mortality in percent but as a fraction such as 2 to 5, 7 to 10, and so on.

From the first part of the table, Graunt concludes that the year 1603 is the greatest plague year. He then asks what year has the greatest total mortality. Since the size of the population is unknown, he uses the number of christenings as a measure of the population size and concludes from the second part of the table that the total mortality is largest and about equal for the years 1603 and 1625. How can this be reconciled with the first conclusion? Graunt proposes to show that the number of plague deaths recorded in 1625 is too small. He notes that in the years before and after 1625, which were years without plague, the number of burials is between 7000 and 8000, whereas the number of nonplague deaths in 1625 is recorded as $54{,}265 - 35{,}417 = 18{,}848$, from which he concludes that about $18{,}000 - 7000 = 11{,}000$ of the ordinary deaths in 1625 ought to have been recorded as plague deaths. [Since a plague death might lead to the isolation of an infected house, Hull (1899, p. 365) suggests that the Searchers may have been bribed to report other causes of death.] With this correction the number of plague deaths in 1625 becomes about 46,000, which is 85% of the number of deaths for the whole year, that is, about the same plague mortality as in 1603. This is just one example of Graunt's critical appraisal of the data.

In the fifth edition of his book Graunt gives the total number of burials and the plague deaths in 1665 as 97,306 and 68,596, respectively. From his table of burials for the preceding years it follows that there was nearly no plague in the four years 1661–1664 and that the average number of burials was about 15,000. Using Graunt's line of reasoning, we see that the number of plague deaths should be increased by about 14,000, which means that the plague mortality becomes about 85% of the total, just as in 1603 and 1625.

Graunt himself has not made this analysis, perhaps for the reason which he in another context tells the reader concerning the analysis of the bills: "conceiving it will be more pleasure and satisfaction [for the readers] to do it themselves, than to receive it from another hand."

In the third edition of his book (1665, p. 106), Graunt estimates the population of London in 1661 at 403,000. (In the first edition he had found 460,000 by cruder methods.) With a yearly increase of about 2%, the population size in 1665 was about 430,000, which gives an uncorrected plague mortality in 1665 of 16% and, adding the 14,000 unrecorded plague deaths, a corrected plague mortality of 19%. The total mortality is 23%. Graunt writes that "about one fifth part of the whole people died in the great Plague-years" and that two-fifths fled to the country. This statement may be found as No. 36 in the Index to all editions of his book; he has obviously spared the reader the "multiloquious deductions."

In the third edition Graunt gave a table of the weekly deaths and plague deaths for the six great plague years. It is odd that he did not give this table in the first edition, since the bills were instituted with the specific purpose to be able to follow (and predict) the course of a plague epidemic. Perhaps he thought that this was the best known and most discussed information in the bills, and he therefore omitted it. He notes that the maximum of the epidemic curve appears in the fall (August–October), that the duration of the maximum varies, and that the ratio of the increasing to the decreasing period is about 3 to 2. Finally, he observes that only five times has there been a doubling of the plague deaths from one week to the next.

From the fact that the number of christenings before and two years after the great plague years are about equal, he concludes that London is fully repeopled within two years, essentially by migration from the country.

The year 1665 was the last great plague year in London. During the latter part of the century, the plague disappeared from England and a little later from the Continent as well.

7.3   JOHN GRAUNT AND HIS *OBSERVATIONS MADE UPON THE BILLS OF MORTALITY*

John Graunt (1620–1674) was the son of a London draper, and after serving an apprenticeship he joined in his father's business, which he eventually took over. He received an ordinary education in English learning and studied Latin and French on his own in the morning before business hours. He soon became a respected London citizen and held a number of important offices in the Draper's Company, in the Ward offices, and in the Council of the City. He was a captain and later became a major of the trained band of the

City. He also had many acquaintances in cultural and scientific circles. About 1650 his influence was great enough to procure a professorship of music at Gresham College for his friend William Petty, assistant professor of anatomy at Oxford.

The publication of his *Natural and Political Observations Made upon the Bills of Mortality* in 1662 resulted in his election to the newly established Royal Society. His book was a success, and a second edition appeared before the end of the year. The great plague in 1665 renewed the interest in the book, and a third edition with new data added was published and reprinted in 1665. A fifth edition was published in 1676 after Graunt's death with "Some further Observations of Major John Graunt."

The great fire in 1666 destroyed his property in London, and he seems to have been in increasing economic troubles for the rest of his life. His difficulties were also enhanced by his change of religion in his later years. Originally a Puritan, he became for many years a Socinian (Unitarian), and finally a Roman Catholic, which forced him to give up his civil and military offices.

There has been much discussion of whether Petty wrote the whole or part of the *Observations*. Glass (1964) has summarized the discussion as follows: "neither direct testimony nor internal evidence furnishes much support for the contention that Petty contributed in any substantial measure to Graunt's *Observations*." However, Cullen (1975, p. 5) argues that "Graunt may have written the body of the work but the general framework and the whole conception were Petty's."

A full account of the little that is known about Graunt's life has been given by Glass (1964).

Graunt's book is a small volume of 85 pages plus two dedications and an index, "the whole Pamphlet, not two hours reading," as Graunt writes in his dedication to Lord Roberts. This is true if one reads the book as an entertainment, mainly to learn about the many new factual results in demography and vital statistics which Graunt deduces from the bills. However, if one is interested in learning the new ideas about statistical analysis embodied in the book, the reading will take much longer. His statistical methods are seldom pronounced directly but are to be found in his examples.

His program for descriptive statistical analysis is announced in the paragraph we have chosen as epigraph. First he stresses the *reduction of data* from "several great confused Volumes into a few perspicuous Tables," and next his *statistical analysis of the tables* is presented in "a few succinct Paragraphs, without any long Series of multiloquious Deductions." This admirable program has ever since been a goal for any statistical office.

Graunt's fascinating book should be read by anyone interested in the history of statistics.

Graunt's book begins with a very useful Index containing 106 propositions proved or discussed in the text. Next follow two dedications and a preface. The 12 chapter headings are as follows:

1. Of the Bills of Mortality, their beginning, and progress.
2. General Observations upon the Casualties.
3. Of Particular Casualities.
4. Of the Plague.
5. Other Observations upon the Plague, and Casualties.
6. Of the Sickliness, Healthfulness, and Fruitfulness of Seasons.
7. Of the difference between Burials, and Christnings.
8. Of the difference between the numbers of Males, and Females.
9. Of the growth of the City.
10. Of the Inequality of Parishes.
11. Of the number of Inhabitants.
12. Of the Country Bills.
The Conclusion.

It will be seen that besides the London bills he had access to the bills of a country parish, namely, Romsey in Hampshire.

At the end of the book he gives eight tables on which all his results depend. The contents of the tables may be described as follows.

1. Number of burials in London for each year from 1629 to 1636 and from 1647 to 1660 classified according to 81 causes of death. Subtotals are given for five groups of four years each, namely, 1629–1632, 1633–1636, 1647–1650, 1651–1654, 1655–1658, and totals are given for the corresponding 20 years. The total number of burials for the 20 years is 229, 250. This is a large table to which Graunt adds the following note: "That the 10 years between 1636 and 1647 are omitted as containing nothing Extraordinary, and as not consistent with the Incapacity of a Sheet."

2a. Number of burials and christenings in London for each year from 1604 to 1664. Burials are divided into plague deaths and other deaths, the latter being divided among three groups of parishes. Totals are given for every eight years.

2b. Number of burials and christenings in seven out-parishes for each year from 1636 to 1659, with information on plague deaths from 1636 to 1648.

3. Number of burials and christenings in London for each year from 1629 to 1664 for males and females. Totals for every eight years.

4. Number of weddings, christenings, and burials in Romsey for each year from 1569 to 1658 for males and females. Totals for every ten years.

5. A similar table for Tiverton from 1560 to 1664.

6. A similar table for Cranbrooke from 1560 to 1649.

7. Total number of deaths and number of plague deaths in London for each week in 1592, 1603, 1625, 1630, 1636, and 1665.

8. Weddings, christenings, and burials in Paris for each month of the years 1670–1672.

Tables 5, 6, and 7 were added in the third edition, and Table 8 in the fifth edition.

On the title page Graunt calls himself "Citizen of London," and in several places of the book he hints at his lack of academic education and his use of shop arithmetic instead of mathematics. There is no doubt, however, that he was perfectly aware of the originality of his approach and his results. He writes to Lord Roberts that "to offer any thing like what is already in other Books, were but to derogate from your Lordship's learning." In the preface he writes, "Moreover finding some Truths, and not-commonly-believed Opinions, to arise from my Meditations upon these neglected Papers, I proceeded further, to consider what benefit the knowledge of the same would bring to the World." In the conclusion he states, "And there is pleasure in doing something new, though never so little, without pestering the World with voluminous Transcriptions."

In accordance with the title of his book, *Natural and Political Observations*, Graunt wrote two dedications. The first is addressed to John Lord Roberts, also called Robartes, Lord Privy Seal and, later, the first Earl of Radnor. Graunt gives a summary of the results in his book concerning "Government, Religion, Trade and Growth" most interesting and useful for a politician. He intimates that his Lordship may know some of these results already, but that the new thing is that they are now *deduced* from the bills of mortality.

The second dedication is to Sir Robert Moray, President of the Royal Society, and to the rest of that Honourable Society. Referring to a work of Francis Bacon, Graunt points out that his work is on natural history, since it concerns "the Air, Countries, Seasons, Fruitfulness, Health, Diseases, Longevity, and the proportion between the Sex and Ages of Mankind." As such it naturally falls under the sphere of interest of the Society.

The data given in the bills are unreliable, incomplete, and often inadequate as a basis for answering the questions raised by Graunt. For example, there is no information on the size of the population and on the age distribution of the dead and the living. Nevertheless, Graunt succeeds in answering a great many questions in demography and vital statistics by shrewd argumentation. He has a remarkable ability to support his conclusions by several independent arguments based on various aspects of the data.

His reasoning is often based on the idea of the stability of statistical ratios; sometimes he demonstrates the stability from the data, in other cases, he uses it as a reasonable hypothesis.

The main impression of his book is one of strong logical consistency of arguments and results. It is no wonder, however, that certain shortcomings occur; we shall comment later on those of a methodological character. It is also easy to find discrepancies between the numbers given in the text and the tables and to find small errors in his arithmetic. We shall not dwell on these blemishes, which are of minor importance compared to the sound statistical principles embodied in his work.

His data are given as a number of time series. Nobody would analyze such data today without plotting them. In Graunt's time, however, the graphical analysis of statistical data had not been invented, so Graunt analyzed his data by looking carefully at the variation of the series of numbers. To help in his evaluation of trends, he computed totals for periods of four, eight, or ten years in the different series.

We have already pointed out that he begins with the history and provenance of the data. We shall next give some comments and examples of his main methods and results.

7.4 GRAUNT'S APPRAISAL OF THE DATA

Graunt begins by a discussion of the reliability of the classification according to cause of death. He says that even if the Searchers may be ignorant and careless, their reports are sufficiently accurate for many purposes, particularly if certain corrections are carried out. We have already seen in §7.2 how by studying the course of the series of burials, Graunt reaches the conclusion that the number of burials are not to be trusted in plague years and that "a fourth part more die of the plague than are set down." For many deaths, however, the cause of death is obvious, and in case of doubt the Searchers usually report the opinion of the physician who treated the patient. Nevertheless, misclassifications occur but "it matters not to many of our purposes, whether the Disease were exactly the same, as Physicians define it in their Books." As an example of underrecording, Graunt mentions the number of deaths from French pox (syphilis). Comparing the reports from several parishes, Graunt concludes that in most parishes syphilis is misclassified as consumption, ulcers, or sores, because the family of the deceased has bribed the Searchers. He concludes that it is important to report the deaths from many causes and many districts so as to be able to check the accuracy by making comparisons.

In his comparisons of the mortality of various causes of death over the period of investigation from 1604 to 1661, Graunt sometimes uses the yearly number of burials (exclusive of the plague deaths) and sometimes the yearly number of christenings as a measure of the population at risk. He thus tacitly assumes constant death and birth rates. How suitable are the recorded numbers of burials and christenings for this purpose? Studying the table of burials and christenings in nonplague years, Graunt observes that the ratio of christenings to burials has decreased from about 1 to 1 before 1642, to about 2 to 3 in 1648 and to about 1 to 2 in 1659, the reason being a change of religious opinion on baptizing infants after the Civil Wars and the change of government. Thus, the number of christenings after 1642 does not adequately represent the number of births. As further evidence of this conclusion, Graunt presents the comparisons shown in the following table:

Number of Deaths and Christenings in London

YEAR	ABORTIONS AND STILLBIRTHS	WOMEN DYING IN CHILDBED	ORDINARY BURIALS	CHRISTENINGS
1631	410	112	8288	8524
1659	421	226	14,720	5670

Source: Graunt, 1662, Chap. 3.

Graunt takes the larger ratio of stillbirths to christenings in 1659 than in 1631 as an indication of a decrease in the number of christenings in relation to the number of births; he thus tacitly assumes that the stillbirth rate has not increased. However, he remarks that the numbers of abortions and stillbirths are rather unreliable, so he turns to the numbers of women dying in childbed, which he considers to be more accurately recorded. Assuming a constant maternal mortality rate, he concludes "that the true number of the Christnings Anno 1659 is above double to the 5,670 set down in our Bills, that is about 11,500, and then the Christnings will come near the same proportion to the Burials, as has been observed in former times." This is typical of Graunt's line of reasoning; he supports his conclusion in three different ways, all based on numerical data. His general conclusion is that the christenings may be considered proportional to the population size before

1642, whereas thereafter the burials should be used. In principle he prefers the christenings to the burials because some people die in London without living there permanently.

7.5   PROPORTIONAL MORTALITY BY CAUSE OF DEATH

From his basic table of burials, classified according to causes of death, Graunt derives the following summary in Chap. 2:

CAUSE OF DEATH	NUMBER OF BURIALS
Plague	16,000
Children's diseases	77,000
Aged	16,000
"Chronical" diseases	70,000
"Epidemical" diseases	50,000
Total	229,000

He defines the "chronical" (endemic) diseases as those which bear a constant proportion to the whole number of burials, whereas the proportion of deaths due to "epidemical" diseases varies from year to year.

He gives a list of 30 diseases and casualties with very low proportional mortality, for most of them less than 1/200 of the total mortality. He writes that many persons live in great fear and apprehension of these diseases and that his table may make them better understand the (small) hazard they are in.

For a great number of diseases he gives in Chap. 3 a detailed analysis of the trend in mortality taking the increasing population into account. As an example, consider the number of deaths due to "Stone," as shown in the following table:

YEAR	STONE	BURIALS
1631–1635	254	47,757
1656–1650	250	68,712

To make life easy for the reader, he has obviously selected two 5-year periods with nearly the same numbers of deaths from Stone. He then remarks that because of the increasing population, the mortality has really decreased.

Graunt faces the same difficulties as we do today in medical statistics:

Diagnoses are improved, and new diseases appear so that the classification changes. For example, in the beginning of the observational period, Livergrown, Spleen, and Rickets are in one group, whereas from 1634, Rickets and, from 1647, Spleen are classified separately. A new disease, Stopping of the Stomach, occurs in 1636. To illustrate Graunt's discussion of these problems we shall give the number of burials for the five 4-year periods from his basic table for several diseases. In his own discussion he uses the yearly numbers.

Number of Burials for Various Diseases Excerpted from Graunt's Basic Table

DISEASE	1629–1632	1633–1636	1647–1650	1651–1654	1655–1658
Livergrown, etc.	392	356	213	269	191
Rickets	0	113	780	1190	1598
Stomach	0	6	121	295	247
Rising of the Lights	309	220	777	585	809
Scurvy	33	34	94	132	300
Gout	14	24	35	25	36

Graunt begins by asking whether Rickets actually first appeared about 1634 or whether a disease which had long been in existence did then first become named. Comparing the first two lines of the table, he concludes that Rickets is a new disease with an increasing mortality, since the rate of increase in the number of burials far exceeds the rate of increase in the total number of burials. (During the period of observation, the total number of burials increased about 50%.) A similar conclusion holds for Stopping of the Stomach. He surmises that the increase in Rising of the Lights may be connected with the two other diseases just mentioned.

He also concludes that the mortality of Scurvy is increasing, whereas Gout does not increase. He also discusses the mortality of several other diseases.

7.6   THE STABILITY OF STATISTICAL RATIOS

The stability of the ratio of the numbers of male and female births and deaths is investigated both for London and for Romsey. Graunt refers the reader. to the tables, from which the stability is obvious, and he calculates the ratios only for the whole period of observation. We have calculated the ratios of the totals given in his tables, as shown in the table below.

Ratio of Number of Males to Number of Females

LONDON			ROMSEY		
PERIOD	CHRISTENINGS	BURIALS	PERIOD	CHRISTENINGS	BURIALS
1629–1636	1.072	1.113	1569–1578	1.03	0.97
1637–1640	1.073	1.149	1579–1588	1.06	0.95
1641–1648	1.063	1.093	1589–1598	1.25	1.19
1649–1656	1.095	1.065	1599–1608	0.97	1.14
1657–1660	1.069	1.093	1609–1618	1.16	0.88
			1619–1628	0.99	1.00
Total	1.074	1.098	1629–1638	1.01	0.99
			1639–1648	0.98	0.98
			1649–1658	1.11	0.99
			Total	1.06	1.00

To help the reader prove that most of the variation may be explained as random binomial variation, we have given the average "sample sizes" in the following table:

Average Number of Christenings and Burials for Males and Females

	CHRISTENINGS		BURIALS	
	MALES	FEMALES	MALES	FEMALES
London per 8 years	33,758	31,430	49,738	45,297
Romsey per 10 years	362	343	293	293

Graunt concludes that in Romsey there were born 15 females for 16 males, whereas in London there were 13 for 14, which shows "that London is somewhat more apt to produce Males than the Country." Perhaps he feels a little unsure of this conclusion, since he adds that "it is possible, that in some other places there are more Females born than Males; which, upon this variation of proportion, I again recommend to the examination of the curious."

This is the first time that the near equality of the numbers of males and females and the stability of the sex ratio at birth have been statistically demonstrated.

Graunt speculates on the reasons for this phenomenon, and he vacilates between religious and practical explanations. He remarks that the higher death rate of adult men and the greater emigration reduce the surplus of men and "bringeth the business but to such a pass, that every Woman may have an Husband, without the allowance of Polygamy." It follows "that the Christian Religion, prohibiting Polygamy, is more agreeable to the Law of

Nature" than Islam. Furthermore, "It is a Blessing to Mankind, that by this overplus of Males there is this natural Bar to Polygamy; for in such a state Women could not live in that parity and equality of expense with their Husbands, as now, and here they do."

Graunt expresses the wish "that Travellers would enquire whether it be the same in other Countries." This wish has certainly been fulfilled. The number of investigations on this problem since the time of Graunt is overwhelming.

It is remarkable that Graunt also investigates the stability of the fluctuations in the time series of christenings and burials. For Romsey he has grouped his data in decades, and in Chap. 12 he tabulates the greatest and least numbers within each of the nine decades, as shown in the following table:

Greatest and Least Yearly Numbers of Burials and Christenings in Romsey

	BURIALS			CHRISTENINGS		
DECADE	MAXIMUM	MINIMUM	RATIO	MAXIMUM	MINIMUM	RATIO
1569–1578	66	34	1.9	70	50	1.4
1579–1588	87	39	2.2	90	45	2.0
1589–1598	117	38	3.1	71	52	1.4
1599–1608	53	30	1.8	93	60	1.6
1609–1618	116	51	2.3	87	61	1.4
1619–1628	89	50	1.8	85	63	1.4
1629–1638	156	35	4.5	103	66	1.6
1639–1648	137	46	3.0	87	62	1.4
1649–1658	80	28	2.9	86	52	1.7

Graunt does not tabulates the ratios but uses some of them in the discussion in the text.

For comparison we have constructed a similar table for London:

Greatest and Least Yearly Numbers of Burials and Christenings in London

	BURIALS[a]			CHRISTENINGS[b]		
DECADE	MAXIMUM	MINIMUM	RATIO	MAXIMUM	MINIMUM	RATIO
1609–1618	9596	6716	1.43	7985	6388	1.25
1619–1628	12,199	7401	1.65	8564	6701	1.28
1629–1638	13,261	8288	1.60	10,311	8524	1.21
1639–1648	12,216	9283	1.32	—	—	—
1649–1658	14,979	8749	1.71	7050	5612	1.26

[a]The number of burials in 1625 has been corrected for plaguedeaths.
[b]The number of christenings in the decade 1639–1648 is omitted for reasons discussed previously.

Graunt states "that the proportion between the greatest and the least Mortalities in the Country are far greater then at London." He gives some examples of this fact and concludes that at Romsey the ratio is generally above 2, whereas in London it is below 2. Graunt is naturally puzzled by this result and tries to explain it in terms of the fresh air in the country compared to the polluted air of London, in accordance with the epidemic theory of the time.

Graunt remarks that the ratios are smaller for the births than for the deaths, but still far greater in Romsey than in London. He does not comment further.

It is clear that this problem is too difficult for Graunt. He understands that trends are more easily discernable from totals than from individual years, but he has no clear conception of the relation between the size of the random variation and the size of the sample. Let us comment on Graunt's numbers. In Romsey the population is nearly constant, and the average yearly numbers of burials and christenings are 58 and 70, respectively. Considering these numbers as the means of two Poisson processes, the coefficients of variation become 13% and 12%, respectively. It follows that the variation of the ratios for the christenings is not much larger than for a Poisson process, whereas the ratios for the burials vary considerably more. This is easy to explain, since the occurrence of epidemics creates a larger variation in the deaths than in the births. In London the population is increasing at a rate of about 18% per decade, which explains a factor of 1.18 in the ratios. The average yearly number of burials and christenings is 10,058 and 7768, respectively, with corresponding coefficients of variation of 1.00% and 1.13%. The 18% increase within a decade, combined with the coefficient of variation, may explain the variation of the ratios for the christenings but not for the burials. It is, however, clear that the larger random variation of the yearly numbers in Romsey plays a large role in explaining that the ratios in Romsey are larger than in London.

7.7   A TEST OF THE HYPOTHESIS "THAT THE MORE SICKLY THE YEAR IS, THE LESS FERTILE OF BIRTHS"

The London bills show that the number of christenings is small in plague years, since many pregnant women die or flee to the country. In continuation of this observation Graunt formulates the refined hypothesis that "sickly" years are less fertile than healthy years. To prove this proposition from the London bills he first excludes the plague years, which he defines as years with more than 200 deaths from plague. This reduces his data from 58 to 34 years. He then makes the shrewd remark that "we may not call that a more sickly year, wherein more die, because such excess of Burials may proceed from increase and access of People to the City onely." [If he had also applied this observation to his definition of a plague year, he would have allowed the limit of 200 deaths to increase over time (population), but this weakness in his definition is not significant.] He defines a "sickly" year as a year wherein the burials exceed those of both the preceding and the subsequent year; that is, he uses the local maxima of the time series of burials for his definition. Implicitly healthy years are defined by means of the local minima. Similarly, he uses local maxima and minima in the time series of christenings to define fertile and less fertile years. His idea is to prove the hypothesis by means of the coincidence (correlation) between the local maxima of the series of burials and the local minima of the series of christenings. He does not give this analysis in detail but writes "that upon view of the Table, [the hypothesis] will be found true, except in a very few cases, where sometimes the precedent, and sometimes the subsequent years vary a little, but never both together." The proposed test is ingenious, but his conclusion doubtful.

His method of analysis of the data from Romsey is simpler because the population size is nearly constant. He therefore compares the deviations from the average number of burials, 58 per year, with the corresponding deviations from the average number of christenings, 70 per year, and pronounces that "you shall finde, that where fewer than 58 died, more than 70 were born. Having given you a few instances thereof, I shall remit you to the Tables for the general proof of this Assertion." His idea of a negative correlation is perfectly clear, but for once his common sense fails him in the execution of the test. He fools himself and the reader by selecting as examples some years where the data support his hypothesis, namely, four years with a high number of christenings and a low number of burials, and four years with a high number of burials and a low number of christenings. In the first group of four years the average number of burials is 41, that is, considerably lower than the total average of 58. However, if we include all the years with a large number of christenings (larger than 86, say), the average number of burials becomes 60. Similarly, we find for the group of four years with a large number of burials an average number of christenings of 58, that is, lower than the total average of 70; but if we include all the years with a large number of burials (larger than 87, say), the average becomes 68. Hence, the data do not support Graunt's hypothesis. This was to be expected in view of the conclusion reached in the previous section that the variation of the christenings is nearly a random binomial variation.

7.8 ON THE NUMBER OF INHABITANTS

Since the death and birth rates play an important role in Graunt's discussion of the population size and of migrations, we shall summarize his results here.

In London "I find by telling the number of families in some parishes within the walls, that 3 [persons] out of 11 families per annum have died." He assumes that the average family size is eight, namely, "the Man, and his Wife, three Children, and three Servants, or Lodgers." The death rate then becomes 3/88 or 34 per 1000. (Instead of 1/29.3, he sometimes uses 1/30 or 1/32 without any explanation.) In most other cases Graunt carefully presents the data on which his estimates are based. It is unfortunate that in this important case he does not report how many families and family members he counted.

We have already mentioned that in the third edition Graunt estimates the population size in 1661 at 403,000. Setting this number in relation to the average number of burials in 1660–1662, we get a death rate of 35 per 1000, in good agreement with his estimate of 3/88.

The average population size for Romsey is estimated at 2700 and the average number of deaths 58, which gives a death rate of 22 per 1000. He attributes the great difference in urban and rural mortality to the unhealthiness of London, with its polluted air and overcrowding.

Since the number of christenings in London is about 12/13 of the number of burials, the birth rate becomes 31 per 1000.

In Romsey the birth rate is estimated at 70/2700 or 26 per 1000.

At Graunt's time it was generally believed that the population of London amounted to about 1 million people. Graunt tells that one day an Alderman of the City asserted that the increase in population from before the plague in 1625 to 1661 was 2 million, and upon this provocation Graunt decided to endeavor to get a little nearer to the truth.

Characteristically, he estimates the number of families in London in three different ways, namely, from the births, the burials, and the number of houses.

Since the number of burials at the time was about 13,000 per year he estimates the number of births at 12,000, even if the number of christenings is considerably smaller for reasons previously mentioned. From the consideration that childbearing women "have scare more than one Childe in two years," he estimates the number of childbearing women at 24,000 and the number of families at 48,000, because he imagines that the number of women between 16 and 76 years is about double the number of women of childbearing age.

Second, from his estimated death rate of three persons per eleven families and the 13,000 yearly deaths, he again finds 48,000 as the number of families.

Third, from a map of London and from the estimated area occupied by a house, he guesses that there are 54 families per 100 square yards, of which there are 220 within the walls. Since the yearly number of burials within the walls is about one-fourth the total number, he again gets 48,000 families.

Of course this agreement is almost too good to be true. On the other hand, Graunt is disarmingly honest; he displays his data frankly and every step in his (bold) reasonings, so that it is easy for the reader to criticize the result or to make improvements if he so desires. It should also be remembered that this is the first attempt to estimate the population of London on an objective, statistical basis.

As mentioned previously, he estimates the average family size to be eight, which gives him a population in London of 384,000 for the area corresponding to the original bills. However, he also has the information that the burials in seven suburb parishes amount to about one-fifth the burials in London, which leads him to 460,000 as the population of Greater London.

In 1665 Graunt got access to a census of the population of London taken in 1631, and he thus got the possibility of checking his estimate. The census covering only the central part of London gave a population of 130,000. Using as usual the ratio of number of burials as a measure of change in population size, Graunt corrects the census result by two factors, the first estimates the increase from 1631 to 1661 and the second the increase of area. His result is

$$130,000 \times \frac{11}{8} \times \frac{9}{4} = 403,000,$$

but he is deplorably vague about how he arrives at the correction factors.

To estimate the growth of the different parts of London, Graunt gives the following figures:

Number of Burials in Different Districts of London

YEAR	97 PARISHES	16 PARISHES	10 OUT-PARISHES	TOTAL
1605	2014	2974	960	5948
1659	3431	6988	4301	14,720
Ratio	1.7	2.3	4.5	2.5

The 97 parishes within the walls have the smallest rate of growth because there is almost no more space for new buildings. Then follows the 16 parishes immediately outside the walls and finally the 10 outer parishes, where the greatest increase has taken place. Graunt describes in some detail this development and points out the economic consequences of this westward movement of London.

Since the number of deaths in London exceeds the number of births, Graunt faces the difficult problem of explaining the growth by means of migration from the country. His information on population movements in the country is limited to the number of burials, christenings, and weddings from 1569 to 1658 in a parish containing the market town Romsey in Hampshire "being a place neither famous for Longevity and healthfulness, nor for the contrary." (Romsey is Petty's birthplace.)

To estimate the population size he first proceeds as for London. Guessing at a death rate of $1/4$ per family, a little smaller than the $3/11$ for London, and using the average yearly number of deaths, he finds $58 \times 4 = 232$ families and $232 \times 8 = 1856$ persons. Comparing the number of families with the number of houses, and the number of people with the number of communicants, he discards these estimates as too low. From an ordinary number of communicants of 1500 and from the supposition that there are nearly as many under 16 years old as above, Graunt concludes that there are about 2700 people in Romsey.

The excess of christenings over burials in Romsey during the 90 years is 1059, or about 12 per year. This is the tiny basis for his speculations on growth and migration. He begins by distributing the surplus of about 1100 people in three categories: about 300 have remained in the parish; it is probable that between 300 and 400 went to London; and it is known that about 400 went overseas. He supposes that Romsey has grown from 2400 to 2700 during the 90 years.

He also estimates the population of England and Wales at about 6,440,000 people. First, he remarks that the whole population equals 14 times the population of London, $14 \times 460,000 = 6,440,000$, since London bears $1/15$ of the whole tax. Second, he notes that there live 220 persons per square mile at Romsey and guesses that for the whole country, the density is about $3/4$ of that. Since the whole area is 39,000 square miles, he obtains the estimate $220 \times (3/4) \times 39,000 = 6,400,000$. Third, he guesses that the average number of people per parish is about 600, and multiplying by 10,000, the number of parishes, he gets 6,000,000.

The problem is to get a consistent whole out of this information on the population of England, London, and Romsey. The most reliable estimate of the migration from the country to London is obtained from the growth of London itself. He considers the 40-year period from 1622 to 1661 in which the population has doubled. This follows from various investigations of the increase in the number of burials and is in agreement with the previous demonstration of an increase by a factor of 2.5 in 54 years. For the 40 years the increase has thus been 230,000 to reach the 460,000 in 1661. To this must be added the excess of deaths over births. Because the recording of christenings during the period in question is unreliable, he uses the excess 33,000 for the period 1604–1643. His conclusion is that immigration has been about 250,000, or about 6000 per year. With a death rate of $3/88$ this gives a yearly increase in the number of burials of about 200, which is in agreement with the bills. (The numbers given in his text on p. 43 do not agree with the numbers in the tables.)

Turning to the direct estimation of the rate of migration, he first observes that the population in the country increases by about 1/7 in 40 years "as we shall hereafter prove doth happen in the Countrie [Romsey]." However, he does not return to this question. Perhaps he started from the increase $1059 \times (40/90) = 470$ in relation to 2700, which is about 1/6, and for some reason changed this estimate to 1/7. With a population outside London of 5,980,000, this gives an increase of 854,000 of which about 1/3, i.e., 285,000, migrates to London according to the experience at Romsey. He also uses the same data in a loose way to estimate the yearly migration as 1 out of every 900 inhabitants, which gives a yearly number of 6600. He concludes that there is reasonable agreement between the estimates of migration based on the Romsey and the London data.

To improve this analysis Graunt realized the necessity of getting more data on the population outside London. He therefore designed a representative sample of seven parishes, but he only succeeded in getting information from two besides Romsey. In the Appendix to the third and following editions he wrote, "I have here inserted two other Country-Bills, the one of Cranbrook in Kent, the other of Tiverton in Devonshire, which with that of Hampshire [Romsey], lying about the midway between them, give us a view of the most Easterly, Southerly, and Westerly parts of England. I have endeavoured to procure the like account from...thereby to have a view of seven countries most differently situated." One would expect that he would have used these new data to improve the analysis carried out in the first edition, but he leaves this task to the reader.

In the fifth edition he included a table of the monthly burials in Paris for the years 1670–1672, which he used for a discussion of the population size of Paris compared to that of London.

7.9 GRAUNT'S LIFE TABLE

Graunt naturally wanted to answer the important political question of how many "fighting men" (men between 16 and 56 years of age) there were in London. He therefore had to construct an age distribution of the living population from his data on the number of deaths according to causes, a seemingly impossible task. He hit upon the idea of estimating childhood and old age mortality from the data and then filing in the gap by guesswork. He thus constructed a life table showing how a cohort of 100 newborn infants dies out. This is perhaps the most famous and most discussed passage in Graunt's book, and we shall therefore quote it extensively.

Having premised these general Advertisements, our first Observation upon the Casualties shall be, that in twenty Years there dying of all diseases and Casualties, 229,250, that 71,124 dyed of the Thrush, Convulsion, Rickets, Teeths, and Worms; and as Abortives, Chrysomes, Infants, Livergrown, and Overlaid; that is to say, that about 1/3 of the whole died of those diseases, which we guess did all light upon Children under four or five Years old.

There died also of the Small-Pox, Swine-Pox, and Measles, and of Worms without Convulsions, 12,210, of which number we suppose likewise, that about 1/2 might be Children under six Years old. Now, if we consider that 16 of the said 229 thousand died of that extraordinary and grand Casualty the Plague, we shall finde that about thirty six per centum of all quick conceptions, died before six years old (Graunt, 1662, p. 15).

His estimate of the mortality for the age group 0–6 years is thus found to be

$$\frac{71,124 + 6105}{229,250 - 16,000} = 0.36.$$

Graunt includes the 8559 abortions and stillborn among the deaths, and he therefore writes about the mortality of "all quick conceptions." If these deaths are excluded, the death rate becomes 0.32.

Old-age mortality is estimated from the 15,757 deaths classified as "aged" in relation to the total number 229,250, which gives 7%. (He does not mention the plague deaths in this connection.) Then he writes on p. 18, "Onely the question is, what number of Years the Searchers called Aged, which I conceive must be the same, that David calls so, viz. 70." However, in his description of the causes of death on p. 13 he writes, "whether men were Aged, that is to say, above sixty years old, or thereabouts, when they died." He is thus rather inconsistent, and as we shall see, he finally decides that only 3% are alive at 66 years of age and 1% at 76.

The life table is derived on pp. 61–62:

Whereas we have found, that of 100 quick Conceptions about 36 of them die before they be six years old, and that perhaps but one surviveth 76, we, having seven Decads between six and 76, we sought six mean proportional numbers between 64, the remainder, living at six years, and the one, which survives 76, and finde, that the numbers following are practically near enough to the truth; for men do not die in exact Proportions, nor in Fractions; from whence arises this Table following.

Viz. of 100 there dies within		The fourth	6
the first six years	36	The next	4
The next ten years, or Decad	24	The next	3
The second Decad	15	The next	2
The third Decad	09	The next	1

From whence it follows, that of the said 100 conceived there remains alive at six years end 64.

At Sixteen years end	40	At Fifty six	6
At Twenty six	25	At Sixty six	3
At Thirty six	16	At Seventy six	1
At Forty six	10	At Eighty	0

There has been much discussion of how Graunt constructed his life table. Evidently he began with the two values (6, 64) and (76, 1), and then he mysteriously "sought six mean proportional numbers" to fill in the gap. Most authors have suggested that he used a geometrical series for the living. The simplest explanation has been given by Westergaard (1901, p. 31), who writes that for the first five decades Graunt used a mortality rate of about 3/8 and thereafter a somewhat increasing mortality. Glass (1950) has given an explanation in terms of the deaths rather than the survivors.

Let l_x denote the number of survivors at age x. The following table shows Graunt's values for l_x and the corresponding values for a constant mortality rate of 3/8 per decade. The two columns agree, except for the last three values. We have also shown the mortality rate for each age interval according to Graunt's table.

Graunt's Life Table and some Supplementary Calculations and Data[a]

GRAUNT (1662)				GENEVA (1601–1700)		
AGE x	l_x	$64(5/8)^y$	DEATH RATE[b]	AGE x	l_x	DEATH RATE[b]
0	100	—	0.36	1	100.0	0.43
6	64	64	0.375	7	57.4	0.16
16	40	40	0.375	17	48.1	0.16
26	25	25	0.36	27	40.5	0.18
36	16	16	0.375	37	33.4	0.22
46	10	10	0.40	47	26.0	0.29
56	6	6	0.50	57	18.4	0.39
66	3	4	0.67	67	11.2	0.57
76	1	2	1.00	77	4.8	0.98
80	0	2		87	0.1	

[a] $y = (x - 6)/10$.

[b] The death rate is the difference between consecutive l's divided by the first l.

How realistic is Graunt's table? Hull (1899, p. 386) has given the distribution of 53,783 deaths according to age in Geneva for the years 1601–1700 taken from a paper by E. Mallet in 1837, and from these data we have calculated the life table (assuming a stationary population) and the mortality rates given in the table above. It will be seen that the death rates in Geneva for the age interval 6–36 years are only about half of the death rates assumed by Graunt, which indicates that Graunt's assumption is unrealistic.

Graunt did not grasp all the implications of his new concept. He did not know the relationship between the life table and the age distribution of the corresponding stationary population, and he even made a mistake in the use of the table for its intended purpose, namely to find the number of men between 16 and 56 years of age. Referring to the table he writes, "There are therefore of Aged between 16 and 56 the number of 40, less by six, viz. 34." However, 34 is the proportion dying between 16 and 56, not the proportion living, as pointed out by Westergaard (1932, p. 23). Graunt goes on to find the number of fighting men in London from the total population, using the previously derived ratio of males to females and the 34% found above.

We mentioned earlier two arguments that depend on the age distribution and were used by Graunt to estimate the population sizes of London and Romsey. For London, he assumes that the number of women between 16 and 40 and between 40 and 76 are about equal, and for Romsey he assumes that the number of people under 16 years of age is a little smaller than over 16. If he had confronted these statements with results derived from his life table, even by his flawed method, he would have perhaps realized that the number of deaths alloted to ages betwen 6 and 46 is much too large. Of course the whole problem was obscured by migration, and it is odd that Graunt did not comment on the different age distributions of the rapidly growing London and the nearly stationary Romsey.

7.10   CONCLUDING REMARKS ABOUT GRAUNT'S OBSERVATIONS

Graunt's book had immense influence. Bills of mortality similar to the London bills were introduced in other cities, for example, Paris in 1667.

Graunt's methods of statistical analysis were adopted by Petty, King, and Davenant in England; by Vauban in France; by Struyck in the Netherlands; and somewhat later by Süssmilch in Germany. Ultimately, these endeavors led to the establishment of governmental statistical offices.

Graunt's investigation on the stability of the sex ratio was continued by Arbuthnott and Nicholas Bernoulli. His life table was given a probabilistic interpretation by the brothers Huygens; improved life tables were constructed by de Witt in the Netherlands and by Halley in England and used for the computation of life annuities. The life table became a basic tool in medical statistics, demography, and actuarial science.

There exists a large literature about Graunt's work. We refer to Hull (1899), Westergaard (1932), Willcox (1937), Greenwood (1941–43), Sutherland (1963), Glass (1964), and K. Pearson (1978).

Kreager (1988) throws new light on Graunt. He traces Graunt's method back to Bacon and relates his statistical technique to common bookkeeping. He also explains many of Graunt's remarks in terms of the prevailing religious and mercantile opinions.

7.11 WILLIAM PETTY AND POLITICAL ARITHMETIC

William Petty (1623–1687) was the son of a clothier in Romsey in Hampshire, where the precocious boy received his basic education, including some mathematics and Latin and Greek. After a roving life that included studies in French, mathematics, astronomy, and navigation at Caen in France; several years in the Royal Navy; and during the Civil Wars several more years of study, particularly medicine, at Utrecht, Leiden, Amsterdam, and Paris, he came back to England in 1647, got his doctor's degree in medicine, and became Professor of Anatomy at Oxford in 1650. It was about that time that he made the acquaintance of Graunt, who helped him to become Gresham Professor of Music, a post he really never took over. The decisive break in his career came in 1652, when he was appointed physician to the army in Ireland and to the Lieutenant-General Henry Cromwell. From 1655 to 1656, Petty undertook a survey of Ireland as a basis for the distribution of forfeited lands to army officers and soldiers, and he also became a member of the commission that distributed the lands. Petty himself took over a great deal of land and was later charged with dishonesty, which led to many lawsuits. By the end of his appointment in 1659, the initially poor university professor had become a great landowner and a wealthy man. Returning to England, he joined a group of natural scientists who met at Gresham College and eventually founded the Royal Society in 1662. At the Restoration in 1660 one would have expected Petty to have great difficulty, but he soon gained the King's favor; he was kinghted in 1661 and appointed Surveyor-General of Ireland. He involved himself in many practical, economic, and political problems in Ireland, where he spent a large part of his life.

Petty's abundant energy also resulted in a great number of reports, essays, and books. Hull (1899, Preface) writes;

The writings of Sir William Petty may be roughly divided into three classes. The first relates to his activities as surveyor of forfeited lands in Ireland under the Protectorate; its present interest is chiefly biographical. The second includes his papers on medicine and on certain mathematical, physical and mechanical subjects. These are now forgotten. The third class comprises his economic and statistical writings. The merit of these has been freely recognized. No writer on the history of political economy who touches the seventeenth century at all has failed to praise them.

Petty is one of the founders of the English school of political economy. His economic writings are responses to actual problems, such as tax problems caused by the Restoration and by the Dutch war, money problems in connection with the recoinage project, assessment of the wealth of England as compared to France as an argument for making Charles II independent of Louis XIV, the economic problems of Ireland to support his arguments for reforms. With his background in mathematics and medicine, he coined the terms "political arithmetick" and "political anatomy" used in the titles of his books.

Petty was greatly influenced by Bacon and Graunt. In the preface of his *Political Arithmetick*, written in 1676 but first published in 1690, he writes;

The Method I take to do this, is not yet very usual; for instead of using only comparative and superlative Words, and intellectual Arguments, I have taken the course (as a Specimen of the Political Arithmetick I have long aimed at) to express my self in Terms of *Number, Weight,* or *Measure*; to use only Arguments of Sense, and to consider only such Causes, as have visible Foundations in Nature.

The importance of the works of Graunt and Petty for the establishment of official statistical offices has been described by Hull (1899, p. LXVI). Hull also points out that statistics rests on the enumeration of a large number of items and the construction of tables by *addition* of the observations into adequate categories; however, Petty's political arithmetic is based on scanty and imperfect data from which by *multiplication* he derives information on phenomena correlated with the observed, using the stability of statistical ratios and often simply guessing at these ratios.

Petty was less critical in his assessment of data and methods than Graunt, and he extended his analysis to many more topics about which less was know. Since Petty did not contribute new methods of statistical analysis, we shall not give an account of the many *Essays in Political Arithmetick*, which he published in the 1680s.

The sketch above is essentially based on the book by Hull (1899), which also contains reprints of the most important of Petty's papers with notes by Hull. More recent investigations of Petty's life and works and on the further development of political arithmetic in the 17th and 18th centuries have been given by Westergaard (1932), Greenwood (1941–1943), and K. Pearson (1978).

CHAPTER 8

The Probabilistic Interpretation of Graunt's Life Table

There are thus two different concepts: the expectation or the value of the future age of a person, and the age at which he has an equal chance to survive or not. The first is for the calculation of life annuities, and the other for wagering.

—*CHRISTIAAN HUYGENS, 1669*

8.1 THE CORRESPONDENCE OF THE BROTHERS HUYGENS, 1669

Presumably, Graunt did not know anything about probability theory, and in particular he did not know Huygens' work. In March 1662, Sir Robert Moray set a copy of Graunt's book to Huygens, who politely thanked him for the gift and expressed his admiration of Graunt's ingenuity in general terms.

In 1669 Lodewijk (Ludwig) Huygens (1631–1699), Christiaan's younger brother, began a correspondence with Christiaan on the expectation of life and the usefulness of Graunt's table for calculating values of life annuities. They never got to life annuities; this was left to de Witt and Hudde two years later, as we shall see next in Chap. 9. They had, however, an interesting correspondence about the expected and the median lifetime, which was not published until 1895 in Huygens' *Oeuvres*, Vol. 6. (In the following, we shall cite page numbers only when referring to this volume.)

In Lodewijk's first letter (p. 483) he writes that he has made "a table of the remaining lifetime for persons of any given age" based on Graunt's table and states that it will be useful for the evaluation of life annuities. He adds that "the question is to what age a newly conceived child will naturally live" and asks the same question for persons of any other age. He states that Christiaan, who at the time was 40 years old, will live to about $56\frac{1}{2}$ years of age. Without disclosing his method of calculation, he asks Christiaan to make similar calculations for comparison.

Lodewijk's formulations are somewhat vague, and Christiaan does not seem to be much interested. In his short reply he states that to get "exact" results, one needs a life table with the number of deaths for each year of age. Further, he considers the life table as defining a game of chance and says that one may bet 4 to 3 on the event that a person aged 16 will live at age 36. (This should have been 24 to 16 or 3 to 2 instead.) Finally, he asks to see Lodewijk's calculations.

In his reply Lodewijk gives the numbers shown in Table 8.1.1.

Within each interval Lodewijk assumes a uniform distribution of the number of deaths so that t_x represents the average lifetime for each of the d_x deaths. He explains how to calculate the average age at death and the corresponding expectation of life (without using the word "expectation") as

$$\bar{t}_x = \sum_{i=x}^{76} t_i d_i \bigg/ \sum_{i=x}^{76} d_i = \sum_{i=x}^{76} t_i d_i \bigg/ l_x, \qquad i \text{ and } x = 0, 6, 16, \ldots, 76,$$

and $\bar{e}_x = \bar{t}_x - x$. For example, for $x = 0$ he says that the total lifetime of 1822 years should be divided equally among the 100 persons, which gives 18.22 years as the "age" for each of them.

Table 8.1.1. Calculation of expectations of life by Lodewijk Huygens in 1669[a]

Age x	Number of survivors l_x	Number of deaths d_x	Midpoint of age interval t_x	$t_x d_x$	Accumlation of $t_x d_x$ from below	Average age at death $\bar{t}_x$	Expectation of life $\bar{e}_x$
0	100	36	3	108	1822	18.22	18.22
6	64	24	11	264	1714	26.78	20.78
16	40	15	21	315	1450	36.25	20.25
26	25	9	31	279	1135	45.40	19.40
36	16	6	41	246	856	53.50	17.50
46	10	4	51	204	610	61.00	15.00
56	6	3	61	183	406	67.67	11.67
66	3	2	71	142	223	74.33	8.33
76	1	1	81	81	81	81.00	5.00
86	0						0.00

[a] Lodewijk does not use the mathematical symbols used here, he only gives the numerical results with verbal explanations.

Source: Huygens' *Oeuvres*, Vol. 6, p. 516.

Perhaps confused by Christiaan's letter he adds the (wrong) comment that a person aged 6 and a person aged 16 have about the same chance of living another 20 years.

In his reply (pp. 524–532 and 537–539), Christiaan repeats Lodewijk's calculations, finds the same results, and gives the correct interpretation.

At the time the only vocabulary available for discussing probability theory was that of games of chance. Therefore Christiaan considers the life table as defining a lottery with 100 tickets, 36 tickets having the value 3, 24 having the value 11, etc. He states that the expectation of life has been calculated according to the rule given in his treatise.

In his description of the life table Huygens considers the remaining lifetime for a person aged x, T_x, say, as a random variable. Instead of the (cumulative) distribution function $\Pr\{T_x \leqslant t\}$ or its complement $\Pr\{T_x > t\}$, he uses the odds and the corresponding bets in a fair game to characterize the distribution. Thus he states that the number of chances that a person aged 16 will die before age 36 equals 24 and that the number of chances that he will die after age 36 equals 16, so that in a fair game one should bet 16 to 24, or 2 to 3, on the event that the person dies before age 36.

Introducing the probability that a person aged x will survive t years as $_tp_x = l_{x+t}/l_x$ and the complementary probability $_tq_x = 1 - {_tp_x} = (l_x - l_{x+t})/l_x$, it will be seen that Huygens uses the odds $_tq_x/_tp_x = (l_x - l_{x+t})/l_{x+t}$ for the description of the probability distribution instead of $_tp_x$, as we do today.

He then turns to the special case where there is an even chance of dying before or after t years, that is, $l_{x+t} = \frac{1}{2}l_x$, and solves this equation for t, the *median* remaining lifetime as it is called today, sometimes also called the probable lifetime. He explains carefully the distinction between the expectation and the median and states that for a newborn child, the median equals about 11 years, whereas the expectation equals 18 years.

He also takes the remarkable step of considering the life table as a continuous distribution and gives a graph such as the one shown in Fig. 8.1.1. This is the first graph of (the complement of) a distribution function. He points out that the median lifetime for a newborn child may be found on the graph as the abscissa corresponding to an ordinate of 50, and he generalizes to the median remaining lifetime for any given age as shown for a person of age 20, for whom the median remaining lifetime is about 16 years. He does not comment on the fact that the graph shows a nearly exponential decrease of l_x, whereas his calculation of $\bar{e}_x$ assumes a linear decrease between the given points.

Finally, Huygens discusses joint-life expectations. As an example he mentions the case of a man aged 56 who marries a woman aged 16 and asks for the expected value of the time they are both living, the expectation of the shortest life, and the expected value of the time of the longest life. He

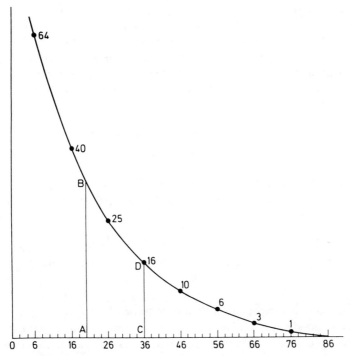

Fig. 8.1.1 Huygens' graph of Graunt's life table and his determination of the median remaining lifetime. For a person aged A, the median remaining lifetime is given by AC, since the ordinate CD is half of the ordinate AB.

does not carry out the calculations for this example but, for simplicity, only for a couple both aged 16.

To describe this situation Huygens considers a lottery containing $l_{16} = 40$ tickets having the values $t_x - 16$, with the corresponding number of chances equal to d_x. The two persons, (x) and (y), say, each draw successively and independently (with replacement) a ticket and the values of T_x and T_y, the remaining lifetimes, are recorded. To find the expectation of $T = \max\{T_x, T_y\}$, Huygens uses a conditional argument based on the idea that if $T_x = t$, it follows that $T = t$ for $T_y \leqslant t$, and $T = T_y$ for $T_y > t$.

Suppose that $T_x = 15$, i.e., (x) dies in the age interval $(26, 36)$, and combine this information with the following distribution of T_y:

T_y	5	15	25	35	45	55	65
d_y	15	9	6	4	3	2	1

To derive the distribution of T, Huygens remarks that the only difficulty occurs for $T_y = T_x = 15$. He distributes the nine deaths uniformly over the ten-year interval and notes that there are 4.5 chances for $T_y \leqslant 15$ and 4.5 chances for $T_y \in (16, 20)$, which he simplifies by using the midpoint $T_y = 18$. For $T_y \geqslant 25$, he uses the shortcut to take the expected lifetime from Lodewijk's table, which gives $53.50 - 16 = 37.50$ for the 16 cases considered. Hence, the distribution of T becomes

$$
\begin{array}{c c c c}
T & 15 & 18 & 37.50 \\
d & 19.5 & 4.5 & 16
\end{array}
$$

which gives the (conditional) expectation $E\{T | T_x = 15\} = 24.3$.

In this way, he calculates all the conditional expectations and obtains the following table:

$$
\begin{array}{c c c c c c c c}
t_x - 16 & 5 & 15 & 25 & 35 & 45 & 55 & 65 \\
E\{T | T_x = t_x - 16\} & 20.3 & 24.3 & 30.2 & 37.6 & 46.1 & 55.3 & 65.0 \\
d_x & 15 & 9 & 6 & 4 & 3 & 2 & 1
\end{array}
$$

The unconditional expectation then becomes $E\{T\} = 29.22$.

This is a good example of an early application of the fundamental principle that the expectation $E\{T\}$ may be found as the expectation of the conditional expectation $E\{T | T_x\}$.

Huygens also indicates how to obtain the expectation of the shortest life.

The step taken by the brothers Huygens in calculating the expectation of life is a very essential one, since the evaluation of a life annuity only requires that T be replaced by the value of an annuity certain of duration T.

For other discussions on the present topic we refer the reader to Kohli and van der Waerden (1975) and Hacking (1975).

8.2 NICHOLAS BERNOULLI'S THESIS, 1709

The results of the brothers Huygens remained unknown outside Holland, but their ideas were in the air. In 1666 a review of Graunt's book containing the life table was published in *Journal des Sçavans*. Based on this, James Bernoulli in a publication of 1686 quoted Graunt's table and stated without proof that the odds for a person aged 16 to die before a person aged 56 against the opposite event are 59:101. In 1709 Nicholas Bernoulli (1687–1759), a nephew of James Betnoulli, took up the same problems as the brothers Huygens and published similar solutions in his thesis *De Usu Artis Conjectandi in Jure* (The Use of the Art of Conjecturing in Law). It seems that neither James nor Nicholas had seen Graunt's book, which is odd in view of its wide circulation and its importance to Nicholas' thesis. There are abundant references to juridical literature in the thesis. For probability theory the only references are to Huygens' treatise (1657) and to the unpublished manuscript of *Ars Conjectandi* by James Bernoulli. The life tables of de Witt (1671) and Halley (1694) are not mentioned. Nicholas Bernoulli published a summary of his thesis in *Acta Eruditorum*, Supplementa, 4, 1711, pp. 159–170.

The title of Chap. 3 of the thesis is "Of an absent person to be presumed dead." After a discussion of the legal meanings of the word "absence," Bernoulli turns to the case where an absent person is "someone of whom it is not known where he is and whether he is alive." After a certain time of absence, the person may be presumed dead by order of the court. Bernoulli quotes many prominent jurists who hold widely varying opinions regarding the length of time of absence necessary to be presumed dead, namely, from 5 and up to 30 years with an "outlier" of 100 years. Bernoulli proposes to solve the problem by means of probability theory and Graunt's life table, which makes it possible to find "in what time it is twice, thrice, four times, etc. as likely that someone be dead than that he still be alive." He then gives the solution for the case where it is twice as probable that the absent person is dead than living; that is, he solves the equation $l_{x+t} = l_x/3$ with respect to t for $x = 0, 6, 16, \ldots, 76$, using linear interpolation in Graunt's table and rounding $l_x/3$ to integers. His results are as follows:

x	0	6	16	26	36	46	56	66	76
t	$20\frac{2}{3}$	$24\frac{4}{9}$	25	25	$23\frac{1}{3}$	20	15	10	$6\frac{2}{3}$

Bernoulli concludes that "if someone should have started his absence at the twentieth or thirtieth year of his life, for example, and has been absent for 25 years, and nothing has been heard from him through that time, the Judge would be able to declare him dead and to grant his goods to his nearest relatives without caution."

It will be seen that Bernoulli, like Huygens, considers the life table to define a probability distribution and that he reasons by means of the odds $_t q_x / _t p_x$.

In Chap. 2 of his thesis, Nicholas Bernoulli discusses the "estimation of the length of human life," in particular the expectation of life, the median lifetime, joint-life expectations and the probability of survivorships.

To prove James Bernoulli's statement mentioned above, Nicholas uses a conditional argument, which may be expressed by the formula

$$\Pr\{T_x < T_y\} = \sum_{i=0}^{\infty} \Pr\{T_x < T_y \mid T_y \in C_i\} \Pr\{T_y \in C_i\},$$

where C_i denotes the ten-year interval $[10i, 10(i + 1)], i = 0, 1, \ldots$. As an example, consider the term for $i = 1$, which means that the person aged 56 dies in the age interval $(66, 76)$. The probability of this event equals $d_{66}/l_{56} = 2/6$. The conditional probability is found by noting that there are $d_{16} = 15$ chances for a person aged 16 to die in the first decade, $d_{26} = 9$ chances to die in the second decade, and $l_{36} = 16$ chances to die later, which gives the probability $(15 \times 1 + 9 \times \frac{1}{2} + 16 \times 0)/40 = 39/80$. For the second decade, Bernoulli distributes the number of chances equally between the two persons because "neither has a stronger hope than the other of surviving." The final result becomes

$$\tfrac{1}{6}(3 \times \tfrac{15}{80} + 2 \times \tfrac{39}{80} + 1 \times \tfrac{54}{80}) = \tfrac{59}{160}.$$

It will be seen that Bernoulli, like Huygens, quite naturally uses conditional expectations.

In accordance with James Bernoulli, Nicholas states that the probability of surviving a certain age cannot be found by *a priori* reasoning as in a game of chance but that it may be estimated from observation, for example, by observing how many of 300 men will survive a period of 10 years. He then quotes Graunt's life table, which he believes "has been observed from collation of very many catalogues of this sort [of deceased persons]."

He begins by calculating the expectation of life by the same method as Lodewijk Huygens and finds the same results as shown in Table 8.1.1. He notes that the expectations may be obtained in an easier way by backward recursion and carries out the calculations by the formula

$$\bar{e}_x = \frac{1}{l_x}\{5d_x + l_{x+10}(10 + \bar{e}_{x+10})\}, \qquad x = 6, 16, \ldots, 76,$$

$$\bar{e}_0 = \frac{1}{l_0}\{3d_0 + l_6(6 + \bar{e}_6)\}.$$

In all of his calculations, Bernoulli assumes that the deaths are uniformly distributed within the age intervals given in Graunt's table, and he therefore interpolates linearly to intermediate values of l_x.

Like Christiaan Huygens, he clearly distinguishes between the mean and the median remaining lifetime. As an example, he shows by linear interpolation that the median lifetime for a newborn child equals $11\frac{5}{6}$ years, whereas the expectation equals $18\frac{11}{50}$ years. He does not give a graph of the life table.

The most remarkable of Bernoulli's results in Chap. 2 concerns the expected lifetime of the longest living among several lives. He begins by proving the following lemma.

Lemma. Let it be equally likely that a person dies at any moment within a time interval of length a and consider the distribution of b deaths within this interval. The expected lifetime of the longest living within the interval equals $ab/(b + 1)$.

In modern terminology this means that he considers a random variable with a uniform distribution over the interval $(0, a)$ and asks for the expectation of the largest of b observations.

For $b = 1$, the expected lifetime is obviously $a/2$.

For $b \geqslant 2$, Bernoulli divides the interval into n equal parts of length m, $nm = a$, and lets n tend to infinity. He speaks of these intervals as moments.

Suppose that $b = 2$ and that the largest lifetime equals xm, say, $x = 1, 2, \ldots, n$. It follows that there are x cases (samples of two observations), giving a largest lifetime of xm, since the first death may occur at anyone of the moments $1, 2, \ldots, x$. The expectation of the longest living then becomes

$$\sum_{x=1}^{n}(xm)x \bigg/ \sum_{x=1}^{n} x = \frac{a}{n}\sum_{x=1}^{n} x^2 \bigg/ \sum_{x=1}^{n} x.$$

For $n \to \infty$, Bernoulli replaces the sums by integrals and gets

$$\frac{a}{n}\int_0^n x^2\, dx \bigg/ \int_0^n x\, dx = \frac{2}{3}a.$$

Bernoulli gives similar proofs for $b = 3$ and 4, from which the general proof follows. The basic problem is to find the number of cases in which $b - 1$ deaths may be distributed on the x moments (an occupancy problem in modern terminology). However, this problem had been solved by James Bernoulli in *Ars Conjectandi*, and Nicholas simply gives the results

$$\binom{x + b - 2}{b - 1} \quad \text{and} \quad \sum_{x=1}^{n}\binom{x + b - 2}{b - 1} = \binom{n + b - 1}{b}$$

without comment. Nicholas' proof implies that the general solution is

$$\sum_{x=1}^{n}(xm)\binom{x + b - 2}{b - 1} \bigg/ \sum_{x=1}^{n}\binom{x + b - 2}{b - 1},$$

which tends to

$$\frac{a}{n}\int_0^n \frac{x^b}{(b - 1)!}\, dx \bigg/ \int_0^n \frac{x^{b-1}}{(b - 1)!}\, dx = \frac{ab}{b + 1}.$$

He supplements this (combinatorial) proof by a geometrical one based on the idea that the expectation of a random variable x with distribution $y(x)$ equals the center of gravity of the figure under the curve $y(x)$ over the interval $(0, a)$; that is, the expectation equals

$$\frac{\int_0^a xy(x)\,dx}{\int_0^a y(x)\,dx}.$$

If the longest living dies at time x, then $y(x)$ will be proportional to x^{b-1}, and the expectation becomes

$$\frac{\int_0^a x^b\,dx}{\int_0^a x^{b-1}\,dx} = \frac{ab}{b+1}.$$

Bernoulli then turns to the calculation of the expected lifetime of the longest of two (independent) lives (x) and (y). We shall use the notation from Table 8.1.1 so that x and $x + i$, say, take on the values $0, 6, 16, \ldots, 86$ only. The distribution of (T_x, T_y) is given by all possible products of the form $d_{x+i}d_{y+j}/l_x l_y$, where $l_x = d_x + d_{x+10} + \cdots + d_{76}$ for $x = 6, 16, \ldots, 76$ and $l_0 = d_0 + l_6$. For $j < i$, we have $T_y < T_x$, so that the longest lifetime equals $i + 5$. Summing d_{y+j} for all j less than i, we get $l_y - l_{y+i}$. For $i = j$, (x) and (y) die in the same interval, and the lemma then gives the longest lifetime as $i + \frac{2}{3} \times 10 = i + 6\frac{2}{3}$, apart from the case where $x = y = i = 0$, which gives $\frac{2}{3} \times 6 = 4$. If both x and y are larger than or equal to 6, the expectation becomes

$$\frac{1}{l_x l_y}\left[\sum_{i=0}^{\infty}(i + 6\tfrac{2}{3})d_{x+i}d_{y+i} + \sum_{i=10}^{\infty}(i + 5)d_{x+i}(l_y - l_{y+i}) + \sum_{i=10}^{\infty}(i + 5)d_{y+i}(l_x - l_{x+i})\right].$$

The necessary modification when at least one age equals zero is obvious. The formula becomes simpler for $x = y$.

Bernoulli does not use the symbols above but gives two numerical examples from which the formula follows. For $x = y = 0$, he finds the expectation equal to 27.8238 years; we shall leave the calculation to the reader as an exercise. For $x = 16$ and $y = 46$, he calculates the expectation as follows:

$$(4 \times 15 \times 6\tfrac{2}{3} + 9 \times 4 \times 15 + 15 \times 3 \times 15 + 3 \times 9 \times 16\tfrac{2}{3} + 6 \times 7 \times 25$$
$$+ 24 \times 2 \times 25 + 2 \times 6 \times 26\tfrac{2}{3} + 4 \times 9 \times 35 + 30 \times 1 \times 35 + 1 \times 4 \times 36\tfrac{2}{3}$$
$$+ 3 \times 10 \times 45 + 2 \times 10 \times 55 + 1 \times 10 \times 65)/40 \times 10 = 25\tfrac{23}{48}.$$

The reader should use Bernoulli's method to find the expected lifetime of the longest of two lives for $x = y = 16$ and thus compare the methods of Huygens and Bernoulli.

Bernoulli notes that his method may be used for any number of lives.

Bernoulli remarks that the assumption of a uniform distribution of the number of deaths over such large intervals as in Graunt's table is unsatisfactory. He has therefore asked a friend to provide some observations from a town in Switzerland on "the ages of nearly 2000 men in the same year, some living and some having died." Unfortunately, he publishes neither the data nor the method of analysis. He only gives a table of $\bar{e}_x$ for $x = 0, 5, 10, \ldots, 85$, the first five expectations being 27, 38, 37, 33, and 30 years, that is, considerably higher than the results from Graunt's table. He writes that the reason for this difference is unclear. It may be that the number of observations is too small to give reliable results, but it may also be, as he suspects, that "in our Switzerland, perchance because of more temperate life or better constitution of air, men more frequently succeed in arriving at a very great age than in France, where those observations which are contained in the *Ephemerides* (*Journal des Sçavans*) were perhaps made." He recommends that accurate parish records of the ages of the deceased be kept in future.

We shall discuss the remainder of Bernoulli's thesis in §§9.1, 9.2, and 21.1. A summary of the thesis with commentaries has been given by Kohli (1975b). The quotations given above are from Drucker's translation (1976).

Part of the present chapter has previously been published by Hald (1987).

CHAPTER 9

The Early History of Life Insurance Mathematics

From these Considerations I have formed the adjoyned Table, whose Uses are manifold, and give a more just Idea of the State and Condition of Mankind, than any thing yet extant that I know of. It exhibits the Number of People in the City of Breslaw of all Ages, from the Birth to extream Old Age, and thereby shews the Chances of Mortality at all Ages, and likewise how to make a certain Estimate of the value of Annuities for Lives, which hitherto has been only done by an imaginary Valuation. Also the Chances that there are that a Person of any Age proposed does live to any other Age given; with many more, as I shall hereafter shew.

—HALLEY, 1694

9.1 THE BACKGROUND

Bequests of maintenances, usufructs, and life incomes have occurred since ancient times. The necessity of evaluating such legacies, whether in kind or in money, arose in ancient Rome, when the Falcidian law, passed in the year 40 B.C., ruled that the heir (or heirs) to an estate should receive not less than one-quarter of the total property left by the testator. Hence, if the testator had given legacies amounting to more than three-quarters, they had to be reduced proportionately, and this also applied to the bequest of life incomes.

The Roman jurist D. Ulpian (?–228) devised a table for the legal conversion of a life annuity to an annuity certain:

Ulpian's Conversion Table

Age of annuitant in years										
0–19	20–24	25–29	30–34	35–39	40	⋯	49	50–54	55–59	60–
Duration of annuity certain in years										
30	28	25	22	20	19	⋯	10	9	7	5

The table is to be found in *Corpus Iuris Civilis*, Digesta, XXXV, 2, 68; the section in question has been quoted (in Latin) by Greenwood (1940).

The interpretation of Ulpian's table has been much discussed. Some authors hold that the table gives the expectation of life and that it to some extent is based on observation. Greenwood (1940), however, points out that the object was to protect the interests of the legal heir, and therefore the valuation is chosen too high. It is particularly unrealistic to suppose that the life expectancy of persons between zero and 19 years of age equals 30 years in view of the high mortality of children. Therefore, Greenwood is of the opinion that the 30 years is not an estimate of the expectation of life, but simply the *legal* maximum valuation of any usufruct. Greenwood further suggests that Ulpian rather arbitrarily may have chosen five years as the last value in the table and then inserted (interpolated to) the intermediate values, so that they generally agree with values previously used.

The approximation to a life annuity obtained as an annuity certain with a duration equal to the annuitant's expectation of life is sometimes called Ulpian's approximation, even if there is no such rule explicitly stated in the *Digest*.

Nicholas Bernoulli (1709, Chap. 5) discusses the application of Ulpian's table and the "Falcidian fourth" in his own time. He criticizes the interpretations made by some lawyers who have not understood the probabilistic nature of Ulpian's rule. Bernoulli considers the numbers in Ulpian's table to be the expectations of life, and he explains how to find the value of the corresponding annuity certain. As an example, he takes a person aged 24, who has a legacy of 10 ducats per year. According to Ulpian this has to be valued by means of the duration of 28 years, and some lawyers therefore use 280 ducats as the corresponding capital. Bernoulli, however, presents a table (in his Chap. 4) of the values of an annuity certain of one per year at 5% interest for all durations from 1 to 100 years. From this table it follows that the correct evaluation leads to a capital of 148.98 ducats.

Finally, Bernoulli remarks that the expectations of life given by Ulpian do not agree with his own results based on Graunt's life table. In particular, the highest expectation of life found by Bernoulli is 21 years, which is considerably less than the 30 years given by Ulpian. This disagreement does not, however, lead Bernoulli to reflect on the provenance and meaning of Ulpian's numbers. Bernoulli concludes "that reason can best enter into this case of the Falcidian law if legacies of this sort are estimated according to the value and prices of life incomes, as we have determined them in the preceding chapter," and he then gives a numerical example of this procedure.

Bernoulli (1709, Chap. 4) also gives some information on "the purchase of yearly incomes for life." First, he discusses whether such a purchase can be considered legal. The background for this question is that temporary life assurances had been declared illegal in several countries, because "men are not included within the terms of merchandise, that a free man cannot be bought and sold, and that a free man cannot be given a value." Bernoulli rejects these arguments and says that a fair contract may be based on the life table. For life annuities, however, most lawyers agree that such contracts are legal. This is also supported by Canon law, which permits a monastery to grant a man a yearly income in return for a gift.

In the Middle Ages it became common for states and towns to raise funds by selling annuities. The price depended on the prevailing rate of interest but was generally independent of the age of the annuitant, which naturally caused the buyers to chose healthy children as nominees. The price varied greatly from town to town and over time; often, it was about half the price of a perpetual annuity. Bernoulli gives many examples of prices of single life annuities varying from six to twelve years' purchase and also examples of prices of annuities for two or more lives. The following example from London in 1704 is quoted by Bernoulli:

> Yesterday the Queen issued an order for the sale of life annuities; in two hours £10,000 were subscribed and to date £100,000. These are the conditions for the annuities; if someone wants 10 pounds of yearly income with a possible extension to two succeeding persons, then he pays 90 pounds for one life, 110 pounds for two lives, 120 pounds for three lives following one another; and if one wishes to have 14 pounds yearly for 99 years, one pays 210 pounds for it.

After "having related all those opinions which are constructed without foundation,"Bernoulli concludes (like de Witt and Halley before him) that the price has to depend on the age and health of the annuitant. As an example, he quotes the prices demanded by the Magistrate of Amsterdam in 1672 as shown in the following table:

Price of an Annuity of 100 Florins Yearly at Amsterdam in 1672

Age of annuitant in years

1–19	20–29	30–39	40–44	45–49	50–54	55–59	60–64	65–69	70–74	75–

Price in florins

1000	950	900	850	800	750	675	600	500	400	300

In 1673 the prices were increased to 800 florins for all over 45 years of age. Bernoulli finds that these prices are in reasonably good agreement with those calculated by himself. Later, we shall discuss his method of calculation.

The price of a life annuity was usually quoted as "14 years' purchase," say, which means that the price was 14 times the yearly payment.

De Witt (1671) states that the current price of an annuity at Amsterdam in 1671 was 14 years' purchase at a time when the rate of interest was 4%. He further says that most of the annuities sold previously at a time with a higher rate of interest were obtained at nine years' purchase. He then presents a method for the correct evaluation of an annuity, which we shall discuss in the following section. He finds that the price of an annuity for a three-year-old child ought to be at least 16 years' purchase.

It is of course essential to know the relation between the value of an annuity and the rate of interest. De Witt (1671) and Halley (1694) calculated values only for one rate of interest. A simple solution was first given by de Moivre (1725) under the assumption that $l_x = 86 - x, 10 \leqslant x \leqslant 86$. Let the yearly rate of interest be i and let a_x denote the value of a life annuity of one per year for a life aged x. Then de Moivre proved that

$$ia_x = 1 - \frac{(1 + i)a_{\overline{86 - x}|}}{(86 - x)}, \qquad 10 \leqslant x < 86,$$

where $a_{\overline{n}|}$ denotes the value of an annuity certain of one payable in n years (see §25.5). Some values of ia_{10} have been given in the following table:

Ratio of Value of Annuity to Value of Perpetuity according to de Moivre's formula for a life aged 10 years

100i	4	6	8	10	12
ia_{10}	0.675	0.770	0.823	0.855	0.877
a_{10}	16.9	12.8	10.3	8.6	7.3

It will be seen that a price of nine years' purchase is advantageous to the purchaser when the rate of interest is a little below 10% and that at 4%, a price of 14 years' purchase is advantageous.

Another scheme for raising money to the state was invented by Lorenzo Tonti (1630–1695), a Neapolitan banker living most of his life in Paris. Tonti proposed to Cardinal Mazarin that funds might be raised by selling shares of a given amount to a number of persons of about the same age and that this group of persons afterward, as long as anyone was alive, should share the interest of the fund. Let the yearly rate of interest be i and the number of persons entering into the arrangement at age x be l_x; then the interest rate obtained after t years will be il_x/l_{x+t}. This is the simplest form of a "tontine." The state gets immediately access to the fund, and in return the state guarantees the payment of a constant yearly annuity to be divided equally among the surviving members of the group and terminating with the death of the last survivor. It is clear that the advertisement of such a project must contain a table of l_x as an essential element.

Cardinal Mazarin did not succeed in carrying out the idea in 1653 because of opposition of the Parliament. The first state tontine in France was established in 1689. Bernoulli gives a detailed description of its rules.

Tonti came much nearer to realizing his idea in Denmark in 1653. Poul Klingenberg (1615–1690) was born in Hamburg, where he got his commercial training. In 1652 he moved to Copenhagen; he became Postmaster General in 1654 and spent the rest of his life in Copenhagen as a member of the civil service. Klingenberg met Tonti in Amsterdam in 1652 and became so knowledgeable about Tonti's scheme that he was able to convince the Danish King Frederik III to try it out. Klingenberg and the King wanted to make subscription to the scheme compulsory for certain persons, but this was rejected by the Parliament. The rules of the tontine were described in a booklet of 49 pages, printed in 750 copies, 450 in Danish and 300 in German. In May 1653 the scheme was advertised all over the Monarchy, but the project was abandoned because only about 5% of the stipulated capital was subscribed.

The project is of interest because it contains the first published life table. Whether it is due to Tonti or Klingenberg is unknown, but as already indicated, Tonti must have had a life table as the basis for his project, so it is presumably his.

The main rules of the State tontine of 1653 are as follows. The total capital amounted to 1,600,000 rix-dollars with a yearly annuity of 80,000 rix-dollars, guaranteed by the King and the Parliament by reserving certain taxes for this purpose. Each share of the capital was set at 100 rix-dollars, and each subscriber could buy several shares. The subscribers were divided into eight classes according to age, the first being children below eight years of age, the next from eight to sixteen, and so on until the last from 56 years and above, the capital for each class being 200,000 rix-dollars. The advertising material contained a life table for each class giving the expected number of survivors for every year. The table for the first class begins with 2000 persons between zero and eight years of age. For brevity, let us set $l_4 = 2000$. The following table gives some values of l_{4+t} [see Iversen (1910) for more details]:

Table of l_{4+t} from Klingenberg's Life Table of 1653

t	0	5	10	15	20	25	30	40	50	60	70
l_{4+t}	2000	1700	1400	1150	900	700	500	250	115	35	2

In 1670 a similar tontine was established by the town of Kampen in Holland, and the administrator Jacob van Dael published the following life table [an English translation of the pamphlet has been given by Hendriks (1853, §29)]:

Van Dael's Life Table from 1670

x	1	12	24	36	48	60	61	⋯	66	68	70	72	74	76	80
l_x	400	200	100	50	25	12	11	⋯	6	5	4	3	2	1	0

Tontines became a popular means of raising funds for many towns and states, and many variants of the simple tontine described above were created. When private firms also established tontines, some of them with disastrous results for the subscribers, the idea was discredited. Du Pasquier (1910) has given a historical account of the tontines.

From an actuarial point of view, the tontines are of interest because of their life tables. The first tables used were presumably guesswork, but they were at least as good (realistic) as Graunt's table. Supposedly, mortality has been overvalued to make the tontines look attractive. To compare the three life tables we have found $l_1 = 93$ and $l_4 = 74$ by interpolation in Graunt's table, and starting from these values we have calculated l_x by means of Klingenberg's and van Dael's tables. By interpolation we have also found the death rates $q_x = (l_x - l_{x+1})/l_x$, as shown in the following table:

Comparison of Three Early Life Tables

	KLINGENBERG (1653)		GRAUNT (1662)		VAN DAEL (1670)	
x	l_x	$1000q_x$	l_x	$1000q_x$	l_x	$1000q_x$
0	—	—	100	—	—	—
1	—	—	93	—	93	—
4	74	—	74	—	77	—
6	70	32	64	46	68	61
16	48	39	40	46	37	56
26	30	49	25	44	21	56
36	16	69	16	46	12	56
46	8	69	10	50	7	56
56	4	95	6	67	4	59
66	1	222	3	104	1	87
76	0.0	—	1	—	0.2	—
80			0			

There exist many books on the history of insurance. For the early history we refer to Trenerry (1926) and for a general history until 1914 to Braun (1925).

A history of mortality from prehistoric times through the Middle Ages has been given by Acsádi and Nemeskéri (1970).

9.2 JAN DE WITT AND HIS REPORT ON THE VALUE OF LIFE ANNUITIES, 1671

When the Northern Provinces of the Netherlands won their independence from Spain in the late 16th century, a period of exceptional growth in power and wealth began. During the Golden Age of the Dutch Republic in the 17th century many outstanding contributions to the arts and sciences were made. In our field of interest we find three men, Jan de Witt, Jan Hudde, and Christiaan Huygens, of the same age who received the same education in law and mathematics. As young men they were inspired by the works of Descartes (who lived in Holland at the time) through their mathematics teacher Frans van Schooten, who induced them to write original papers on mathematics, which he published under their own names as appendices to his *Exercitationum Mathematicarum* (1657) and his edition of Descartes' *Géométrie* (1659–1661). They also made fundamental contributions to probability theory and insurance mathematics. De Witt became the leading statesman of the Republic; Hudde became burgomaster of Amsterdam; and

Huygens became the leading natural philosopher in Europe during the time period between Descartes and Newton.

Jan (Johan) de Witt (1625–1672) was born into an old burgher-regent family. After studies at the University of Leiden and a traveling tour of Europe he settled in the Hague as an advocate. His most important mathematical work, from about 1650, gives a systematic treatment of the conics by means of analytical geometry. In 1650 he began his remarkable political career as pensionary of Dordrecht (secretary of the town council) and leader of the town deputation in the States of Holland. In 1653 he became grand pensionary of Holland (prime minister) at the age of 28. As leader of the Republican Party, he was in opposition to the Prince of Orange.

It was a period of fierce commercial and maritime competition between the two rapidly growing powers, England and Holland, leading to many disputes and eventually to war. De Witt conducted his policy with great diplomatic skill and consolidated the finances of the state in the relatively peaceful period between 1654 and 1665. The second Anglo-Dutch war 1665–1667 tapped the Republic for much of its sea power and economic strength with heavy taxation as a result. An impending war against France made it necessary to build up the army, and de Witt proposed to raise funds by selling annuities. In a report to the States General in 1671 he showed how to calculate the value of an annuity. When in 1672 France invaded the Republic, de Witt resigned and was replaced by the Stadholder William III. De Witt was murdered by a mob in 1672.

Jan (Johannes) Hudde (1628–1704) spent most of his life as a politician and civil servant. In 1672 William III chose him as one of the burgomasters of Amsterdam, a post he had for 21 years, intermittently being chancellor of the admiralty. His most important mathematical papers are on the solution of algebraic equations and on the maxima, minima, and tangents of algebraic curves, in which he continues the investigations of Descartes and Fermat, respectively. In 1665 he had a correspondence with Huygens on games of chance (discussed in §6.2). In 1671 he checked de Witt's paper, which led to correspondence with de Witt. About the same time he had correspondence with Huygens on the mortality of annuitants.

De Witt wrote his *Waerdye van Lyf-Renten Naer proportie van Los-Renten* (Value of Life Annuities in Proportion to Redeemable Annuities) in 1671 to the States General, and it was published in the *Resolution van de Heeren Staten van Holland en West-Friesland* in the same year. Some copies were also printed for private circulation. At the time the rate of interest was 4%, and life annuities were sold at 14 years' purchase. De Witt proved that the price ought to be at least 16 years' purchase for a child aged 3, and from his report it is easy to calculate the prices for other ages of entrance.

De Witt begins his report by defining the expectation of a random variable and explaining its calculation. He closely follows the definitions and proofs of Huygens (1657). He next turns to the assumptions on mortality. He divides the age interval from 3 to 80 years into four periods, namely, (3, 53), (53, 63), (63, 73), and (73, 80), and makes two assumptions: (1) within each period "it is not more likely that this man should die in the first half-year of a given year, than in the second half"; and (2) that the "chance of dying" (*die apparentie of dat hazardt van sterven*) in a given year of the second period is not more than 3/2 times the chance of dying in a given year of the first period, and that the corresponding factors for the third and fourth periods are 2 and 3. As an example he takes a man aged 40 and another aged 58 to whom the factor 3/2 applies. He says that it is a fair contract, if the one aged 40 were to inherit 2000 florins in case the one aged 58 should die within six months in return for an inheritance of 3000 florins to the one aged 58 in case the 40 year old dies within six months. Hence, if the one aged 58 dies, the amount to be paid is only 2/3 of the amount to be paid if the one aged 40 dies.

From the first assumption de Witt deduces that within each of the four periods each half-year of life is equally mortal. Combining this with the second assumption, he obtains the distribution of the number of chances of dying in each half-year. His reasoning is somewhat obscure and has given rise to much discussion. Here we shall only present his result, which is stated very clearly.

It is natural for de Witt, like Huygens, to consider *the distribution of the number of deaths* as the fundamental probability distribution. Since annuities were paid in half-yearly instalments, he uses the half-year as time unit. He specifies the number of chances (deaths) in each half-year of the four age intervals as 1, 2/3, 1/2, and 1/3, respectively, the total number of chances thus being

$$1 \times 100 + \tfrac{2}{3} \times 20 + \tfrac{1}{2} \times 20 + \tfrac{1}{3} \times 14 = 128.$$

An annuitant dying in the $(t + 1)$st half-year will get the half-yearly payment, 1, say, at the end of each of the t half-years. The present value of these payments is

$$a_{\overline{t}|} = \sum_{k=1}^{t} (1 + i)^{-k}, \qquad \text{where} \quad 1 + i = 1.04^{1/2}.$$

De Witt therefore gives a table of $a_{\overline{t}|}$ for $t = 1, 2, \ldots, 200$; actually, he tabulates

$10^8 a_{\overline{\eta}|}$ to avoid decimals. Finally, he calculates the expectation of $a_{\overline{\eta}|}$ as

$$E\{a_{\overline{\eta}|}\} = \frac{1}{128}\left\{\sum_1^{99} a_{\overline{\eta}|} + \frac{2}{3}\sum_{100}^{119} a_{\overline{\eta}|} + \frac{1}{2}\sum_{120}^{139} a_{\overline{\eta}|} + \frac{1}{3}\sum_{140}^{153} a_{\overline{\eta}|}\right\} = 16.001607.$$

De Witt points out that the value of the annuity really is somewhat higher than 16 florins for several reasons, which he has not taken into account in his assumptions on the mortality. First, there is the effect of selection; the buyer of an annuity will appoint a nominee "in full health, and with a manifest likelihood of prolonged existence," which will lead to a low mortality in the first few half-years. Second, the corresponding increase in value of the annuity is larger than may be thought, because:

one half-year of life, at the commencement of and shortly after the purchase of the life annuity, is of greater value to the annuitant, with respect to the price of such a purchase, than eighteen half-years during which the person upon whom the annuity is purchased might live after the said purchase, from the age, for example, of 70 to 79 years.

Third, there is a certain probability of living to age 80 and above, and this has been neglected. De Witt says that if the span of life is extended to 100 years, the value of the annuity will increase only by 0.7 florins, and if we further assure the heirs of the annuitant of a perpetual annuity, the increase in value will be only 0.5 florins. Fourth, the effect of a higher mortality than assumed may also be evaluated. If the factors 2 and 3 for the last two intervals of age are changed to 3 and 5, respectively, the value of the annuity will decrease only by 0.3 florins. Taking all these considerations into account de Witt concludes that the value of an annuity for a nominee aged 3 will certainly be above 16 years' purchase. He further adds that annuities are exempt from taxes.

This is certainly an ingenious analysis of the effects of selection and changes of mortality on the value of an annuity.

At the request of de Witt, Hudde certified that the method used was mathematically correct and that the numerical result of a price of at least 16 years' purchase also was correct, provided that the calculations had been carried out without error. The basic table of annuities certain was certified by the two bookkeepers of the States General.

In a supplement to his report de Witt says that it may be difficult for the untrained to follow his theoretical reasoning, and he has therefore provided an empirical proof based on the registers of annuitants for Holland and West Friesland. He has extracted for some thousand of cases the age of the nominee at the purchase of the annuity and the duration until death. For each annuitant he has calculated the present value of the payments received and then found the average for "a fair number of young lives. This being calculated upon considerably more than a hundred different classes, each class consisting of about one hundred persons," it has invariably been found that the average value exceeds 16 florins and for many classes higher values have been found, even up to 18 florins. Unfortunately, these calculations have not been preserved.

Hudde began his work on the mortality statistics of annuitants in the spring of 1671. In a letter to Huygens (Huygens' *Oeuvres*, Vol. 7, p. 59), he states that he has found some preliminary results quite different from those of Graunt. This seems to indicate that Hudde knew of the correspondence of the brothers Huygens in 1669. The day after the presentation of de Witt's report to the States General, Hudde sent his final table of mortality statistics to de Witt, and three weeks later he sent a copy to Huygens (Huygens' *Oeuvres*, Vol. 7, pp. 95–98). A correspondence with de Witt ensued. Unfortunately, Hudde's letters to de Witt are not preserved; we only have de Witt's letters to Hudde [see Hendriks (1853, §28)].

De Witt and Hudde discuss three topics: (1) the value of an annuity evaluated from the data on annuitants; (2) the mortality of annuitants; and (3) the calculation of annuities on two or more lives.

Hudde's table is similar to that described by de Witt in his Supplement. It contains the age at purchase of 1495 annuitants from the register at Amsterdam and the duration in years of the annuity, all annuities having been bought in the period 1586–1590. For the 796 persons who were between 1 and 10 years of age at purchase, Hudde finds the average value of the corresponding annuities certain to 17.6 florins, whereas de Witt for the same age class finds 17.9 florins based on the registers at The Hague. Hudde adds that the average value for all the annuitants is 16.6 florins, which covers persons between 1 and 50 years of age at purchase.

The data tabulated by Hudde are ideal for the calculation of death rates according to age. It does not seem that Hudde has made such calculations, but a remark from de Witt indicates that Hudde proposed to replace de Witt's hypothetical distribution of deaths by the simpler hypothesis of a uniform distribution over the age interval from 6 to 86 years. De Witt, however, analyses the data (whether his own or Hudde's), probably by counting the number of survivors aged 50, 55, 60, ..., and finds "that they [annuitants at 50 and over] die almost exactly, at least without any sensible difference," as shown in the following table. For comparison we have added the death rates calculated by Iversen (1910) from Hudde's data and also the rates following from de Witt's original hypothesis and from Hudde's hypothesis.

Probability of Dying within Five Years

Age	50	55	60	65	70	75	80	85
From Hudde's								
data	0.160	0.205	0.259	0.322	0.468	0.591	0.704	0.813
De Witt	1/6	1/5	1/4	1/3	1/2	3/5	2/3	7/9
De Witt's								
hypothesis	0.255	0.263	0.321	0.394	0.565	1.000		
Hudde's								
hypothesis	0.139	0.161	0.192	0.238	0.313	0.455	0.833	1.000

It will be seen that the death rates found by de Witt agree remarkably well with the rates found from Hudde's data and that they are somewhat lower than the rates following from the hypothesis in de Witt's report. De Witt notes that "contrary to my expectation" there is a discrepancy between the data and his original hypothesis and indicates that he will investigate the matter further, also for ages below 50. He does not, however, return to this problem in the letters to Hudde.

De Witt also presents a general method for finding the value of an annuity based on the last survivor of several lives. He considers as an example a group of eight young lives of the same age and with the remaining lifetimes 7, 15, 24, 33, 41, 50, 59, and 68 years. From a table of the eight annuities certain payable yearly he finds the average 17.2 florins. He has obviously chosen a nearly uniform distribution of the eight deaths such that the average value of the annuities certain equals the average previously found from his registers. The value of an annuity on the last survivor of the eight lives equals an annuity certain of duration 68 years, which gives 23.3 florins, and the values of annuities on a number of lives between one and eight will lie between 17.2 and 23.3 florins.

For m lives, $2 \leqslant m \leqslant 8$, there are $\binom{8}{m}$ combinations of the eight ages and to each combination a corresponding annuity certain of longest duration. The number of combinations for each of the possible durations, beginning with the shortest, is

$$\binom{m-1}{0} \quad \binom{m}{1} \quad \binom{m+1}{2} \quad \cdots \quad \binom{7}{8-m}.$$

De Witt does not give this formula, but his table leaves no doubt about these binomial coefficients. The average of the annuities certain with the binomial coefficients as weights gives the value sought. For two lives de Witt shows that the value becomes 20.8 florins, and he further explains how the averages may be found by successive summations of the annuities certain, instead of by multiplication, because of the special structure of the weights.

That de Witt was ahead of his time may be demonstrated by a comparison of his method with those used by Nicholas Bernoulli (1709) 38 years later. Bernoulli begins by setting the value of an annuity equal to the value of an annuity certain of duration equal to the expectation of life. He therefore gives a table of annuities certain for durations of 1 to 100 years at a rate of interest of 5%, and by means of this table and his previously calculated expectations of life, according to Graunt's table, he tabulates the values of annuities for the usual ages. Then he writes, "Indeed while I write these things, I notice that the value of these incomes is not correctly calculated by supposing that the return will last as many years as someone is supposed probably to live." He afterwards explains that the correct method is to calculate the expected value of the annuity certain using the probability distribution of the deaths. Since Graunt's table gives the deaths for ten-year intervals only, he approximates the sum for each interval by the total number of deaths multiplied by the average value of the annuities certain. Finally, he tabulates the value of the annuities. For example, for $x = 16$ his first method gives a value of 12.6, whereas his second (correct) method gives 10.6. It is odd that Bernoulli does not interpolate to yearly values of the number of survivors in the life table to get a better approximation. As remarked in the previous section, he finds these values in good agreement with the prices demanded at Amsterdam in 1672.

Finally, we shall comment on de Witt's work using modern notation and terminology.

Let T denote the remaining lifetime for the life (x) with probability distribution d_{x+t}/l_x, $t = 0, 1, \ldots$, where $d_x = 0$ for $x \geqslant \omega$ and $l_x = d_x + d_{x+1} + \cdots + d_{\omega-1}$. De Witt's formula for the value of an annuity may then be written as

$$a_x = E\{a_{\overline{T}|}\} = \frac{1}{l_x} \sum_{t=1}^{\omega-x-1} a_{\overline{t}|} d_{x+t}.$$

The corresponding formula for the expectation of life,

$$e_x = E\{T\} = \frac{1}{l_x} \sum_{t=1}^{\omega-x-1} t d_{x+t}$$

had already been used (with a small modification) by the brothers Huygens, as explained in §8.1. Whether de Witt knew about this, perhaps through Hudde, is unknown. It is more likely that de Witt's theoretical approach was inspired by his empirical investigations in which he calculated the arithmetic mean in the distribution of $a_{\overline{t}|}$ for the deceased annuitants. In a letter to

Hudde on the calculation of joint-life annuities, de Witt writes that he could "establish thereon an argument *a priori*, although I have found it *a posteriori*, like in almost all inventions."
    Inserting

$$a_{\overline{t}|} = \sum_{k=1}^{t} (1 + i)^{-k}$$

into de Witt's formula and changing the order of summation, we get

$$a_x = \frac{1}{l_x} \sum_{t=1}^{\omega - x - 1} (1 + i)^{-t} l_{x+t},$$

which is Halley's formula (1694) for the value of an annuity. De Witt does not give this formula, but his remarks on the contributions of the first few years, compared to many years later in life, indicate that he was familiar with this form of the formula.
    De Witt and Hudde did not construct life tables, possibly because their calculations only required the distribution of the deaths. Eneström (1896, 1898) has pointed out that the life table corresponding to the values of d_x used by de Witt in his report is a piecewise linear function:

$$l_x = \begin{cases} 128 - 2(x - 3), & 3 \leqslant x \leqslant 53, \\ 28 - \frac{4}{3}(x - 53), & 53 \leqslant x \leqslant 63, \\ 14\frac{2}{3} - (x - 63), & 63 \leqslant x \leqslant 73, \\ 4\frac{2}{3} - \frac{2}{3}(x - 73), & 73 \leqslant x \leqslant 80. \end{cases}$$

Eneström has further pointed out that the death rate $q_x = d_x/l_x$ increases in each of the four intervals but that q_x jumps from a higher to a lower value at the end of each interval; for example, $q_{52} = 0.067$ and $q_{53} = 0.048$, the new level being approximately 2/3 times the foregoing in this case. This is of course unreasonable, and one may wonder why de Witt had not observed or commented on this fact.
    This leads to the crucial question about de Witt's determination of the distribution of the deaths. In his second assumption on the mortality it is assumed that the death rate at age 58 equals 3/2 times the death rate at 40; that is, he assumes that the death rates are increasing. However, in the following he never argues in terms of death rates; he only uses the chances of dying for a three-year-old nominee, and he assumes that the number

of deaths at age 58 equals 2/3 times the number of deaths at 40. He does not explain how to obtain this result from his assumption.

Looking at the formula for a_x it will be seen that an essential simplification of the calculations is obtained by keeping d_x constant over suitably chosen intervals. It also has the further advantage that the calculation of a_x as a function of x becomes very simple. Such considerations may have induced de Witt to chose a constant value of d_x within each of the four intervals.

Today we are inclined to read de Witt's report as a mathematical paper. It is, however, a prime minister's attempt to convince the States General that the price of annuities should be raised from 14 to 16 years' purchase. As other prime ministers in critical situations, de Witt was short of time and money. He had to work out his report in a very short time, and he had presumably no hope of getting the price raised to more than 16 years' purchase. This may explain the inconsistencies in his paper. We shall support this point of view by means of a small calculation which might easily have been carried out by de Witt himself.

The average values of the annuities certain for the four intervals follow directly from de Witt's table. Hence, it is easy to find the value of the annuity for different distributions of deaths, as shown in the table below for (1) the distribution corresponding to de Witt's assumption with the factors 1, 3/2, 2, and 3; (2) a uniform distribution; and (3) the distribution actually used by de Witt with the factors 1, 2/3, 1/2, and 1/3.

Value of Annuity for Three Different Distributions of Deaths

INTERVAL	AVERAGE VALUE OF ANNUITY CERTAIN	NUMBER OF DEATHS		
0–99	14.076	100×1	100	100×1
100–119	22.280	$20 \times \frac{3}{2}$	20	$20 \times \frac{2}{3}$
120–139	23.243	20×2	20	$20 \times \frac{1}{2}$
140–153	23.816	14×3	14	$14 \times \frac{1}{3}$
Total		212	154	128
Value of annuity		18.90	17.22	16.00

It will be seen that de Witt's original assumption would have led to a value of 18.90 florins. Presumably, de Witt realized that he had no chance of getting an increase from 14 to 19 accepted. Furthermore, a price of 19 was higher than the one following from his empirical investigations. He could then have turned to a uniform distribution, which would have given a price of 17.22, in good agreement with his empirical results. Instead, he chose factors, reciprocal of the original ones, leading to the (desired) result 16.00.

De Witt was both a great statesman and a great mathematician.

Hudde's data has been analyzed by Westergaard (1901, pp. 270–273) and Iversen (1910) who have calculated the corresponding life table. Westergaard has also studied the mortality over time and shown the effects of the plague in Hudde's data. Plotting the life table calculated by Iversen it will be seen that l_x is nearly linear, namely, $l_x = 111 - 1.37x$ for $15 \leqslant x \leqslant 75$. For $x \leqslant 15$, selection makes the death rates smaller. Of course, Hudde's data comprise mainly healthy and wealthy persons, so that the death rates are not representative for the population in general.

De Witt's considerations on the five-yearly death rates are essential and might have led him to revise his mortality assumptions.

De Witt's method of calculating the value of an annuity on the last survivor of two lives is rather primitive compared to Huygens' method of finding the corresponding expectation of life. This seems to indicate that de Witt did not know the details of the correspondence of the brothers Huygens.

As mentioned before, the city of Amsterdam took the remarkable step in 1672 to offer annuities at a price dependent on the age of the annuitant. The price was, however, very low (at most 10 years' purchase) compared to the results found by de Witt and Hudde. The explanation for this lower price may be the need for money, an increase in the rate of interest and perhaps also the fear of inflation following the war.

It has often been asserted that de Witt's report is very scarce. However, according to van Brakel (1976) one edition existed for the general public and at least three printings for the States General and the administration. De Witt's method is mentioned by Struyck (1740) in his discussion of life annuities, and his report is mentioned in Gouraud's *History* (1848).

De Witt's method was superseded by Halley's (1694), which became very popular, and from the middle of the 18th century, de Witt's paper was forgotten until Hendriks found it and provided an English translation with commentaries in 1852–1853.

Besides the references given we refer to Algemeene Maatschappij (1898, 1900); Braun (1925); van Haaften (1925); Chateleux and Rooijen (1937); Kohli and van der Waerden (1975); and Seal (1980).

9.3 HALLEY AND HIS LIFE TABLE WITH ITS SEVEN USES, 1694

Edmond Halley (1656–1742) was the son of a wealthy citizen of London. He received a good classical education and, at the same time, he studied astronomy, navigation, and mathematics and carried out astronomical observations with instruments provided by his father. At seventeen years old he went to Oxford to continue his studies and his observations. Without taking his degree in Oxford he followed his interest in astronomy and went to the island of St. Helena to make observations of stars which were too near the south pole to be visible in Europe. After his return in 1678 he published a catalog of the positions of 341 southern stars, which was of importance to the navigation of the many new trade routes to the southern hemisphere. On the basis of this work he was elected a fellow of the Royal Society in 1678 at the age of 22. He served the Society well as assistant secretary and as editor of the *Philosophical Transactions* for several years, and he helped Newton with the *Principia*. In 1704 he became professor of geometry at Oxford, a post he held until 1720; when 63 years old, he took over the post of Astronomer Royal at Greenwich, in which position he remained until his death.

Halley was a scientist of unusual versatility and energy. His main works are on astronomy and geophysics, but he also made contributions to physics, mathematics, demography, and insurance mathematics. Furthermore, he edited the mathematical works of Apollonius and several other Greek mathematicians in Greek with a Latin translation, and he combined his astronomical and classical knowledge to help historians to date important events in ancient times.

In geophysics he made important contributions to the theory of terrestrial magnetism, and on a two years' voyage as captain of a small ship he traversed the Atlantic and made a chart of the variation of the compass "in all those seas where the English Navigators were acquainted."

Halley's many observations of positions of the stars, the planets, the moon, and the comets led him to important practical and theoretical results. He discovered the proper motions of the stars, which hitherto had been considered fixed. He proposed to use transits of Venus across the sun for a better determination of the sun's distance from the earth. He improved the tables of the moon's orbit to make determinations of the longitude at sea more precise.

Using Newton's theory he worked out the orbits of 24 comets. Noting that the orbits of the comets of 1531, 1607, and 1682 were similar, he conjectured that these comets were one and the same object moving periodically in an elliptical orbit, and he predicted the next appearance of this comet, today known as Halley's comet.

To get a better understanding of Halley's paper on demography and insurance mathematics, we shall introduce some fundamental notions from population theory. The life table gives the number of survivors l_x at age x from a number of births l_0. Consider now a population where (1) the number of births in each calendar year is l_0; (2) the mortality as defined by the life table is constant over time; and (3) there is no migration. This is the *stationary* population corresponding to the given life table. The problem is to find the age distribution of the living in the stationary population and the size of the population.

The lifetime of a person may be depicted in a coordinate system with age as abscissa and calendar time as ordinate, a Lexis (1875) diagram, as shown in Fig. 9.3.1. For each person a lifeline is drawn beginning at the date of birth on the ordinate axis and continued diagonally until the date of death, the corresponding abscissas being zero and the age at death. In the diagram a lifeline has been shown ending with death at the age 1.5 years. The number of lifelines beginning in each calendar year equals l_0 of which l_x will cross the vertical line x years later. The number of deaths d_x between x and $x + 1$ equals $l_x - l_{x+1}$ and is represented by the number of lifelines stopping in the corresponding parallelogram. This number may be divided into the number of deaths before (b) and after (a) the first of January, $d_x = d_x^b + d_x^a$, say. The number of persons between x and $x + 1$ years of age, L_x, equals the number of lines crossing the corresponding horizontal line so that

$$L_x = l_x - d_x^b = l_{x+1} + d_x^a = \tfrac{1}{2}(l_x + l_{x+1}) + \tfrac{1}{2}(d_x^a - d_x^b).$$

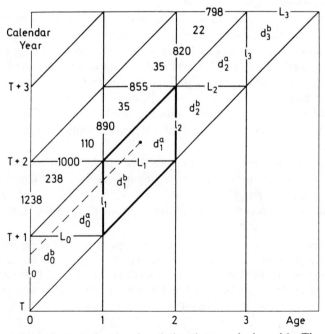

Fig. 9.3.1. A Lexis diagram showing the relations between l_x, d_x, and L_x. The numerical values of L_x are from Halley's life table.

If the number of deaths are nearly uniformly distributed over the parallelogram, we obtain the approximation $L_x \simeq \frac{1}{2}(l_x + l_{x+1}) \simeq l_{x+1/2}$, which is usually sufficiently accurate, apart from very small and very large values of x. The population size is obtained by adding all the L's.

The background for Halley's work is as follows. At Breslau in Silesia registers of births and deaths according to sex and age had been kept since the end of the 16th century. A prominent evangelical pastor and scientist, Caspar Neumann (1648–1715), used the lists from 1687 and the following years in his attempts to fight popular superstitions about the influence on health of the phases of the moon and the climacteric ages, i.e., ages divisible by seven and nine. Neumann sent his results to Leibniz, who in 1689 informed Justell, secretary of the Royal Society at London, of Neumann's researches. After the works of Graunt and Petty, members of the Royal Society had been waiting for observations suitable for the construction of a life table, and Justell therefore wrote to Neumann who responded by sending his observations for each of the years 1687–1691. The Society asked Halley to analyze the data, and in 1693 Halley presented his paper *An Estimate of the Degrees of the Mortality of Mankind, drawn from curious Tables of the Births and Funerals at the City of Breslaw; with an Attempt to ascertain the Price of Annuities upon Lives.*

Halley begins by referring to Graunt and Petty and points out that they themselves had been aware of the defects of their data, namely, that the population size was unknown, that the ages of the deceased were unknown, and that the populations of London and Dublin were increasing because of migration. These defects seem to a great extent to be alleviated in the bills of mortality at Breslau "wherein both the Ages and Sexes of all that die are monthly delivered, and compared with the number of Births," which exceeds the number of deaths only to a small extent. The only thing missing is the population size.

Halley does not present Neumann's data in detail; he merely gives the average yearly numbers without distinguishing between the sexes. The average numbers of births and deaths are 1238 and 1174, respectively, so that there is an increase in the population of about 5% per year. Halley writes,

I will suppose the people of Breslaw to be encreased by 1238 births annually. Of these it appears by the same Tables, that 348 do die yearly in the first Year of their Age, and that but 890 do arrive at a full Year's Age; and likewise, that 198 do die in the five years between 1 and 6 compleat, taken at a Medium; so that but 692 of the persons born do survive Six whole Years.

Following Graunt's idea, Halley begins by constructing a life table with $l_0 = 1238, l_1 = 890$, and $l_6 = 692$. For ages above six, he gives the information that of the whole People of Breslaw there die yearly, as in the following Table, wherein the upper Line shews the Ages, and the next under it the Number of Persons of that Age dying yearly. And where no Figure is placed over, it is to be understood of those that die between the Ages of the preceding and consequent Column [see Table 9.3.1].

Note that age seven, say, in Table 9.3.1 indicates the seventh year of life, i.e., the age interval from six to seven. Halley's table contains the first published empirical distribution of deaths according to age.

Table 9.3.1. The Average Yearly Number of Deaths at Breslau for the Period 1687–1691

Age	7	8	9	·	14	·	18	·	21	·	27	28	·	35
Deaths	11	11	6	$5\frac{1}{2}$	2	$3\frac{1}{2}$	5	6	$4\frac{1}{2}$	$6\frac{1}{2}$	9	8	7	7

Age	36	·	42	·	45	·	49	·	54	55	56	·	63	·	70
Deaths	8	$9\frac{1}{2}$	8	9	7	7	10	$10\frac{1}{2}^a$	11	9	9	10	12	$9\frac{1}{2}$	14

Age	71	72	·	77	·	81	·	84	·	90	91	·	98	99	100
Deaths	9	11	$9\frac{1}{2}$	6	7	3	4	2	1	1	1	1^a	0	$\frac{1}{5}$	$\frac{3}{5}$

^{a}These two numbers are missing in Halley's table, possibly due to printer's errors, (see Knapp, 1874, p. 127).
Source: E. Halley, *Phil. Trans.*, 1694, Vol. 17, p. 599.

Halley's analysis of these numbers is rather brief. He considers four age intervals: 9–25, 25–50, 50–70, and above 70. He observes that within the first interval there is a minimum number of deaths about age 14. He does not believe in the reality of this minimum for two reasons. First, it may be "attributed to Chance, as are the other irregularities in the Series of Ages, which would rectifie themselves, were the number of Years much more considerable, as 20 instead of 5." Second, experience from Christ Church Hospital in London indicates that the death rate in this age group is about 1%, and since the number of people of age 14 in Breslau is about 600, this would lead to an expected number of deaths of about six. He therefore uses six as the minimum number of deaths in the first interval. For the two following intervals he uses a gradual increase of the number of deaths from 7 to 11, and for the last interval a gradual decrease from 11 to 0.

Without further explanation he writes; "From these Considerations I have formed the adjoyned Table" (see Table 9.3.2), and "This Table does shew the number of Persons that are living in the Age current annexed thereto".

Halley's table gives L_x for $x = 0, 1, \ldots, 83$, and the sums of seven consecutive L's, i.e., the number of persons in seven-year age intervals and also the number

Table 9.3.2. Halley's Life Table[a]

Age Curt.	Per-sons	Age Curt.	Per-sons	Age Curt.	Per-sons	Age Curt.	Per-sons	Age Curt.	Per-sons	Age Curt.	Per-sons	Age Curt.	Per-sons
1	1000	8	680	15	628	22	586	29	539	36	481	7	5547
2	855	9	670	16	622	23	579	30	531	37	472	14	4584
3	798	10	661	17	616	24	573	31	523	38	463	21	4270
4	760	11	653	18	610	25	567	32	515	39	454	28	3964
5	732	12	646	19	604	26	560	33	507	40	445	35	3604
6	710	13	640	20	598	27	553	34	499	41	436	42	3178
7	692	14	634	21	592	28	546	35	490	42	427	49	2709
												56	2194

Age Curt.	Per-sons	Age Curt.	Per-sons	Age Curt.	Per-sons	Age Curt.	Per-sons	Age Curt.	Per-sons	Age Curt.	Per-sons	Age Curt.	Per-sons
43	417	50	346	57	272	64	202	71	131	78	58	63	1694
44	407	51	335	58	262	65	192	72	120	79	49	70	1204
45	397	52	324	59	252	66	182	73	109	80	41	77	692
46	387	53	313	60	242	67	172	74	98	81	34	84	253
47	377	54	302	61	232	68	162	75	88	82	28	100	107
48	367	55	292	62	222	69	152	76	78	83	23	Total	34000
49	357	56	282	63	212	70	142	77	68	84	20		

[a] E. Halley, *Phil. Trans.*, 1694, Vol. 17, p. 600. Age current equal to x means the age interval $x - 1$ to x.

of persons between 84 and 100, the total sum, 34,000, being the population size.

A thorough discussion of the demographic part of Halley's paper, with an attempt to reconstruct his method of finding the L's, has been given by Böckh (1893), which we shall follow. Presumably, Halley used the monthly observations to find L_0, which he then for convenience rounds to 1000. From $l_0 = 1238$, $L_0 = 1000$, and $l_1 = 890$, it follows that $d_0^b = 238$ and $d_0^a = 110$, which seems to be a reasonable distribution of the number of deaths in the first year (see Fig. 9.3.1). Halley must also have calculated $l_2, \ldots, l_6$ from the data even if he only informs us that $l_6 = 692$; perhaps he only calculated l_6 and found the remaining values by interpolation. Assuming an even distribution of the deaths within each age interval he may have found L_x to be $\frac{1}{2}(l_x + l_{x+1})$ for $x = 1, \ldots, 5$, as indicated in Fig. 9.3.1.

The crucial point, however, is the relation between the numbers in Tables 9.3.1 and 9.3.2. If Halley had assumed that the population of Breslau was stationary, he could have calculated the life table by summation of the successive numbers of deaths. He would then have obtained $l_0 = 1174$. By choosing $l_0 = 1238$, he had to increase the number of deaths in Table 9.3.1 by $1238-1174 = 64$. He seems to have adjusted the numbers in the table in three ways: (1) he increased the yearly number of deaths in the interval from 9 to 25 to a minimum of 6 for the reasons already mentioned; (2) he carried out a smoothing operation to reduce chance variations; and (3) he combined the smoothing with a distribution of the 64 deaths, using the restriction that the yearly number of deaths should be between 6 and 11 for the age interval 9 to 70 and gradually decreasing thereafter.

Böckh has carried out this adjustment and given a table of the resulting values of l_x, the first eight being 1238, 890, 820, 776, 744, 720, 700, and 685. We note that Halley's adjustments also have affected l_6, originally given as 692, such that the adjusted value becomes 700 with the result that $L_6 = 692$, which of course has caused a great deal of confusion.

Halley's adjustments mean that the differences in his table of L_x are nearly constant for considerable sections of the table; they increase from 6 to 11 for $6 \leqslant x \leqslant 53$, nearly all of them equal 10 for $54 \leqslant x \leqslant 77$, and then they decrease to zero. Of course, theoretically, d_x varies continuously, but because of rounding to integers, Halley reaches the artificial result that d_x is piecewise constant.

In Fig. 9.3.2 we have shown the distribution of 1000 deaths according to Halley's table and for comparison the distributions corresponding to the tables of Graunt and de Witt. Although Halley's d's are constant within rather short intervals, we have shown the distribution as continuous.

At the end of his paper Halley briefly discusses whether the life table and the results derived from it may be considered "universal." He notes that the

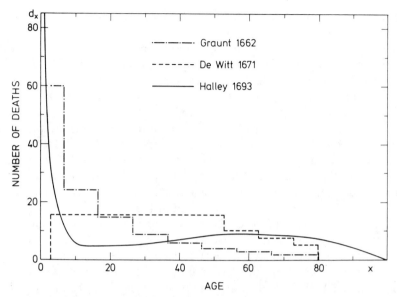

Fig. 9.3.2. Distributions of the number of deaths according to the life tables by Graunt, de Witt, and Halley: $\sum d_x = 1000$; Halley's $d_0 = 281$.

crude death rate in Breslau is 1174/34,000, or 35 per 1000, which is about the same as Petty (Graunt) found for London, and that infant mortality also is about the same as found by Graunt. He concludes that "there cannot perhaps be one better place [than Breslau] proposed for a Standard." He does not comment on the great differences between Graunt's life table and his own, neither does he discuss the changes in mortality and population composition due to the plague and other epidemic diseases. These problems have been discussed by Westergaard (1901, pp. 34–39).

We shall now briefly recount the seven ways Halley used his table.

(1) Like Graunt, he finds the "Proportion of Men able to bear Arms." Summing L_x from 18 to 56 and dividing by 2, he gets 9027 men, so that the required proportion is 9/34, "which may perhaps pass for a Rule for all other places."

He also finds that there are about 7000 women above 16 and under 45 years of age and wonders why the yearly number of births is only 1238. He recommends "an effectual Care to provide for the Subsistence of the Poor, by finding them Employments, whereby they may earn their Bread, without being chargeable to the Publick" as a means for increasing the birth rate.

(2) He uses the odds $L_{x+t}/(L_x - L_{x+t})$ as a measure of the "differing degrees of Mortality, or rather Vitality in all Ages."

(3) He defines the median remaining lifetime for a life (x) as the solution of the equation $L_{x+t} = \frac{1}{2}L_x$.

(4) He remarks that the price of a term assurance may be found from the odds calculated under item (2).

(5) On the valuation of annuities upon lives Halley writes "that the Purchaser ought to pay for only such a part of the value of the Annuity, that he has Chances that he is living; and this ought to be computed yearly, and the Sum of all those yearly values being added together, will amount to the value of the Annuity for the Life of the Person proposed." Halley's formula for the value of an annuity thus becomes

$$\frac{1}{L_x} \sum_{t=1}^{\omega - x - 1} (1 + i)^{-t} L_{x+t}.$$

Halley carefully explains that the present value of 1 payable after t years equals $(1 + i)^{-t}$ and that this value has to be multiplied by the chance that (x) is living at time t. Halley presents a table of 1.06^{-t} for $t = 1(1)40(5)100$ and a table of the corresponding values of annuities for every fifth year of age.

Value of an Annuity at 6% Interest According to Halley

Age	1	5	10	20	30	40	50	60	70
Value of annuity	10.28	13.40	13.44	12.78	11.72	10.57	9.21	7.60	5.32

Halley remarks that the English government sells annuities at a price of seven years' purchase, which is to great advantage for the purchasers. Halley's paper did not affect the government, which continued to sell annuities cheaply and at a price independent of the age of the annuitant.

(6) The value of an annuity dependent on two lives, (x) and (y), say, may be found by a similar calculation, since "the number of Chances of each single Life, found in the Table, being multiplied together, become the Chances of Two Lives."

Let us denote the number of deaths in Halley's table between x and $x + t$ by $D_{x,t}$ so that $D_{x,t} = L_x - L_{x+t}$. Halley then finds the number of chances corresponding to the four possible combinations of survival and death by expanding $L_x L_y$ as

$$L_x L_y = (L_{x+t} + D_{x,t})(L_{y+t} + D_{y,t})$$
$$= L_{x+t}L_{y+t} + L_{x+t}D_{y,t} + L_{y+t}D_{x,t} + D_{x,t}D_{y,t},$$

and he gives a pictorial representation of this relation by means of the four areas in a rectangle, as shown in Fig. 9.3.3. Halley does not give the formulae for the corresponding annuities, but it is implied that the formula for the joint-life annuity is

$$\frac{1}{L_x L_y} \sum_{t=1}^{\omega - x - 1} (1 + i)^{-t} L_{x+t} L_{y+t} \qquad \text{for} \quad x \leqslant y,$$

with analogous formulae holding for the three remaining cases.

(7) The results of the previous section are generalized to three lives. After a discussion of the meaning of the eight terms of the product $L_x L_y L_z$, Halley gives the value of the reversion of the younger life (x) after the two elder, (y) and (z), as

$$\frac{1}{L_x L_y L_z} \sum_{t=1}^{\omega - x - 1} (1 + i)^{-t} L_{x+t} D_{y,t} D_{z,t}.$$

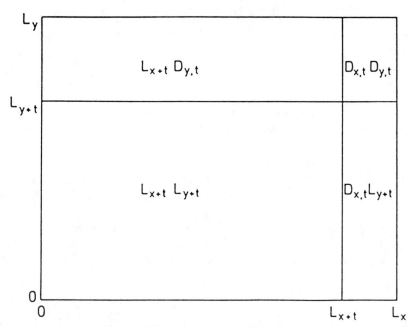

Fig. 9.3.3. Halley's pictorial representation of the number of chances for survival and death for two independent lives. The notation is of a later date.

He shows that it will be expedient to introduce $L_{x+t} = L_x - D_{x,t}$ and calculate

$$\frac{1}{L_x L_y L_z} \sum_{t=1}^{\omega-x-1} (1+i)^{-t} D_{x,t} D_{y,t} D_{z,t}.$$

The result may then be subtracted from the main term, which may be expressed as an annuity certain. In an example he calculates only the first six terms of the sum above because of the laborious calculations.

Halley does not discuss the expectation of life, but the formulae in question follow immediately from his formulae for annuities by setting $i = 0$. It will be seen that his method is more direct and easier to understand than Huygens' method for joint lives. Halley's method was rediscovered by N. Bernoulli in 1709, as discussed in §8.2.

The demographic concepts introduced by Halley were not understood by his contemporaries, possibly because Halley's explanations were somewhat unsatisfactory.

De Witt used the distribution of deaths as starting point, and he therefore calculated the value of an annuity as the expectation of the payments made to the deceased annuitants, that is, as $\sum a_{\bar{t}|} d_{x+t}/l_x$. Working independently of de Witt, Halley used the distribution of the survivors as the starting point, and he therefore introduced the expectation of the payments made to the living, that is, $(1+i)^{-t} L_{x+t}/L_x$, and found the value of an annuity as the sum of all these expectations. Halley's paper became of great importance to actuarial science. The expectation $(1+i)^{-t} l_{x+t}/l_x$, today called a pure endowment, became a fundamental quantity in life insurance and has remained so ever since.

Halley may have written his paper because as an editor, he needed papers of high quality for the *Phil. Trans.*; he never returned to this topic. It is a singular proof of his versatility that he produced this seminal paper in a field rather far from his main interests.

For supplementary reading we refer to Graetzer (1883), Greenwood (1941–1943), and K. Pearson (1978). Daston (1987) has given a sociological analysis of the interplay between probability and insurance business. Part of the present chapter has previously been published by Hald (1987).

9.4 PROBLEMS

1. Compare the life tables of Graunt, de Witt, and Halley by means of graphs of l_x, d_x, and q_x.

2. For

$$a_x = E\{a_{\overline{T}|}\} = \frac{1}{l_x} \sum_{t=1}^{\omega-x-1} a_{\overline{t}|}d_{x+t}$$

$$e_x = E\{T\} = \frac{1}{l_x} \sum_{t=1}^{\omega-x-1} td_{x+t}$$

prove that $a_x < a_{\overline{e_x}|}$.

3. Compare e_x, the curtate expectation of life, and $\bar{e}_x$, the complete expectation of life, and show that $\bar{e}_x \simeq e_x + \frac{1}{2}$ as a first approximation.

3. Calculate the expected payment according to age for each survivor in the Danish State tontine of 1653.

5. Set $l_x = 86 - x$ for $x \leqslant 86$ and $l_x = 0$ for $x > 86$, as suggested by Hudde. Use de Witt's formula to find a_x and show that $a_{50} = 11.3$ at 4% interest and yearly instalments. This result is given by de Witt in a letter to Hudde. Solve the same problem for $l_x = 111 - 1.37x$.

6. Assume that (x) and (y) are independent and set $l_{xy+t} = l_{x+t}l_{y+t}$ for $t \geqslant 0$. Let $\bar{e}_{xy}$ denote the joint expectation of life. Show that

$$\bar{e}_{xy} = \frac{1}{l_{xy}} \int_0^\infty t\,\delta_{xy+t}\,dt = \frac{1}{l_{xy}} \int_0^\infty l_{xy+t}\,dt,$$

where $\delta_{xy+t} = d(-l_{xy+t})/dt$.

Replacing L by l in Halley's formula, we get the value of the joint-life annuity,

$$a_{xy} = \frac{1}{l_{xy}} \sum_{t=1}^{\omega-x-1} (1+i)^{-t}l_{xy+t}, \qquad x \leqslant y.$$

Setting $i = 0$, we get the curtate expectation of life e_{xy}.

7. Following Halley, let us write

$$l_x l_y = (l_{x+t} + d_{xt})(l_{y+t} + d_{yt}).$$

Use this formula to prove that the probability that at least one of the lives will survive t years is

$$1 - {}_tq_x\,{}_tq_y = {}_tp_x\,{}_tp_y + {}_tp_x\,{}_tq_y + {}_tp_y\,{}_tq_x$$

$$= {}_tp_x + {}_tp_y - {}_tp_{xy}, $$

where ${}_tp_x = l_{x+t}/l_x,\,{}_tq_x = 1 - {}_tp_x$, etc.

8. From Problem 7 it follows that the expectation of life for the last survivor equals $\bar{e}_x + \bar{e}_y - \bar{e}_{xy}$. Find this expectation for $x = y = 16$ from Graunt's life table and compare with Huygens' method and result.

   Hint: Consider l_{xy+t} as a life table and compute $\bar{e}_{xy}$. The result is $20.25 + 20.25 - 11.42 = 29.08$. This computation is based on the assumption that l_{xy+t} is linear between the given points, which is at variance with the assumption that l_{x+t} is linear in each interval. Hence, improve the computation by using a quadratic function for l_{xy+t}.

9. Following Halley, the expectation in Problem 8 may also be found as

$$\int_0^{\omega - x} (1 - {}_tq_x\,{}_tq_y)\,dt \qquad \text{for} \quad x \leqslant y.$$

   Evaluate this integral by the trapezoidal rule and show that the result for $x = y = 16$ becomes $70 - 40.92 = 29.08$, in agreement with the result in Problem 8.

10. Write out the various forms of the value of an annuity on the last survivor of three lives.

CHAPTER 10

Mathematical Models and Statistical Methods in Astronomy from Hipparchus to Kepler and Galileo

10.1 OBSERVATIONAL ERRORS AND METHODS OF ESTIMATION IN ANTIQUITY AND THE MIDDLE AGES

Astronomy was the most important and the most advanced field of applied mathematics from antiquity until the 18th century. Observational and mathematical astronomy give the first examples of parametric model building and the fitting of models to data. In this sense, astronomers are the first mathematical statisticians, and it seems therefore natural to begin the history of mathematical statistics with a sketch of the history of mathematical models in astronomy. Problems in astronomy gradually led to the principle of the arithmetic mean and to various methods of estimation in parametric models, culminating with the method of least squares.

The crude instruments used by astronomers in antiquity and the Middle Ages could lead to large systematic and random errors. By planning their observations astronomers tried to balance positive and negative systematic errors. They seem not to have developed fixed rules for taking averages of observations to estimate the true value. If they made several observations of the same object they usually selected the "best" as estimator of the true value, the "best" being defined from such criteria as the occurrence of good observational conditions, special care having been exerted, and so on. Sheynin (1973) has made a survey of these problems with many quotations and references.

144

A probabilistic theory of errors and a corresponding theory of estimation did not emerge until the 18th century. Nevertheless, parameters had been successfully estimated by "primitive" methods for more than 2000 years.

Examples are to be found in estimating the length of periods of revolution of the heavenly bodies. The simplest model assumes that the sun revolves with constant speed in a circular orbit around the earth, which means that the angular position of the sun is a *linear* function of time. To estimate the length of the (tropical) year, Hipparchus about 135 B.C. uses a simple ratio estimate (as we would say today) of the slope; that is, he observes two points on the line and uses the ratio of the differences of the coordinates as the estimator. He also realizes that the error of the estimate depends on the distance between the two points. He uses the date of the summer solstice of the year 280 B.C., observed by Aristarchus of Samos, and his own observation for the year 135 B.C. The difference in time expressed in days and hours divided by the number of revolutions, $280 - 135 = 145$ gives him $365\frac{1}{4}$ days minus $(1/300)$th of a day as estimate of the length of a year. The correction of $(1/300)$th of a day was an improvement of the then current estimate of $365\frac{1}{4}$ days. Berry (1898, p. 55) writes,

> It is interesting to note as an illustration of his scientific method that he discusses with some care the possible error of the observations, and concludes that the time of a solstice may be erroneous to the extent of about $\frac{3}{4}$ day, while that of an equinox may be expected to be within $\frac{1}{4}$ day of the truth. In the illustration given, this would indicate a possible error of $1\frac{1}{2}$ days in a period of 145 years, or about 15 minutes in a year. Actually his estimate of the length of the year is about six minutes too great, and the error is thus much less than that which he indicated as possible.

Ptolemy checked this result by a similar computation based on an observation by Hipparchus of the autumn equinox and a corresponding observation of his own 285 years later and found the same estimate as Hipparchus, see Plackett (1958). Ptolemy also discussed several observations of the vernal and autumnal equinoxes and concluded that the length of the year is constant.

Since the parameters of the model could usually be identified with or related to specific observable phenomena, the natural method of estimation was to form the equations of condition (as they were later to be called) by equating predicted and observed values and to solve for the parameters. It was also natural to use only as many equations as there were parameters. In the linear case exemplified above we have $ax_1 + b = t_1$ and $ax_2 + b = t_2$, which give the ratio estimate $a = (t_2 - t_1)/(x_2 - x_1)$. A constant systematic error in the observation of t will disappear in the ratio estimate. We shall return to parametric models and estimation in §10.4.

10.2   PLANNING OF OBSERVATIONS AND DATA ANALYSIS BY TYCHO BRAHE

The *Almagest* by Ptolemy about A.D. 150 contained a catalog of the positions of 1028 stars, tables of important astronomical constants, and parameters for the planetary motions. Such tables depend on both direct observation and the theory of the motion of celestial bodies employed to reduce the observed positions to geocentric coordinates. Several Arab and European astronomers revised these tables during the Middle Ages. They corrected computational errors and mathematical inconsistencies, but they did not essentially reduce the uncertainties due to observational errors. Copernicus (1543) gave a star catalog based on Ptolemy's catalog corrected by a few new observations and arranged so as to avoid the effects of the precession of the equinoxes. Erasmus Reinhold published in 1551 his *Prussian Tables* containing revised tables of the positions and the motions of the celestial bodies calculated from the Copernican model.

In the construction of his heliocentric model, Copernicus used data covering the large time interval from Hipparchus about 150 B.C. to his own time. These data were naturally somewhat inconsistent and encumbered with many errors that misled Copernicus into introducing some long-term changes in his parameters. The Ptolemaic and Copernican models fitted the data about equally well, and many astronomers therefore realized the importance of providing new and more accurate observations as a basis for further development of models of the universe.

The Danish astronomer Tycho Brahe (1546–1601) became the leading observational astronomer in the latter part of the 16th century. Supported by the Danish king, he built a magnificent observatory on the island Hven, where he educated a large number of assistants and demonstrated his instruments and methods to many visitors. Very early in life he set himself an immense task: to redetermine the positions of the celestial bodies with far greater accuracy than previous observers and thus to create a new empirical foundation for revising the existing mathematical models.

He constructed better and larger instruments with finer gradations. He mounted some of the instruments on sturdy supports and built an underground observatory to avoid the effects of wind, temperature, and other outside disturbances. He regularly checked his instruments against each other to control systematic errors.

He trained his assistants in observational procedures and had them independently observe the same phenomenon at the same time to check each other's results.

He began a program of regular observation of the same celestial bodies that extended over a period of nearly 25 years.

In modern terms, he controlled both systematic and random errors by measuring the variations between instruments, between observers, and over time. His methods served as a model for other observatories and scientific laboratories.

He collected his observations in a catalog of 1000 stars. In most cases the positions were determined with an error of less than 2′, and for eight of his nine "reference stars," the error was less than 1′. This was a great improvement for observations made with the naked eye.

We shall now discuss two recent investigations of Tycho's work in more detail. The first is due to Plackett (1958). Tycho knew that his observations were affected by systematic errors due to parallax and refraction. In his attempts to determine the right ascension of the star α Arietis, he therefore combined his observations in pairs based on objective astronomical criteria of symmetry so that the two observations might be supposed to have systematic errors of the same size but of opposite sign. For the years 1582–1588 he collected 12 such pairs. To eliminate the systematic errors he calculated the arithmetic mean of each pair. For example, the first pair gave 26°4′16″ and 25°56′23″ with a mean of 26°0′20″. (Plackett presents a complete table with the date for each observation, from which it follows that the systematic errors, the differences divided by 2, vary from about 0′ to 8′, the example above with a systematic error of about 4′ being typical.) Deducting 26° from each mean, Tycho's results expressed in seconds of arc are as follows:

$$20, \ 38, \ 18, \ 32, \ 42, \ 37, \ 27, \ 29, \ 14, \ 4, \ 28, \ 39.$$

Tycho supplemented the twelve means with three single observations equal to 44″, 32″, and 30″. He gave the final result of his analysis as 26°0′30″ without explaining how he arrived at this number, and he referred this result to the end of the year 1585, which is nearly the midpoint of the observational period. By modern methods the position is calculated to be 26°0′45″.

This is a fine example of carefully planned observations combined with refined data analysis using the arithmetic mean to eliminate systematic errors and choosing a central value as an estimate of true position to reduce the effects of random errors.

For comparison with Tycho's result we note that the mean and standard deviation of the 12 means are 26°0′27″ and 11.5″, respectively, so that the standard error of the mean is 3.3″. This indicates that the mean (and also Tycho's value) is significantly smaller than the modern value, if we assume that the latter is without error. The difference is, however, very small. Multiplying the standard deviation by $\sqrt{2}$ we obtain 16″ as the standard deviation of a single observation, which shows the great accuracy of Tycho's observational method.

One may wonder why the practice of calculating the arithmetic mean of several independent, equally good observations did not become firmly established before the 18th century. The concept of the arithmetic mean had been known in other contexts long before, for example, the center of gravity. Part of the explanation may be that since new observations were taken with the purpose of improving previous results, the observations were not considered equally good.

The early history of the arithmetic mean has been given by Eisenhart (1974). Kepler's few attempts of statistical considerations have been described by Sheynin (1975).

To investigate the accuracy of the instruments used by Tycho Brahe, Wesley (1978) computed the positions of 20 stars at Tycho's times with an error negligible compared to the errors of Tycho's observations. These computed values may thus be considered the true values, and the difference between Tycho's observations and the true values demonstrate the combined effects of systematic and random errors. For example, for the declination of the star Tauri the average error equals 60″, and the standard deviation of the errors equals 15″ for observations carried out with the mural quadrant. Wesley states that the number of observations is at least eight, so we are not able to give the standard error of the average, but we can say that it is smaller than $15''/\sqrt{8} = 5''$. This means, however, that the larger part of the average represents systematic errors. To test Tycho's assertion that his instruments gave consistent results, Wesley has calculated the average error for five other instruments and found the results 39″, 52″, 44″, 105″, and 59″, which is in agreement with the first result if we take the slightly larger standard deviations and the smaller number of observations into account. Wesley finds similar results for the other 19 stars and for a few more instruments as well. He also concludes that the average errors do not change greatly from year to year, presumably because of Tycho's constant checking of each instrument against all the others.

Tycho Brahe succeeded in carrying out his observational program. He also proposed a new mathematical model, a compromise between the Ptolemaic and the Copernican models; however, this was shortly superseded by Kepler's model. In his *Astronomia Nova* from 1609 Kepler writes,

> Since the divine goodness has given to us in Tycho Brahe a most careful observer, from whose observations the error of 8′ is shewn in this calculation,... it is right that we should with gratitude recognise and make use of this gift of God.... For if I could have treated 8′ of longitude as negligible I should have already corrected sufficiently the hypothesis... discovered in chapter XVI. But as they could not be neglected, these 8′ alone have led the way towards the complete reformation of astronomy, and have been made the subject-matter of a great part of this work. (Berry, 1898, p. 184.)

10.3 GALILEO'S STATISTICAL ANALYSIS OF ASTRONOMICAL DATA, 1632

In his works on physics Galileo describes many experiments to support and elucidate his mathematically formulated physical laws. He did not, however, publish his experimental results, and therefore there has been some discussion on whether he actually carried out these experiments. Undoubtedly, some of the experiments were imaginary, or "thought experiments," in which he appealed to generally acknowledged experiences and to common sense; in other cases, as documented from his notebooks, he did carry out experiments.

Drake (1978, p. 89; 1985) and others have analyzed Galileo's experiments with a ball rolling down an inclined plane placed on a table, the ball being deflected horizontally until it hits the floor. Galileo records the vertical distance of fall (above the table) and the horizontal distance traversed (from the table to the end of the trajectory). Assuming that horizontal distance is proportional to the square root of vertical distance, he estimates the constant from one pair of observations, calculates the expected value of the other horizontal distances, and notes the deviations.

In his *Dialogo* Galileo (1632) gives a detailed statistical analysis of observations on the new star of 1572. He was provoked by S. Chiaramonti, professor of philosophy at Pisa, who published the *De tribus novis stellis quae annis 1572, 1600, 1604 comparuere* (On the Three New Stars Which Appeared in the Years 1572, 1600, 1604) in 1628 to prove that the new star was sublunar, in opposition to Tycho Brahe and some other astronomers, who held that it was situated among the fixed stars. As mentioned earlier in §2.3, the question of the immutability of the heavens was a very serious one, both philosophically and theologically. Tycho wrote a small book on the new star in 1573, and in his posthumously published works in 1602 he returned to a detailed discussion and interpretation of the observations made by himself and other astronomers; a summary has been given by Dreyer (1890, Chaps. 3 and 8). The *Dialogo* has been translated into German by E. Strauss (1891) and into English by S. Drake (1967), see the references under Galileo (1632), and both translators have provided many useful notes. The following page references are to Drake's translation. The whole analysis takes up pp. 280–318.

Galileo gives an informal discussion of random errors on pp. 281–282 and 287–293. He does not discuss systematic errors, nor does he use the terms "random" and "distribution"; rather, he writes about "observational errors." However, his description of the properties of these errors leaves no doubt that he is discussing what today is called the distribution of random errors. His discussion may be summarized as follows:

1. There is only one number which gives the distance of the star from the center of the earth, the true distance.
2. All observations are encumbered with errors, due to the observer, the instruments, and the other observational conditions.
3. The observations are distributed symmetrically about the true value; that is, the errors are distributed symmetrically about zero.
4. Small errors occur more frequently than large errors.
5. The calculated distance is a function of the direct angular observations such that small adjustments of the observations may result in a large adjustment of the distance.

We do not know whether the properties of observational errors as described by Galileo were generally known among astronomers of the time; at any rate they were not taken into account by Chiaramonti. These properties are the foundation on which mathematical error theory later was built.

Based on these suppositions, Galileo discusses how to compare the two hypotheses about distance. Comparing the observations with the two hypothetical values he says that *the most probable hypothesis is the one which requires the smallest corrections of the observations.* He writes on p. 290, "Then these observers being capable, and having erred for all that, and their errors needing to be corrected in order for us to get the best possible information from their observations, it will be appropriate for us to apply the minimum amendments and smallest corrections that we can," and on p. 293, "Those observations must be called the more exact, or the less in error, which by the addition or subtraction of the fewest minutes restore the star to a possible position. And among the possible places, the actual place must be believed to be that in which there concur the greatest number of distances, calculated on the most exact observations."

Galileo's error theory has been discussed by Maistrov (1974, pp. 32–34) who reached similar results to those stated above.

The crucial question is, What does Galileo mean by "the smallest corrections"? He does not answer explicitly, but from his analysis of the data it follows that he uses *the sum of the absolute deviations from the hypothetical value* as his criterion.

We shall use the notation indicated in Fig. 10.3.1.

The data adopted by Galileo from Chiaramonti's book are given in Table 10.3.1.

Unfortunately, Galileo does not offer a critical evaluation of the data. He only remarks that repeated observations of the polar altitude by the same observer may result in variations of a minute or so, or even of many minutes, and that Tycho and the Landgrave are known to be among the best observers.

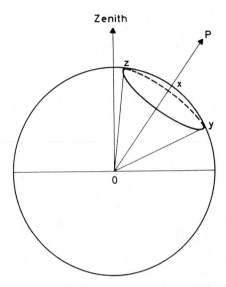

Fig. 10.3.1. The point O represents the center of the earth, the circle the celestial sphere, x the altitude of the polar star P, and y and z the minimum and the maximum altitudes of the new star corresponding to the daily rotation of the earth.

Table 10.3.1. Observations of the Star of 1572

Observer	Altitude of the pole, x	Minimum altitude of the star, y	Maximum altitude of the star, z
1a Tycho	55°58′	27°57′	84°00′
1b Tycho		27°45′	—
2a Camerarius	52°24′	24°28′	80°30′
2b Camerarius		24°20′	80°27′
2c Camerarius		24°17′	80°26′
3 Peucer	51°54′	23°33′	79°56′
4 The Landgrave	51°18′	23°03′	79°30′
5 Reinhold	51°18′	23°02′	79°30′
6 Busch	51°10′	22°40′	79°20′
7 Gemma	50°50′	—	79°45′
8 Ursinus	49°24′	22°	79°
9a Hainzel	48°22′	20°09′40″	76°34′
9b Hainzel		20°09′30″	76°33′45″
9c Hainzel		20°09′20″	76°35′
10 Hagek	48°22′	20°15′	—
11 Muñoz	39°30′	11°30′	67°30′
12 Maurolycus	38°30′	—	62°

Source: Galileo, *Dialogo*, 1632, pp. 294–295.

In his notes, Strauss (1891) points out that Chiaramonti has obtained his data from Tycho's book (1602) apart from Maurolycus' observations, which have been obtained from another publication. Tycho himself remarks that his first observations 27°57′ is unreliable, so that only the second should be used. Tycho also remarks that Reinhold has copied the observations of the Landgrave. Galileo speaks of 13 astronomers, but two of them, Peucer and Schuler, have used the same data. Chiaramonti and Galileo therefore treat the data as 12 independent sets.

Looking at the data in Table 10.3.1, it is clear that they are of somewhat varying quality. Some observations are recorded in whole degrees only, whereas others are given to an accuracy of 10″; it is obvious, however, that the three observations by Hainzel cannot be independent. Looking at the differences between the maximum altitude of the star and the polar altitude, it will be seen that the maximum altitude given by Maurolycus is an outlier being at least 4° too small. Chiaramonti seems uneasy about Maurolycus' observation but concludes that since Maurolycus is the Bishop at Messina, his observation ought to be reliable. Galileo does not comment on this matter.

The plotting of observations in a coordinate system had not yet been invented at the time of Galileo. Nevertheless, we shall plot the data to give

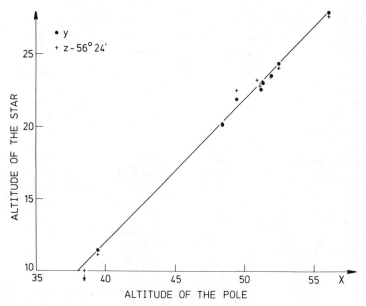

Fig. 10.3.2. The observed altitude of the star plotted against the polar altitude. The slope of the line equals unity. The altitude corresponding to the arrow is 5°36′.

the reader a better understanding of the following discussion (see Fig. 10.3.2.) For each observer we have used the mean of the altitudes. It will be seen that the minimum and maximum altitudes are nearly linear functions of the polar altitude. (The constant $56°24'$ is the average difference between the maximum and the minimum altitudes.) Galileo explains this phenomenon on p. 282, "For if it [the star] was placed in the firmament among the other fixed stars, its meridian altitudes when taken at different elevations of the pole would have to differ among themselves in the same way as did these polar elevations." This means that the true values of x, y, and z satisfy the relations $x - y = z - x = \alpha$, say, α being a constant, if the new star is placed among the fixed stars. It follows that the true values of y and z are linear functions of the true values of x, the slope of the line being unity, as shown in Fig. 10.3.2.

On p. 311 Galileo remarks "that if the new star, or some other phenomenon, is close to the earth and is turning in the diurnal motion about the pole, it will show itself more distant from the pole when it is below the pole on the meridian than when above it."

To simplify the exposition we shall introduce some symbols and formulae. Galileo, however, proceeds by numerical examples only.

Let the three altitudes of the ith observation be denoted by (x_i, y_i, z_i). In Table 10.3.1 the observations were ordered according to decreasing values of x_i, the polar altitude. Consider the comparison of two observations, as shown in the following table:

x_i	y_i	$x_i - y_i$
x_j	y_j	$x_j - y_j$
$x_i - x_j$	$y_i - y_j$	p_{ij}

Assuming that $d_{ij} = x_i - x_j > 0$, the parallax of the star in relation to the two observational places i and j is found to be

$$p_{ij} = (y_i - y_j) - (x_i - x_j).$$

Another estimate of the parallax is obtained by replacing y by z. In the following we shall always give the parallax in minutes.

It will be seen that $p_{ij} \gtreqless 0$ corresponds to

$$\frac{y_i - y_j}{x_i - x_j} \gtreqless 1,$$

so that $p_{ij} > 0$ if the slope of the line from (x_i, y_i) to (x_j, y_j) is larger than 1. If follows from Fig. 10.3.2 that the majority of the p's vary closely around zero.

According to the hypothesis of Tycho and Galileo the true value of the parallax is zero, whereas by Chiaramonti's hypothesis the true value is positive.

The distance of the star from the earth may be computed from the minimum (or maximum) altitudes of the star observed from two different places on the earth, located on the same circle of longitude and with different latitudes, i.e., different altitudes of the pole. The distance, expressed in multiples of the radius of the earth, *from observer j to the star*, is estimated by means of the formula

$$\frac{2\sin(\tfrac{1}{2}d_{ij})\sin(180 - y_i + \tfrac{1}{2}d_{ij})}{\sin p_{ij}}.$$

The distance from the center of the earth, r_{ij}, say, is obtained by adding 1. A geometrical proof of the distance formula is given by Galileo on p. 297; an analogous figure and proof using the notation above is shown in Fig. 10.3.3. Chiaramonti uses a similar formula.

As an example, consider the computation of distance based on observations 1a and 10, as follows:

Observation	x	y
1a	55°58′	27°57′
10	48°22′	20°15′
Difference	7°36′	7°42′

The parallax becomes $p = 6'$, and the distance formula gives

$$r = 1 + \frac{2\sin 3°48'\sin 155°51'}{\sin 6'} = 32.$$

This is an example of Chiaramonti's (and Galileo's) computations of distances. Today such a computation is easy, but at the time of Galileo it required a great deal of labor, as can be seen from his book. It is therefore understandable that they did not carry out all the computations that the data admit.

If there had been only one observation for each of the 12 astronomers, $\binom{12}{2} = 66$ parallaxes and distances could have been computed. However, 10 of the astronomers have observed the minimum altitude, and 11 have

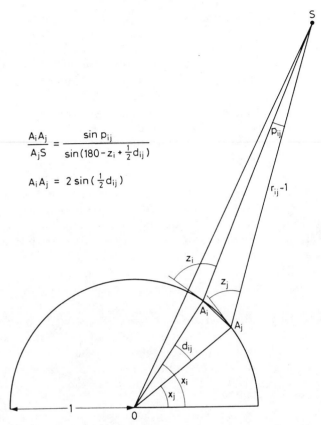

Fig. 10.3.3. Galileo's proof of the distance formula: O represents the center of the earth, S the position of the star, and A_i the position of astronomer i.

observed the maximum altitude, so that it is possible to make

$$\binom{10}{2} + \binom{11}{2} = 100$$

pairs. Furthermore, in some cases several observations are recorded for both y and z, which makes the number of possible pairs considerably larger.

Chiaramonti writes that to investigate all pairs would be a nearly impossible task, and he therefore limits himself to demonstrating that it is possible to find 12 pairs, that lead to sublunar distances using observations from all 12 astronomers. However, one of his pairs gives a negative parallax, which by a correction he makes positive; like Galileo we shall disregard this result.

Besides the computation of $r = 32$ given above, Chiaramonti gives 10 more, leading to distances between 1.2 and 32 (see Table 10.3.2.) He concludes that the new star is sublunar, since the distance to the moon according to Ptolemy is at least 33 radii. Chiaramonti rejects the pairs of observations giving negative parallaxes because they place the star in "impossible positions," that is, beyond the fixed stars. Galileo, however, says that such observations are not at variance with the model of the universe because the negative parallaxes can be explained as due to observational errors. Galileo writes on p. 293 that "the most trifling errors made by the observer with his instrument will change the location from finite and possible to infinite and impossible" and that, conversely, "in these calculations made from the observations which would put the star infinitely distant, the addition or subtraction of one single minute would often restore it to a possible location."

After having criticized the rejection of negative parallaxes, Galileo points out that Chiaramonti has also omitted five pairs of observations giving distances beyond the moon and, furthermore, five pairs with parallaxes equal to zero.

To simplify the investigation of Chiaramonti's hypothesis, that the true value of r lies between 1 and 33, Galileo proposes to replace the composite hypothesis by the simple hypothesis that the true value of r equals 32. He chooses 32 because it is the largest value found by Chiaramonti, and it is just below the upper limit of the sublunar sphere and thus the "most favorable" to Chiaramonti.

Galileo's own hypothesis implies that the true value of r equals infinity. He can only handle the distribution of r by remarking that most of the observations lead to distances which are infinite or "beyond." This fact naturally supports Galileo's hypothesis, but he wants more convincing arguments. He therefore turns to the distribution of the parallaxes, which under his hypothesis has a true value equal to zero and properties corresponding to those described in his error theory. He uses the quantity $A_0 = \sum |p_{ij}|$ as a measure of agreement between the observations and his hypothesis.

Unfortunately, the simple alternative hypothesis becomes composite when expressed in terms of parallax. To find the value of the parallax, c_{ij}, say, corresponding to a distance of 32, Galileo solves the distance equation for p_{ij}, which leads to the equation

$$\sin c_{ij} = \frac{2 \sin(\tfrac{1}{2} d_{ij}) \sin(180 - y_i + \tfrac{1}{2} d_{ij})}{31}$$

Table 10.3.2. Values of p_{ij}, r_{ij}, and c_{ij} for the 20 Pairs of Observations

	Chiaramonti's pairs[a,d]					Galileo's pairs[a,e]			
i,j	p_{ij}	r_{ij}	c_{ij}	$\|p_{ij}-c_{ij}\|$	i,j	p_{ij}	r_{ij}	c_{ij}	$\|p_{ij}-c_{ij}\|$
9*,12*	282.5	3	20	262.5	1b,4	2	61	3.9	1.9
1a,9c	10	19	6	4	2a,11	4	61	7.7	3.7
1a,4	14	10	4	10	1*,11*	2	479	30.8	28.8
7*,9*	42.5	4	5	37.5	3*,11*	2	358	23.0	21.0
2c,4	8	4	1	7	4*,9b*	0.25	716	5.6	5.3
8,10	43	1.5	1	42	2c*,3*	0	∞	1.0	1.0
4,6[b]	15	1.2	0.1	14.9	4*,9a*	0	∞	5.6	5.6
11*,12*	270	1.2	2	268	1*,3*	0	∞	7.8	7.8
7*,11*	55	13	20	35	5*,9a*	0	∞	5.6	5.6
8*,11*	96	7	20	76	2c 10	0	∞	3.0	3.0
Total	836		79.1	756.9[c]		10.25		94.0	83.7

[a](*) Maximum altitude has been used.
[b]In (4, 6) Galileo writes 10″, but in his calculations on p. 306 he uses 6″.
[c]According to Drake (1967, p. 485), the sum 756′ would have been reduced to 658′ if Galileo had chosen a distance of 7 radii instead of 32.
[d]Note that Chiaramonti uses observation 12 twice and 1a twice.
[e]Calculation of the c_{ij} for Galileo's 10 pairs is due to me.

for the determination of c_{ij}. Galileo does not give this equation explicitly but proceeds by trial and error to find an approximation to c_{ij}. His measure of agreement for Chiaramonti's hypothesis thus becomes $A_c = \sum |p_{ij} - c_{ij}|$.

Galileo first considers the ten pairs of observations with nonnegative parallaxes which most support his own hypothesis (see Table 10.3.2). They consist of five pairs with positive parallaxes and a distance beyond the moon; the parallaxes are between $\frac{1}{4}'$ and 4′, and the distances between 61 and 716 radii. The sum of the five parallaxes equals 10′. The remaining five pairs have parallaxes equal to zero. Hence, the ten pairs contribute 10′ to A_0.

Next, Galileo considers the ten pairs which most support Chiaramonti's hypothesis, as mentioned above. He calculates the corresponding values of c_{ij} and finds that they contribute 757′ to A_c. The calculations are shown in Table 10.3.2.

From these results Galileo concludes that the ten pairs which most support Chiaramonti's hypothesis require a total correction of 757′ to be adjusted to a common distance of 32 radii, whereas the ten pairs which most support his own hypothesis only require a total correction of 10′ to place the star in the firmament, and he regards this comparison as strong evidence for his own hypothesis, see pp. 307–308.

Galileo does not stop with the comparison of the corrections based on the 20 pairs with nonnegative parallax. He points out that one should use all the observations. He writes on p. 308, "Of the remaining combinations that can be made of observations taken by all these astronomers, those which imply the star to be infinitely high are much more numerous—about thirty more—than those which upon calculation place the star beneath the moon."

Galileo may mean that each pair of astronomers should contribute only one estimated distance. The 12 astronomers will thus give 66 distances, which presumably is the number Galileo has in mind, but he does not tell how to select the 66 pairs among the many more possible pairs. We have summarized Galileo's analysis in Table 10.3.3 and Fig. 10.3.4.

The three pairs leading to a subterranean position have been listed by Galileo as $(4^*, 7^*)$, $(5^*, 7^*)$, and $(6^*, 7^*)$; they have $x_i > x_j$ and $z_i < z_j$. The following $11 + 5 + 5$ pairs are the 20 pairs listed in Table 10.3.2 and the pair $(1a, 10)$ with a parallax of $6'$ and a distance of 32 radii.

Galileo does not evaluate the contributions to A_0 and A_c of the 45 pairs with negative parallaxes. He writes, however, on p. 309 "that the corrections to be applied to observations which give the star as at an infinite distance will, in drawing it down, bring it first and with least amendment into the

Table 10.3.3. Distributions of Distances and Parallaxes for Galileo's Pairs of Observations

Estimated position of star	Value of p_{ij}	No. of pairs
Subterranean	$p < 0$	3
Sublunar	$p > 0$	11
Between the moon and the firmament	$p > 0$	5
Firmament	$p = 0$	5
Beyond the firmament	$p < 0$	42

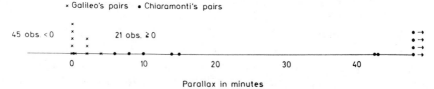

Fig. 10.3.4. The distribution of Galileo's 66 parallaxes. The four arrows indicate the four largest values in Table 10.3.2.

firmament rather than below the moon." This means that the negative parallax is first adjusted to zero by adding $|p_{ij}|$ and then it is further adjusted by adding c_{ij}. The contribution to A_0 is thus $\sum |p_{ij}|$, whereas the contribution to A_c becomes $\sum |p_{ij}| + \sum c_{ij}$, the summation being over the 42 pairs. Galileo does not discuss whether this contribution is sufficient to make A_c larger than A_0. It seems to him that the evidence presented in favor of his hypothesis is overwhelming. Furthermore, independent of the statistical analysis above, he gives other arguments in favor of his hypothesis, see pp. 310–318.

Chiaramonti (1633, 1643) naturally replied to Galileo's criticism, however, nothing of statistical interest seems to be involved. Riccioli (1651) gives a detailed discussion of the whole controversy and concludes that "It is more probable that the star of the year 1572 and 1573 has been above rather than below the moon" (Riccioli, 1651, Vol. 1, Part 2, p. 165). This seems to end the discussion, which is natural enough in view of the rapid progress of astronomy in the second half of the 17th century.

We shall now comment further on Galileo's analysis. Let us first consider the evaluation of A_0 and A_c for the pairs in Table 10.3.3, except for the three subterranean pairs. The contributions from the 21 pairs are listed in Table 10.3.2. How could Galileo easily find an approximate value of $\sum c_{ij}$ for the 42 pairs? From his calculation of c_{ij} for Chiaramonti's 10 pairs it will be seen that the average value of c_{ij} is about 8', so it is therefore tempting to use the approximation $42 \times 8 = 336$. (Galileo might have used a central value other than the mean if he had concerned himself with this problem.) Denoting the unknown sum of the 42 $|p_{ij}|$'s by a, say, we get the following evaluation:

$$A_0 = 836 + 6 + 10 + a = 852 + a,$$

$$A_c = 757 + 0 + 84 + a + 336 = 1177 + a,$$

so that A_0 is somewhat smaller than A_c.

Today it is well known that the value of c minimizing $\sum |p_{ij} - c|$ is the median of the distribution of p_{ij}. Since the majority of the p's are negative, this means that the minimum is obtained for a value of c slightly less than zero, and it follows that A_0 is smaller than A_c for positive c's.

In his criticism of Chiaramonti's analysis Galileo naturally used the same data and the same method for determining the distances by pairing the observations, but he worked directly with the parallaxes instead of the distances. However, the pairing introduces correlations that make the statistical analysis unnecessarily difficult. Today one would use regression analysis of the independent observations to test the hypothesis, as indicated in Fig. 10.3.2.

The statistical lesson to be drawn from Galileo's discussion is that a comparison of a mathematical model with data may be misleading unless a theory of observational errors is taken into account.

Even though Galileo did not state his method unambigously and did not carry out all the numerical details, his statistical analysis still contains the rudiments of a theory for comparing hypotheses using the sum of the absolute deviations as criterion. In view of the wide circulation of Galileo's book it is odd that his method of analysis was not discussed by other astronomers. It may be that the controversy involved soon ceased to be of interest, and this section of Galileo's book became less important. The method of least absolute deviations was not taken up again by astronomers until Boscovich made use of it more than a century later.

A large part of the present section is reprinted with permission from my 1986 paper in the *International Statistical Review*. This paper also contains two regression analyses of Galileo's data.

10.4 MATHEMATICAL MODELS IN ASTRONOMY FROM PTOLEMY TO KEPLER

An enormous literature exists about the history of mathematical models in astronomy from Hipparchus to Newton; some expositions stress the astronomical details, others the philosophical implications. We shall try to sketch the history from a statistical point of view with respect to model building and goodness of fit. This would have been impossible for us but for the paper by Riddell (1980), Parameter Disposition in Pre-Newtonian Planetary Theories, on which the following exposition is based.

Mathematical astronomy in antiquity culminated with the works of Hipparchus about 150 B.C. and Ptolemy about A.D. 150. The *Almagest* by Ptolemy became the undisputed authority for about 1400 years until the work of Copernicus. Ptolemy inherited from earlier Greek astronomers and philosophers the idea that the revolutions of the heavenly bodies about the earth are describable in terms of uniform circular motions and on that assumption Ptolemy constructed his ingenious epicyclic model. His purpose was to predict the heavenly phenomena as accurately as possible, and he was perfectly aware of the fact that the models were descriptive and did not reflect physical realities. He stressed the importance of using simple models.

The observations are the positions of the sun, the moon, and the planets referred to a coordinate system with origin O at the center of the earth. The angular positions are characterized by the longitude λ, measured eastward along the ecliptic from the spring equinox, and the latitude, measured along circles at right angles to the ecliptic. We shall write S for the sun and P for the planet in question. The positions are observed at different times, and time t will be the independent variable in the system.

We shall illustrate the main ideas of the models by giving numerical results for the longitude of the planet Mars. Similar results on the latitude and for the other planets, the sun, and the moon may be found in Riddell's paper.

Ptolemy's model may be conceived as consisting of three stages; the third-order model is shown in Fig. 10.4.1. The second-order model is obtained by letting the points E and O coincide with M, the center of the circle.

In the first-order model, the planet moves with constant angular speed on a circle so that the longitude is a linear function of t. The model thus contains three parameters, the radius R, which is unidentifiable, and the two angular parameters, which are independent of R and easily estimated from observations.

Because of the retrogressions of the planets as seen from the earth, the circular model is obviously wrong. To deal with this phenomenon a second-order model is constructed in which P moves with constant speed on a small circle, the epicycle, with center C, which again moves with constant speed on a larger circle, the deferent, with center at O. The three parameters of the deferent are thus supplemented by three parameters of the epicycle, namely, the radius r and the two angular parameters. To match the phenomena, C must travel around O, and P must travel around C in the same direction once in a siderial year in such a way that the line CP is always parallel to the line from the origin to the (mean) sun. This is effected by introducing two linear relations between the four angular parameters.

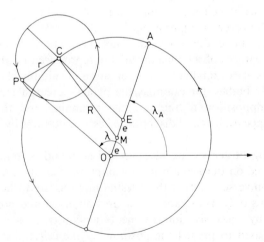

Fig. 10.4.1. Ptolemy's third-order model for the longitude λ of an outer planet.

Hence the six parameters are reduced to four, of which the two angular parameters and $\rho = r/R$ are estimated from the observations, whereas R is unidentifiable.

The *goodness of fit* is specified by Riddell by means of the maximum discrepancy between the observed and the computed longitude. Riddell (1980, p. 97) writes, "The reader is emphatically cautioned that our error estimates are in many cases rather rough, and are meant as an indication of orders of magnitude in comparisons with other similar estimates, not as exact maximum errors." For Mars the maximum error for the first-order model is $52°$, which is reduced to $13°$ for the second-order model, corresponding to an increase in the number of estimable parameters from two to three.

Since discrepancies of $13°$ are considerably larger than the observational error, Ptolemy constructs a third-order model by letting the center of the deferent, M, say, be different from O. He introduces three new parameters, the two distances $|OM|$ and $|ME|$ and the longitude of the apogee A, the point on the deferent farthest from O. Ptolemy makes the point E, the equant, the center of equal motion; that is, he assumes that the angle AEC (rather than the angle AMC) increases uniformly with time.

For all the planets except Mercury, Ptolemy makes the two distances equal, which means that he uses $e = |OM| = |ME|$, the eccentricity, as parameter. The longitude of A and the eccentricity are estimated from the observations. The number of estimable parameters is thus increased by two, and the error decreases from $13°$ to $4°$, which is not much greater than the observational error.

Taking the latitude also into account, the complete third-order model for the motion of Mars is generated by means of sixteen parameters subject to seven restrictions, which leaves nine parameters to be estimated from the observations; however, one parameter is unidentifiable.

As indicated above, the restrictions have been chosen such that certain characteristic features of the phenomena are reproduced by the model. The free parameters are estimated from observation of suitably chosen positions of the heavenly bodies, for example, the planet's greatest elongation from the sun, the opposition to the sun, the beginning and the end of the retrocession, appearances and disappearances, occultations by the moon and eclipses.

In evaluating Ptolemy's ingenious model it should be remembered that his work is based on the two a priori assumptions that the earth is the fixed center of the universe and that the celestial motions are to be composed of circular motions only. He steadily compares observed and predicted values and succeeds by successive improvements to construct a complex model which can be used to predict the positions of the celestial bodies with an accuracy nearly equal to the observational error. It is an important feature of the model that it is composed of linear angular motions which make it computationally rather simple and therefore easy to use in the many practical applications.

As more observations accrued during subsequent centuries, small systematic discrepancies between observed and predicted values were noted. Arab and European astronomers in the Middle Ages thus improved upon Ptolemy's model by adding epicycles to epicycles, eccentrics to eccentrics, eccentrics with moving centers, and so on. The mathematical simplicity and beauty of the model thereby disappeared. The basic structure of the model was not questioned until the work of Copernicus in 1543.

The mathematical complexities of Ptolemy's model, in particular, his assumption about the uniform motion in relation to the equant rather than the center of the deferent, caused Copernicus to give up the assumption of the earth as the fixed center of the universe and to assume instead that the earth and the (other) planets revolved around the sun. By this fundamental innovation he was able to describe the complicated motion of a planet as seen from the earth by means of two independent circular motions of the earth and the planet around the sun. Copernicus also postulated that the apparent diurnal motion of the heavens is due in reality to the rotation of the earth.

For the longitudinal motion of the planets, Copernicus begins with P's uniform circular motion around $\bar{S}$ with radius R_P, where the mean sun $\bar{S}$ is defined as the center of the earth's circular orbit. To find the motion of P in relation to O it is necessary to combine the motion of P and O around $\bar{S}$. Geometrically the resulting motion may be described by the deferent-epicycle model with $R = R_P$ and $r = R_O$, so that Copernicus' first-order model is equivalent to Ptolemy's second-order model, with a maximum error of 13° for Mars. However, in the heliocentric model the radius is estimable.

Copernicus did not question the assumption of compound circular motions. His second-order model makes use of a construction similar to that shown in Fig. 10.4.1 but with O replaced by $\bar{S}$, so that P moves on an epicycle, and $\bar{S}$ has an eccentricity in relation to the center of the deferent. The model has six estimable parameters and leads to a maximum error of 4°, just as for Ptolemy's third-order model. This means that the Copernican model gives essentially the same predictions as the Ptolemaic model.

Copernicus did not make many observations himself but relied on the observations given in the *Almagest* and on observations made by his contemporaries. On his estimation of parameters Riddell (p. 124) writes, "Copernicus generally takes over Ptolemaic parameter values and adjusts them *ad hoc* where necessary."

Comparing the Copernican and Ptolemaic models we conclude that they are about equal in mathematical and computational complexity, that they use about the same number of parameters, and they give nearly the same predictive accuracy. What, then, are the advantages of the Copernican model?

First, by choosing the sun as center of the universe, Copernicus gives a natural explanation of the retrogradations of the planets. Similarly, he explains a large number of other phenomena and points out that his model gives a more natural and harmonious picture of the universe.

Second, the assumption of the diurnal rotation of the earth explains the diurnal motion of the stars.

Third, Copernicus demonstrates the natural order of the heavenly bodies and shows that the planetary periods of revolution T are increasing with distance R from the sun. Copernicus estimates T and R for each of the five known planets relative to the values for the earth.

Tycho Brahe did not believe in the motions of the earth as proposed by Copernicus. In 1588 he published a model of his own based on the idea that the sun is the center of the orbits of the five planets and that the earth is the center of the orbits of the sun and the moon. He also supposed that the sphere of the fixed stars revolves around the earth in 24 hours, so that the position of the earth as stationary at the center of the universe was preserved. Conceptually his system was different from the Copernican model, but mathematically the two models were equivalent. He intended to determine the parameters by means of his own observations, but he died before he completed this project.

In 1597 Tycho left Denmark, and in 1599 he settled in Prague as Imperial Mathematician at the court of the Emperor Rudolph II. Johannes Kepler (1571–1630) worked as assistant to Tycho from the beginning of the year 1600, and when Tycho died in 1601, Kepler inherited his position and his observations.

Like Copernicus, Kepler was a great mathematical astronomer, and he now had in his possession the observations necessary to revise the Copernican model. Furthermore, Kepler had a physical theory about the planetary motions which guided him in his search for a simpler and more precise model.

Kepler derived the following three laws:

1. The orbit of a planet is an ellipse with the sun in one focus.
2. The radius vector from the sun to a planet sweeps out equal areas in equal times. This law is also called the area law.
3. The squares of the times of revolution of any two planets around the sun are in the same proportion as the cubes of their mean distances from the sun; that is T^2/R^3 is the same for all the planets, where T is the planet's siderial period, and R is the length of the major semiaxis.

In a much simplified version, Kepler's physical hypothesis about the motion of the sun and the planets is as follows. A magnetic force emanates from the rotating sun, and this force spreads in the plane of the ecliptic and produces a vortex which would carry the planets in circular orbits around the sun were they not, like the earth, magnetic. Because of the interaction of the magnetic forces, the orbits become oval. As the emanating force diminishes with the distance from the sun Kepler assumes that the velocity of a planet varies inversely as the distance from the sun. The inverse distance law is incorrect but it nevertheless led Kepler to the area law.

Based on Tycho Brahe's observations, Kepler resolved the apparent motion of Mars around the earth into the orbits of Mars and the earth around the sun. Kepler began his model building with two incorrect assumptions: (1) that the orbit of Mars is circular with the sun displaced from the center; and (2) that the inverse distance law holds. In his application of the inverse distance law to find the angular velocity of Mars around the center of the circular orbit, a problem which really requires calculus for its solution, he made various mathematical approximations, which luckily led him to the area law. He proceeded to use the area law to compute Mars' orbit and then found deviations from the observations up to 8'. Since he knew that Tycho's observations were more accurate than that, he discarded the circular hypothesis and turned to an egg-shaped oval instead. After much computational work, he finally found that an ellipse gave the best fit.

Kepler published his results in the *Astronomia Nova* (1609), which contains his first two laws. They were based on his results for Mars only, but he postulated that the same laws were valid for all the planets.

Kepler makes the *real* sun the center of the planetary system and thus draws the full consequence of the Copernican hypothesis. Copernicus had used the *mean* sun as center, but Kepler states that the mean sun is just a point dependent on the earth's orbit, and a physical explanation of the motion requires a real body in the center.

Ignoring the small eccentricity of the sun, Kepler's first-order model assumes that all the planets (including the earth) move with uniform speed around the real sun. The angular motion is described by the linear relation $\varphi = 2\pi f t + \varphi_0$, so that the period of revolution equals $T = 1/f$. The parameters are f, φ_0, and R, the radius of the circle. The first-order models of Kepler and Copernicus are thus the same, except for the substitution of the real sun for the mean sun.

In the second-order model, the circle is transformed to an ellipse, and the uniform speed is changed to a variable speed by means of the area law. The reader should draw a figure of Kepler's model based on an ellipse with major axis of length $2R$ ending in the aphelium A and with S in the focus farthest from A. This transformation requires two new parameters, the eccentricity and the longitude of the major axis λ_A. According to the area law the planet P moves such that the area, M, say, of the segment ASP of the ellipse increases at a constant rate, which leads to the equation

$$M(t) = \pi R^2 f(t - t_A),$$

where t_A, the time for P's passage through A, may be expressed by f and φ_0. Hence, the area law does not introduce new parameters. The second-order model contains five estimable parameters and leads to an astoundingly small maximum error of $10'$.

A comparison of the three models is shown in Table 10.4.1.

Table 10.4.1 Number of Parameters and Maximum Error for Models of the Motion of Mars

| Model | Longitudes | | | Latitudes | Totals |
	First order	Second order	Third order		
Ptolemy	$(3, 0, 2)^a\,52°$	$(6, 2, 3)\,13°$	$(9, 3, 5)\,4°$	$(7, 4, 3)\,2°$	$(16, 7, 8)$
Copernicus	$(3, 0, 3)\ \ 13°$	$(8, 2, 6)\,4°$		$(8, 5, 3)\,2°$	$(16, 7, 9)$
Kepler	$(3, 0, 3)\ \ 13°$	$(5, 0, 5)\,10'$		$(2, 0, 2)\,15'$	$(7, 0, 7)$

[a]The three numbers in the parentheses are the number of parameters generating the model, the number of restrictions, and the number of estimable parameters, respectively.

Source: Riddell (1980).

Each line in Table 10.4.1 shows that the increasing complexity of the model, expressed by the increasing number of parameters, improves the goodness of fit. Comparison of the Ptolemaic and Copernican models on the one hand with Kepler's on the other shows that an even greater improvement is obtained with a smaller number of parameters by means of a fundamental change of the mathematical form of the model. Table 10.4.1 strikingly demonstrates the simplicity and the strength of Kepler's model compared with previous models. Furthermore, Kepler's model treats all the planets in the same way in relation to the sun.

In one respect, however, Kepler's model is more involved than the other two, namely, with respect to the computation of the heliocentric longitude of P. To find the angle ASP corresponding to the area $M(t)$, it is necessary to solve Kepler's equation

$$\varphi + \frac{e}{R} \sin \varphi = 2\pi f(t - t_A),$$

with respect to φ for any given t. Kepler prepared tables from which the solution may be obtained by interpolation.

In his indefatigable search for mathematical regularities, Kepler also investigated the relationship between T and R for the five planets, as indicated earlier by Copernicus. By trial and error, Kepler succeeded in finding his third law. It takes great skill to find a nonlinear relation between two variables without a plot of the observations. Furthermore, it requires good intuition and great luck when it turns out that this empirical relation, based on five observations, is identical with the theoretical one later derived from a general theory about the phenomena observed. Kepler gave his result in the *Harmonice Mundi* (The Harmony of the World), 1619, as shown in Table 10.4.2.

Table 10.4.2. Mean Planetary Periods T and Distances R Relative to the Earth

Planet	T	$T^{2/3}$	R
Mercury	245,714	392	388
Venus	612,220	721	724
Earth	1,000,000	1000	1000
Mars	1,878,483	1523	1524
Jupiter	11,877,400	5206	5200
Saturn	29,539,960	9556	9510

Source: Kepler (1619), *Harmonice Mundi*, Ges. Werke, Bd. 6, pp. 311 and 358.

No wonder that Kepler ecstatically wrote about the mathematical harmony of the Divine Design, and like the Pythagoreans he found a correspondence between musical harmonies and the velocities of the planets which he transformed into a "music of the spheres."

In 1627 Kepler published the *Rudolphine Tables* containing tables and formulae for the computation of the positions of the heavenly bodies based on Tycho's observations and his own mathematical model.

Kepler's laws are empirical in the sense that they represent the observations with an accuracy corresponding to the observational error, but they are not deduced from axioms based on a physical theory of motion. The solution of this problem was given by Newton in his *Principia* (1687).

The story told above reveals some principles for empirical model building of importance for any scientist and particularly for statisticians as professional model builders. The Gauss linear model, which is the most important tool of the statistician, may be compared to Ptolemy's model in the sense that it is flexible almost without limit; the introduction of more and more parameters leads to a good description but not necessarily to an understanding of the fundamental relations between the quantities involved.

For a statistician, Kepler's principles of model building are instructive and valuable even today. They may be summarized as follows:

1. Model building is a sequential process proceeding from simpler to more complex models.
2. The complexity of a model depends both on its mathematical form and the number of parameters.
3. Models should be simple, aesthetic, and harmonious in mathematical respects.
4. Models should be based on a physical theory about the phenomena.
5. The ultimate criterion for the choice among models is the agreement with the observations, the goodness of fit.
6. These principles should interact on the various stages of the model-building process. This means that a many-parameter model built on one mathematical form may be superseded by a model with fewer parameters based upon another mathematical form.

For supplementary reading we refer to Kuhn (1957), Wilson (1968), and Aiton (1969).

10.5 PROBLEMS

1. *Regression Analysis of Galileo's Data.* In the following table we have given Galileo's data except for the outliers 1a and 12 and observation 5, which is a copy of 4. Furthermore, we have replaced the three observations given by astronomers 2 and 9 by their means:

	Altitude (degrees)		
Observer	x	y	z
1	55.97	27.75	84.00
2	52.40	24.36	80.46
3	51.90	23.55	79.93
4	51.30	23.05	79.50
6	51.17	22.67	79.33
7	50.83	—	79.75
8	49.40	22	79
9	48.37	20.16	76.57
10	48.37	20.25	—
11	39.50	11.50	67.50

Suppose that $y_i = \alpha_1 + \beta_1 x_i + \epsilon_{1i}$ and $z_i = \alpha_2 + \beta_2 x_i + \epsilon_{2i}$, where the ϵ's are independent random variables with mean zero. It follows that

$$p_{ij} = y_i - y_j - (x_i - x_j) = (\beta_1 - 1)(x_i - x_j) + (\epsilon_{1i} - \epsilon_{1j}),$$

an analogous expression being valid for the z's. Galileo's hypothesis may therefore be tested by testing the hypotheses $\beta_1 = 1$ and $\beta_2 = 1$.

Show that the two slopes are not significantly different; estimate the common slope and show that it does not differ significantly from 1. Setting $\beta_1 = \beta_2 = 1$ in the model above, show that the means and variances of $x - y$ and $z - x$ do not differ significantly. Estimate the common mean and variance and interpret the result. Test the linearity of the regression line.

The nonlinear regression for $r = 32$ has been discussed by Hald (1986).

2. Find the two distributions of p_{ij} and calculate the mean, the median, and $\sum |p_{ij}|$.

3. Consider Galileo's distribution of $n = n_1 + n_2 + n_3$ parallaxes, where n_1 is the number of negative parallaxes, n_2 the number of p's such that $0 \leqslant p_{ij} < c_{ij}$, and n_3 the number of p's such that $p_{ij} \geqslant c_{ij}$. Let

$$\sum = \sum_1 + \sum_2 + \sum_3$$

be the corresponding sums.

Prove that $A_0 < A_c$ if and only if

$$2 \sum_2 p_{ij} < \sum_2 c_{ij} + \sum_1 c_{ij} - \sum_3 c_{ij}$$

and give an interpretation of this criterion.

4. Plot Kepler's values of (R, T) on double-logarithmic paper and fit a straight line to the points. Having found a slope slightly different from $\frac{3}{2}$, would you dare to round it to $\frac{3}{2}$ and announce the result as a natural law for planetary motion?

CHAPTER 11

The Newtonian Revolution in Mathematics and Science

11.1 INTRODUCTION

One would have imagined that the publication in 1657 of Huygens' treatise on probability as part of van Schooten's widely circulated text on mathematics would have caused many mathematicians to take up this new subject. That this did not happen was perhaps due to the fact that the mathematical world was fully occupied with developing the calculus. Furthermore, probability theory did not yet have any important application in science, only in life insurance.

The leading scientists at the time were Huygens, Newton, and Leibniz. All three exerted a strong influence on the four leading probabilists in the beginning of the 18th century, James and Nicholas Bernoulli, Montmort, and de Moivre, whose works we shall discuss in detail in the following chapters. As a background we shall first sketch some features of the Newtonian revolution in mathematics and science.

Karl Pearson (1926) writes about the influence of Newton:

De Moivre expanded the Newtonian theology and directed statistics into the new channel down which it flowed for nearly a century. The causes which led de Moivre to his "Approximatio" or Bayes to his theorem were more theological and sociological than purely mathematical, and until one recognises that the post-Newtonian English mathematicians were more influenced by Newton's theology than by his mathematics, the history of science in the 18th century—in particular that of the scientists who were members of the Royal Society—must remain obscure.

This statement is, of course, extreme and provocative. It is true that theological considerations and beliefs played a large role in the general philosophical outlook at Newton's time, but the development of mathematics and science, the creation of new scientific institutions furthering communication among scientists, and the impact of society on science were even more important.

In the beginning of the century, religious intolerance combined with economic and political interests led to persecutions and wars; such as the Thirty Years' War (1618–1638) in Germany, the Civil Wars in the 1640s in England, and the persecution of the Huguenots in France that culminated in the Revocation of the Edict of Nantes in 1685. Many refugees found a new home in the Dutch Republic. After the Revolution of 1688 in England, religious freedom was guaranteed for the dominant Protestant sects by the Act of Toleration. Parallel to this political development ran a philosophical discussion of the relation of religion to science.

For centuries the Christian religion had dominated the natural sciences but during the 17th century this situation gradually changed. The advancing knowledge of natural phenomena left smaller room for explanation by means of Divine providence and miracles. The importance of Descartes' mechanical philosophy in this respect is evident. In England most of the scientists were devout Protestants who considered the study of nature to be part of their religious duty to reveal the harmony of the world as created by God. From the order and wonders observed in nature, scientists concluded that the world had been created by a supreme being, thus inferring the existence of God. They wrote of God as the supreme clockmaker and "very well skilled in mechanics and geometry;" they referred to nature as God's handiwork and considered the Book of Nature comparable to the Book of Relevations. Gradually, their attempts to provide a rational foundation of the Christian religion led them to a natural or rational religion, with little room left for the original spiritual side of Christianity. An analysis of the relation of science to religion in 17th century England has been given by Westfall (1958), a more comprehensive study is due to Shapiro (1983). The discussion on natural theology, probability, and statistics still goes on, as can be seen from contributions of Bartholomew (1984, 1988).

An essential improvement in communication among scientists occurred in the latter part of the 17th century through the creation of scientific societies and journals. Most societies began as informal private gatherings of scientists and gradually developed into formal societies or academies. In Rome the Accademia dei Lincei (The Academy of the Lynx-like) had only a short life in the beginning of the century. The Accademia del Cimento (The Academy of Experiments) founded in 1657 in Florence with the purpose of continuing the experimental science of Galileo and Torricelli and financed by the Grand Duke of Tuscany had to give up after about ten years. These first attempts were, however, soon to be followed by the creation of national academies which still exist today. The Royal Society of London for Improving Natural Knowledge was founded in 1660 by a group of scientists and other interested individuals; the Royal Society was independent of government subsidies. In Paris the Académie Royale des Sciences, founded in 1666, consisted of a few members selected and salaried by the government. The Berlin Academy in 1700 and the St. Petersburg Academy in 1724 followed the French model, whereas the American Philosophical Society in 1769 followed the English. In most of the smaller European countries, academies of science were founded during the first half of the 18th century.

The well-established science of astronomy obtained state-supported observatories, such as the Observatoire de Paris in 1669 and the Royal Observatory at Greenwich in 1675.

The academies were influenced by Bacon's ideas on collective research and the usefulness of science. The program of the Royal Society was to improve natural knowledge by data collection and by experiments and thereby to create a natural philosophy. During the early years of the Society, the meetings were largely occupied with the performance and discussion of experiments, a task neglected by the universities. The Society also sponsored and published treatises and books on natural philosophy, the most famous being Newton's *Principia* in 1687.

Of great importance for the dissemination of knowledge and new results was the publication of journals by the societies. Previously, new results had been communicated by letter to small groups of fellow scientists. To secure wider circulation a few persons had established themselves as intermediaries who received manuscripts and copied them for the information of their correspondents. This practice was very useful but liable to produce misunderstandings and priority disputes. It continued for some time after the appearance of the first journals.

In 1665 the Royal Society began to publish the *Philosophical Transactions*; the French *Journal de Sçavans* began the same year. About the same time the Paris Academy began the publication of several journals. The *Acta Eruditorum* was founded by Leibniz in 1682.

Hence, by the end of the 17th century an international community of mathematicians and natural scientists had been established.

11.2   THE NEWTONIAN REVOLUTION

Isaac Newton (1642–1727) entered Trinity College at Cambridge University in 1661. He taught himself mathematics by reading van Schooten's second edition of Descartes' *Geometry*, van Schootens' textbooks, and the works of

Vieta and Wallis. He also studied astronomy and physics, in particular, the theory of motion and the theory of light.

Because of the plague the university was closed for about two years, 1665–1667, which Newton spent in his family home at Woolsthorpe, Lincolnshire, continuing his studies, but now also producing extraordinary results of his own. Returning to Cambridge he got his master's degree, and in 1669 he succeeded Barrow as professor of mathematics, a post he held until 1696 when he became warden and in 1699, master of the Mint in London. He was president of the Royal Society from 1703 until his death.

After having mastered mathematics, astronomy, and optics, about 1670 he turned with equal fervor to the more difficult subjects of alchemy and theology. For about ten years he devoted most of his time to these topics, although he occasionally returned to mathematics and optics to rewrite and complete some of his earlier papers.

For a man of Newton's intellect, theology had to be studied intensely and systematically from the original sources. He came to the conclusion that the Scriptures had been corrupted to support Trinitarianism from the time of Athanasius in the fourth century, and he became an Arian, rejecting the Trinity. His conviction, if publicized, would have cost him his position, and he therefore kept it secret. Another subject which he studied intensively was the prophesies, the Book of Daniel and the Book of Revelations, presenting new interpretations and comparing these with the historical facts.

After he moved to London in 1696 he again took up theological studies, which became one of his major occupations for the rest of his life. He left a large number of manuscripts at his death, and two of the less provocative were published as *The Chronology of Ancient Kingdoms Amended* (1728) and *Observations upon the Prophesies* (1733). The greater part of his manuscripts on theology were first studied in the 20th century.

Nearly all Newton's papers on mathematics and natural philosophy were completed and circulated as manuscripts at the last moment to prove his priority. This procedure unavoidably led to serious conflicts, the most notable being with Hooke, Flamsteed, and Leibniz.

Newton's contributions to the development of pure and applied mathematics, astronomy, optics, and the philosophy of science were revolutionary. The story of the Newtonian revolution has been told many times; the latest and most comprehensive version is by Westfall (1980). We shall comment only slightly on the calculus, celestial mechanics and the philosophy of science.

In 1665 Newton discovered the infinite series for the binomial with fractional exponent. This result was of fundamental importance for his first exposition of the theory of infinite series and the calculus, which was circulated in manuscript form in 1669 as *De Analysi per Aequationes Numero*

Terminorum Infinitas (On Analysis by Means of Equations with an Infinite Number of Terms); it was not published until 1711. Besides the binomial theorem, this paper contains the definition of the rate of change of a function as the limit of the difference quotient and the fundamental theorem of the calculus that integration and differentiation are inverse operations. It also contains many examples of derivatives, integrals, and infinite series for algebraic and transcendental functions.

In 1671 Newton wrote another exposition of his calculus, *Tractatus de Methodis Serierum et Fluxionum* (A Treatise of the Methods of Series and Fluxions); it was not published until 1736. Inspired by the theory of motion, he now considered his variables (fluents) as functions of time and introduced their rates of change under the name of fluxions. (He later introduced the dot notation for fluxions.) He also introduced derivatives of higher order (fluxions of fluxions) and partial derivatives, and he proved the rules for differentiation of products and quotients and of implicit functions. He applied the method of fluxions to finding tangents, maxima and minima, points of inflection, and curvatures. Besides finding the relation between the fluxions for a given relation between two fluents, he also studied the problem of finding the relation between the fluents given a relation between the fluxions; that is, he studied and solved differential equations.

The first published version of Newton's calculus was given in his main work *Philosophiae Naturalis Principia Mathematica* (The Mathematical Principles of Natural Philosophy), 1687, in which he applied the calculus to problems in physics and celestial mechanics. However, in most cases the proofs were given a geometrical formulation, following the tradition from Ptolemy to Kepler; only occasionally did he use the methods from *De Analysi* and *Methodis Fluxionum*.

Gottfried Wilhelm Leibniz (1646–1716) developed his form of the calculus in 1675 and published it for the first time in 1684, i.e., before the publication of the *Principia*. Leibniz was aware of Newton's previous work but did not refer to Newton, which led to the famous priority dispute. Leibniz introduced the notation that is still in use: dx, dy/dx, and the integral sign as an elongated S for summa. Much of Leibniz' work was incomplete, but it inspired James and John Bernoulli to write a large number of papers that contributed essentially to the clarification and the rapid development of the calculus on the Continent.

After having worked on infinite series and the calculus, Newton turned to studying the theory of motion in the period 1665–1666. For uniform circular motion he found that the force by which a body endeavors to recede from the center is proportional to the square of the velocity and inversely proportional to the radius; that is $f \propto v^2/r$. As usual, he did not publish his result. Huygens found the same result independently and published it in 1673, calling the force a centrifugal force.

According to Newton's own account written many years later, he combined his formula for the centrifugal force with Kepler's third law, according to which T^2/r^3 is constant, where T denotes the time of revolution. Since $v = 2\pi r/T$ he found that $f \propto 1/r^2$, the inverse square law. "I deduced that the forces which keep the Planets in their orbs must be reciprocally as their distances from the centers about which they revolve", see Berry, (1898, p. 212).

This is the first result on the way to Newton's law of gravitation, which asserts that every particle of matter in the universe attracts every other particle with a force varying directly as the product of their masses and inversely as the square of the distance between them.

Provoked by an impending priority dispute with Robert Hooke (1635–1702) and challenged by a question by Halley, Newton wrote a nine-page manuscript *De Motu Corporum in Gyrum* (On the Motion of Bodies in an Orbit) in 1684. Urged on by Halley, Newton wrote the *Principia* in about two year's time between 1684 and 1687. Deeply impressed by Newton's results Halley, with unique magnanimity, devoted his time and money to the publication of the *Principia*. With the help of Roger Cotes (1682–1716), a thoroughly revised edition was published in 1713; a third edition with a few alterations was published in 1726.

At the time the overriding problem in celestial mechanics was the derivation of Kepler's laws from fundamental physical principles on motion. This is just one of the many problems solved by Newton in the *Principia*. Based on three axioms, today called Newton's three laws of motion, he builds up a theory of motion of point masses in a central force field. As a special case, he shows that Kepler's laws hold for the motion of a single point mass under the attraction of a central force. [As pointed out by Riddell (1980), we may thus add a line to Table 10.4.1 containing Newton's model with a total parameter index of (6, 0, 6).] For two or more bodies, Newton proves that Kepler's laws must be modified due to perturbations caused by interaction of the bodies involved; however, because of observational errors, Kepler (fortunately) did not notice these perturbations.

After having developed a mathematical theory of motion, Newton compares the theory with the astronomical observations. He verifies Kepler's third law from three sets of observations: for the satellites of Jupiter, for the satellites of Saturn, and for the planets around the sun, respectively. For the third case, he uses Kepler's data, see Table 10.4.2, and a similar set of data due to the astronomer Boulliau. Newton calculates the expected distance from the sun for each planet as function of the periodic time by means of Kepler's law and compares the calculated and the observed distances; he notes that the differences are insignificant but does not carry out a statistical analysis.

Having thus shown that the motion of the heavenly bodies conforms to mathematical laws, Newton shows that the force of attraction used in the mathematical theory may be identified with the force of gravity. The proof is the famous earth–moon test of the law of gravitation in which Newton demonstrates that the acceleration of the moon toward the earth equals the acceleration of a body at the surface of the earth due to gravity when the inverse square law is taken into account.

Finally, Newton gradually constructs his "system of the world" by combining experiments, observations, and theory to explain a large number of astronomical phenomena, such as the masses and densities of the planets, the axes and eccentricities of the orbits, the perturbations caused by the interactions of several planets, the shape of the earth, the dependence of weight on the position of the earth, the motion of the moon, the tides, and the orbits of comets.

Newton proves that Descartes' explanation of the motion of the planets by means of vortices is irreconcilable with Kepler's third law. With regard to his own explanation by means of universal gravitation, he remarks that "to us it is enough that gravity does really exist, and act according to laws which we have explained, and abundantly serves to account for all the motions of the celestial bodies, and our sea."

Newton was also an ingenious experimenter who designed and carried out many experiments of great accuracy, not merely the pendulum experiments reported in the *Principia*, but many others reported in his *Opticks* (1704), among them his "crucial experiment" proving the heterogeneity of light.

With the *Principia* Newton created a model for the construction and exposition of scientific theories that has been followed ever since. The idea of explaining a large number of seemingly unconnected facts from a few simple laws, and after verification using these laws to predict and explain further facts, came to exert a profound influence on the development of science and philosophy. Newton characterized this method as "The Method of Analysis and Synthesis;" today it is called the inductive–deductive method. The "Newtonian style" has been discussed by Cohen (1980).

11.3 NEWTON'S INTERPOLATION FORMULA

Consider a function and its successive forward differences defined as

$$\Delta f(x) = f(x + 1) - f(x),$$
$$\Delta^k f(x) = \Delta^{k-1} f(x + 1) - \Delta^{k-1} f(x), \qquad k = 2, 3, \dots.$$

In an unpublished manuscript from about 1611, Thomas Harriot

(1560–1621), mathematician and astronomer at Oxford, constructed a table of differences for a function tabulated at unit intervals and used it to find the values of $f(x)$ for more finely spaced arguments by interpolation. In modern notation Harriot's interpolation formula may be written

$$f(a + h) = f(a) + \binom{h}{1}\Delta f(a) + \binom{h}{2}\Delta^2 f(a) + \cdots + \binom{h}{5}\Delta^5 f(a).$$

Setting $h = \frac{1}{5}, \frac{2}{5}, \frac{3}{5}$, and $\frac{4}{5}$, Harriot used this formula to subtabulate $f(x)$ from a to $a + 1$. Of course he also used the simpler form of the formula obtained under the assumption that differences of order higher than 2, say, are zero. Briggs (1624) also used this method of quadratic interpolation to subtabulate $\log x$ to intervals of one-tenth the original tabulation. Without proof Briggs gave a method for constructing the difference table for unit intervals from a given difference table for five-unit intervals, a method that has become the standard method of subtabulation. A more detailed account of the early history of interpolation has been given by Goldstine (1977).

About 1670 it was generally known that the nth difference of a polynomial of the nth degree is constant, and several mathematicians developed formulae for polynomial interpolation similar to the one found by Harriot and Briggs. Most of Newton's work on interpolation was carried out during 1675–1676 but not published until 1687 in *Principia*, Book 3, Lemma 5, where he states the interpolation problem as follows: "To find a parabolic curve which shall pass through any given points." Without proof he gives what today are called the Gregory–Newton interpolation formula and Newton's interpolation formula with divided differences. The proofs were not published until 1711 in his *Methodus Differentialis*.

During the decade before his work on interpolation Newton had studied the theory of infinite series, and it was therefore natural for him to represent the given function as a power series

$$f(A + x) = a + bx + cx^2 + dx^3 + \cdots,$$

say. If the series converges it will often be sufficiently accurate for practical purposes to take only a finite number of terms into account, in particular if A is chosen such that x is small.

Hence, if $n + 1$ values of a function $f(x)$ are given, we may write $f(x) = f_n(x) + r_n(x)$, where $f_n(x)$ is a polynomial of the nth degree to be determined from the given values, and $r_n(x)$ denotes the remainder term. The problem is to devise a practical method for calculating $f_n(x)$ for any given value of x in the interval defined by the $n + 1$ given arguments and also to find an upper bound for $r_n(x)$. We shall sketch Newton's solution as given in the *Methodus Differentialis*.

Let $f(x)$ be given for $x = a_0, a_1, \ldots, a_n$, and define the divided differences of f as

$$f(a_0, a_1) = \frac{f(a_1) - f(a_0)}{a_1 - a_0},$$

$$f(a_0, a_1, \ldots, a_k, a_{k+1}) = \frac{f(a_1, \ldots, a_{k+1}) - f(a_0, \ldots, a_k)}{a_{k+1} - a_0}, \qquad k = 1, 2, \ldots, ;$$

see also the arrangement in the following table:

Table of Divided Differences According to Newton

x	$f(x)$	$f(.,.)$	$f(.,.,.)$	$f(.,.,.,.)$
a_0	$f(a_0)$			
		$f(a_0, a_1)$		
a_1	$f(a_1)$		$f(a_0, a_1, a_2)$	
		$f(a_1, a_2)$		$f(a_0, a_1, a_2, a_3)$
a_2	$f(a_2)$		$f(a_1, a_2, a_3)$	
		$f(a_2, a_3)$		
a_3	$f(a_3)$			

The name "divided difference" and the notation used here are of a later date than Newton's.

Setting $f(x) = a + bx + cx^2 + \cdots$, Newton first proves that the first differences between the values of the function are divisible by the differences between the arguments and that the differences between the divided differences of the first order are divisible by the differences between every second argument, and so on. Assuming that $f(x)$ is a polynomial of the fourth degree, he presents a table of divided differences containing such results as

$$f(a_0, a_1) = b + c(a_0 + a_1) + d(a_0^2 + a_0 a_1 + a_1^2)$$
$$+ e(a_0^3 + a_0^2 a_1 + a_0 a_1^2 + a_1^3),$$

$$f(a_0, a_1, a_2) = c + d(a_0 + a_1 + a_2) + e(a_0^2 + a_0 a_1 + a_1^2 + a_0 a_2 + a_1 a_2 + a_2^2),$$

$$f(a_0, a_1, a_2, a_3) = d + e(a_0 + a_1 + a_2 + a_3),$$

$$f(a_0, a_1, a_2, a_3, a_4) = e.$$

He points out that the last of the divided differences will be equal to the last of the coefficients in the polynomial and that the remaining coefficients may be found by means of the remaining divided differences by working backward in the difference table, which is obvious from the formulae given above. He concludes that when these coefficients are determined, the curve passing through the end-points of the ordinates has been found.

Newton does not write the polynomial explicitly in terms of the divided differences as we do today; however, if we follow his directions and begin with the divided difference of order $n + 1$, we have by definition that

$$f(x, a_0, a_1, \ldots, a_{n-1}) = f(a_0, a_1, \ldots, a_n) + (x - a_n)f(x, a_0, a_1, \ldots, a_n).$$

Using this for backward recursion, we end with

$$f(x, a_0) = f(a_0, a_1) + (x - a_1)f(x, a_0, a_1),$$
$$f(x) = f(a_0) + (x - a_0)f(x, a_0).$$

By substitution we obtain the identity

$$f(x) = f(a_0) + (x - a_0)f(a_0, a_1) + (x - a_0)(x - a_1)f(a_0, a_1, a_2) + \cdots$$
$$+ (x - a_0)(x - a_1)\cdots(x - a_{n-1})f(a_0, a_1, \ldots, a_n)$$
$$+ (x - a_0)(x - a_1)\cdots(x - a_n)f(x, a_0, a_1, \ldots, a_n),$$

which is Newton's interpolation formula in modern notation. The first $n + 1$ terms give the interpolation polynomial, and the last term is the remainder. If $f(x)$ is a polynomial of order n, the remainder disappears.

In the special case with equidistant arguments, Newton introduces the more convenient central differences, $\delta f(x) = f(x + \frac{1}{2}) - f(x - \frac{1}{2})$, instead of divided differences. He leaves it to the reader to express the coefficients in terms of these differences; following the same procedure as above, the resulting formulae are known today as the Newton–Stirling and the Newton–Bessel interpolation formulae for n odd and even, respectively.

Let $f(x)$ be given for $x = a + ih, i = 0, 1, \ldots, n$, and set $(x - a)/h = y$. It follows from Newton's interpolation formula that

$$f_n(x) = f_n(a + yh) = f(a) + \binom{y}{1}\Delta f(a) + \binom{y}{2}\Delta^2 f(a) + \cdots + \binom{y}{n}\Delta^n f(a),$$

where $\Delta f(a) = f(a + h) - f(a)$. This result was found independently by James Gregory (1638–1675) in 1670 and by Newton in 1676. It is, of course, just a slight generalization of Harriot's formula.

Newton points out that the area under any curve may be found approximately by integration of the corresponding interpolation polynomial; as an example he gives the three-eighths rule

$$\int_0^3 f(x)\, dx \cong \frac{3}{8}[f(0) + 3f(1) + 3f(2) + f(3)].$$

Newton's idea was taken up by Cotes who approximated the integral by a linear combination of $n + 1$ equally spaced values of the integrand. In 1722 he published the coefficients in these Newton–Cotes formulae for $n = 2$, $4, \ldots, 10$; they were much used by de Moivre and Simpson to approximate integrals in probability theory and insurance mathematics.

In an unpublished manuscript Newton gave another version of his interpolation formula using "adjusted differences," the adjusted difference of order n being equal to $n!$ times the divided difference. The resulting formula becomes directly comparable to the formulae with equidistant arguments, see the detailed discussion by Fraser (1927).

Newton presents many examples of the evaluation of the remainder term by considering the maximum value of the first neglected term of the series.

Newton clearly states that the interpolation polynomial may be given many forms depending on the various applications. He also hints at a more general theory of interpolation using functions other than polynomials.

Fraser (1928, p. 71) evaluates Newton's work as follows: "Modern workers have struggled up to the level reached by Newton, and the twentieth century will no doubt see extensions and developments of the subject of interpolation beyond the boundaries marked by Newton 250 years ago." This statement is correct if we add the qualification "polynomial" to interpolation. The essential progress made since Newton consists in investigations of the remainder term and "optimum" spacings of the arguments.

Fraser has, however, overlooked one important invention from 1906, namely, Thiele's interpolation formula with "reciprocal differences," which uses rational functions instead of polynomials. Other types of functions, among them trigonometric functions, have also been used. In the 1930s the extension of polynomial interpolation to osculatory interpolation was much discussed, and this later developed into spline interpolation.

The history of Newton's interpolation formula has been discussed by Fraser (1928) and by Whiteside in his edition of Newton's mathematical papers.

After having stated his general interpolation formula in the *Principia*, Newton continues with a lemma entitled, "Certain observed places of a comet being given, to find the place of the same at any intermediate given time." He does not give a numerical example; he only mentions that if five observations are given, the interpolation formula may be used to find a polynomial of the fourth degree through the given points. Neither does he give numerical examples in his other publications on interpolation. However, in *Newton's Mathematical Papers*, 1976, Vol. 7, 682–687, Whiteside reports and comments upon a numerical example found on one of Newton's worksheets from 1695. As shown in the following table the data consist of five observations on the path of a comet, the longitude of the fourth observation having been chosen as origin:

Observations of the Longitude x and Latitude $f(x)$ (in Degrees) of a Comet at Five Dates Between December 21, 1680 and January 13, 1681[a]

x	$f(x)$	$10f_1$	$10^2 f_2$	$10^3 f_3$	$10^4 f_4$
−63.6922	21.75833				
		2.2589197			
−40.4178	27.01583		−0.388216		
		0.6089636		0.00192787	
−21.1912	28.18666		−0.375937		0.000011127
		−0.9104911		0.00201785	
0.0000	26.25722		−0.364316		
		−2.3081702			
17.1733	22.29333				

[a]The divided difference of order i is denoted by f_i.

Newton writes the result in a form which is equivalent to

$$f(x) = f_0 + f_1 p_1(x) + \cdots + f_4 p_4(x),$$

where the f's are the descending differences in the table, and

$$p_i(x) = (x - a_0)(x - a_1) \cdots (x - a_{i-1}), \quad i = 1, 2, \ldots.$$

As a check on the calculation of the difference table he uses the formula to find $f(a_4)$, which is found to agree with the tabular value. Newton's goal is to find the slope of the path of the comet for $x = a_2$. Noting that $p_i = (x - a_{i-1})p_{i-1}$ so that

$$p_i' = p_{i-1} + (x - a_{i-1})p_{i-1}',$$

he obtains a simple expression for $f'(x)$, and, finally, he calculates $f'(a_2) = -0.01218842$. (Actually, Newton finds the result 0.01216933 because of a small error in his divided difference of the fourth order.)

Newton does not discuss the observational errors of the data, nor does he discuss the accuracy of the result obtained. He seems not to distinguish between interpolation on a mathematically given function and observational data, like many other scientists in the following time. From a statistical point of view Newton's method is unsatisfactory; the reader should analyze Newton's data, taking both the astronomical model and observational error into account.

About 1670 Newton developed an iterative method for finding the root of an equation based on previous work by Vieta. The method was further improved by Joseph Raphson (1648–1715) in 1690, and is now known as the Newton–Raphson method. Newton's basic idea was to linearize the solution of polynomial equations by setting $x^n = (a + \epsilon)^n \cong a^n + na^{n-1}\epsilon$, where a is a first approximation to the root, and then solve for ϵ. Obviously this corresponds to using the first two terms of Taylor's expansion. In general, let a_i be an approximation to the root of $f(x) = 0$. By means of Taylor's formula we have

$$f(a_i) + (x - a_i)f'(a_i) \cong 0.$$

Solving for x, we get a new approximation, a_{i+1}, say, of the form

$$a_{i+1} = a_i - \frac{f(a_i)}{f'(a_i)}, \qquad i = 1, 2, \ldots,$$

which is the modern formulation of the Newton–Raphson formula. The method and its generalization to functions of several variables has many important applications, for example, solving Kepler's equation, finding the interest rate in equations involving annuities, and solving the likelihood equations in statistical estimation problems.

The conditions of convergence were not discussed until much later.

CHAPTER 12

Miscellaneous Contributions between 1657 and 1708

The Reader may here observe the Force of Numbers, which can be successfully applied, even to those things, which one would imagine are subject to no Rules. There are very few things which we know, which are not capable of being reduc'd to a Mathematical Reasoning; and when they cannot, it's a sign our Knowldege of them is very small and confused;.... I believe the Calculation of the Quantity of Probability might be improved to a very useful and pleasant Speculation, and applied to a great many events which are accidental, besides those of Games.

—JOHN ARBUTHNOTT, 1692

12.1 PUBLICATION OF WORKS FROM BEFORE 1657

In the period between the publication of Huygens' treatise in 1657 and Montmort's essay in 1708, some important works from before 1657 were published for the first time.

In 1660 the Dutch version of van Schooten's *Mathematical Exercises* containing Huygens' treatise was published, see §6.1. Cardano's *Liber de Ludo Aleae* was published in 1663 in Volume 1 of his collected works, see §4.3. Pascal's treatise on the arithmetic triangle and its uses was published in 1665, see §5.1. Pascal's letters to Fermat in 1654 were published in the 1679 edition of Fermat's works, see §5.1.

12.2 NEW CONTRIBUTIONS PUBLISHED BETWEEN 1657 AND 1708

The contributions are ordered chronologically.

Leibniz' *Dissertatio de Arte Combinatoria* was published in 1666. It is more a treatise on symbolic logic than on combinatorics, and Todhunter (p. 32) remarks that "The mathematical treatment of the subject of combinations is far inferior to that given by Pascal."

According to Todhunter (pp. 44–46), a textbook on mathematics, *Mathesis Biceps*, by John Caramuel was published in 1670. It contains a section on combinatorics giving well-known results; a reprint of Huygens' treatise, attributed to the Danish astronomer Longomontanus; and some attempts of Caramuel to solve elementary problems in games of chance which, however, do not go beyond the results in Huygens' treatise.

The French mathematician Joseph Saveur gave in *Journal des Sçavans*, 1679, some formulae without proofs for the advantage of the banker at Bassette, which at the time was a fashionable card game, see Todhunter (pp. 46–47).

According to Stigler (1988) Thomas Strode published *A Short Treatise of the Combinations, Elections, Permutations and Composition of Quantities; Illustrated by Several Examples, with a New Speculation of The Differences of the Powers of Numbers* in London 1678. Strode wrote in his preface that he had found many combinatorial results without knowing Pascal's work, but during the printing of his book he got hold of Pascal's treatise and took the opportunity to make some revisions and additions. Besides the properties of the binomial coefficients he found the result that the number of permutations of n elements consisting of k classes of r_i identical elements in the ith class equals $n!/r_1! \cdots r_k!$, where $r_1 + \cdots + r_k = n$. He also tabulated the distribution of the total number of points thrown by two, three, or four dice, his derivation being the same as that used by Cardano and Galileo for two and three dice. Generalizing to n dice each having f faces, he proved that the number of chances for getting a total of s points equals $\binom{s-1}{n-1}$ for $s < n + f$. It seems that all of Strode's results were given previously by Mersenne (1636), see §5.2.

The *Algebra* by Wallis from 1685 contains a section on elementary combinatorics (see Todhunter, pp. 34–36; and David, 1962).

In 1685 James Bernoulli proposed the following two problems for solution in the *Journal des Sçavans*:

1. Players A and B play with a die on the condition that he who first throws an ace wins. Player A throws once, then B throws once; thereafter A throws two times in succession and then B two times; then A throws three times in succession and B also three times, and so on.

2. Alternatively, A throws once, then B two times in succession, then A three times, then B four times, and so on until one of them wins. What is the ratio of their chances?

The two problems are generalizations of Huygens' Problem 1.

After having waited in vain for a solution for five years, Bernoulli finally published his own solution without proof in the *Acta Eruditorum*, 1690, giving A's probability of winning as the infinite series

$$P_1 = 1 + q^2 + q^6 + q^{12} + q^{20} + \cdots - q - q^4 - q^9 - q^{16} - \cdots,$$

$$P_2 = 1 + q^3 + q^{10} + q^{21} + q^{36} + \cdots - q - q^6 - q^{15} - q^{28} - \cdots,$$

where $q = \frac{5}{6}$. Bernoulli's paper is important because here for the first time we have the solution of a probability problem expressed as an infinite series. This presumably induced Montmort (1708, pp. 157–158) and de Moivre (1712, pp. 233–234) to develop the same method of solution of Huygens' Problem 1, a method they also used on many other occasions. Bernoulli gave the proof in *Ars Conjectandi* (1713, pp. 54–60); we shall return to this in §§14.2 and 15.3.

Later in 1690 Leibniz published a solution in *Acta Eruditorum* in the form

$$P_1 = (1 - q)(1 + q^2 + q^3 + q^6 + q^7 + q^8 + q^{12} + q^{13} + q^{14} + q^{15} + \cdots),$$

which he reduced to Bernoulli's result, and he gave a similar solution of the second problem. Both solutions have been discussed by Biermann (1957) and Mora-Charles (1986).

An English translation of Huygens' treatise by John Arbuthnott was published in 1692 under the title *Of the Laws of Chance, or, a Method of Calculation of the Hazards of Game, Plainly demonstrated, And applied to Games as present most in Use*. Besides the translation it contains a preface and some examples of the application of Huygens' method to simple card and dice games, see Todhunter (pp. 48–53). Most interesting is the preface from which we shall give some quotations showing that Arbuthnott had similar ideas about the definition and applicability of probability as later expressed by James Bernoulli (1713):

> It is impossible for a Die, with such determin'd force and direction, not to fall on such a determin'd side, only I don't know the force and direction which makes it fall on such a determin'd side, and therefore I call that Chance, which is nothing but want of art;... .

Further,

> as I have hinted already, all the Politicks in the World, are nothing else but a kind of Analysis of the Quantity of Probability in casual Events, and a good

Politician signifies no more, but one who is dextrous at such Calculations; only the Principles which are made use of in the Solution of such Problems, can't be studied in a Closet, but acquir'd by the Observation of Mankind.

We have quoted Arbuthnott's great expectations for the force of mathematical reasoning in the epigraph. Finally, he writes

There is likewise a Calculation of the Quantity of Probability founded on Experience, to be made use of in Wagers about any thing; it is odds, if a Woman is with Child, but it shall be a Boy; and if you would know the just odds, you must consider the Proportion in the Bills that the Males bear to the Females.. . .

He returned to this problem in a paper 1712, see §17.1. He also gives other examples of relative frequencies from population statistics, which clearly shows the influence of Graunt's book and testifies to his awareness of the usefulness of probability theory in this field of application.

In 1699 the Scottish mathematician John Craig published a short book *Theologiae Christianae Principia Mathematica* (Mathematical Principles of Christian Theology) on the credibility of human testimony. Many critical comments and interpretations indicate the difficulty of understanding his theory, which nevertheless inspired many of the great French probabilists in their works on a theory of testimony. The following is based on the comments by K. Pearson (1978, pp. 465–467). Pearson first praises Craig for his originality in devising a mathematical theory on this subject. He stresses the soundness of Craig's ideas that the reliability of testimonies decreases with the number of successive witnesses, the lapse of time, and the distance from the occurrence of the event under investigation. Craig's formula for the credibility of the historical evidence of an event gives the probability as

$$P = x + (m-1)s + \frac{kT^2}{t^2} + \frac{qD^2}{d^2},$$

where x is "the probability that the first witness transmits"; m the number of witnesses in the chain; and s the suspicion created by a transfer; t the interval elapsed; d the distance of the event; and k, T, q, and D constants. The suspicion is given a negative value. Pearson remarks that Craig's formula "is perfectly arbitrary, as are also Craig's numerical values for the constants." Pearson also objects to the formula because the three determinants of the probability enter additively instead of multiplicatively. Craig uses his formula to deduce that 1454 years after the writing of his book, all written testimony for the life of Christ will have disappeared in terms of credibility.

Referring to Craig, but developing his own formula by the usual probabilistic arguments, Laplace uses the decreasing evidence for historical events to demolish the conclusion of Pascal's wager. The eternal happy life (or the infinite number of happy lives) promised by the Scriptures does not have a finite probability as assumed by Pascal. In his *Essai* (1814), Laplace writes, "The expression for the probability of their testimony thus becomes infinitely small. Multiplying it by the infinite number of happy lives promised, infinity disappears from the product that expresses the expectation resulting from this promise, which destroys the argument of Pascal." Pearson (1978, pp. 677–678) presents further details of Laplace's calculations.

After having been subjected to heavy criticism for nearly three centuries, Stigler (1986) attempts to rehabilitate Craig by giving a completely new and surprising interpretation of his theory. Stigler points out that there did not exist a unique definition of probability at the time and that Craig's probability concept differs from that used by Huygens and others measured on a scale from 0 to 1. Stigler suggests that Craig's P equals the likelihood ratio of the present available evidence in relation to the historical hypothesis. Furthermore, Stigler considers t and d to be fixed units of time and distance and uses T and D as variables which lead to a log likelihood ratio as a linear function of m and a quadratic function of T and D. We refer the reader to Stigler's paper for further details.

A comparison of Stigler's testimony with that of Pearson and his predecessors offers a striking example of the difficulty of forming a mathematical theory of the credibility of testimony.

Another paper on the same topic entitled *A Calculation of the Credibility of Human Testimony* was published anonymously in *Phil. Trans.* 1700. According to Stigler (1988) the author is Bishop George Hooper. The author speaks about "moral certitude absolute" and "moral certitude incompleat," which has "several Degrees to be estimated by the Proportion it bears to the Absolute." A similar terminology was later employed by J. Bernoulli (1713). According to K. Pearson (1978, pp. 466–469), the author derives the probability that a report is true when it is transmitted by single successive reporters who are equally credible. He also finds the probability that at least one report is true among several independent concurrent testimonies. Finally, he discusses the truth of either oral or written transmission over time. In all cases he solves the problems by means of the multiplication theorem for probabilities of independent events. The results of this paper have been discussed by Shafer (1978) and Zabell (1988).

It should be noted that the account given above is based on secondary sources; the reader who wants to be sure of the interpretations should therefore read the originals himself.

Let us digress for a moment to the realm of philosophy and consider the discussion on the concept of (subjective) probability by John Locke in his

Essay Concerning Human Understanding (1690), Book 4, Chapter 15, Of Probability and Chapter 16, Of the Degree of Assent, from which the following sentences have been pieced together:

Probability is the appearance of agreement upon fallible proofs. It is to supply the want of knowledge. Being that which makes us presume things to be true before we know them to be so. The grounds of probability are two; conformity with our own experience, or the testimony of other's experience. Probability, then being to supply the defect of our knowledge, and to guide us where that fails, is always conversant about propositions whereof we have no certainty, but only some inducement to receive them for true. The grounds of it are, in short, these two following: First, The conformity of any thing with our own knowledge, observation and experience. Secondly, The testimony of others, vouching their observation and experience. In the testimony of others, is to be considered, (1) The number. (2) The integrity. (3) The skill of the witnesses. (4) The design of the author, where it is the testimony out of a book cited. (5) The consistence of the parts and circumstances of the relation. (6) Contrary testimonies.... Probability is either matter-of-fact or speculation.

By "matter-of-fact," Locke is thinking mainly of natural laws for which no causal explanation has been given. He writes, "These probabilities rise so near to certainty that they govern our thoughts as absolutely, and influence all our actions as fully, as the most evident demonstration; and in what concerns us, we make little or no difference between them and certain knowledge. Our belief thus grounded rises to assurance."

On testimony, Locke writes, "Unquestionable testimony and experience for the most part produce confidence. Fair testimony, and the nature of the thing indifferent, produce also confident belief. Experience and testimonies clashing, infinitely vary the degrees of probability. Traditional testimonies, the farther removed, the less their proof." Presumably, Craig was influenced by Locke's philosophy of testimony and probability.

Locke does not give numerical values of probabilities, nor does he indicate a probability calculus.

How much of this development was common knowledge at the end of the period? This question is easily answered because Montmort in the preface to the second edition of his *Essay* (1713) wrote the first history of probability and statistics in which he covers everything that has been mentioned above, except for Arbuthnott's translation of Huygens' treatise and Strode's treatise. Montmort also mentions contributions to the theory of combinations by J. Prestet and A. Tacquet. Montmort states that the contributions of Pascal to combinatorics and probability theory are far superior to all the others and that his own book is a continuation of Pascal's work.

12.3   CONTRIBUTIONS DURING THE PERIOD PUBLISHED AFTER 1708

The most important unpublished works from the period in question are by Huygens and Bernoulli. We have discussed Huygens' work on probability after 1657 in §§6.3 and 6.4 and his use of Graunt's life table in §8.1. James Bernoulli's work on *Ars Conjectandi* can be followed through his mathematical diary, the *Meditationes*, which has been published in Volume 3 of Bernoulli's collected works (1975). We shall return to Bernoulli's work in Chapters 14–16.

Newton does not seem to have been interested in probability and statistics and did not make essential contributions to these fields. In an unpublished note from his time as a student he extends Huygens' Proposition 3 to the case with an irrational number of chances. He considers a ball dropped at random on a circle divided into two areas in the ratio $2: \sqrt{5}$ and gives the expectation as $(2a + b\sqrt{5})/(2 + \sqrt{5})$. He also mentions that examples of dice games may be generalized by using a die with unequal sides. In 1693 Samuel Pepys asked Newton to solve the following problem: What is the probability of throwing at least one six with 6 dice, at least two sixes with 12 dice and at least three sixes with 18 dice? Newton gave a simple algorithm for the solution corresponding to one, two, and three terms of the tail of the binomial with $p = \frac{1}{6}$. Pepys might have found similar problems with their solutions in Arbuthnott's translation (1692) of Huygens' treatise; Newton does not refer to Huygens' work.

We mentioned earlier Newton's use of his interpolation formula to fit a polynomial to observations, see §11.3. In his *Chronology* (1728), Newton presents a statistical analysis of the length of reigns of successive kings in many countries using the mean as estimator. Finding that the mean reign in most cases is about 20 years, he rejects historians' assertions, that rulers of some ancient kingdoms reigned about 40 years on average and explains that the discrepancy is due to confusion about generation and reign; he was thus able to reconstruct some ancient chronologies, see K. Pearson (1928) and Stigler (1977).

More details on Newton's work have been given by David (1962), Sheynin, (1971), and Gani (1981).

Leibniz did not make any direct contribution to probability theory, but in his unpublished manuscripts and his correspondence with other scientists, he had considerable influence on the philosophy of probability and the scope of its applications. Opinions differ, however; Hacking (1975) stresses his importance, whereas Schneider (1980, 1981, 1984) is more reserved. The unpublished notes on combinatorics by Leibniz have been discussed by Knobloch (1974).

12.4 A NOTE ON DATA ANALYSIS

In the period considered, no statistical theory emerged. We have already mentioned successful attempts to fit equations to data in astronomy and mechanics, but these results were obtained without using formal methods of statistical estimation and tests of goodness of fit. A much-needed survey of data analysis before 1750 has been given by Plackett (1988), who concludes,

> Sciences as diverse as mechanics, pneumatics and physiology employed models involving a single unknown parameter, estimated by selecting a single pair of measurements. Models were confirmed by the comparison of observed and theoretical values, residuals were calculated, and outliers noted. But the sciences considered were variously affected by limitations of experimental technique, and there was no systematic advance on procedures of data analysis.

A survey of the developments in political arithmetic has been presented by Westergaard (1932, Chapter 4).

CHAPTER 13

The Great Leap Forward, 1708–1718: A Survey

In 1708 he [Montmort] published his work on Chances, where with the courage of Columbus he revealed a new world to mathematicians.
—TODHUNTER, 1865

13.1 A LIST OF PUBLICATIONS

The beginnings of a theory of probability in the 1650s did not lead to the same rapid development as was the case for the calculus; for about 50 years, no essential new results were published. This period of stagnation ended in 1708 with the publication of Montmort's *Essay*. There followed a decade of hectic activity and competition, as attested by the following list of publications:

1708. P. R. de Montmort, *Essay d'Analyse sur les Jeux de Hazard*. Paris.

1709. Nicholas Bernoulli, *De Usu Artis Conjectandi in Jure*. Basel.

1712. J. Arbuthnott, An Argument for Divine Providence, taken from the constant Regularity observed in the Births of both Sexes. *Phil. Trans.* London.

1712. A. de Moivre, De Mensura Sortis. *Phil. Trans.* London.

1713. James Bernoulli, *Ars Conjectandi*. Basel.

1713. P. R. de Montmort, *Essay*. 2nd. ed. Paris.

1716. N. Struyck, *Uytreekening der Kanssen in het spelen, door de Arithmetica en Algebra, beneevens eene Verhandeling van Looterijen en Interest*. Amsterdam.

1717. Nicholas Bernoulli, Solutio Generalis Problematis XV propositi à
D. de Moivre, in tractatu de Mensura Sortis. *Phil. Trans.* London.
1717. A. de Moivre, Solutio generalis altera præcedentis Problematis. *Phil. Trans.* London.
1718. A. de Moivre, *The Doctrine of Chances.* London.

After the breakthrough of probability theory in this decade, the mathematical world had at its disposal four excellent textbooks in French, Latin, Dutch, and English, respectively. The three texts by Bernoulli, Montmort, and de Moivre contained so many new ideas, methods, and problems that it took nearly a century to digest and develop them. The second edition of the *Doctrine of Chances* (1738) became the standard text for the remaining part of the century and was superseded only in 1812 by Laplace's *Théorie Analytique des Probabilités.*

James Bernoulli died in 1705 and left a nearly finished manuscript of *Ars Conjectandi.* In a eulogy by Saurin (1706) a summary of the table of contents was published, but no indication of Bernoulli's proofs was given. According to Kohli (1975a), it follows from Bernoulli's mathematical diary that he had found the main results before 1690, but he was still working on his book when he died. Due to quarrels within the Bernoulli family, *Ars Conjectandi* was not published until 1713.

The assignment of priority to the many new definitions, methods, theorems, problems, and solutions is therefore a difficult matter. Many of James Bernoulli's results were found independently by Montmort (1708) and de Moivre (1712) and published before Bernoulli's book (1713). Moreover, the second edition of Montmort's book (1713) contains a correspondence between Montmort and Nicholas Bernoulli, a nephew of James, with many new results of great importance for the following development.

13.2 METHODS AND RESULTS

We shall now present a survey of the methods used and the results obtained by James Bernoulli, Montmort, Nicholas Bernoulli, and de Moivre. In the following chapters we shall study some of their results in detail.

Notations and Definitions

As did their forerunners, they defined probability as the number of favorable cases (chances) divided by the number of equally likely cases (chances), and they considered only discrete probability spaces. Probabilities were written

$a/(a + b), c/(c + d)$, etc., and only occasionally did they introduce a single letter symbol for a probability. The formulae were therefore cumbersome to read and write, the more so because indices not yet were in common use.

The classical probability concept based on symmetry considerations in pure games of chance proved to be insufficient in games where the outcome also depends on the skill of the players. James Bernoulli therefore introduced a distinction between probabilities which could be calculated *a priori* (deductively, from symmetry) and *a posteriori* (inductively, from relative frequencies). A posteriori probabilities were also used in demography and insurance.

Bernoulli also introduced a new and fundamentally different concept of probability, a *subjective* probability as a measure of the degree of belief in a statement or a proposition about things or events, however, this concept only became important in probability theory in the latter part of the century.

Elementary Rules of the Calculus of Probability

The addition and multiplication rules were derived from the classical definition of probability by the same arguments as those used today for a finite probability space, which means that the same rules hold for probabilities and relative frequencies. The authors clearly distinguished between independent and dependent events and illustrated this distinction by drawings with and without replacement from a pack of cards or from an urn with differently colored balls, or by comparison of dice and card games. The multiplication theorem was usually formulated for independent events only and then used with the adequate conditional probabilities for dependent events.

Methods of Solution

The classical method of enumeration of favorable and equally likely cases by combinatorial methods, as used by Pascal and Fermat, was developed much further. Each of the three books contains a section on combinatorics.

Beginning with simple examples, complete and, more often, incomplete induction was used when direct proofs were too difficult.

Recursion by means of difference equations, the method used by Pascal and Huygens, was used to solve more complicated problems. In many cases only a numerical solution was obtained, but in some cases they succeeded in solving the difference equation by ad hoc methods. Finally, de Moivre found more general results by means of his theory for the summation of recurring series.

The method of inclusion and exclusion was used as a matter of course to solve many complicated problems. It was used to derive the compound probability theorem for exchangeable events by de Moivre in 1718.

To find the probability of winning in an infinite number of independent and periodic trials Huygens' analytic method was used.

Using the newly developed theory of infinite series, the probability of winning at each trial of a possibly infinite number of trials was found, and the total probability of winning was obtained by summation.

Conditional probabilities and expectations were used to analyze two-stage games of chance.

Generating functions or corresponding algorithms were invented; however, de Moivre's proof in which a generating function was used for the first time was not published until 1730.

The first continuous distribution, the uniform distribution, was introduced as a limiting case of a discrete distribution, and by means of the calculus the expectation of the largest observation in a sample was found by Nicholas Bernoulli.

Limiting processes were used to derive the Poisson approximation and the law of large numbers for a binomially distributed random variable.

Problems Solved

The binomial and the hypergeometric distributions and the corresponding multivariate distributions were derived by combinatorial methods, and the waiting time distributions were found.

The general solution of the classical problem of points was obtained, and various generalizations to bowl games and tennis were formulated and solved.

Huygens' five problems were solved by several methods.

The distribution of the total number of points obtained by throwing any given number of dice was found and tabulated.

The binomial equation $B(c, n, p) = \frac{1}{2}$ was solved by means of the Poisson approximation, and np was tabulated as function of c.

The expectation of the player in a large number of popular dice games, card games, and lotteries was calculated.

The problem of coincidences (matches, rencontres) was formulated and solved.

Various versions of the occupancy problem for a given number of balls distributed at random in a given number of cells, formulated as a problem in dicing, were solved.

The probability distribution for the duration of play in a circular tournament (Waldegrave's problem) was derived from the corresponding difference equation by Nicholas Bernoulli.

The probability distribution for the duration of play in the ruin problem, a random walk with absorbing barriers, was found in various ways by Nicholas Bernoulli and de Moivre.

In a discussion with Montmort and Nicholas Bernoulli, Waldegrave found the optimal mixed strategy in the game Her by using the minimax principle; they did not, however, clearly recognize this as a fundamental new principle.

James Bernoulli proved his fundamental theorem on the convergence of a binomially distributed relative frequency to the true value (the weak law of large numbers), and as part of his theorem he found an upper bound for the tail probability of the binomial distribution. Sharpening this proof, Nicholas Bernoulli found an approximation to the tail probability, which he used to test the stability of the sex ratio based on Arbuthnott's data.

The problems were formulated in terms of games of chance. Today we know that many of these problems are of great significance outside this context and they have therefore become standard problems in textbooks on probability.

James Bernoulli and Montmort died rather young, and Nicholas Bernoulli did not have much time for research in probability theory once he became professor of law. It therefore became the fate of de Moivre to fulfill the work so splendidly begun in this decade. This he did in the second edition of the *Doctrine of Chances* (1738), the third edition (1756), and numerous editions of his *Annuities on Lives* from 1725 onward.

CHAPTER 14

New Solutions to Old Problems, 1708–1718

14.1 THE PROBLEM OF POINTS

The problem of points for players of equal skill was solved by Pascal and Fermat in 1654, as related in §5.3. We shall discuss here the solutions for players of unequal skill using notation similar to that in §5.3.

Montmort (1708, pp. 165–178) reprints Pascal's letter to Fermat of 24 August 1654, and on this basis he discusses the problem of points for two and three players without progressing further than Pascal and Fermat. James Bernoulli (1713, pp. 106–110) likewise refers to the Pascal–Fermat correspondence and discusses the combinatorial solution. They both give the formula (5.3.4) for two players and do not attempt to solve the problem for players of different skill.

Montmort sent a copy of his *Essay* to John Bernoulli, who responded with a letter of 17 March 1710, reprinted in Montmort (1713, pp. 283–298). John Bernoulli remarks that the solution of the problem of points for any value of p is obtained from the expansion of $(p + q)^{a+b-1}$, the sum of the last b terms giving A's probability of winning, and the sum of the first a terms B's probability of winning. He gives the solution as

$$e(a, b) = \sum_{i=a}^{a+b-1} \binom{a+b-1}{i} p^i q^{a+b-1-i}, \tag{1}$$

which is the generalization of (5.3.4) obtained by replacing $(\frac{1}{2})^{a+b-1}$ by $p^i q^{a+b-1-i}$.

The same solution is given independently by de Moivre (1712 and 1718, Problem 2).

In 1713 (pp. 244–246), Montmort added three pages to his previous discussion. He first gives John Bernoulli's solution. Next he uses the binomial to derive a new form of the solution. Beginning with A's probability of getting a successes in n trials,

$$\binom{n}{a} p^a q^{n-a},$$

he notes that A's probability of getting the ath success at the $(n+1)$st trial is

$$\binom{n+1}{a} p^a q^{n+1-a} - q \binom{n}{a} p^a q^{n-a} = \binom{n}{a-1} p^a q^{n+1-a}.$$

Setting $n + 1 = a + i, i = 0, 1, \ldots, b - 1$, and summing he gets

$$e(a, b) = p^a \sum_{i=0}^{b-1} \binom{a-1+i}{a-1} q^i, \tag{2}$$

which is the generalization of (5.3.9) with $(\frac{1}{2})^{a+i}$ replaced by $p^a q^i$. To prove that the two expressions for $e(a, b)$ give the same result, Montmort multiplies q^i in (2) by $(q + p)^{b-1-i}$, and expanding the binomial he reduces the resulting double sum to the form (1). He gives the proof only for $a = 3$ and $b = 5$.

Today, the distribution

$$\binom{a-1+i}{a-1} p^a q^i, \qquad i = 0, 1, \ldots, \tag{3}$$

is called the *binomial waiting-time distribution*. It is usually derived directly as the probability of getting $a - 1$ successes in $a - 1 + i$ trials, times the probability of getting a success in the next trial. It is also called the *negative binomial distribution* because it may be found by expansion of $p^a(1 - q)^{-a}$.

In an example with three players of equal skill, Montmort (1708) first gives the solution based on the multinomial distribution and next the corresponding solution based on waiting times, without expressing his results in formulae. However, this was done independently by Montmort (1713, pp. 242, 353, 371) and de Moivre (1712 and 1718, Problem 8).

Let k players, $A_1, \ldots, A_k$, need points $a_1, \ldots, a_k$, respectively, to win, $\sum a_i = n$, say, and let $p_1, \ldots, p_k, \sum p_i = 1$, respresent the skill of the players. Since the play will end after at most $n + 1 - k$ games, the solution may be obtained from the multinomial

$$(p_1 + \cdots + p_k)^{n+1-k} = \sum C(x_1, \ldots, x_k) p_1^{x_1} \cdots p_k^{x_k},$$

where $x_1 + \cdots + x_k = n + 1 - k$. If $x_1 \geq a_1$ and all the other x's are smaller than the corresponding a's, then the term in question is part of P_1, A_1's probability of winning. For the remaining terms the coefficient C has to be split up between the players, depending on the *permutations* of the $n + 1 - k$ p's. If p_1 occurs at least a_1 times *before* p_2 occurs a_2 or more times and so on, then the term belongs to P_1. It will be seen that this procedure is identical to the one explained by Fermat in his letter to Pascal of 25 September 1654. As remarked by Montmort the work of counting the number of permutations belonging to each of the players makes the method unpractical for large values of $\{a_i\}$.

The solution was given a more practical form by de Moivre (1730; 1738, Problems 6 and 69; 1756, Problem 6) by means of the waiting-time argument, which leads to

$$P_1 = \sum \frac{(a_1 - 1 + x_2 + \cdots + x_k)!}{(a_1 - 1)! \, x_2! \cdots x_k!} \, p_1^{a_1} \, p_2^{x_2} \cdots p_k^{x_k}, \tag{4}$$

where the summation extends over all integer values of x_i from 0 to $a_i - 1$ for $i = 2, \ldots, k$. For $k = 2$, the solution becomes identical to Montmort's, given by (2).

14.2 SOLUTIONS OF HUYGENS' FIVE PROBLEMS

Huygens' five problems came to play an important part in the development of probability theory because all the leading probabilists felt obliged to solve them and to generalize them. They used the same two methods that Huygens used, the method of recursion and Huygens' analytical method from his Proposition 14. Further, they introduced infinite series, as shown by Bernoulli in 1690; combinatorial methods; explicit solution of a difference equation; and a martingale argument. We shall discuss the solutions given by James Bernoulli (1713, pp. 49–71, 144–149), Montmort (1708, pp. 156–165; 1713, pp. 216–223), Nicholas Bernoulli in Montmort (1713, pp. 309–311), de Moivre (1712, Problems 9, 11–14; 1718, Problems 7, 10, 11, 20), and Struyck (1716, pp. 32–45, 46–48, 58–64, 90–96, 108–116). Huygens' own solutions were discussed in §6.3.

Huygens' First Problem

The order of the players in A BB AA BB AA, and so on. The probability of winning in a single trial is p_1 for A and p_2 for B, $p_1 + q_1 = p_2 + q_2 = 1$.

Spinoza (1687), James Bernoulli, and Montmort all give the solution by means of Huygens' analytical method.

Bernoulli also gives the solution as an infinite series, using the same reasoning as in his commentary to the analogous problem in Huygens' Proposition 14. He gives A's probability of winning as

$$P_A = p_1 + q_1 q_2^2 p_1 + q_1^2 q_2^2 p_1 + q_1^3 q_2^4 p_1 + q_1^4 q_2^4 p_1 + \cdots.$$

Taking every second term together, Bernoulli gets two geometric series with the common quotient $q_1^2 q_2^2$, which leads to the sum

$$P_A = \frac{p_1(1 + q_1 q_2^2)}{1 - q_1^2 q_2^2},$$

in agreement with the result previously found. Similarly, Bernoulli finds P_B and notes that $P_A + P_B = 1$. De Moivre and Struyck derive the same result by essentially the same method.

Montmort does not give the infinite series for Huygens' problem but turns to the more difficult problem in which the order of the players is A BB AA BBB AAA, etc. He gives the solution as

$$P_A = p_1 + q_1 q_2^2 p_1 + q_1^2 q_2^2 p_1 + q_1^3 q_2^5 p_1 + q_1^4 q_2^5 p_1 + q_1^5 q_2^5 p_1 + \cdots,$$

and remarks that it is difficult to find the sum. For $p_1 = p_2 = p$, he gets the special result

$$P_A = 1 - q + q^3 - q^5 + q^8 - q^{11} + \cdots,$$

which is obviously the solution of a problem analogous to the two posed by Bernoulli in 1685 and solved by him in 1690, see §12.2. Montmort does not refer to Bernoulli in 1708; however, in the preface to the 1713 edition of his book, he mentions the two papers by Bernoulli.

Bernoulli stresses that his method may also be used for nonperiodic cases, and as an example he considers the following play:

Player	A	B	A	B	$\cdots$
Number of trials	k_1	k_2	k_3	k_4	$\cdots$
Cumulative number	m_1	m_2	m_3	m_4	$\cdots$

He assumes that the probability of success in a single trial is p for both A and B and that the play ends as soon as success occurs. He notes that the probability of winning in a series of k trials will contain the expression

$$p + qp + q^2p + \cdots + q^{k-1}p = 1 - q^k,$$

so that

$$P_A = 1 - q^{k_1} + q^{m_2}(1 - q^{k_3}) + q^{m_4}(1 - q^{k_5}) + \cdots$$
$$= 1 - q^{m_1} + q^{m_2} - q^{m_3} + q^{m_4} - \cdots.$$

This result is also given by Montmort in 1713. Bernoulli's two results from 1690 are special cases of the formula above.

Struyck, who had the advantage of reading the books by Bernoulli and Montmort, made a further generalization by considering the following play:

Player	A	B	C	A	B	C	A	B	C	⋯
Number of trials	a	b	c	d	e	f	d	e	f	⋯

The play is periodic after the first three turns. We therefore introduce the quotient

$$r = \frac{q_1^a q_2^b q_3^c}{1 - q_1^d q_2^e q_3^f}.$$

Using Bernoulli's method, Struyck finds

$$P_A = 1 - q_1^a + (1 - q_1^d)r,$$
$$P_B = q_1^a - q_1^a q_2^b + (q_1^d - q_1^d q_2^e)r,$$
$$P_C = q_1^a q_2^b - q_1^a q_2^b q_3^c + (q_1^d q_2^e - q_1^d q_2^e q_3^f)r,$$

which contain all the previous results. Struyck remarks that the extension of the formulae to more than three players follows easily from the simple structure of the expressions. He gives six examples, among them the solution to Huygens' first problem and the solutions to Bernoulli's two problems from 1685.

Huygens' Second Problem

The order of the players is ABC ABC, etc., and they draw chips from a bowl containing a white and b black chips; the one who first draws a white chip wins.

James Bernoulli points out that the problem may be interpreted in three different ways: (1) The chips may be replaced in the bowl after each drawing; (2) the drawings are without replacement; (3) each player has a bowl with a white and b black chips from which he draws. Further, more than three players may be introduced. We shall consider the first two interpretations and leave the solution of the third to the reader.

For drawings with replacement, Bernoulli and Montmort first give Huygens' analytical solution. Montmort remarks that the solution may be obtained as an infinite series but does not give the result. This is done by Bernoulli, de Moivre, and Struyck, who state that the probability of winning in the next drawing after k unsuccessful drawings equals

$$e_{k+1} = \left(\frac{b}{n}\right)^k \frac{a}{n}, \qquad n = a+b, \quad k = 0, 1, \dots .$$

By summation of every third term they get for $a/n = p$,

$$P_A = \frac{p}{1-q^3}, \qquad P_B = qP_A, \qquad P_C = q^2 P_A, \qquad P_A + P_B + P_C = 1.$$

For drawings without replacement they find similarly,

$$e_{k+1} = \frac{b^{(k)}}{n^{(k)}} \frac{a}{n-k}, \qquad k = 0, 1, \dots, b,$$

from which they get the probabilities required by summation.

Huygens' Third Problem

James Bernoulli first solves the problem by recursion, like Huygens. He then remarks, like Montmort and Struyck, that the natural way of solving this problem is by combinatorics and gives the solution as $10^4/\binom{40}{4}$.

Huygens' Fourth Problem

All four authors solve the fourth problem by combinatorics and give the solution as

$$\binom{4}{3}\binom{8}{4}\bigg/\binom{12}{7}.$$

Bernoulli and de Moivre also give the general hypergeometric probability

$$\binom{a}{k}\binom{n-a}{m-k} \Big/ \binom{n}{m}.$$

They also remark that the probability of getting at least k white chips is obtained by summation of hypergeometric probabilities.

Huygens' Fifth Problem

Let A have m counters and B n counters at the beginning of the play, and let their probabilities of winning in each trial be p and $q = 1 - p$, respectively. After each trial the winner gets a counter from the loser, and the play continues until one of the players is ruined. What is the probability of A being ruined?

This is the general formulation of the Gamber's Ruin problem as given by James Bernoulli and de Moivre; Huygens' fifth problem corresponds to $m = n = 12$.

We shall first discuss the problem for $m = n$. Huygens solved it for $m = 1, 2, 3, 4, 6$ and concluded by analogy that the general solution is $P_A:P_B = p^m:q^m$, where P_A denotes the probability that A wins, see §6.3. In a letter to Huygens in 1665, Hudde solved the problem for $m = 1, 2, 3$ by the same method as Huygens, see Huygens' *Oeuvres*, Vol. 5, pp. 470–471.

In the *Meditationes* Bernoulli presents a numerical solution based on the recursion formula $e(x) = [5e(x + 1) + 9e(x - 1)]/14$, where $e(x)$ denotes A's expectation when he has x counters, $x = 1, 2, \ldots, 23, e(0) = 0$, and $e(24) = 1$. By substitution he solves the equations successively and finds that the ratio of $e(12)$ and $1 - e(12)$ agrees with the answer given by Huygens. Bernoulli did not publish this procedure in the *Ars Conjectandi*; independently, Montmort (1708, pp. 162–165; 1713, pp. 222–223) used the same method.

In *Ars Conjectandi* Bernoulli considers the general recursion formula

$$e(x) = pe(x + 1) + qe(x - 1), \quad x = 1, 2, \ldots, 2m - 1, \quad e(0) = 0, \quad e(2m) = 1.$$

He solves the equations for $m = 2$ and 3, and then he states that the ratio of the expectations of A and B is as $p^m:q^m$, which may be shown by induction. Bernoulli's proof is thus as incomplete as the proofs given by Huygens and Hudde.

Finally, Bernoulli gives the solution of the general problem as

$$P_A:P_B = (p^n q^m - p^{m+n}):(q^{m+n} - p^n q^m) \quad \text{for} \quad m \neq n,$$

and as $m:n$ for $p = q$. He leaves the proof to the reader without any indication of his method.

A complete proof based on the difference equation was first given by Struyck (1716, pp. 108–109). He finds the explicit solution of the difference equation first for $m = n$ and then for $m \neq n$. In modern notation his ideas may be expressed as follows. The difference equation

$$e(x) = pe(x + 1) + qe(x - 1), \qquad x = 1, 2, \dots, m + n - 1,$$
$$e(0) = 0, \quad e(m + n) = 1,$$

may be written as

$$pe(x + 1) = (p + q)e(x) - qe(x - 1),$$

so that

$$e(x + 1) - e(x) = \frac{q}{p}[e(x) - e(x - 1)],$$

which leads to

$$e(x + 1) - e(x) = \left(\frac{q}{p}\right)^x e(1),$$

and

$$e(m) = \sum_0^{m-1} [e(x + 1) - e(x)] = \frac{[1 - (q/p)^m]e(1)}{1 - (q/p)},$$

and an analogous expression for $e(m + n)$. Using $e(m + n) = 1$, Struyck gets the explicit solution

$$P_A = e(m) = \frac{1 - (q/p)^m}{1 - (q/p)^{m+n}},$$

and a similar expression for P_B, in agreement with the results previously given by de Moivre and Bernoulli.

The first solution published was that by de Moivre (1712, pp. 227–228; 1718, pp. 23–24) who used a completely different method. De Moivre's proof depends on an ingenious trick. Suppose that A's counters are numbered from 1 to m and B's from $m + 1$ to $m + n$ and that counter x is given the fictitious value $(q/p)^x$. Suppose further that in each trial only consecutively numbered counters are used such that in the first trial A's stake is $(q/p)^m$ and B's is $(q/p)^{m+1}$. If A wins he gets counter $m + 1$ from B, and in the next trial A's stake is therefore $(q/p)^{m+1}$ and B's is $(q/p)^{m+2}$ and so on. Hence, in any trial A's expectation will be $p(q/p)^{x+1} - q(q/p)^x = 0$, and B's expectation will similarly be 0. Since they have the same expectation in every trial, their total expectation must also be the same. However, the total expectation equals the probability of winning times the total amount won at the end of the play, so that

$$P_A\left[\left(\frac{q}{p}\right)^{m+1} + \cdots + \left(\frac{q}{p}\right)^{m+n}\right] = P_B\left[\frac{q}{p} + \cdots + \left(\frac{q}{p}\right)^m\right].$$

Assuming that $P_A + P_B = 1$, the solution is easily found. De Moivre remarks that "The supposition that any counter is to the following as p to q does not change the probabilities of winning; therefore supposing that the counters have the same value, the probabilities of winning will still be in that same ratio which we have determined."

In a letter of 26 February 1711 from Nicholas Bernoulli to Montmort, Nicholas found the probability that the play ends with the ruins of one of the players after at most x games, and he derived the solution of Huygens' problem by letting x tend to infinity. His letter is published in Montmort's *Essay* (1713, pp. 309–311), see §20.3.

For other discussion of the history of this problem we refer the reader to Thatcher (1957), Kohli (1975b), Edwards (1983), and Seneta (1983).

14.3 TO FIND THE NUMBER OF CHANCES OF THROWING s POINTS WITH n DICE, EACH HAVING ƒ FACES

This problem was solved for two and three ordinary dice by enumeration by Cardano and Galileo (see §§4.3 and 4.4) and by Huygens in his introductory remarks to Proposition 10. The enumeration becomes rather complicated for more than three dice and must have required a great deal of labor for Montmort (1708, pp. 141–143; 1713, pp. 203–205) who published a table of the distribution of points for two to nine ordinary dice. He did not disclose his method, but in a letter to John Bernoulli of 15 November 1710 (see Montmort, 1713, p. 307), he gave without proof the formula for finding the distribution for $f = 6$ in such a form that it is easy to generalize to any f. The general formula with a combinatorial proof was published by Montmort (1713, pp. 46–50). In the meantime, de Moivre (1712, pp. 220–22; 1718, pp. 17–19) had published the same formula without proof; he gave the proof by means of a generating function in his *Miscellanea Analytica* (1730, pp. 191–197). James Bernoulli (1713, pp. 23–25) gave an algorithm for finding the distribution and used it for tabulating the distribution for two to six ordinary dice. The same algorithm was given by Montmort (1713, pp. 51–55). We shall present these results, beginning with Bernoulli, continuing with Montmort and ending with de Moivre, although this was not the order of publication.

Besides the classical methods of combinatorics two new methods of proof were used, namely, the method of inclusion and exclusion and the method of generating functions.

From a statistical point of view, the problem is to find the distribution of the sum of n observations from a discrete rectangular distribution.

James Bernoulli's Algorithm

Bernoulli gives his algorithm in the form of a table where he introduces one die at a time. For one die there is one chance for each of the six points. Combining two dice there are $6^2 = 36$ chances to be distributed over the possible sums from 2 to 12. The distribution may be found as shown in Table 14.3.1.

As Bernoulli remarks, the method of construction is obvious; each face of the first die is successively combined with all the faces of the second die, and the 36 chances are distributed over the two-way table taking the sum of the points into account. By summation of the number of chances in each column, the distribution of the number of chances corresponding to the sum of points is found.

For three dice the number of chances is $6^3 = 216$, which should be

---

Table 14.3.1. Bernoulli's and Montmort's Algorithm for Finding the Number of Chances of Throwing a Given Number of Points with Two Dice

		2	3	4	5	6	7	8	9	10	11	12
		1	2	3	4	5	6					
1		1	1	1	1	1	1					
2			1	1	1	1	1	1				
3				1	1	1	1	1	1			
4					1	1	1	1	1	1		
5						1	1	1	1	1	1	
6							1	1	1	1	1	1
No. of chances		1	2	3	4	5	6	5	4	3	2	1

Sum of points for two dice (header spanning columns 2–12)

Table 14.3.2. Algorithm for Finding the Number of Chances of Throwing a Given Number of Points with Three Dice

	Sum of points for three dice							
	3	4	5	6	7	8	9	10
	2	3	4	5	6	7	8	9
1	1	2	3	4	5	6	5	4
2		1	2	3	4	5	6	5
3			1	2	3	4	5	6
4				1	2	3	4	5
5					1	2	3	4
6						1	2	3
No of chances	1	3	6	10	15	21	25	27

distributed over the possible sums from 3 to 18. Combining the distribution of the 36 chances for two dice with the six chances for the third die in a table analogous to Table 14.3.1 above, Bernoulli finds the result shown in Table 14.3.2.

We have abridged Bernoulli's table because the second half is symmetric with the first. In this manner Bernoulli derives the number of chances for up to six dice.

Montmort's Combinatorial Solution

Let x_i denote the number of points thrown by die i, $x_i = 1, 2, \ldots, f$. The total number of points then becomes

$$s_n = x_1 + \cdots + x_n, \qquad s_n = n, n+1, \ldots, nf.$$

Let the "reduced" number of points be

$$r_n = s_n - n + 1, \qquad r_n = 1, 2, \ldots, n(f-1) + 1.$$

Montmort's formula for the number of chances of getting s points may then be written as

$$
\begin{aligned}
N(s; n, f) &= \sum_{i=0}^{[(s-n)/f]} (-1)^i \binom{n}{i} \binom{s - if - 1}{r - if - 1} \\
&= \sum_{i=0}^{[(s-n)/f]} (-1)^i \binom{n}{i} \binom{s - if - 1}{n - 1}.
\end{aligned}
\tag{1}
$$

Montmort first proves that

$$N(s; n, f) = \binom{s-1}{r-1} = \binom{s-1}{n-1} \qquad \text{for } f \geqslant r = s - n + 1. \qquad (2)$$

He considers the cases for $n = 1, 2,$ and 3, and by complete enumeration he shows that the number of chances may be found by successive addition in the same way as the figurate numbers so that (2) holds.

In Table 14.3.3 we have given an abridged version of Montmort's table for $n = 3$. The table shows the formation of each sum by listing the values of the three x's in increasing order of magnitude. In counting the number of chances it is thus necessary to take the number of permutations, 1, 3, and 6 respectively, into account.

Table 14.3.3. **For Given Values of $s = x_1 + x_2 + x_3$, the Table Contains the Compositions (x_3, x_2, x_1) in Increasing Order of Magnitude and the Corresponding Number of Chances**

$s =$	3	4	5	6	7	8	9	10	11	12
	111	112	113	114	115	116	117	118	119	11, 10
			122	123	124	125	126	127	128	129
					133	134	135	136	137	138
							144	145	146	147
									155	156
				222	223	224	225	226	227	228
						233	234	235	236	237
								244	245	246
										255
							333	334	335	336
									344	345
										444
N_0	1	3	6	10	15	21	28	36	45	55
N_1					3	9	18	30	45	63
N_2									3	9
N	1	3	6	10	12	12	10	6	3	1

Source: Montmort (1713, p. 48).

The row indicated by N_0 shows the number of chances when there are *no restrictions on the x's*, so that for each s, the largest value of any x equals $s - 2$. For each s, the value of N_0 is obtained by adding the number of permutations of the compositions listed in the corresponding column.

The row indicated by N is obtained by *excluding* all compositions containing *one or more* numbers larger than 4, as indicated by the triangular domains above the step-lines. This means that *all the x's are restricted* to the numbers $1, \ldots, 4$, and N therefore gives the number of chances of throwing s points with 3 four-sided dice.

The problem is to prove that N may also be obtained from N_0 by applying the corrections N_1 and N_2, in agreement with (1), which shows that

$$N(s; 3, 4) = \binom{s-1}{2} - 3\binom{s-5}{2} + 3\binom{s-9}{2} = N_0 - N_1 + N_2. \tag{3}$$

It will be seen that N_1 is obtained simply by multiplying N_0 by 3 and moving four places to the right, and that, similarly, N_2 is obtained by multiplying N_0 by 3 and moving eight places to the right.

Let us first *restrict one and only one of the x's*, x_1, say, to take on the values 1 to 4. Considering $s = 12$, the compositions which should be excluded are shown in Table 14.3.4.

Table 14.3.4. The Compositions (x_3, x_2) Giving
$s = x_1 + x_2 + x_3 = 12$ for x_1 Larger than 4

x_1	10	9	8	7	6	5
$s - x_1$	2	3	4	5	6	7
	11	12	13	14	15	16
			22	23	24	25
					33	34
No. of chances	1	2	3	4	5	6

Montmort does not include a table exactly like this, but Table 14.3.4 is actually only part of his table for $n = 2$. It will be seen that the corresponding number of chances equals the figurate numbers of order 2, their sum being 21. This fact is explained by Montmort in the following way: To obtain 12 points with $x_1 > 4$ and (x_2, x_3) free from restrictions is obviously the same as to obtain $12 - 4 = 8$ points without any restrictions on the x's. However, it follows from (2) (see also Table 14.3.3) for $s = 8$, that the corresponding number of chances equals 21, namely, the figurate number of order 3, which, as shown in Table 14.3.4, is obtained as the sum of the figurate numbers of order 2. Since one die may be selected among the three in three ways, we have to deduct $3 \times 21 = 63$ if only one of the x's is restricted at a time (see N_1 in Table 14.3.3 for $s = 12$).

Comparing Tables 14.3.3 and 14.3.4, it will be seen that we have deducted too much; instead of 3×21 cases we should only have deducted 3×18 cases.

The difference is due to the compositions which contain *two* x-values larger than 4, namely, 156, 165, and 255. Correcting for this error we have to exclude the number of cases corresponding to 165, which is included in Table 14.3.4 but not in Table 14.3.3, so that the 21 should be reduced to 19, and we should further count 255 only once instead of twice, which reduces the 19 to 18. Hence, we have to add $3(1 + 2) = 9$, as shown in Table 14.3.3 as N_2 for $s = 12$. Montmort explains this result as follows: To obtain 12 points with $x_1 > 4$, $x_2 > 4$, and no restrictions on x_3 is the same as to obtain $12 - 2 \times 4 = 4$ points without any restrictions on the x's. It follows from (2) (see also Table 14.3.3 for $s = 4$) that the corresponding number of chances equals 3. Since the two x's under restriction may be chosen in three ways, we have to multiply by 3.

The proof given by Montmort follows the reasoning in the example. Montmort (1713, p. 50) writes,

Suppose that the dice are denoted by the letters A, B, C, D, E, etc. and that one of the dice, for example A, has all its face values larger than f; it is then evident that this die A together with the other dice B, C, D, etc. have to give, besides f points which with certainty are included on one of the faces of A, further $s - f$ points. Hence, one has to take the number of cases expressing the number of ways of getting $s - f$ points and multiply by the number of dice, because it may be either A or B or C, etc. which is chosen to have face values larger than f. But it may happen that we have deducted too much, namely in the cases where two dice have been chosen to have all their face values larger than f and it is thus necessary to add the number of cases of getting $s - 2f$ points multiplied by $n(n - 1)/2$, because two dice may be selected.

Montmort continues this way of reasoning up to $s - 4f$ points.

The above analysis of Montmort's proof is essentially due to Henny (1975).

It will be seen that Montmort uses what today is called *the method of inclusion and exclusion* in his proof. He does not single out this method in his reasoning but employs it as a matter of course; he also uses it in his proof of the number of coincidences, see §19.2

Neither Montmort nor de Moivre discusses the properties of the function $N(s; n, f)$ apart from noting its symmetry around $s = n(f + 1)/2$. Montmort gives a complete table of the terms of $N(s; 8, 6)$ from which the construction may be seen. The distribution is obviously a linear combination of $[(s - n)/f]$ polynomials of degree $n - 1$, a new term being added each time s increases by f.

As mentioned before, Montmort tabulates the distribution for $n = 2, 3, \ldots, 9$ and $f = 6$. He does not comment on the shape of the distribution for increasing values of n. Since this is one of the first numerical examples illustrating the central limit theorem (with hindsight)we have plotted four of Montmort's distributions in Fig. 14.3.1. We have not normalized the graphs as we would do today by taking the standard deviation into account.

De Moivre's Algebraic Solution. Generating Functions

De Moivre (1738, p. 37) writes, "Although, as I have said before, the Demonstration of this Lemma may be had from my *Miscellanea*; yet I have

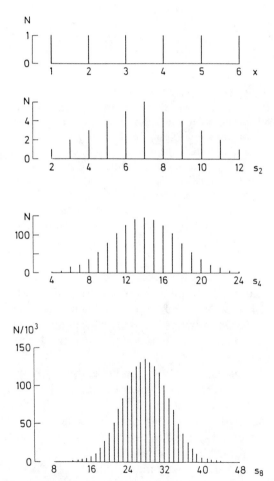

Fig. 14.3.1. Graphs of Montmort's distributions of sums of points by throwing one, two, four, and eight dice; N indicates the number of chances.

thought fit, at the desire of some Friends, to transfer it to this place." He then proceeds to prove (1) by the method of generating functions, as we would say today.

In the proof de Moivre considers a die with t faces marked 1, t^2 faces marked 2, and so on up to t^f faces marked f, so that the total number of faces equals $t + t^2 + \cdots + t^f$. Using the multiplication rule, the number of chances for the various outcomes of a throw with n dice is given by the terms of the product $(t + t^2 + \cdots + t^f)^n$. Since we are interested only in the sum of the points, we have to add all products with the same exponent, s, say, to find the number of chances of getting the sum s; that is, we have to find the coefficient of t^s, which de Moivre does as follows:

$$(t + t^2 + \cdots + t^f)^n = t^n(1 - t^f)^n(1 - t)^{-n}$$

$$= t^n \sum_{i=0}^{n} (-1)^i \binom{n}{i} t^{if} \sum_{j=0}^{\infty} \binom{n+j-1}{j} t^j. \qquad (4)$$

The coefficient of t^s is obtained by setting $if + j = s - n, i = 0, 1, \ldots, [(s - n)/f]$, which gives

$$\binom{n+j-1}{j} = \binom{s-if-1}{s-if-n} = \binom{s-if-1}{n-1},$$

so that the number of chances becomes $N(s; n, f)t^s$. Setting $t = 1$, the proof is completed.

It will be seen that $(t + t^2 + \cdots + t^f)^n$ for $n = 2$ and 3 and $t = 1$ is the algebraic equivalent of the algorithm given in Tables 14.3.1 and 14.3.2.

De Moivre's ingenious method was taken up by Simpson and Lagrange and used to derive the distribution of the sum of independent errors, each having a discrete probability distribution.

Laplace gave the method its name and its modern formulation. Consider a series of real numbers, $p_0, p_1, \ldots$, and the corresponding power series $g(t) = \sum p_x t^x$. If the power series converges in an interval for t, then $g(t)$ is called the generating function of the series $\{p_x\}$. Thus, there exists a one-to-one correspondence between the function $g(t)$ and the series $\{p_x\}$. If p_x denotes the probability that a random variable takes on the value x and $\sum p_x = 1$, then $g(t)$ converges for $|t| \leq 1$. Today t is considered a dummy variable and is not considered to be a number of chances (or a probability) as it was by de Moivre.

From this definition it is easy to prove that the generating function of s_n, the sum of n independent random variables, equals the product of the generating functions of the n components. The proof is essentially the same as that given by de Moivre. Let X and Y be independent random variables with probability distributions p_{1x} and p_{2y}, respectively, $(x, y) = 0, 1, \ldots$, and the generating functions $g_1(t)$ and $g_2(t)$. The probability that the random variable $Z = X + Y$ takes on the value z is

$$p_z = \sum_{x=0}^{z} p_{1x} p_{2,z-x}, \qquad z = 0, 1, \ldots, \tag{5}$$

and the generating function equals $g(t) = \sum p_z t^z$. It follows that

$$
\begin{aligned}
g_1(t) g_2(t) &= \sum_{x=0}^{\infty} p_{1x} t^x \sum_{y=0}^{\infty} p_{2y} t^y \\
&= \sum_{z=0}^{\infty} \sum_{x=0}^{z} p_{1x} p_{2,z-x} t^z = g(t).
\end{aligned}
$$

By induction it may be proved that

$$g_{s_n}(t) = g_1(t) \cdots g_n(t). \tag{6}$$

The early history of generating functions has been recounted by Seal (1949).

Generalizations

Montmort (1713, pp. 52–55) derives an algorithm for finding the number of chances of throwing s points with n dice having different numbers of faces, $f_1, \ldots, f_n$. This is an extension of the algorithm given in Tables 14.3.1 and 14.3.2. De Moivre does not consider this problem; however, it is obvious that the number of chances may be found as the coefficient of t^s in the expansion of the generating function

$$\prod_{i=1}^{n} (t + t^2 + \cdots + t^{f_i}),$$

which corresponds to Montmort's algorithm.

Montmort and Bernoulli consider two other interpretations (applications) of the algorithm, namely, to find the number of combinations with restricted repetitions and the number of divisors of a given integer.

Let $a_1, \ldots, a_n$ denote n different elements or n different prime numbers larger than 1. The number of m-combinations with repetitions wherein a_i is restricted to occur at most f_i times may be found by the algorithm above. Bernoulli (1713, p. 123) gives a numerical example similar to that given by Montmort.

Let the given number be

$$N = a_1^{f_1} \cdots a_n^{f_n}.$$

All divisors of N may be written in the same form as N with f_i replaced by d_i, say, $0 \leqslant d_i \leqslant f_i$ and $\sum d_i = d$. The algorithm then gives the number of divisors of dimension $d = s - n$, including the divisor 1 corresponding to $d = 0$.

Bernoulli (1713, p. 135) also derives an algorithm for finding the number of m-permutations with restricted repetitions. The equivalent generating function is

$$\prod_{i=1}^{n} \left(1 + t + \frac{t^2}{2!} + \cdots + \frac{t^{f_i}}{f_i!} \right)$$

and the number of m-permutations is found to be the coefficient of $t^m/m!$ in the expansion of the generating function.

Finally, Montmort (1713, pp. 59–62) considers the difficult problem of finding the number of chances for getting the sum s by drawing n cards without replacement from a pack of cards consisting of k suits with f cards numbered from 1 to f in each suit. He first gives an algorithm for solving this problem for a single suit, and then he shows by an example how to use this for three suits of ten cards each. He does not give a formula for the solution analogous to (1). A similar example is given by Bernoulli (1713, pp. 169–174).

14.4 TO FIND THE NUMBER OF TRIALS GIVING AN EVEN CHANCE OF GETTING AT LEAST c SUCCESSES. THE POISSON APPROXIMATION

The problem of finding the number of trials that gives an even chance of getting at least one success had been discussed by Cardano, de Méré, Pascal, Huygens, Bernoulli, and Montmort, and the equation $q^n = \frac{1}{2}$ had been solved, first numerically for some important cases and later in general by logarithms giving $n = (\ln 2)/(-\ln q)$.

Huygens and Bernoulli had also discussed the equation

$$q^n + npq^{n-1} = \frac{1}{2},$$

and obtained a numerical solution for $p = \frac{1}{6}$.

Furthermore, Bernoulli (1713, pp. 38–44) had considered the general problem to determine n such that $\Pr\{x \geqslant c\} = \Pr\{x \leqslant c - 1\}$, which means that the equation

$$\Pr\{x \leqslant c - 1\} = \sum_{x=0}^{c-1} \binom{n}{x} p^x q^{n-x} = \frac{1}{2} \tag{1}$$

has to be solved with respect to n for given c and p; however, Bernoulli did not attempt to solve this equation for c larger than 2.

Independently, de Moivre (1712, Problems 5–7) had formulated the same equation and given an approximate solution.

De Moivre's Solution for $p \to 0$ and $np/q = m$

De Moivre remarks that the solution for $p = \frac{1}{2}$ is $n = 2c - 1$. This follows from the symmetry of the binomial distribution for $p = \frac{1}{2}$.

De Moivre first considers the equation $q^n = \frac{1}{2}$ for $p \to 0$. Introducing $p/q = 1/r$, he writes the equation as

$$q^{-n} = \left(1 + \frac{1}{r}\right)^n = 1 + \binom{n}{1} r^{-1} + \binom{n}{2} r^{-2} + \cdots = 2. \tag{2}$$

For $p \to 0$ we have $r \to \infty$, and (2) shows that $n \to \infty$. Setting $n/r = m$, say, de Moivre finds

$$\left(1 + \frac{1}{r}\right)^n \to 1 + m + \frac{m^2}{2!} + \cdots. \tag{3}$$

The right-hand side is obviously equal to e^m, but since this notation had not yet been invented, de Moivre had to write "the number of which the hyperbolic logarithm is m." In modern notation we get $e^m = 2$, with the solution $m = \ln 2 = 0.693$, as given by de Moivre.

Proceeding similarly in the general case we have from (1)

$$q^n \sum_{x=0}^{c-1} \binom{n}{x} r^{-x} = \frac{1}{2}, \tag{4}$$

which de Moivre writes as

$$\left(1 + \frac{1}{r}\right)^n = 2\left[1 + \binom{n}{1} r^{-1} + \binom{n}{2} r^{-2} + \cdots + \binom{n}{c-1} r^{-c+1}\right]. \tag{5}$$

Rather than give the limiting form of (5),

$$e^m = 2\left[1 + m + \frac{m^2}{2!} + \cdots + \frac{m^{c-1}}{(x-1)!}\right], \tag{6}$$

de Moivre takes logarithms and gives the result as

$$m = \ln 2 + \ln\left[1 + m + \frac{m^2}{2!} + \cdots + \frac{m^{c-1}}{(c-1)!}\right]. \tag{7}$$

He solves this equation numerically for $c = 1, \ldots, 6$ as shown in the following table:

De Moivre's Table of $m = np/q$

c	1	2	3	4	5	6
$p = \frac{1}{2}$	1	3	5	7	9	11
$p \to 0$	0.693	1.678	2.675	3.672	4.670	5.668

Source: De Moivre (1712, p. 225).

De Moivre's results for $p \to 0$ deviate at most 0.002 from the correct solution. He implies that m is an increasing function of p, so that the tabular values are the lower and upper limits for m.

From the table he concludes that $m \cong c - \frac{1}{2}$ for large values of c.

He gives some examples of applications to dice playing, all of the following type: "To find in how many throws of Three Dice one may undertake to throw Three Aces [at least] twice." (*Answer*: $m = n/215 = 1.678$; i.e., $n = 360.7$.)

With small modifications the proof from 1712 is repeated in all three editions of the *Doctrine of Chances* (1718, Problems 5–7; 1738 and 1756, Problems 3–5). Some further examples (lottery drawings) and remarks have been added.

Since de Moivre concentrates his exposition on the problem at hand, i.e., on the solution of (1), he does not explicitly state the result that

$$\sum_{x=0}^{c-1}\binom{n}{x}p^x q^{n-x} \to e^{-m}\sum_{x=0}^{c-1}\frac{m^x}{x!} \quad \text{for} \quad p \to 0, \quad n \to \infty, \quad m = \frac{np}{q}, \tag{8}$$

which follows immediately by comparison of (1) and (7). This result is known today as Poisson's limit of the binomial distribution or the Poisson distribution.

Today it is well known that (1) has the asymptotic solution

$$n + \frac{1}{3} \sim \frac{c - \frac{1}{3}}{p} \quad \text{or} \quad m \sim \frac{c - \frac{1}{3}(1 + p)}{q}, \tag{9}$$

and that (6) has the corresponding solution $m \sim c - \frac{1}{3}$ and not $c - \frac{1}{2}$, as stated by de Moivre.

Simpson (1740, pp. 33–37) gives a more detailed analysis of (5). He finds an expansion of $m - c$ in terms of c and tabulates m to four decimal places up to $c = 10$. From his table he concludes that $m \sim c - 0.3$ for $p \to 0$. Since $m = 2c - 1$ for $p = \frac{1}{2}$, he states that

$$(c - 0.3) + \frac{p}{q}(c - 0.7)$$

will be a good approximation to m for $0 \leqslant p \leqslant \frac{1}{2}$.

Struyck's Randomized Number of Trials

In the solution of a similar problem, Struyck (1716, p. 65; see also p. 94) finds that $n = 3.2$ and makes the following remark: "One may make the objection that it is impossible to make 1/5 trial. It is, however, possible to put 5 tickets in a bag, one of them being marked. After having made 3 trials the player draws a ticket at random from the bag and if he draws the ticket marked he is allowed to make another trial, otherwise he has to stop." This idea is due to Montmort, see §18.5.

De Moivre does not consider this artifice; he just chooses the nearest integer and says that the chance of winning then is slightly larger (smaller) than $\frac{1}{2}$.

It will be seen that Struyck is embedding the original experiment in a randomized (two-stage) experiment. His idea is the same as the technique used today for randomized critical regions in significance testing. In the present problem c is given, and we have to randomize for n; in significance testing n is given, and we have to randomize for c.

Poisson's Proof

In his exposition of the limiting forms of the binomial distribution, Poisson first treats the case with fixed p and $n \to \infty$, which leads to the normal distribution, as previously proved by de Moivre and Laplace. Using waiting-time reasoning, as in (14.1.3), Poisson begins by proving the relation

$$\sum_{x=0}^{c} \binom{n}{x} p^x q^{n-x} = q^{n-c} \sum_{x=0}^{c} \binom{n-c+x-1}{x} p^x,$$

because the negative binomial distribution on the right-hand side is a more convenient starting point for the mathematical technique he is going to use. After having carried out the proof for fixed p, he goes on to consider the case with $p \to 0$, $n \to \infty$, and $m = np$ fixed (Poisson, 1837, pp. 205–207).

For fixed c he finds

$$q^{n-c} = \left(1 - \frac{m}{n}\right)^n \left(1 - \frac{m}{n}\right)^{-c} \to e^{-m}$$

$$\binom{n-c+x-1}{x} p^x = \frac{(n-c+x-1)^{(x)} m^x}{n^x} \frac{m^x}{x!} \to \frac{m^x}{x!},$$

so that

$$\sum_{x=0}^{c} \binom{n}{x} p^x q^{n-x} \to e^{-m} \sum_{x=0}^{c} \frac{m^x}{x!}.$$

This proof is in principle not different from de Moivre's, which was (or ought to have been) known to Poisson. However, Poisson had the explicit objective of deriving a limiting form of the binomial, he writes the distribution function explicitly, and he states that it gives the probability that an event, whose chance on each trial is the very small fraction m/n, will not happen more than c times in a large number n of trials.

An English translation of the relevant pages from Poisson's book has been given by Stigler (1982).

The same method of proof may be used directly on the binomial, which is the procedure commonly used today.

Poisson does not give any applications apart from the one given by de Moivre for $c = 1$, see Poisson (1837, pp. 40–41). The Poisson distribution was not much used before Bortkiewicz (1898) expounded its mathematical properties and its statistical usefulness. Uspensky (1937, pp. 135–136) has given an evaluation of the error in using the Poisson distribution function as approximation to the binomial. A comprehensive survey of the Poisson distribution and its generalizations, including a section on its history, is due to Haight (1967).

14.5 PROBLEMS

1. Let

$$R_m^n = \binom{n+m-1}{m}$$

denote the number of m-combinations with repetitions among n elements, and let the n elements be divided into two groups of a and b elements, $a + b = n$. Prove by direct reasoning that

$$\sum_x R_x^a R_{m-x}^b = R_m^n.$$

Give an algebraic proof by comparing the coefficients of t^m in the two expansions of

$$(1+t)^{-a}(1+t)^{-b} = (1+t)^{-n}.$$

2. Use Montmort's idea, indicated in §14.1, and the result in Problem 1 to prove that the expressions (14.1.1) and (14.1.2) are identical.

3. Consider the relation between the number of successes, c, say, in a given time interval and the waiting time until the occurrence of the cth success. Use this relation to see that

$$\sum_{x=c+1}^{n} \binom{n}{x} p^x q^{n-x} = \sum_{t=c+1}^{n} \binom{t-1}{c} p^{c+1} q^{t-c-1}$$

$$= \sum_{y=0}^{n-c-1} \binom{c+y}{y} p^{c+1} q^y.$$

4. Prove that the positive integer s may be written as a sum of n positive integers in $\binom{s-1}{n-1}$ ways. *Hint:* Write s as a sum of 1's. To break this sum into n parts we have to remove $n-1$ plus signs among the $s-1$.

5. Generalize the Gambler's Ruin problem to three players, assuming that the winner of each game gets a counter from each of the losers.

6. Tabulate the number of chances of throwing $s = 8, 9, \ldots, 48$ points with eight ordinary dice, see Montmort (1713, p. 46).

7. Find the number of permutations with restrictions of the letters $a^4b^3c^2$ by the Bernoulli–Montmort algorithm and by a corresponding generating function, see Bernoulli (1713, p. 135) and Montmort (1713, p. 56).

8. From 10 cards numbered $1, 2, \ldots, 10$ are drawn n cards without replacement, $n = 1, 2, \ldots, 10$. Find the number of chances of getting s points, $s = n(n + 1)/2, n(n + 1)/2 + 1, \ldots, n(21 - n)/2$, see Montmort (1713, pp. 59–62). Generalize this problem and find the solution.

9. From three suits of ten cards each are drawn three cards without replacement. Find the number of chances of getting s points, $s = 3, 4, \ldots, 30$, see Montmort (1713, p. 62). Generalize this problem and find the solution.

10. To find in how many throws of six dice one may undertake to throw fifteen points twice. *Answer*: About 45 throws, see de Moivre (1712, p. 224).

11. In a lottery consisting of a great number of tickets, where the blanks are to the prizes as 50 to 1, find how many tickets a person ought to take to expect at least five prizes? See Simpson (1740, p. 36). *Answer*: $n = 239$ according to Simpson's approximation. Compare with the exact solution.

CHAPTER 15

James Bernoulli and
Ars Conjectandi, 1713

*To conjecture about something is to measure its probability.
The Art of Conjecturing or the Stochastic Art is therefore
defined as the art of measuring as exactly as possible the
probabilities of things so that in our judgments and actions
we can always choose or follow that which seems to be better,
more satisfactory, safer and more considered. In this alone
consists all the wisdom of the Philosopher and the prudence of
the Statesman.*

—JAMES BERNOULLI, 1713

15.1 JAMES, JOHN, AND NICHOLAS BERNOULLI

Among the members of the Bernoulli family at Basel were many prominent
merchants, politicians, artists, jurists, mathematicians, and scientists. Four
of them, James, John, Nicholas, and Daniel contributed to probability theory;
we shall discuss the lives of the first three, and in particular that of James.
A partial pedigree of the Bernoulli family is

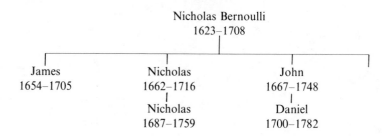

James (Jakob, Jacques) Bernoulli (1654–1705) studied philosophy and theology at Basel and got his degree in theology in 1676. At the same time he studied mathematics and astronomy. He then spent four years traveling as tutor and scholar in Switzerland and France. In Paris he studied the works of Descartes and his followers. It was possibly during his stay at Paris that he became interested in probability theory; it was only 13 years after the publication of Pascal's *Traité*, and Huygens was still a member of the Academy of Sciences. After his return to Basel in 1681 he published a paper in which he tried to derive the path of a comet from fundamental laws and thus to predict its occurrences. This shows him taking sides with the astronomers against some theologians who viewed comets as warnings from God.

He left Basel again and spent two years traveling in the Netherlands and England to study mathematics and science. He met Hudde in Amsterdam; Flamsteed, Boyle, and Hooke in England; and attended a meeting of the Royal Society in 1682.

Returning to Basel in 1683 he lectured on experimental physics; perfected himself in the mathematics of Descartes, Wallis, and Barrow; and wrote some papers on mathematics to the *Acta Eruditorum*, among them the previously mentioned paper of 1685 on a problem in probability.

In 1687 he became professor of mathematics at the university of Basel. He was an excellent teacher and had many students, among them his younger brother John and his nephew Nicholas.

In 1684 Leibniz started to publish his papers on the calculus; they were difficult to understand because of the new concepts and their lack of clarity. Bernoulli worked hard to understand and master the new method, and from about 1690 he and his brother John made a large number of important contributions to the development of the calculus so that they, after Newton and Leibniz, may be regarded as the creators of the calculus.

Starting about 1690 the Bernoullis corresponded regularly with Leibniz and Huygens. Together with Leibniz they developed the rules of differentiation and integration in the form used today. The mathematization of science going on from Galileo and Kepler to Huygens now got the tool it needed for a unified treatment. The old problems of mathematical physics, which had been solved by ad hoc methods, could now be formulated as special cases of differential equations and solved by routine methods, and many new and more difficult problems could be formulated and solved.

James Bernoulli made important contributions to the theory of differential equations and mechanics, the calculus of variations, the theory of infinite series, and the theory of probability. To be more specific, James gave the formula for the radius of curvature in terms of the derivatives, and he developed a theory of evolutes. He studied many special curves corresponding to physical problems, such as the cycloid, the catenary, the sail curve, the lemniscate, the "elastic" curve, and the caustic by refraction. Using polar coordinates he discussed the parabolic and logarithmic spirals, which led him to an elliptic integral for the curve length. He solved the differential equation $y' = f(x)y + g(x)y^n$, which bears his name. In his works on the brachistochrone problem (the curve of the quickest descent) and the isoperimetric problem (the closed curve bounding the maximum area among curves with a given perimeter), he laid the foundation of the calculus of variation. All this work was carried out in strong competition with his brother John, Leibniz, and Huygens. We shall discuss his contributions to probability theory in the following sections.

John (Johann, Jean) Bernoulli (1667–1748) studied medicine in accordance with his father's wishes and got his degree in 1690. At the same time he studied mathematics, guided by his brother. His paper in the *Acta Eruditorum* in 1691 on the catenary showed him to be an accomplished mathematician mastering the new calculus. During 1691–1692 he lived in Paris, where mathematicians not yet were fully acquainted with the calculus. Marquis de l'Hospital, one of the leading mathematicians, engaged Bernoulli to teach him the new methods, and based on these lessons and Bernoulli's notes, de l'Hospital wrote the first textbook on the differential calculus, *Analyse des Infiniment Petits* (1696), which was widely circulated.

In 1695 John became professor of mathematics at Groningen at the recommendation of Huygens. He remained there until 1705, when he succeeded James as professor of mathematics at Basel.

For many years John worked on the same problems as James, and he often gave alternative and more elegant solutions. He wanted to prove his independence and superiority and was rather unscrupulous in his tactics, both in his published papers and in his private correspondence. James retaliated by pointing out the superficiality of some of John's solutions, and the dispute became rather disgraceful and created great bitterness between the brothers. John was also deeply involved in the priority dispute between Newton and Leibniz, see Westfall (1980, pp. 760–776). In his later years John worked on problems in astronomy, mechanics, hydrodynamics, and experimental physics. He only made minor contributions to probability theory; we shall discuss his correspondence with Montmort in §18.4.

A survey of the mathematical works of the brothers Bernoulli has been given by Fleckenstein (1949). Biographies have been written by K. Pearson (1978) and David (1962).

Nicholas (Nikolaus) Bernoulli (1687–1759) studied mathematics and law at Basel. He was taught mathematics by his uncles, first by James until 1705 and then by John. He took his master's degree under James in 1704 on the theory of infinite series. In 1709 he got his doctor's degree in law with the thesis *De Usu Artis Conjectandi in Jure*, part of which we have already discussed in §8.2. During 1712–1713 he traveled in England, the Netherlands, and France. In London he met de Moivre and Newton (see Westfall, 1980, pp. 741–744), and in Paris he met Montmort with whom he had been in correspondence on probability problems since 1710. They became friends, and Nicholas spent three months at the country estate of Montmort working on the second edition of Montmort's *Essay* (1713). In 1716 Nicholas became professor of mathematics at Padua. Returning to Basel he became professor of logic in 1722 and professor of law in 1731. He continued his mathematical work on infinite series and differential equations and corresponded with Leibniz and Euler. We shall discuss his work on probability theory, statistics, and life insurance mathematics in the following chapters. Biographies have been written by Fleckenstein (1973) and Youshkevitch (1986b).

15.2 *ARS CONJECTANDI*

James Bernoulli's preliminary work on probability, which ultimately resulted in the *Ars Conjectandi* (The Art of Conjecturing, 1713), may be followed in his *Meditationes* between 1684 and 1690. He begins by solving Huygens' problems; continues with some remarks and examples on the possibility of employing probability calculus to problems other than games of chance; and finally discusses the convergence of binomially distributed relative frequencies and proves his famous theorem. During the 1690s there are no notes on probability; presumably he was fully occupied with his mathematical work. It was, however, a long way from his notes in the *Meditationes* to the manuscript of the *Ars Conjectandi*. It can be seen from his correspondence with Leibniz that he worked on the manuscript during the last two years of his life.

When Bernoulli died in 1705 the manuscript was nearly completed, but due to quarrels within the Bernoulli family it was not published until 1713. Leibniz advised the widow to let Jakob Hermann, a pupil of Bernoulli's, sort out and order the manuscript. Through Hermann an outline of the main contents became known and was mentioned in the eulogies of Bernoulli, the best being the one by Saurin (1706), who closed his eulogy by recommending that *Ars Conjectandi* should be published under the supervision of John Bernoulli. However, due to the widow's mistrust of John this was out of the question. Montmort, who had published his *Essay* in 1708, wrote to John in 1710 and asked him to communicate an offer of publishing *Ars Conjectandi* at his expense, since he was interested in learning James's results before publishing the second edition of his *Essay*. The offer was, however, rejected. Nicholas, who had read the manuscript when he studied under James, had used it to his own advantage in his thesis, and he also profited from it in his discussions with Montmort. Under the pressure of Montmort and other mathematicians, *Ars Conjectandi* was finally published in Basel in 1713 with a short preface by Nicholas. The preface has been translated into English by David (1962). Nicholas writes that the printers asked him to complete the manuscript but that "I was too young and inexperienced to know how to complete it.... I advised the printers to give it to the public as the author left it." More details on the publication history have been given by Kohli (1975a).

Ars Conjectandi consists of four parts:

1. The treatise *De Ratiociniis in Ludo Aleae* by Huygens with annotations by James Bernoulli, pp. 2–71.
2. The doctrine of permutations and combinations, pp. 72–137.
3. The use of the preceding doctrines on various games of chance and dice games, pp. 138–209.
4. The use and application of the preceding doctrines on civil, moral, and economic affairs, pp. 210–239.

An appendix contains the *Lettre à un Amy, sur les Parties du Jeu de Paume* (Letter to a friend on the points in the game of tennis), 35 pp.

Ars Conjectandi has been reprinted in 1968 in *Editions Culture et Civilisation*, and in 1975 in *Die Werke von Jakob Bernoulli*, Vol. 3. A German translation with a good historical introduction and many special notes on historical and mathematical problems is due to Haussner (1899). Part 1 has been translated into French by Vastel (1801) and into Italian, with many commentaries, by Dupont and Roero (1984). The first three chapters of Part 2 have been reprinted and translated into English by Maseres (1795). Part 4 has been translated into Russian by J. V. Uspensky, with a preface by Markov (1913), and into English by Bing Sung (1966), with a preface by Dempster. The Russian translation has been reprinted in 1986 with new commentaries.

In the present chapter we shall discuss the contents of *Ars Conjectandi*, expect for Bernoulli's proof of his main theorem, which is discussed in Chapter 16. When quoting from *Ars Conjectandi* we shall cite page numbers only; we have generally used the translation by Bing Sung.

At the risk of some repetition later on, we shall begin with a survey of Bernoulli's work.

In the first three parts of *Ars Conjectandi* Bernoulli generalizes the doctrine of chances as previously developed by Pascal and Huygens. He presents an ingenious commentary on Huygens' treatise and a unified, generalized treatment of combinatorics in the spirit of Pascal but with no reference to

Pascal's treatise, which he presumably did not know. He clearly states the multiplication rule for the case of independence; he derives the binomial distribution; and he finds the probability of winning a game with a possibly infinite number of trials as the sum of an infinite series. Finally, he demonstrates the power of his methods by solving 24 problems taken from popular games of chance. The first 209 pages of *Ars Conjectandi* constitute an excellent textbook that consolidates and generalizes the existing calculus of chances. It is a pedagogical masterpiece with a clear formulation of theorems supported by elaborate proofs both in abstract form and by means of numerical examples.

In Part 4, the last 30 pages of *Ars Conjectandi*, Bernoulli takes up new and fundamental problems of probability theory and its applications. Instead of the old concept of chance based on symmetry, he introduces a new concept of probability, defined as a measure of our knowledge of the truth of a proposition, "a degree of certainty." This is a revolutionary step because probability in this sense relates to propositions and not directly to events. Bernoulli thinks that this subjective probability is universally applicable. He tries to give rules for the measurement and combination of such probabilities but does not succeed outside the already known fields of games of chance and stable relative frequencies.

Bernoulli writes that it is a well-known fact that the relative frequency of an event will be nearer to the truth if based on many rather than few observations; however, no one has proved the corresponding property of relative frequencies in probability theory. He therefore proves "*the law of large numbers*," which we shall present in a modern and slightly more general formulation that Bernoulli's .

Consider n independent trials, each with probability p of success, and let s_n be the number of binomially distributed successes and $h_n = s_n/n$ the relative frequency. Then for given positive numbers ϵ and δ,

$$\Pr\{|h_n - p| \leqslant \epsilon\} > 1 - \delta \quad \text{for} \quad n > n(p, \epsilon, \delta),$$

which means that the absolute value of the difference $h_n - p$ will be smaller than ϵ, with a probability tending to 1 for n tending to infinity. The name, the law of large numbers, is due to Poisson (1837, p. 7).

This is the first limit theorem proved in probability theory, and it is of fundamental importance for statistical estimation theory. It was Bernoulli's intention to derive an interval estimate of p, but he does not indicate how to use his theorem for this purpose.

Bernoulli was a great probabilist and mathematical statistician, but he was no applied statistician. He never analyzed a set of observations nor did he comment on the analyses made by Graunt and Halley.

15.3 BERNOULLI'S COMMENTARY ON HUYGENS' TREATISE

In Part 1 of *Ars Conjectandi* Bernoulli improves and supplements Huygens' reasoning, replaces Huygens' numerical results by formulae, generalizes the problems, and provide new methods of solution. The commentary is about four times as long as the original text.

Considering the repetition of a game of chance, Bernoulli stresses that the probability of winning in a single game is constant, that is, independent of the outcomes of the previous games. In recognition of his achievements, repeated independent trials are today called Bernoulli trials when there are only two possible outcomes for each trial (success and failure) and the probabilities of these outcomes remain the same for all the trials.

Bernoulli gives the following alternative proof of Huygens' basic Proposition 3. Suppose that a lottery consists of p tickets, each with the prize a, and q tickets, each with the prize b. Let $p + q$ players each buy a ticket so that their total winnings are $pa + qb$. Since all players have the same expectation, its value must be $(pa + qb)/(p + q)$.

Bernoulli remarks that the word "expectation" is used in a technical sense in probability theory. He also states that the rule for finding the expectation is the same as the rule for finding the average price of a product obtained by mixing several components of different prices.

In several instances Huygens finds the expectation of one player and obtains the expectation of the other by subtraction. Bernoulli points out that this is permissible only if the total stake is to be divided between the two players; that is, when the corresponding events are disjoint and complementary. As an example he supposes that two convicts, sentenced to death, are allowed to play with a die on their lives so that the one who throws the larger number of points will be reprieved and that both will be reprieved if they throw the same number of points. The expectation of each convict is $7/12$ of a life, which shows that the expectation of the one cannot be found by subtracting the expectation of the other from unity. The reason is of course that there are cases in which both of them will be reprieved. To avoid such errors Bernoulli normally finds the probabilities of winning for both A and B and checks that $P_A + P_B = 1$. He does not formulate the addition rule for probabilities explicitly but uses it as a matter of course.

He formulates the multiplication rule for independent events as follows. Consider a series of trials with different probabilities of success, $p_1, p_2, \ldots$. The probability of getting a series of successes and failures in a given order is the product of the corresponding probabilities, $p_1 p_2 p_3 q_4 q_5 p_6 \ldots$, say.

For a series of n trials with the same probability of success, p, say, the probability of m successes and $n - m$ failures in a given order thus becomes $p^m q^{n-m}$. If the order of successes and failures is immaterial, the probability of m successes and $n - m$ failures becomes $\binom{n}{m}p^m q^{n-m}$, since there are $\binom{n}{m}$ different orders of m successes and $n - m$ failures. Bernoulli's derivation of the binomial distribution is thus the one used today.

In his generalization of Huygens' Propositions 10–12, Bernoulli considers the probability of getting at least m successes in n trials. Let the number of successes be x and the number of failures $y, x + y = n$, and let A win if $x \geqslant m$. Let B's expectation (probability of winning) be $e(m, n)$. Bernoulli finds $e(m, n)$ by the recursion

$$e(m, n) = pe(m - 1, n - 1) + qe(m, n - 1), \qquad m = 1, 2, \ldots, n - 1,$$

$e(0, n) = 0$, and $e(n, n) = 1 - p^n = (p + q)^n - p^n$. He does not give this formula explicitly but tabulates the probabilities by recursion for $n = 1, 2, \ldots, 6$. He concludes (by incomplete induction) that

$$e(m, n) = q^n + \binom{n}{1}pq^{n-1} + \binom{n}{2}p^2 q^{n-2} + \cdots + \binom{n}{m-1}p^{m-1}q^{n-m+1}.$$

He remarks that this result may also be obtained directly by adding the m probabilities of the outcomes in question.

Let us introduce the notation

$$b(x, n, p) = \binom{n}{x}p^x q^{n-x}, \qquad n = 1, 2, \ldots, \quad x = 0, 1, \ldots, n, \quad 0 < p < 1,$$

$$B(c, n, p) = b(0, n, p) + b(1, n, p) + \cdots + b(c, n, p), \qquad c = 0, 1, \ldots, n.$$

The formula above for B's probability of winning may then be written

$$1 - B(n - m, n, q) = B(m - 1, n, p),$$

corresponding to $y \geqslant n - m + 1$ and $x \leqslant m - 1$, respectively.

In his commentary on the problem of points, Huygens' Propositions 4–9, Bernoulli only extends Huygens' tabular solution. He returns to the problem in Part 2, pp. 107–112, but does not go beyond the solutions given by Pascal and Fermat. He refers to Pascal's letters to Fermat, published in 1679, and says that Pascal was not able to find the general solution. Had Bernoulli known of Pascal's treatise, he would have found there a solution with a proof more elegant than his own. It is strange that Bernoulli did not consider the problem of points for $p \neq \frac{1}{2}$.

In connection with his discussion of Huygens' problems of dicing, Bernoulli gives an algorithm for finding the probability of the number of points obtained by throwing n dice, which has been presented in §14.3.

The most important of Bernoulli's contributions to Part 1 is found in his comments on Proposition 14 about the possibly infinite series of trials in the order BA BA BA, etc. Making use of the addition and multiplication rules and the theory of infinite series, he gives an alternative to Huygens' method and points out that his method is more general. He considers infinitely many players in succession, each having one throw. All even-numbered players have probability p_1 of winning a throw, and the odd-numbered players have probability p_2. The condition for player number 4, say, to win is that the preceding players have been unsuccessful and that he himself has a success; the probability of this event is $q_2 q_1 q_2 p_1$. Bernoulli then lists the probabilities of winning for each player as follows:

Player No.	1	2	3	4	5	6	$\cdots$
Probability	p_2	$q_2 p_1$	$q_2 q_1 p_2$	$q_2^2 q_1 p_1$	$q_2^2 q_1^2 p_2$	$q_2^3 q_1^2 p_1$	$\cdots$
Player	B	A	B	A	B	A	$\cdots$

The probability of winning for A is obviously the sum of the probabilities for the even-numbered players and analogously for B. Summing the two geometric series, Bernoulli finds

$$P_A = p_1 q_2 [1 + q_1 q_2 + (q_1 q_2)^2 + \cdots] = \frac{p_1 q_2}{1 - q_1 q_2},$$

$$P_B = p_2 [1 + q_1 q_2 + (q_1 q_2)^2 + \cdots] = \frac{p_2}{1 - q_1 q_2}.$$

It seems that Bernoulli has overlooked the relationship between his method and the method of recursion. The recursion formula (6.2.4) leads to an infinite series for e_0, which is the same as Bernoulli's result for P_A given above.

Bernoulli uses his method to solve the problem he posed in 1685 and Huygens' Problems 1 and 2. He solves Huygens' Problems 3 and 4 by combinatorial methods and discusses various interpretations. He states the solution of the Gambler's Ruin problem without giving a satisfactory proof. All of these problems have been discussed earlier in §14.2.

15.4   BERNOULLI'S COMBINATORIAL ANALYSIS AND HIS FORMULA FOR THE SUMS OF POWERS OF INTEGERS

Bernoulli mentions the importance of combinatorial analysis for many problems in philosophy, history, medicine, and politics; the solution of such problems depends on conjecture, which in turn depends on a combination of causes. Great mathematicians, such as van Schooten, Leibniz, Wallis, and Prestet, have therefore worked on combinatorial analysis, and Bernoulli is now to supplement this work with his own contributions giving a systematic exposition of the whole theory, since this has not been done before.

It is remarkable that Bernoulli does not mention Pascal, whose *Traité* (1665) contains what Bernoulli states is missing, see §5.2. On the other hand, even if Bernoulli had known of Pascal's treatise, Bernoulli's exposition is still worthwhile because it is more comprehensive and because his discussions on permutations, combinations, and figurate numbers are neatly integrated and given with elaborate explanations and improved notation. Bernoulli's work became the most popular text on combinatorics in the 18th century.

According to Haussner (1899) it was Bernoulli who introduced the term "permutation" in combinatorial theory. The term "combination" had been used by both Pascal and Leibniz; however, according to Stigler (1987) Strode used the word "permutation" in the title of his book in 1678.

We shall use the following terminology. Let there be given n elements. A permutation is an ordered arrangement of the n elements. An m-permutation is an ordered selection or arrangement of m of these elements. An m-combination is a selection of m elements without regard to order. Instead of m-permutations, Bernoulli uses the term m-combination with its permutations, which later became known as "variations." Montmort (1708) used the term "arrangement" for permutation. We shall use the notation C_m^n for the number of m-combinations, $n!$ for the product $n(n-1)\cdots 1$, and $n^{(m)}$ for the product $n(n-1)\cdots(n-m+1)$; Bernoulli does not use these abbreviations.

Part 2 of *Ars Conjectandi* consists of nine chapters on the following topics: (1) Permutations. (2) The number of combinations of all classes. (3) The number of combinations of a particular class, the figurate numbers and their properties, the sums of powers of integers. (4) Properties of C_m^n, the hypergeometric distribution, the problem of points for $p = \frac{1}{2}$. (5) Combinations with repetitions. (6) Combinations with restricted repetitions. (7) Variations without repetitions. (8) Variations with repetitions. (9) Variations with restricted repetitions.

By induction Bernoulli first proves that the number of permutations of n different elements equals $n!$ and that the number of permutations of n elements consisting of k classes with r_i identical elements in the ith class equals $n!/r_1!\cdots r_k!$, where $r_1 + \cdots + r_k = n$. He does not at this stage discuss the number of m-permutations but continues with the theory of combinations.

Bernoulli begins by listing all the combinations of $n = 1, \ldots, 5$ elements, adding one element at a time, and counting the additional number of m-combinations for each n. We shall denote this number by a_{nm}. Here is an abridged version of Bernoulli's table:

Table of Combinations

n	m			
	1	2	3	4
1	a			
2	b	ab		
3	c	ac bc	abc	
4	d	ad bd cd	abd acd bcd	$abcd$

Source: Bernoulli, Ars Conjectandi, p. 83.

Table of a_{nm}: The Number of Additional Combinations

n	m			
	1	2	3	4
1	1	0	0	0
2	1	1	0	0
3	1	2	1	0
4	1	3	3	1

Source: Bernoulli, Ars Conjectandi, p. 87.

From this construction Bernoulli observes that a_{nm} is given by

$$a_{n1} = 1; \quad a_{nm} = 0 \quad \text{for} \quad n < m; \quad a_{nm} = \sum_{i=1}^{n-1} a_{i,m-1}. \tag{1}$$

He remarks that the theory of combinations thus surprisingly leads to the figurate numbers, which are defined by the relations above, and that these numbers have been previously studied by several mathematicians, among them Faulhaber, Remmelin, Wallis, Mercator, and Prestet, but they have not succeeded in giving satisfactory proofs of the properties of these numbers.

From the construction of the table of combinations it also follows that the total number of combinations equals

$$C_m^n = \sum_{i=1}^{n} a_{im} = a_{n+1,m+1}. \tag{2}$$

By inspection of the table of a_{nm}, Bernoulli records 12 "wonderful properties" of these numbers, among them the symmetry property, the relation to the binomial coefficients, the recursion formula

$$a_{nm} = a_{n-1,m} + a_{n-1,m-1}, \tag{3}$$

and, finally,

$$\frac{a_{n+1,m+1}}{a_{nm}} = \frac{n}{m}. \tag{4}$$

These properties are of course the same as those listed by Pascal (see §5.2).

Like Pascal, Bernoulli notes that the crucial problem is to prove (4), which is the link between the additive and the multiplicative forms of the a's. Bernoulli first gives a rather lengthy proof; he then states that based on this, his brother John has constructed a more elegant proof of the following lemma.

Lemma. Assuming that $\sum_{i=1}^{n} a_{im}/na_{nm} = 1/r$ for all n, it follows that

$$\sum_{i=1}^{n+1} \frac{a_{i,m+1}}{(n+1)a_{n+1,m+1}} = \frac{1}{r+1}.$$

In the proof Bernoulli makes repeated use of (1) combined with the assumptions given in the lemma, as shown here:

$$\sum_{i=1}^{n+1} a_{i,m+1} = \sum_{i=1}^{n} a_{im} + \sum_{i=1}^{n-1} a_{im} + \cdots + \sum_{i=1}^{m} a_{im}, \qquad \text{from (1),}$$

$$= \frac{na_{nm} + (n-1)a_{n-1,m} + \cdots + na_{mm}}{r}, \qquad \text{from the assumption,}$$

$$= \frac{n\sum_{i=1}^{n} a_{im} - (a_{n-1,m} + 2a_{n-2,m} + \cdots + (n-m)a_{mm})}{r}$$

$$= \frac{na_{n+1,m+1} - \sum_{i=1}^{n} a_{i,m+1}}{r}, \qquad \text{from (1),}$$

which yields

$$ra_{n+1,m+1} + r\sum_{i=1}^{n} a_{i,m+1} = na_{n+1,m+1} - \sum_{i=1}^{n} a_{i,m+1},$$

so that

$$\sum_{i=1}^{n} a_{i,m+1} = \frac{a_{n+1,m+1}(n-r)}{r+1}. \tag{5}$$

Adding $a_{n+1,m+1}$ on both sides the lemma follows.

For $m = 1$ the lemma is obviously true for $r = 1$. Bernoulli then proves by induction that $r = m$, so that

$$a_{n+1,m+1} = \sum_{i=1}^{n} a_{im} = \frac{a_{nm}n}{m},$$ (6)

and using (5) he obtains

$$\sum_{i=m}^{n} a_{im} = \sum_{i=1}^{n} a_{im} = \frac{a_{n+1,m}(n-m+1)}{m}.$$ (7)

Repeated application of this result leads to

$$C_m^n = a_{n+1,m+1} = \frac{n(n-1)\cdots(n-m+1)}{m!}.$$ (8)

By means of these results Bernoulli proves the properties of the binomial coefficients.

Comparison with Pascal's exposition (see §5.2) will show that Pascal bases his proof on recursion (3), whereas Bernoulli uses the more cumbrous recursion (1), which makes Bernoulli's proof more clumsy.

Combining (1) and (8), Bernoulli gets the addition formula for the binomial coefficients

$$\sum_{i=1}^{n} \binom{i-1}{m-1} = \binom{n}{m}.$$ (9)

He states that this relation may be used to find the sum of the powers of the natural numbers. Writing the binomial coefficients as polynomials in i and n, respectively, he tabulates the power-sums as polynomials in n for $m = 1, 2, \ldots, 10$ by recursion. It would have been simpler if, like Pascal, he had derived an explicit recursion formula. Bernoulli does not mention that Faulhaber had previously tabulated these polynomials up to the 17th degree, see §5.2.

Studying the pattern of the coefficients of the polynomials, Bernoulli remarks that it is not necessary to continue the tabulation, since the general formula is obviously,

$$\sum_{i=1}^{n} i^m = \frac{n^{m+1}}{m+1} + \frac{n^m}{2} + \frac{1}{2}\binom{m}{1}B_2 n^{m-1}$$

$$+ \frac{1}{4}\binom{m}{3}B_4 n^{m-3} + \frac{1}{6}\binom{m}{5}B_6 n^{m-5} + \cdots,$$ (10)

the series terminating with the last positive power of n, and $B_2, B_4, \ldots$, denoting what de Moivre and Euler later called the Bernoulli numbers. (Bernoulli does not use the notation $B_2, B_4, B_6, \ldots$; rather, he uses $A, B, C, \ldots$.) He gives the first of the coefficients as $B_2 = 1/6$, $B_4 = -1/30$, $B_6 = 1/42$, $B_8 = -1/30$ and remarks that they may be found by recursion, since for each m the sum of the coefficients of the polynomial in question equals 1. Presumably he has obtained this relation by setting $n = 1$, which gives

$$\frac{1}{2} = \frac{1}{2k+1} + \sum_{i=1}^{k} \frac{1}{2i}\binom{2k}{2i-1} B_{2i}, \qquad k = 1, 2, \ldots. \tag{11}$$

Bernoulli does not explain how he got the idea of introducing the binomial coefficients as factors in the expansion.

As previously stated, the formula for the sums of powers of integers was an important tool for computing the areas of plane figures, i.e., for obtaining approximations to integrals. Bernoulli's result was therefore a major achievement and marks the end of a long search carried on by many mathematicians. It was later derived as a special case of the Euler–Maclaurin formula.

Finally, Bernoulli considers a generalization of the figurate numbers, which he obtains by repeated summations, starting from any number, d, say, in the first column and introducing successively the numbers c, b, and a in the following columns, so that the generalized figurate number of order 4 becomes

$$u_{n4} = a + ba_{n2} + ca_{n3} + da_{n4},$$

or

$$u_{n4} = a + b\binom{n-1}{1} + c\binom{n-1}{2} + d\binom{n-1}{3}. \tag{12}$$

Bernoulli points out that the advantage of writing a polynomial of the third degree in this form is that the sum of any number of equidistant terms may be found by means of (9) so that

$$\sum_{i=1}^{n} u_{i4} = a\binom{n}{1} + b\binom{n}{2} + c\binom{n}{3} + d\binom{n}{4}. \tag{13}$$

He concludes that any series of numbers $u_1, u_2, \ldots$, having constant differences of order m, say, may be written in the form (12) so that the sum of the first n terms may be expressed in the form (13).

Bernoulli does not mention that (12) is a special case of the

Gregory–Newton formula (see §11.3), which gives

$$u_n = u_1 + \binom{n-1}{1} \Delta u_1 + \cdots + \binom{n-1}{m} \Delta^m u_1. \tag{14}$$

Thus, only (13) is new; however, Montmort had independently published this result in *Journal des Sçavans* (1711) and in his *Essay* (1713, pp. 63–67) with a proof similar to Bernoulli's.

Let the n different elements be divided into two groups of m and $n - m$ elements, respectively, and let b elements be selected from among the n in such a way that a elements are selected among the m and $b - a$ among the $n - m$. Bernoulli formulates this problem and finds the corresponding number of combinations to be

$$\binom{m}{a}\binom{n-m}{b-a}.$$

This result leads to the hypergeometric distribution (as it is called today), and Bernoulli uses it to solve Huygens' Problems 3 and 4.

After his exposition of the theory of m-combinations without repetitions, Bernoulli gives an analogous discussion of m-combinations with repetitions; i.e., each of the n elements is allowed to occur up to m times in a selection of m elements, and he proves that the number of m-combinations with repetitions equals

$$\binom{n+m-1}{m}.$$

To find the number of m-permutations, Bernoulli notes that for each m-combination there are $m!$ permutations so that the number of m-permutations equals $C_m^n m! = n^{(m)}$. He does not, as does Montmort, mention that the first element may be chosen in n ways, the second in $n - 1$ ways, and so on. He remarks that the number of m-permutations of n different elements is equal to the number of permutations of n elements of which m elements are identical. He proves that the total number of permutations of order $1, 2, \ldots, n$ may be found recursively by the formula

$$\sum_{m=1}^{n} n^{(m)} = n \left[1 + \sum_{m=1}^{n-1} (n-1)^{(m)} \right].$$

Further, Bernoulli notes that the number of m-permutations with repetitions equals n^m, and their sum equals $(n^m - 1)n/(n - 1)$.

He points out the importance of combinatorics for evaluating expressions of the form $(a_1 + \cdots + a_n)^m$. The number of terms in the expansion equals the number of m-combinations with repetitions, i.e.,

$$\binom{n + m - 1}{m},$$

and the coefficient of any product of m factors equals the corresponding number of permutations so that

$$(a_1 + \cdots + a_n)^m = \sum \frac{m!}{r_1! \cdots r_n!} a_1^{r_1} \cdots a_n^{r_n}, \qquad r_i = 0, 1, \ldots, m,$$

and $r_1 + \cdots + r_n = m$. Such multinomial expansions had been discussed in other contexts in the 1690s by Leibniz, John Bernoulli, and de Moivre.

Finally, Bernoulli also discusses the number of combinations and permutations with restricted numbers of repetitions, as indicated in §14.3.

It is easy to trace the influence of Part 2 of the *Ars Conjectandi* on the exposition of combinatorial analysis in modern textbooks on probability and mathematical statistics.

Edwards (1987) has discussed Bernoulli's combinatorial work in relation to previous results.

15.5   BERNOULLI ON GAMES OF CHANCE

In Parts 1 and 2 Bernoulli develops new tools for solving problems of games of chance and uses them in his analysis of the classical problems previously discussed by Pascal, Fermat, and Huygens. In Part 3 he solves a variety of problems of games of chance not previously analyzed in this way. He formulates and solves 24 problems related to popular games. Most of these problems are not particularly difficult for us and are not of much general interest, but they show the state of the art at the time. We shall limit our discussion to only 5 of the 24 problems in detail. Among the remaining problems we have used six of the most interesting as problems for the reader in §15.8.

For the solution Bernoulli uses the addition and multiplication rules; enumeration of equally likely and favorable cases by combinatorial methods; calculation of expectations, making systematic use of conditional expectations; and recursion.

Problems 14 and 15

Problems 14 and 15 are two-stage dicing problems where the outcome at the first stage determines the game to be carried out at the second stage. A die is thrown, and the number of points obtained determines the number of dice to be used in the following game. Let the outcome of the first throw be x and the number of points obtained by throwing x dice be y_x. The player gets nothing if $y_x < 12$; he gets half the stake if $y_x = 12$; and he gets the whole stake if $y_x > 12$. Find the player's expectation. Let z be a random variable taking on the values $0, \frac{1}{2}$, and 1 corresponding to the three outcomes, so that $\Pr\{z = 0 | x\} = \Pr\{y_x < 12\}$, and so on. These probabilities are easily obtained from Bernoulli's table of the number of chances of getting any number of points by throwing $1, 2, \ldots, 6$ dice. For example, for $x = 4$ the number of chances for $z = 0, \frac{1}{2}$, and 1 are 310, 125, and 861, respectively, with the sum 1296. Bernoulli thus finds that the conditional expectation is

$$\frac{125 \times \frac{1}{2} + 861 \times 1}{1296} = \frac{1847}{2592}.$$

Similarly, he finds the other five conditional expectations and finally the unconditional expectation by averaging over x, each x having the probability $\frac{1}{6}$, which gives $E(z) = 15{,}295/31{,}104$. Bernoulli, like Huygens, uses the formula $E(z) = EE(z|x)$ as a matter of course.

By the same method Bernoulli finds the expected number of points $E(y) = 12\frac{1}{4}$. He notes that it is difficult to explain why the player's expectation is less than $\frac{1}{2}$ in view of the fact that the average number of points is larger than the critical value 12. We shall leave the explanation to the reader.

Using the same method he solves two more problems of the same kind. In the first he uses the condition $x + y_{x-1} < 12$ and in the second the condition $y_x < x^2$.

Problem 16

Cinq et neuf is a dice game which may involve several throws depending on the outcome of the previous throws. Player A throws two dice; B wins if 5 or 9 points are thrown; A wins if he throws 3 or 11 points or a doublet; for the remaining outcomes, A continues to play until either B wins by getting 5 or 9 points, or A wins by getting the same number of points as made him continue.

Considering the first throw, it is easy to see that B wins in 8 cases, A wins in 10 cases, and A continues to play in 18 cases of a total of 36 equally likely cases.

To find A's expectation Bernoulli uses a conditional argument combined with the following lemma: Let A's expectation at the beginning of a game be e, say. Suppose that there are three prizes a, b, and e and that the corresponding number of chances are p, q, and r. Then, $e = (pa + qb)/(p + q)$. The proof is straightforward.

Bernoulli analyzes the continuation cases as follows. Suppose that A gets 4 points in his first throw without getting the doublet $(2, 2)$. The play then continues until A throws either 5 or 9 points, which has probability 8/36, or A throws 4 points, which has probability 3/36. For all other outcomes A is back in the same situation as he was before his last throw. It follows from the lemma that A's expectation equals $(3 \times 1 + 8 \times 0)/11 = 3/11$. In this way Bernoulli finds the following results:

No. of points in first throw	4	6	7	8	10
Probability of continuation	2/36	4/36	6/36	4/36	2/36
Conditional expectation	3/11	5/13	3/7	5/13	3/11

Bernoulli finally calculates A's expectation as

$$[10 \times 1 \times 8 \times 0 + 4 \times 3/11 + 8 \times 5/13 + 6 \times 3/7]/36 = 4189/9009.$$

Montmort (1708, pp. 109–113; 1713, pp. 173–177) discusses a slightly more general version of this game (*Quinquenove*) and uses essentially the same method for its solution.

Problem 19

Problem 19 concerns the banker's expectation in a series of m games in which the banker wins each game with probability $p, p + q = 1$, and $p - q = r > 0$, and the banker's probability of continuing as banker in the next game is $h, h + k = 1$, and $h - k = t > 0$. The winner of each game gets the amount a from the loser. In the first game the banker's expectation is $pa + q(-a) = ra$. Hence in any game in which a player is banker, his expectation is ra, and if he is not the banker his expectation is $(-ra)$. For the first two games the expectation of the player who was banker in the first game is $ra + h(ra) + k(-ra) = ra(1 + t)$. Continuing in this way Bernoulli finds that the banker's expectation is

$$ra(1 + t + t^2 + \cdots + t^{m-1}) = \frac{ra(1 - t^m)}{1 - t} = \frac{ra(1 - t^m)}{2k}.$$

As a warning to the reader against erroneous reasoning, Bernoulli also gives the three wrong answers that he had considered before finding the correct one above.

Problem 20

Problem 20 is about the banker's expectation in a card game named *Bock*. At the beginning of a game each player puts down his stake, and the banker then deals a card to each of the $n - 1$ players and to himself from a pack of $N = sf$ cards consisting of s suits each having f face values marked from 1 to f. If the face value of the banker's card is larger than or equal to the punter's (the banker's adversary), the banker wins the stake; otherwise, he has to pay the same amount as the punter's stake. The banker continues as banker until it happens that he loses to *all* the punters in a game.

Bernoulli first finds the banker's probability of winning over any punter. The banker wins if the two face values are equal or if he has a face value larger than the punter's. Since the total number of cases equals $\binom{sf}{2}$ and the number of cases with two equal face values equals $\binom{s}{2}f$, the difference of these numbers gives the number of cases with different face values, and half of that is the number of cases favorable to the banker. The number of favorable cases is thus equal to

$$\binom{s}{2}f + \frac{1}{4}s^2f(f-1) = \frac{1}{4}sf(sf + s - 2).$$

Dividing by the total number of cases, the banker's probability of winning becomes $\frac{1}{2}(sf + s - 2)/(sf - 1)$.

The banker's probability of continuing as banker depends on the number of players. For $n = 2$ it is equal to the probability of winning as found above. For $n = 3$ Bernoulli gives a detailed derivation, whereas for $n = 4$ he only indicates the proof and leaves the details to the reader. We shall give Bernoulli's proof for $n = 3$ and $s = 4$ and like Bernoulli leave the general proof to the "industrious reader," see Problem 5 in §15.8.

The total number of cases, which equals $\binom{4f}{3}$, is divided into three classes: first, the cases where the three cards have the same face value, which amounts to $\binom{4}{3}f = 4f$ cases; second, the cases where two cards have the same face value and the third a different value, which amounts to $\binom{4}{2}f(4f - 4) = 24f^2 - 24f$; third, the cases where the three cards have different face values, which amounts to the difference between the total number of cases and the two numbers just found and thus becomes $(32f^3 - 96f^2 + 64f)/3$.

The next step is to find the probability that the banker continues as banker given that the outcome is a member of any of the three classes. In the first class the banker always continues. Subdividing the second class into two according to whether the third card has a face value larger or smaller than the two having the same face value, Bernoulli notes that the banker always continues if the outcome is a member of the first subclass, whereas he only has probability $\frac{2}{3}$ for continuing in the second subclass. In the third class the banker also continues with probability $\frac{2}{3}$ because four of the six permutations of the three different face values are favorable to the banker. Weighting the number of cases found above with these probabilities and dividing by the total number of cases, Bernoulli finds that the banker's probability of continuation is $(16f^2 - 3f - 4)/(24f^2 - 18f + 3)$.

For once, the principle of Bernoulli's proof is not completely clear because he gives the results as polynomials in f without stating the intermediate results in terms of binomial coefficients and number of permutations. It is therefore not completely trivial for the reader to derive the general result indicated in Problem 5, §15.8.

After finding the two basic probabilities, Bernoulli considers a series of games and use the result from Problem 19 to find the banker's expectation.

Problem 21

In Problem 21 Bernoulli discusses the banker's advantage in the card game *Bassette*. We shall give a somewhat formalized description of the rules of the game. Consider a pack of $2n$ cards of which k are marked a and $2n - k$ marked b, for example, for $2n = 52$ and $k = 4$. Drawing two cards successively without replacement, the four possible outcomes are ab, ba, aa, and bb. In the first case the banker wins one from the punter; in the second cases he loses one, in the third case he wins one; and in the fourth case the play goes on with the banker drawing another pair of cards from the stock. This is the main rule; like Bernoulli, we shall discuss some modifications later.

To find the banker's expectation, Bernoulli states that we may disregard the outcomes ab with a gain of 1 and ba with a loss of 1 because they have the same probability of occurrence. He therefore considers only the number of chances for aa in a series of n drawings.

For $k = 1$, a doublet is impossible, so that the number of chances for aa is zero.

For $k > 1$, the total number of permutations of the $2n$ letters is $\binom{2n}{k}$. The banker wins if the first drawing of two cards gives aa or if the second drawing gives aa and, so on. The corresponding numbers of permutations are

$$\binom{2n-2}{k-2}, \quad \binom{2n-4}{k-2}, \ldots$$

Hence Bernoulli finds that the banker's expectation is

$$\sum_{i=1}^{n} \binom{2i-2}{k-2} \Big/ \binom{2n}{k}, \qquad k = 2, 3, \ldots, 2n.$$

Bernoulli does not give the general formula but derives the results for $k = 2, 3$, and 4 by reasoning as above. He also finds that the sums of the binomial coefficients are polynomials in n, see Problem 6, §15.8.

To make the banker's expectation positive for $k = 1$ as well, the main rule is amended so that the banker loses nothing if a occurs as the last card. Hence, of the $2n$ permutations, $n-1$ pairs give no contribution to the expectation; one permutation gives a gain of 1; and one gives zero, so that the banker's expectation becomes $1/2n$.

The second modification consists of a reduction in the banker's gain from 1 to $\frac{2}{3}$ when a occurs as the first card at the beginning of the game. The reduction in the banker's expectation thus amounts to $\frac{1}{3}$ times

$$\binom{2n-1}{k-1},$$

which is the number of permutations with a in the first place.

Denoting the banker's expectation by e_k, the result may be written as $e_1 = 1/3n$ and

$$e_k = \left[\sum_{i=1}^{n} \binom{2i-2}{k-2} - \frac{1}{3}\binom{2n-1}{k-1} \right] \Big/ \binom{2n}{k}, \qquad k = 2, 3, \ldots, 2n.$$

Bernoulli gives the results as

$$e_2 = \frac{n+1}{6n^2 - 3n}, \qquad e_3 = \frac{n+1}{4n^2 - 2n}, \qquad e_4 = \frac{4n^2 + n - 6}{12n^3 - 24n^2 + 9n}.$$

He also derives the expectation corresponding to further modification of the rules; however, we shall ignore this because it does not involve any new probabilistic arguments.

Bernoulli refers to the paper by Sauveur in *Journal des Sçavans*, 1679, containing the same results without proof.

The banker's expectation has also been analyzed by Montmort (1708, pp. 65–74; 1713, pp. 144–156), who derives the general formula both by combinatorics and by recursion, and by de Moivre (1718, pp. 32–39).

15.6   BERNOULLI'S LETTER ON THE GAME OF TENNIS

In his *Letter to a Friend on the Game of Tennis*, Bernoulli begins with a summary of his considerations in the *Ars Conjectandi* on the difference between games of chance and games that depend on the skill of the players, on the corresponding determination of probabilities a priori and a posteriori, and on the law of large numbers, which justifies the use of the relative frequency of winning as a measure of the probability of winning. Apart from this short introduction, the letter is really an exercise in probability theory and could well have been included in Part 3 of the *Ars Conjectandi*.

Bernoulli writes that he will not explain the rules of the game because they are well known. The game is more complicated than tennis but with the same scoring rules; a detailed description of the game has been given by Haussner (1899). We shall disregard the rules related to a modification named "chase."

Bernoulli analyzes many problems of tennis. There are, however, no new methods used in his analysis; he keeps strictly to the methods used by Huygens, solving most of the problems by recursion between expectations. The letter is an imposing work, demonstrating Bernoulli's pedagogical qualities, his ability to systematize, and his thoroughness. We shall confine ourselves to a discussion of the main points, leaving out most of the details. It seems that Bernoulli's results have been overlooked by modern writers on the game.

For convenience we shall award a player one point for each play that he wins rather than using Bernoulli's scoring system $(0, 15, 30, 45, \text{game})$. Player A's probability of winning a point will be denoted by p and player B's probability by $q, p + q = 1$. We shall denote the state of a game by the number of points, (i, j), say, won by the two players. The game is won by the player who scores four points before the other player scores more than two points; furthermore, if the game reaches the state $(3, 3)$, the player who first wins two points more than his opponent wins the game.

Using modern terminology, the play may be described as a random walk in two dimensions with absorbing barriers, see Fig. 15.6.1. The random walk starts at $(0, 0)$ and moves one step to the right, with probability p if player A wins, and one step up, with probability q if player B wins. In Fig. 15.6.1 we have cut off the continuation region at the score $(7, 7)$.

Let $g(i, j)$ denote A's probability of winning the game, given that the game is in state (i, j). Since A wins the next points with probability p and loses with probability q, we have

$$g(i,j) = pg(i + 1, j) + qg(i, j + 1). \tag{1}$$

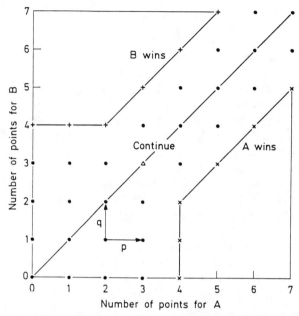

Fig. 15.6.1. A random-walk diagram for a game of tennis limited to seven plays.

This is the fundamental formula which Bernoulli derives and uses to tabulate $g(i,j)$, with the modification that he uses $n = p/q$ as parameter.

Beginning with the state $(3, 3)$ and using the recursion two times, Bernoulli finds

$$g(3,3) = p^2 g(5,3) + 2pqg(4,4) + q^2 g(3,5).$$

Since $g(5,3) = 1$, $g(3,5) = 0$, and $g(4,4) = g(3,3)$, he gets

$$g(3,3) = \frac{p^2}{p^2 + q^2} = \frac{n^2}{n^2 + 1}. \tag{2}$$

Using (1) again, he obtains

$$g(2,3) = pg(3,3) + qg(2,4) = pg(3,3) = \frac{n^3}{n^3 + n^2 + n + 1},$$

and continuing in this manner, he finds $g(i,j)$ for $i \leqslant 3$ and $j \leqslant 3$ and thus solves the problem completely. He tabulates $g(i,j)$ as the ratio of two polynomials of the same degree in $n = p/q$, as shown in the following example:

$$g(0,0) = \frac{n^7 + 5n^6 + 11n^5 + 15n^4}{n^7 + 5n^6 + 11n^5 + 15n^4 + 15n^3 + 11n^2 + 5n + 1}. \tag{3}$$

By means of these formulae, Bernoulli calculates all the values of $g(i,j)$ for $p/q = 1, 2, 3, 4$.

We note that (1) leads to

$$(p^2 + q^2)g(i,j) = p^{4-i}f_{4-j}(p,q), \tag{4}$$

where f denotes a homogeneous polynomial in (p,q) of degree $4 - j$. For example,

$$(p^2 + q^2)g(0,0) = p^4(p^4 + 6p^3q + 16p^2q^2 + 26pq^3 + 15q^4), \tag{5}$$

in agreement with (3). We further note that (4) may be reduced to a simpler form by introducing $x = 4 - i$, $y = 4 - j$, and $d(x,y) = (p^2 + q^2)g(i,j)$. We shall leave that to the reader.

Bernoulli uses his results to determine the size of handicaps to get a fair game. He first asks the question; How many points should be accorded the weaker player for the game to be fair? Suppose that $p/q = 2$. Then, Bernoulli's table shows that $g(0,2) = 208/405 = 0.514$, so that a handicap of two points to B will nearly equalize their chances of winning. Considering the same problem for $p/q = 3$, Bernoulli notes that $g(0,2) = 891/1280 = 0.696$ and $g(0,3) = 243/397 = 0.612$, so that handicaps of two and three are not enough to equalize the chances. He then finds that $g(1,3) = 81/160 = 0.506$, which means that a game starting with one point for A and three points for B will be almost fair.

He next solves the inverse problem: If B has been given a handicap of j points to make the game fair, what does that mean for the relative strength of the players? Obviously one has to solve the equation $g(0,j) = \frac{1}{2}$ with respect to $n = p/q$ for a given value of j. This leads to an algebraic equation in n. For $j = 2$, say, corresponding to the first example above, Bernoulli solves an equation of the sixth degree and finds $n = 1.946$.

Bernoulli also discusses the probability of winning a set of games. He states that because of notational difficulties he will only illustrate this problem by examples. As usual, however, his procedure is very clear and easy to translate to modern notation. Let $s(u,v)$ denote A's probability of winning the set when A and B have already won u and v games, respectively. Bernoulli's procedure corresponds to the recursion formula

$$s(u,v) = g(0,0)s(u+1,v) + [1 - g(0,0)]s(u,v+1), \tag{6}$$

which is analogous to (1) with $g(0,0)$ substituted for p. Bernoulli's difficulties stem from the fact that he does not have a notation for the probabilities which we have denoted by $g(i,j)$ and $s(u,v)$.

Generalizing (6) to the case where the number of points is (i,j) in game (u,v), Bernoulli uses the formula

$$m(u,v;i,j) = g(i,j)s(u+1,v) + [1 - g(i,j)]s(u,v+1) \qquad (7)$$

to find A's probability of winning the set. Bernoulli uses these formulae to discuss the problem of handicaps. We shall report only one of his examples.

Suppose that B has a handicap of "half-45," which in our notation means that in alternate games he has a handicap of two and three points, respectively. The problem is to find the value of $n = p/q$ for which A's probability of winning equals $\frac{1}{2}$. Considering two games in succession, A's probability of winning the first and the second, respectively, equals $g(0,2) = a/(a+b)$ and $g(0,3) = c/(c+d)$, the ratios being Bernoulli's notation. His reasoning may

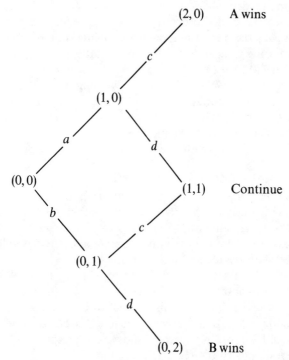

Fig. 15.6.2. The possible states of two games of tennis, with the number of chances of winning and losing.

be illustrated by means of Fig. 15.6.2, where the states refer to the number of games.

Let A's probability of winning in the state $(0, 0)$ and therefore also in the state $(1, 1)$ be denoted by z. By recursion Bernoulli finds that

$$\frac{a[(c + dz)/(c + d)] + b[cz/(c + d)]}{a + b} = z,$$

which leads to $z = ac/(ac + bd)$. It will be seen that the method used by Bernoulli is the same that Huygens used in solving his fifth problem, see §6.3, Fig. 6.3.1. Setting $z = \frac{1}{2}$, Bernoulli finds that $a/b = d/c$ so that

$$\frac{g(0, 2)}{1 - g(0, 2)} = \frac{1 - g(0, 3)}{g(0, 3)},$$

an equation of the 11th degree in n, which according to Bernoulli has the root $n = 2.7$. This root lies between the two roots $n = 1.9$ and $n = 4.2$, corresponding to handicaps of two and three, respectively.

Bernoulli also extends his model by taking into account the fact that the player who serves has a greater probability of winning a point than when he is not serving. Further, he discusses a game with three and four players.

15.7 BERNOULLI'S CONCEPT OF PROBABILITY AND HIS PROGRAM FOR APPLIED PROBABILITY

After having discussed games of chance in the first three parts of *Ars Conjectandi*, Bernoulli turns to a discussion and extension of the probability concept and its applications in Part 4. To understand the following account of Bernoulli's work, we shall comment on the various concepts of probability.

As noted in Chapter 3 there are two types of probability: objective and subjective. Objective probabilities are used for describing properties of chance setups and chance events. We attempt to measure objective probabilities by calculating relative frequencies from experiments or observations, and for this reason objective probabilities are also sometimes called frequentist or statistical probabilities. Subjective probabilities are used for measuring the degree of belief in a statement or a proposition about things or events; they thus refer to our imperfect knowledge or our judgment and not directly to the things or events about which the statements are made. We attempt to measure subjective probability (at least since the 1920s) by finding the odds or the betting ratio at which the person in question is willing to bet on the truth of the statement. It is, of course, confusing that the same word is used for two concepts so fundamentally different.

In philosophical literature the two types of probability are also called *aleatory* and *epistemic*, respectively.

The two main types of probability have been subdivided into several classes, according to the field of application considered and various philosophical opinions of induction. For more than a century there have been heated discussions among statisticians, probabilists, and philosophers about the correct definition of probability.

It is important to note the distinction between probability and chance which was used in the 17th century and before. Probability was related to opinions, propositions, and beliefs, like subjective probability, whereas chance was used to mean objective probability. The distinction between objective and subjective probability was formulated clearly by Cournot (1843, pp. V–VII and 437–440) who also tried to revive the old usage, as is evident from the title of his book, *An Exposition of the Theory of Chances and Probabilities*. Poisson (1837, p. 31) also distinguished between chance and probability.

From a mathematical point of view, however, it is not necessary or even desirable to define probability explicitly. According to the usual axiomatic procedure, probability is an undefined notion, a real number between 0 and 1 satisfying certain rules of operation from which the calculus of probability is developed by deduction. From this point of view any interpretation of probability is admissible if only the axioms are satisfied. It is therefore the duty of the applied probabilist or the statistician to demonstrate that the objects under study obey the rules derived from the axioms. Since the number of chances by definition is additive and the corresponding probability is a proper fraction, and since objective probability is an idealized relative frequency, the axioms have been chosen, implicitly in the beginning and explicitly later on, such that probabilities satisfy the same basic rules of operation as relative frequencies. From the beginning the basic concepts have been additivity and independence.

It is characteristic of games of chance that they may be resolved into a finite number of possible outcomes which are mutually exclusive and equally likely (probable) so that the probability of success may be found as the ratio of the number of favorable cases to the total number of cases. Of course, such calculation of the probability presumes an idealized chance mechanism that can only be approximated by the coins, dice, cards, urns, lotteries, roulettes, etc., actually used. From Cardano to Bernoulli this assumption has been expressed by phrases, such as equal conditions, honest dice, pure chance, equal ease of occurrence, and symmetry.

After Graunt in 1662 had demonstrated the stability of statistical ratios for some demographic phenomena and constructed his life table, the calculus of chances was extended to cover mortality problems, seemingly without any hesitation or qualification. Huygens and his contemporaries called the distribution of deaths according to age the chances of death, even if such chances could not be found a priori but had to be calculated from the observed death rates.

We owe to Bernoulli the important distinction between probabilities which can be calculated a priori (deductively, from considerations of symmetry) and those which can be calculated only a posteriori (inductively, from relative frequencies). (This characterization has nothing to do with a priori and a posteriori distributions used in modern Bayesian terminology.)

In the first chapter of Part 4 Bernoulli defines the concepts of contingency and probability. According to Hacking (1971), similar ideas had previously been expressed by Leibniz; however, Bernoulli gave the first systematic exposition of these fundamental concepts, and his formulations and viewpoints have been influential ever since, as we shall indicate below.

In his discussion it is the theologian and philosopher more than the mathematician who speaks. Like most of his contemporaries, Bernoulli considered the world to be deterministic. The omnipotent and omniscient God leaves nothing to chance; every event is determined by Him, normally by causal laws. Bernoulli therefore states that every event is objectively certain. He then turns to subjective certainty, which is certainty in relation to us and thus a measure of our knowledge. Bernoulli defines probability as follows:

> probability is a degree of certainty and differs from absolute certainty as a part differs from the whole. If, for example, the whole and absolute certainty—which we designate by the letter a or by unity—is supposed to consist of five probabilities or parts, three of which stand for the existence or future existence of some event, the remaining two against its existence, this event is said to have $(3/5)a$ or $3/5$ certainty (p. 211).

For this calculation Bernoulli presumably has Huygens' definition of expectation is mind: There are three cases for a and two cases for zero, so that the expectation becomes $\frac{3}{5}a$.

Likewise, what is considered *contingent* depends on our knowledge. Bernoulli illustrates this by means of three examples: (1) predicting the outcome of throwing a die, (2) forecasting tomorrow's weather, and (3) predicting a planetary eclipse. He stresses that if we could measure the initial conditions accurately and if we knew the physical laws governing the motions involved, then we would be able to predict the happenings in the first two examples just as accurately as in the third. He writes, experience has maintained that eclipses alone are calculated be necessity, but that the fall of a die and the future occurrence of a storm are calculated by contingencies. There is no reason for this fact other than that those data which are supposed to determine later events (and especially such data which are in nature) have nevertheless not been learned well enough by us. If they had been learned well enough by us, the studies of geometry and physics have not been well enough refined so that these effects can be calculated from data in the same way that eclipses can be computed and predicted from the principles of astronomy. Therefore just as much as the fall of a die and the occurrence of a storm, these very eclipses—before astronomy had advanced to such a degree of perfection—needed to be referred among future contingencies. Hence, it follows that what can seem to be to one person at one time a contingent event may be at another time to another person (indeed, to the very same person) a necessary event after its causes have been learned (pp. 212–213).

Laplace (1814, p. ii) held opinions similar to Bernoulli's on determinism and contingency with the modification that he replaced God by "an intelligence which could comprehend all the forces of nature."
In the philosophical literature of today one also finds formulations which can be traced to Bernoulli. Popper (1959, p. 205) writes,

One sometimes hears it said that the movements of the planets obey strict laws, whilst the fall of a die is fortuitous, or subject to chance. In my view the difference lies in the fact that we have so far been able to predict the movements of the planets successfully, but not the individual results of throwing dice. In order to deduce predictions one needs laws and initial conditions; if no suitable laws are available or if the initial conditions cannot be ascertained, the scientific way of prediction breaks down.

Bernoulli also introduces the important concept of *moral certainty*. He writes,

That is *morally certain* whose probability nearly equals the whole certainty, so that a morally certain event cannot be perceived not to happen; on the other hand, that is *morally impossible* which has nearly as much probability as renders the certainty of failure moral certainty. Thus if one thing is considered morally certain which has 999/1000 certainty, another thing will be morally impossible which has only 1/1000 certainty (pp. 211–212).

Because it is still rarely possible to obtain total certainty, necessity and use desire that what is merely morally certain be regarded as absolutely certain. Hence, it would be useful if, by the authority of the magistracy, limits were set up and fixed concerning moral certainty. I mean, if it were fixed whether 99/100 certainty would suffice for producing moral certainty, or whether 999/1000 certainty would be required. Note that then a judge could not be biased, but he would have a guideline which he would continually observe in passing judgment (p. 217).

Today we use *practically certain* rather than morally certain. Most textbooks on statistics contain such considerations as given by Cramér (1946, p. 149): "If E is an event of this type [i.e., having a small probability] and if the experiment $\mathscr{E}$ is performed one single time, it can thus be considered as practically certain that E will not occur".

R. A. Fisher (1925), the statistical magistrate of our time, has made Bernoulli's 99 and 99.9% probability limits, supplemented by the 95% limits, popular among applied statisticians in testing and estimation problems.

The art of conjecture, Bernoulli says, is the art of measuring as exactly as possible the probabilities of things. Furthermore, probability is a degree of certainty and as such a measure of our imperfect knowledge; probability is personal in the sense that it varies from person to person according to his knowledge, and for the same person it may vary with time as his knowledge changes. We shall discuss Bernoulli's attempt to measure this elusive quantity.

According to Bernoulli, probability theory is applicable to all fields of the empirical sciences—physical, moral, and political—and to everyday life as well. All empirical knowledge is probable knowledge only. Examples of fields where certainty reigns are the Scriptures, logic, and mathematics.

Bernoulli's discussion of the measurement question is closely related to his distinction between *a priori* and *a posteriori* probabilities.

With respect to games of chance, Bernoulli states that experience has taught the inventors of such devices how to obtain physical symmetries in their constructions, and we may therefore calculate the probability of a specified outcome as the number of favorable cases divided by the total number of cases, since these are equally likely.

In most other situations, however, it is impossible to apply this procedure because we are not able to find the number of equally likely cases. As examples Bernoulli lists the probability of dying at a certain age, of contracting a certain disease, of changes in the weather, and of winning a game that depends on the intelligence or physical abilities of the players. The outcome of such events depends on causes that are completely hidden from us, and the many possible interactions of these causes will always deceive us.

> But indeed, another way is open to us here by which we may obtain what is sought; and what you cannot deduce *a priori*, you can at least deduce *a posteriori*— i.e. you will be able to make a deduction from the many observed outcomes of similar events. For it may be presumed that every single thing is able to happen and not to happen in as many cases as it was previously observed to have happened or not to have happened in like circumstances (p. 224).

Bernoulli adds

> that this empirical way of determining the number of cases by trials is neither new nor unusual, for the celebrated author of the *Ars Cogitandi* [Arnauld's *Logic*,

1662] a man of great insight and intelligence, prescribes a similar method in Chapter 12 & ff. of the last part of his work; and everyone constantly observes the same thing in daily life (p. 225).

Bernoulli obviously considered the *Ars Conjectandi* (*The Art of Conjecturing*) an extension of the *Ars Cogitandi* (The Art of Thinking), the most popular textbook on logic at his times. Bernoulli stresses that everyone uses the a posteriori method for estimating probabilities in daily life. In view of the fact that Graunt had published his *Observations* with its abundance of examples in the same year as Arnauld's *Logic*, it is surprising that Bernoulli refers only to *Logic*, which contains only some naive reflections of a general nature about a posteriori probabilities without empirical examples of relative frequencies.

How are we to determine the probability in the vast number of situations which do not fall under the two categories mentioned above, for example, in situations in which no repetition is possible?

Bernoulli states that "probabilities are estimated by the *number* and the *weight* of *arguments*" (p. 214) and that arguments are either intrinsic, depending on the thing or event itself, or extrinsic, depending on the authority and testimony of men.

He then lists nine "general rules dictated by common sense":

1. One must not use conjecture in cases where complete certainty is obtainable.

2. One must search for all possible arguments or evidence concerning the case.

3. One must take into account both arguments for and against the case.

4. For a judgment about general events, general arguments are sufficient; for individual events, however, special and individual arguments have to be taken into account.

5. In case of uncertainty, action should be suspended until more information is at hand; however, if circumstances permit no delay, the action that is most suitable, safe, wise, and probable should be chosen.

6. That which can be useful on some occasion and harmful on no occasion is to be preferred to that which is useful and harmful on no occasion.

7. The value of human actions must not be judged by their outcome.

8. In our judgments we must be wary of attributing more weight to a thing than its due and of considering something that is more probable than another to be absolutely certain.

9. This is the rule, already discussed, that states that moral certainty should in practice be considered absolute certainty.

The nine rules are illustrated by examples. According to rule 1, an astronomer should not guess about eclipses when he can predict an eclipse with certainty by calculation. Rule 2 states that we must take into account all available information and the specific circumstances causing the event in question. Rule 4 is illustrated by the distinction between an age-specific death rate and the probability of dying for a specific person of that age, taking all information about his health into account.

It is easy to agree with Bernoulli that these rules are reasonable, but it is difficult to see how they can be of much use in finding the numerical values of probabilities.

Bernoulli goes on to discuss the classification of arguments and the determination of the probability corresponding to a combination of arguments. First, he distinguishes among three types of arguments: those that exist necessarily and indicate the thing contingently; those that exist contingently and indicate the thing necessarily; and those that both exist and indicate the thing contingently. He gives some examples, but it is nevertheless difficult to understand his meaning. Shafer (1978) suggests the following explanation:

> If we think of an argument as consisting of premises and conclusion, then we surely come close to Bernoulli's meaning if we say that the argument exists contingently when the premises do not necessarily hold, and that the argument proves contingently when the premises do not necessarily entail the conclusion.

Another classification introduced by Bernoulli is between *pure* and *mixed* arguments. A pure argument proves a thing with a certain probability without giving a positive probability to the opposite thing, whereas a mixed argument proves a thing with a certain probability and proves the opposite with the complementary probability. Considering several arguments for proving the same thing, Bernoulli lists for each argument the total number of cases, the number of cases proving the thing, and the number of cases not proving (pure argument) or proving the opposite (mixed argument). He then derives the power of the proof of all the arguments combined by means of Huygens' formula for expectations. If all the arguments are mixed, this procedure gives the usual result for the probability of a combination of independent events, whereas special results are obtained if some or all of the arguments are pure because the probabilities of such an argument and its opposite do not add to unity. Bernoulli's discussion does not seem to have influenced contemporary probabilists. This section is rather difficult to understand and has only recently been analyzed in detail by Shafer (1978).

Bernoulli does not, however, give a single example which convincingly demonstrates how to determine a subjective probability numerically. All his examples are artificial and constructed in analogy with the classical definition; he speaks of probabilities instead of the number of chances. This is apparent in connection with his definition of probability: there are three cases for the existence of an event and two cases against, hence the probability of its existence is $\frac{3}{5}$. In another example, Bernoulli discusses a murder case and attempts to combine the arguments for Gracchus being the murderer. He writes,

> From what has been said previously, it is clear that the power which any arguments has depends on a multitude of cases in which it can exist or not exist, in which it can indicate or not indicate the thing, or even in which it can indicate the opposite of the thing. And so, the degree of certainty or the probability which this argument generates can be computed from these cases by the method discussed in the first part just as the fate of gamblers in games of chance are accustomed to be investigated.... Moreover, I assume that all cases are equally possible, or that they all can happen with equal ease; for in other cases discretion must be applied, and any case which occurs rather readily must be counted as many times as it occurs more readily as others (pp. 218–219).

It is little wonder that Bernoulli was unable to solve this problem which has occupied philosophers ever since; however, his indication of a solution, as quoted above, had wide-ranging consequences because in the hands of Laplace, it led to the *principle of insufficient reason*, which asserts that equal probabilities must be assigned to each of several alternatives if there is *no known reason* for preferring one to another.

How did the contemporaries and immediate successors of Bernoulli react to his definition of probability? Montmort, de Moivre, and Struyck limited themselves to discussions of games of chance and mortality statistics so it was not necessary for them to introduce a new concept of probability. However, in the prefaces to their books, both Montmort and de Moivre made some remarks on the concepts of probability, to which we shall return.

Bernoulli envisioned the application of probability theory to "civil, moral and economic affairs," as expressed in the title of Part 4, but he never succeeded in carrying out this program. It has been suggested that he did not complete the manuscript and publish the book because he wanted to include some numerical examples. He wrote to Leibniz and asked him for examples from jurisprudence and to send him a copy of de Witt's treatise, but Leibniz did not help him. It is strange that Leibniz did not refer him to the paper by Halley (1694) with its statistical analysis of the Breslau mortality data and its evaluation of annuities on lives; perhaps Leibniz took it for granted that Bernoulli knew of Halley's paper. Bernoulli might also have gotten the data he needed from Graunt's book or by writing to Caspar Neumann at Breslau or by looking for bills of mortality analogous to the

London bills in his native Basel, or if not in Basel then in Geneva. For incomprehensible reasons he did none of these obvious things, so he never succeeded in getting a set of observations to illustrate his theory.

During 1703–1705, Bernoulli corresponded with Leibniz about the estimation of probabilities a posteriori and about his main theorem. In a letter dated 3 October 1703, Bernoulli explains his results in words nearly identical to those used in *Ars Conjectandi*. In his reply of 3 December 1703, Leibniz is somewhat skeptical. He says that in some political and legal situations, "there is not as much need for fine calculation as there is for the accurate recapitulation of all the circumstances." Presumably, he means that his experiences as a diplomat and jurist show that in such situations it is not possible to find cases where the same circumstances prevail; therefore, a posteriori estimation is impossible. Furthermore, in other situations, even if repetition is possible, there is the difficulty "that happenings which depend upon an infinite number of causes cannot be determined by a finite number of experiments." Mortality, he says, depends on an innumerable number of causes, and furthermore new diseases continually crop up and change the rate of mortality. In his reply of 20 April 1704, Bernoulli claims that probability theory is important to legal affairs with regard to insurance, annuities on lives, marriage contracts, and so on. He again explains that his theorem gives him the means to determine an unknown probability with moral certainty and that in principle there is no difference between the determination of an unknown ratio of white and black balls in an urn and an unknown rate of mortality. That the number of causes of death is very large or even infinite, whereas the number of balls in the urn is finite, does not matter because the ratio of two infinite numbers may be finite. If new diseases occur, the mortality rate changes, and new observations are necessary to estimate the new rate.

They did not quite understand each other. Leibniz's skepticism seems to be based on the practical difficulties of getting a sufficiently long series of observations under the same essential conditions, whereas Bernoulli's main occupation was to explain his important theoretical result.

In a letter from April 1705, four months before Bernoulli died, Leibniz refers to Pascal's *Traité*; Bernoulli did not respond to this letter, neither did he mention Pascal's *Traité* in his book.

Without mentioning Leibniz's name, but under the guise of objections from "some learned men," Bernounlli recounts the discussion in *Ars Conjectandi*, pp. 227–228.

Most of the correspondence has been translated into English by Bing Sung (1966) as an appendix to his translation of *Ars Conjectandi* and into German by Kohli (1975c), who has also given a detailed commentary.

The ideas of subjective probability and the applicability of the calculus of probability to other fields than games of chance had previously been indicated by Locke (1690), Arbuthnott (1692), and Hooper (1700) as mentioned in §12.2. Bernoulli does not refer to these authors. Bernoulli's discussion is much clearer and more comprehensive than the previous ones.

In addition to the papers previously mentioned, we refer to Hacking (1971, 1975), Schneider (1972, 1984), and van der Waerden (1975b, 1975c).

15.8 PROBLEMS FROM *ARS CONJECTANDI* AND BERNOULLI'S LETTER ON TENNIS

Problems from *Ars Conjectandi*, Part 3

1. Six persons participate in a play consisting of five successive games. In the first game, A_1 play against A_2; the winner plays against A_3 in the second game, and so on, until the winner of the fourth game plays against A_6 in the last game. After the first game, where the chances of winning are equal for A_1 and A_2, any player playing against his second opponent has odds 2:1 of winning, against his third opponent 4:1, and so on. Prove that the number of chances of winning for the six participants are 7680:7680:5440:4488:4250:4887 (Bernoulli's Problem 4).

   Generalization. Let $p_{ij}, i < j$, denote the probability that A_i defeats A_j in a single game so that the probability that A_j defeats A_i is $q_{ij} = 1 - p_{ij}$. Find the probabilities of winning for the six players.

2. A pack of 20 cards contains 10 court cards. Three players draw at random seven, seven and six cards, respectively. The one who gets the largest number of court cards wins. If two of the players get the same number of court cards, which is larger than the number obtained by the third player, they divide the stake equally. Show that the number of chances of winning are 142,469:142,469:84,574 (Bernoulli's Problem 9).

   Generalization. Let the pack contain n cards that are distributed randomly among the three players with n_1, n_2, and n_3 cards, respectively. Let the number of court cards be m. The player who gets the largest number of court cards wins. If several players get the same and largest number of court cards, they divide the stake equally. Find the probability of winning for the three players.

3. Throwing a die six times find the probability of getting (a) six different faces (Bernoulli's Problem 11); (b) an ace in the first throw, a deuce in the second, and so on (Bernoulli's Problem 12).

4. An urn contains 32 tickets, four marked 1, four marked 2, and so on, until four marked 8. Four tickets are drawn without replacement. Find the number of chances of getting the different possible sums between 4 and 32 (Bernoulli's Problem 17).

5. *Generalization of Bernoulli's Problem 20.* A random sample of n cards is selected from a pack of $N = sf$ cards. The composition of the sample is given by

$$n = n_1 + \cdots + n_k, \qquad k = 1, \ldots, f,$$
$$1 \leqslant n_1 \leqslant \cdots \leqslant n_k \leqslant s,$$

where n_i denotes the number of cards having the same face value and k the number of different face values. Further, let a_k denote the number of permutations of $(n_1, \ldots, n_k)$.

As pointed out by Haussner (1899), the basic identity to be used for solving Bernoulli's problem is

$$\sum a_k \binom{s}{n_1} \cdots \binom{s}{n_k} \binom{f}{k} = \binom{fs}{n},$$

where the summation is over all values of $(n_1, \ldots, n_k)$ and k, satisfying the relations above.

Let d_k denote the number of permutations of $(n_2, \ldots, n_k)$ for $n_1 = 1$, and let $d_k = 0$ for $n_1 > 1$. Prove that the banker's expectation equals

$$\sum \left(a_k - \frac{d_k}{n} \right) \binom{s}{n_1} \cdots \binom{s}{n_k} \binom{f}{k} \bigg/ \binom{fs}{n}$$

(see Haussner, 1899).

6. Prove that

$$\sum_{i=1}^{n} \binom{2i-2}{2} = \frac{1}{6}(4n^3 - 9n^2 + 5n)$$

(part of Bernoulli's Problem 21).

7. Suppose that A's probability of winning in a single trial is p and that he has to pay the amount 1 to B for each trial. If A wins he gets the amount

m from B, and if he loses a given number of consecutive trials, n, say, he gets the amount n back from B. Prove that A's expectations is

$$m(1 - q^n) - \left(\frac{1 - q^n}{p} - nq^n \right)$$

(Bernoulli's Problem 22).

Problems on the Game of Tennis

8. Use the recursion formula (6.1) to find the polynomials $f_{4-j}(p, q)$ defined by (6.4).

9. Write $d(x, y)$ defined in §15.6 in the form

$$d(x, y) = p^x \sum_{i=0}^{y} p^{y-i} q^i a_i(x, y);$$

derive a recursion formula for $a_i(x, y)$ and tabulate these functions.

10. Prove directly that

$$g(0, 0) = p^4(1 + 4q + 10q^2) + \frac{20p^5 q^3}{1 - 2pq}.$$

Show that this result is identical to (6.3) and (6.5).

11. Let $(u, v) = (1, 2); (i, j) = (1, 3)$; and $p/q = 2$. Show that A's probability of winning the set equals $19{,}031{,}314{,}432/23{,}643{,}278{,}649$ (see Bernoulli's *Letter*, p. 11).

12. Suppose that B's handicap is j points in each game, $j = 1, 2, 3$; however, A prefers to play with B without any handicap in each game and instead to give B a handicap expressed as a number of games, $x - 1$, say, so that A wins the set if besides the $x - 1$ games, he wins two games in succession. Let P denote A's probability of winning a single game. Show that A's probability of winning the set equals

$$\frac{P^{x-1} P^2}{P^2 + Q^2}, \qquad P + Q = 1,$$

and find the value of x, making this probability equal to $\frac{1}{2}$ for $j = 3$. Bernoulli finds $x = 38$ (see his *Letter*, pp. 23–26).

CHAPTER 16

Bernoulli's Theorem

Therefore, this is the problem which I have decided to publish here after I have pondered over it for twenty years. Both its novelty and its great utility, coupled with its just as great difficulty, exceed in weight and value all the other chapters of this doctrine.

—*JAMES BERNOULLI in Ars Conjectandi, 1713*

16.1 BERNOULLI'S FORMULATION OF THE PROBLEM

In Bernoulli's time it was recognized that the equally large (a priori) probabilities attributed to the outcomes of a game of chance constitute a mathematical model from which the probabilities of compound events resulting from the combination of several games can be calculated, essentially by combinatorial methods. When this calculus of chances was applied to other fields, for example, to vital statistics and insurance, relative frequencies were used as estimates of probabilities. As noted by Bernoulli, it is impossible in these applications to identify the number of equally likely cases. Therefore it is necessary to establish a new justification for applying the calculus of chances outside the field of games of chance.

Considering the relative frequency of an event, calculated from observations taken under the same circumstances, Bernoulli states as a fundamental empirical fact that "the more observations that are taken, the less the danger will be of deviating from the truth" (p. 225). He adds that this is well known and that everyone knows that it is not enough to take one or two observations but that a large number must be taken to determine the probability of the event in this way.

257

Bernoulli continues:

Something further must be contemplated here which perhaps no one has thought about till now. It certainly remains to be inquired whether after the number of observations have been increased, the probability is increased of attaining the true ratio between the number of cases in which some event can happen and in which it cannot happen, so that this probability finally exceeds any given degree of certainty; or whether the problem has, so to speak, its own asymptote—that is, whether some degree of certainty is given which one can never exceed (p. 225).

However, lest these remarks be misunderstood, it must be noted that I do not wish for this ratio, which we undertake to determine by trials, to be accepted as precise and accurate (for then the contrary would result from this, and it would be the more unlikely that the true ratio has been found the greater the number of observations made); but rather, the ratio is taken in some approximation, i.e. it is bounded by two limits. Moreover, these limits can be set up as close together as one wishes (p. 226).

In other words, Bernoulli formulates the fundamental question, Does there exist a theoretical counterpart to the empirical fact mentioned above? If so, then we may safely apply the calculus of chances to the fields where stable relative frequencies exist. Hence, Bernoulli approaches the question on the relation between observations and models in statistics in much the same spirit as Newton approached the correspondence between astronomical observations and celestial mechanics.

Let us first formulate Bernoulli's theorem in modern terminology and notation so that it becomes easier to point out the restrictions imposed by Bernoulli. Consider n independent trials, each with probability p for the occurrence of a certain event, and let s_n denote the number of successes. According to probability theory, s_n is binomially distributed. The question is now, What does probability theory tell us about the mathematical object $h_n = s_n/n$, the relative frequency. Let E denote the event $\{|h_n - p| \leqslant \epsilon\}$. Then it can be proved that $\Pr\{E\} > 1 - \delta$ for $n > n(p, \epsilon, \delta)$, where ϵ and δ are any given small positive numbers. This is also expressed by saying that h_n converges in probability to p for $n \to \infty$. Of course, this property of h_n is a purely algebraic property of the binomial distribution, namely, that the ratio of the body (defined by E) to the tail (defined as the complement to E) of the distribution tends to infinity. Hence, the theorem is valid for any interpretation of the probabilities involved provided that the probability concept in question satisfies the axioms leading to the binomial distribution.

In the proof Bernoulli considers a trial with $t = r + s$ equally likely outcomes of which r are favorable so that $p = r/(r + s)$. He proves that h_n converges in probability to p or, in his terminology, that we can be morally certain that h_n does not deviate greatly from p if n is sufficiently large. He also states that h_n is an estimate of p that is just as good as the a priori value if n is infinitely large. Without further discussion he proposes to use the relative frequency of an event as an estimator of p in other fields of application as well, even if this a posteriori calculation of p has been justified only for trials with a finite number of equally likely outcomes.

According to the *Meditationes*, Bernoulli proved his theorem between 1688 and 1690. He began with a numerical example for $p = \frac{1}{2}$, continued with a general proof for $p = \frac{1}{2}$, and finally developed the proof for $p = r/(r + s)$ essentially in the form given in the *Ars Conjectandi* (pp. 228–238); for details of the early proofs we refer to van der Waerden (1975b).

Bernoulli's own formulation of the theorem is as follows: "It must be shown that so many observations can be made that it will be c times more probable than not that the ratio of the number of favourable observations to the total number of observations will be neither larger than $(r + 1)/t$ nor smaller than $(r - 1)/t$ (p. 236)." Note that the inequality

$$\frac{r-1}{t} \leqslant h_n \leqslant \frac{r+1}{t}$$

may be written $|h_n - p| \leqslant 1/t$. Bernoulli's $1/t$ corresponds to ϵ and $1/(c + 1)$ to δ.

Comparing Bernoulli's formulation of the theorem with his previous statement (quoted above) that the ratio is bounded by two limits which can be set up as close together as one wishes, it must have been natural for him to chose the limits as close together as possible, that is, $(r - 1)/t$ and $(r + 1)/t$, since r and t are integers. If he could prove his theorem for these limits, then obviously a similar and stronger result would hold for a set of wider limits; it is also obvious how to modify his proof to cover the case with limits equal to $(r \pm a)/t, a = 1, 2, \ldots$. Stigler (1986, p.67) correctly points out that "Bernoulli's ε was not an arbitrary positive number; it was always taken as $1/(r + s)$," but then he goes on to the dubious conclusion that "His estimation was essentially an attempt to identify a discrete r and s with 'moral certainty'."

16.2   BERNOULLI'S THEOREM, 1713

Bernoulli's Theorem. Let r and s be two positive integers and set $p = r/(r + s)$ and $t = r + s$. For any positive real number c, we have

$$\Pr\left\{|h_n - p| \leqslant \frac{1}{t}\right\} > \frac{c}{c + 1}, \tag{1}$$

for $n = kt$ sufficiently large, i.e., for $k \geqslant k(r,s,c) \vee k(s,r,c)$, where $k(r,s,c)$ is the smallest positive integer satisfying

$$k(r,s,t) \geqslant \frac{m(r+s+1)-s}{r+1}, \tag{2}$$

and m is the smallest positive integer satisfying

$$m \geqslant \frac{\ln[c(s-1)]}{\ln[(r+1)/r]}. \tag{3}$$

Remarks. For given p we may chose t as large as we like so that the interval for h_n becomes arbitrarily small. Apart from rounding to integers, k and thus n are linear functions of $\ln c$, which means that the left-hand side of (1) tends to 1 when c, and thus n, tends to infinity.

Bernoulli's proof is rather lengthy because he does not use indices to indicate the successive terms of a series; he also writes the binomial coefficients as ratios of two products, and he evaluates the tail probabilities both for the left and the right tail, even when one follows from the other. We shall follow Bernoulli's proof closely but take advantage of modern notation.

Proof. Bernoulli chooses n as a multiple of t, $n = kt$, $k = 1, 2, \ldots$, which makes the proof very neat. The expansion of $(p+q)^n$ contains $n+1 = kr + ks + 1$ terms. The expected number of successes $np = kr$ is the "central term"; to the left, there are kr terms and to the right, ks terms. Setting $h_n = s_n/n$, we have to find n such that

$$P_k = \Pr\{|s_n - kr| \leqslant k\} > \frac{c}{c+1},$$

or, equivalently, such that $P_k/(1 - P_k) > c$.

Since $(p+q)^n = (r+s)^n t^{-n}$, we need only consider the terms of $(r+s)^n$. We shall set

$$(r+s)^n = \sum_{x=0}^{n} \binom{n}{x} r^x s^{n-x} = \sum_{i=-kr}^{ks} f_i,$$

say, where

$$f_i = \binom{kr+ks}{kr+i} r^{kr+i} s^{ks-i}, \qquad i = -kr, -kr+1, \ldots, ks.$$

Since f_{-i} for $i = 0, 1, \ldots, kr$ is obtained from f_i for $i = 0, 1, \ldots, ks$ by interchange of r and s, results proved for f_i also hold for f_{-i}, and we shall therefore only give the proof for the right tail.

To prove that P_k, the central term of the series plus k terms to each side, is larger than c times $1 - P_k$, the sum of the $k(r - 1)$ terms of the left tail and the $k(s - 1)$ terms of the right tail, it is sufficient to prove that

$$\sum_1^k f_i \geqslant c \sum_{k+1}^{ks} f_i, \tag{4}$$

for $k \geqslant k(r, s, c)$.

Bernoulli first investigates the ratio

$$\frac{f_i}{f_{i+1}} = \frac{(kr + i + 1)s}{(ks - i)r} = \frac{rs + (i + 1)s/k}{rs - ir/k} > 1, \qquad i = 0, 1, \ldots, ks - 1. \tag{5}$$

From this expression it follows that

(a) f_i is a decreasing function of i for $i \geqslant 0$;
(b) $f_0 = \max\{f_i\}$;
(c) f_i/f_{i+1} is an increasing function of i for $i \geqslant 0$;
(d) $f_0/f_k < f_i/f_{k+i}$ for $i \geqslant 1$.

Partitioning the tail probability into $s - 1$ sums each containing k terms and using property (a), Bernoulli gets the upper bound

$$\sum_{k+1}^{ks} f_i \leqslant (s - 1) \sum_{k+1}^{2k} f_i,$$

and combining this with property (d), he gets

$$\frac{\sum_1^k f_i}{\sum_{k+1}^{ks} f_i} \geqslant \frac{\sum_1^k f_i}{(s - 1) \sum_{k+1}^{2k} f_i} > \frac{f_0/f_k}{s - 1}. \tag{6}$$

To prove (4) it is thus sufficient to prove that

$$\frac{f_0}{f_k} \geqslant c(s - 1).$$

It will be seen that Bernoulli has reduced the problem to evaluate the ratio of two terms of the binomial instead of two sums, or in geometrical terms, the ratio of two ordinates instead of the ratio of two areas.

From (5) it follows that

$$\frac{f_0}{f_k} = \frac{rs+s}{rs-r+(r/k)} \frac{rs+s-(s/k)}{rs-r+(2r/k)} \cdots \frac{rs+(s/k)}{rs}.$$

To find a lower bound for this ratio, Bernoulli states that the k factors lie between $(rs+s)/(rs-r)$ and 1 and that we can therefore choose any fixed number between these limits, $(r+1)/r$, say, so that the first of the factors are larger than this number and the following smaller. By suitable choice of k, the mth factor, $1 \leqslant m \leqslant k$, becomes equal to $(r+1)/r$; that is,

$$\frac{rs+s-(m-1)(s/k)}{rs-r+(mr/k)} = \frac{r+1}{r},$$

which gives the relation between k and m as

$$k = \frac{m(r+s+1)-s}{r+1}. \tag{7}$$

Hence, for this value of k the ratio f_0/f_k consists of m factors larger than or equal to $(r+1)/r$ and $k-m$ factors larger than 1 so that

$$\frac{f_0}{f_k} \geqslant \left(\frac{r+1}{r}\right)^m,$$

and it is thus sufficient to find m from the inequality

$$\left(\frac{r+1}{r}\right)^m \geqslant c(s-1).$$

Solving for m we obtain (3), and by means of (7) we get (2). Finally, k is found as the larger of the two integers $k(r, s, c)$ and $k(s, r, c)$, and n is found as kt. This completes Bernoulli's proof.

Bernoulli gives just one example. Consider n drawings with replacement from an urn with white and black balls in the ratio 3:2. In the theorem we may chose (r, s) as $(30, 20)$, $(300, 200)$, etc., depending on what limits for h_n we want to have. Bernoulli chooses $r = 30$ and $s = 20$ so that

$29/50 \leqslant h_n \leqslant 31/50$. Choosing $c = 1000$, which gives a moral certainty of $1000/1001$ for the inequality to hold, Bernoulli finds for the right tail $m = 301$, $k = 495$, and $n = 24{,}750$, and for the left tail $m = 211$, $k = 511$, and $n = 25{,}550$. Hence, for 25,550 observations, it is at least 1000 times more probable that h_n will fall inside the interval specified than outside.

Bernoulli ends the *Ars conjectandi* by the philosophical remark, inspired by his law of large numbers, that if all events were observed thoughout eternity one would find that everything in the world occurs according to certain causes and laws so that we are bound to recognize a certain necessity even in the seemingly most accidental events.

Bernoulli's theorem is of fundamental importance for statistical estimation theory by giving a theoretical justification for using h_n as estimator of p. He observed that it is necessary to consider an interval, $|h_n - p| \leqslant 1/t$, since the probability of getting $h_n = p$ tends to zero. He does not, however, indicate how to use the theorem to find an interval for p from an observed value of h_n. The theorem assumes that p is known, and the necessary number of observations, $n(p, t, c)$, say, depends on p. If n and h_n are known and p is unknown, one may of course solve the equation $n(h_n, t, c) = n$ with respect to c and thus get an approximation to the lower bound $c/(c + 1)$ for the probability that $|h_n - p| \leqslant 1/t$ and then solve this inequality for p; but there is no hint that Bernoulli had this in mind. The difficulty of formulating a theory of interval estimation may have been one of Bernoulli's reasons for not completing the manuscript.

Bernoulli's proof is remarkable, not only for giving the first limit theorem in probability theory, but also for being completely rigorous. Because of the crude evaluation of the tail probability and of the ratio f_0/f_k, the lower bound to P_k is rather unsatisfactory and leads to a number of observations much too large. That Bernoulli accepted this is perhaps due to the fact that his main objective was to prove that h_n converges in probability to p and only afterward was he led to the extension which requires a determination of n. This is indicated in the *Ars Conjectandi* where he first gives an unsatisfactory proof of the convergence and after this gives the correct proof reported above. His proof was worked out at the latest in 1690. It must have seemed rather unsatisfactory, to Bernoulli himself as well, when he included it in the manuscript 15 years later in view of the fact that the integral calculus had been developed in the meantime. In 1705 it would have been natural to evaluate the areas (sums) by means of integrals instead of limiting ordinates, see §16.5. The need for a revision of the proof may have been another reason for Bernoulli's hesitation to publish.

Bernoulli's theorem may be considered to give a lower bound to P_k or an upper bound to $1 - P_k$ for given n; we only have to solve (2) for given $k = n/t$ with respect to c.

Corollary to Bernoulli's Theorem. For $k = n/t$ we have $P_k > c(k)/[c(k) + 1]$, where $c(k) = c(r, s, k) \wedge c(s, r, k)$, and

$$\ln c(r, s, k) = \frac{k(r + 1) + s}{r + s + 1} \ln \frac{r + 1}{r} - \ln (s - 1). \tag{8}$$

Since $\ln c$ is a linear function of n, it follows that the tail probability tends to zero, at least exponentially when n tends to infinity.

In modern notation Bernoulli's theorem is as follows: for given p, $\epsilon > 0$ and $c > 0$, we have $P_{n\epsilon} > c/(c + 1)$ for $n \geq n(p, \epsilon, c) \vee n(q, \epsilon, c)$, where

$$n(p, \epsilon, c) \geq \frac{(1 + \epsilon)m - q}{\epsilon(p + \epsilon)}, \qquad m \geq \frac{\ln [c(q - \epsilon)/\epsilon]}{\ln [(p + \epsilon)/p]}, \tag{9}$$

and $\epsilon n(p, \epsilon, c)$ and m are the smallest positive integers satisfying the inequalities.

16.3   NICHOLAS BERNOULLI'S THEOREM, 1713

Occasioned by a discussion of the variations in the ratio of male to female births in London, Nicholas Bernoulli wanted to test the hypothesis that the true value of this ratio is as $18:17$. He therefore needed an approximation to the tail probability of the binomial better than the one provided by James Bernoulli's theorem. His results are to be found in a letter of 23 January 1713 to Montmort (1713, pp. 388–394). In the same letter he informed Montmort that *Ars Conjectandi* was in the press at Basel. His proof is based on James's proof; he simply sharpens two of the inequalities involved.

We shall use the same notation as used in §16.2 and set

$$P_d = \Pr\{|s_n - kr| \leq d\},$$

where d is a positive integer.

Nicholas Bernoulli's Theorem. For any $d > 0$,

$$\frac{P_d}{1 - P_d} > \min\left(\frac{f_0}{f_d}, \frac{f_0}{f_{-d}}\right) - 1; \tag{1}$$

that is,

$$P_d > 1 - \max\left(\frac{f_d}{f_0}, \frac{f_{-d}}{f_0}\right). \tag{2}$$

For large values of n in relation to d, we have

$$\frac{f_0}{f_d} \cong \left(\frac{kr+d}{ks-d+1} \frac{kr+1}{kr} \frac{ks}{kr} \right)^{d/2}, \tag{3}$$

and f_0/f_{-d} is obtained from (3) by interchange of r and s.

Proof. As above we need only consider the right tail because the results for the left are obtained by interchanging r and s. Let us define consecutive sums of length d of terms of the binomial by

$$A_m = \sum_{i=1}^{d} f_{md+i}, \qquad m = 0, 1, \ldots, \left[\frac{ks}{d} \right].$$

From the properties of f_i, it follows that

$$\frac{A_m}{A_{m+1}} \geqslant \frac{f_{md+1}}{f_{md+d+1}} > \frac{f_0}{f_d},$$

so that

$$A_m < A_0 \left(\frac{f_d}{f_0} \right)^m, \qquad m \geqslant 1. \tag{4}$$

Hence we have

$$\frac{A_0}{A_1 + \cdots + A_{[ks/d]}} > \frac{f_0}{f_d} - 1. \tag{5}$$

Noting that $t^n P_d$ equals A_0 plus the corresponding sum to the left plus f_0, it follows that (1) holds and thus (2).

It will be seen that (5) is Nicholas' improvement of James's lower bound (2.6). It is obtained by using the terms of the geometric series (4) as upper bounds for the consecutive sums.

Like James, Nicholas has reduced the problem to evaluate f_0/f_d. Instead of seeking a lower bound to this quantity Nicholas finds an approximation. Using (2.5) he first gets

$$\frac{f_0}{f_d} = \frac{kr+d}{ks-d+1} \frac{kr+d-1}{ks-d+2} \cdots \frac{kr+1}{ks} \left(\frac{ks}{kr} \right)^d.$$

He notes that for large values of n, that is, of kr and ks, in relation to d the first d ratios do not differ much. Taking logarithms and replacing the sum of the d logarithms by d times the average value of the first and the last terms, he obtains

$$\frac{f_0}{f_d} \cong \left[\frac{kr+d}{ks-d+1} \frac{kr+1}{ks} \left(\frac{ks}{kr}\right)^2 \right]^{d/2},$$

which gives (3). This completes Nicholas Bernoulli's proof.

It will be seen that Nicholas Bernoulli's theorem consists of two parts, the first giving a lower bound to P_d and the second an approximation to this lower bound. However, the approximate lower bound will normally also be a true lower bound because the difference between P_d and the lower bound is usually much larger than the error of the approximation (3).

Taking $n = 14,000$ as the yearly number of births in London and $p = 18/35$, we have $k = 14,000/35 = 400$, $kr = 7200$, and $ks = 6800$. To find a lower bound to P_d for $d = 163$, Nicholas computes

$$\log \frac{f_0}{f_d} \cong \frac{163}{2} \log \left(\frac{7363}{6638} \frac{7201}{7200} \frac{6800}{7200} \right) = 1.6507,$$

so that $f_0/f_d \cong 44.74$. Similarly, he finds $f_0/f_{-d} \cong 44.58$ and thus $P_d > 0.9776$. De Moivre's (1733) approximation by means of the normal distribution gives $P_d = 0.9942$.

To compare with James's theorem, which Nicholas does not do, let us disregard the requirement that r and s should be integers. Setting $k = 163$, we find $t = n/k = 85.89$ and $r = 44.17$, which leads to $c = 0.15647$ by means of (2.8) and thus $P_k > 0.1353$. This clearly demonstrates the great improvement obtained by Nicholas Bernoulli.

Nicholas' approximation to $\log(f_0/f_d)$ is very easy to compute so that there was no reason for him to investigate the approximation further. Nevertheless, if he had gone one step further and used the logarithmic series for each of the three factors of (3), he would have found that

$$\ln \frac{f_0}{f_d} = \frac{d^2}{2npq} \left(1 + \frac{q-p}{d} + \cdots \right),$$

which converges to $d^2/2npq$ if only d is of the order of $\sqrt{n}$ and $n \to \infty$, a result that was first found by de Moivre (1733) by another method of proof.

Returning to James Bernoulli's proof, it is clear that he realized that the ratio f_i/f_{i+1} is the only manageable quantity for the evaluation of P_k. He evaluated the ratio $P_k/(1 - P_k)$ by means of f_0/f_k, which he expressed as the product of the k ratios f_i/f_{i+1}, $i = 0, 1, \ldots, k - 1$. He really evaluated

$$\log \frac{f_0}{f_k} = \log \frac{f_0}{f_1} + \cdots + \log \frac{f_{k-1}}{f_k},$$

by noting that by a suitable choice of k, the first m terms are all larger than $\log[(r + 1)/r]$ and the last $k - m$ terms are all larger than zero. Nicholas approximated this sum by the number of terms times the average value of the first and the last terms. It is strange that they did not try to approximate the sum by an integral, a procedure that was well known at the time. They had every means for reaching the integral of the normal distribution, but it did not occur to them to work in that direction. We shall return to this problem in §16.5.

The exposition in §§16.2 and 16.3 is based on the paper by Hald (1984).

Let us finally consider how Nicholas' result may be used to solve James's problem, namely, to determine $n = kt$ such that $1 - P_k < 1/(c + 1)$. Looking at the right tail only, it follows from (2) that we have to solve the inequality $f_k/f_0 \leqslant 1/(c + 1)$. The only change necessary in James's proof is thus to replace $c(s - 1)$ by $c + 1$. Nicholas did not himself formulate this obvious corollary.

Corollary to Nicholas Bernoulli's Theorem. James Bernoulli's theorem holds for

$$k(r, s, c) \geqslant \frac{\ln(c + 1)}{\ln[(r + 1)/r]} \frac{r + s + 1}{r + 1} - \frac{s}{r + 1}. \tag{6}$$

An approximate value of k may be found by using (3) to solve the equation $f_0/f_k = c + 1$, which leads to

$$k = \frac{2\ln(c + 1)}{\ln[(r + 1)s/r(s - 1)]}. \tag{7}$$

Returning to James Bernoulli's example with $r = 30$, $t = 50$, and $c = 1000$, James's theorem gives $n \geqslant 25{,}550$, the corollary above gives 17,350, and the approximation (7) gives 8400. The value found from de Moivre's theorem is 6500.

16.4   SOME COMMENTS BY MARKOV, USPENSKY, AND K. PEARSON

The importance of Bernoulli's theorem was immediately recognized by his contemporaries, as is evident from several eulogies on Bernoulli.

Twenty years after the publication of the two proofs of the Bernoullis, de
Moivre showed that the normal distribution gives a good approximation to
the binomial for large values of n; his proof was published in the second
edition of the *Doctrine of Chances* (1738), see §24.5. De Moivre's proof in the
form given by Laplace became a standard item in textbooks on probability,
and since Bernoulli's law of large numbers is easily obtained as a corollary
to de Moivre's theorem, Bernoulli's proof has but seldom been reported.
Exceptions are the books by Markov (1924) and Uspensky (1937).

In 1913, 200 years after the publication of the *Ars Conjectandi*, a translation
of Part 4 into Russian by J. V. Uspensky, edited by A. A. Markov, was
published in St. Petersburg.

Markov (1924, pp. 43–53) writes that he will first report James Bernoulli's
proof and then give a modernized version without the restrictions on p, ϵ,
and n imposed by Bernoulli. However, without referring to Nicholas
Bernoulli, Markov first uses Nicholas' results to prove (3.6) and in two
footnotes he points out how this proof differs from James Bernoulli's.
He then goes through the same steps again in modernized form.

Uspensky (1937, pp. 96–101) writes, "Several proofs of this important
theorem are known which are shorter and simpler but less natural than
Bernoulli's original proof. It is his remarkable proof that we shall reproduce
here in modernized form." He then gives Markov's modernized version of
the combination of the proofs due to James and Nicholas, mentioning neither
Nicholas or Markov. Whether the modernized proof is due to Markov or
Uspensky and whether they knew Nicholas' theorem, we do not know.

In one respect Uspensky goes one step further than his predecessors; his
lower bound for n is independent of p. He notes that

$$\ln \frac{r+1}{r} > \frac{1}{r+1},$$

which combined with (3.6) gives

$$\frac{\ln(c+1)}{\ln[(r+1)/r]} < 1 + (r+1)\ln(c+1),$$

and using the right-hand side as m is (2.7) it follows that James Bernoulli's
theorem holds for

$$k \geqslant 1 + (t+1)\ln(c+1). \tag{1}$$

Uspensky's version of Bernoulli's theorem is as follows: For given $p, \epsilon > 0$, and $\delta > 0$,

$$\Pr\{|h_n - p| \leqslant \epsilon\} > 1 - \delta \quad \text{for} \quad n \geqslant \frac{1+\epsilon}{\epsilon^2} \ln \frac{1}{\delta} + \frac{1}{\epsilon}. \tag{2}$$

This result is obtained from (1) for $\epsilon = 1/t$ and $\delta = 1/(c+1)$. In Bernoulli's example, (1) gives $n \geqslant 17{,}700$ compared with $17{,}350$ from (3.6).

Markov (1912, pp. 135–141; 1924, pp. 104–115) offers a method for computing the tail probability with known limits of error. He uses Stirling's formula for evaluating the bounding ordinate and a continued fraction for the ratio of the tail to the bounding ordinate. He demonstrates his method on Bernoulli's example for $p = \frac{3}{5}$ and setting $n = 6520$, he finds that

$$0.999028 < \Pr\left\{|h_n - p| < \frac{1}{50}\right\} < 0.999044.$$

K. Pearson (1925) has given a detailed discussion of Bernoulli's proof. He points out that an upper bound for the tail probability may be found by means of a dominating geometric series. Fundamentally he uses the same idea as Nicholas Bernoulli but applies it to the individual terms. In our notation this analysis may be written as follows:

$$\sum_{d+1}^{ks} f_i = f_d\left\{\frac{f_{d+1}}{f_d} + \frac{f_{d+2}}{f_{d+1}}\frac{f_{d+1}}{f_d} + \cdots\right\}$$

$$< f_d \sum_{i=1}^{\infty} \left(\frac{f_{d+1}}{f_d}\right)^i,$$

which gives

$$\sum_{d+1}^{ks} f_i < f_d \frac{(ks-d)p}{d+q} < \frac{f_d(ks-d)p}{d}. \tag{3}$$

Pearson gives this result only for $d = k$. In modern notation Pearson's inequality becomes

$$\sum_{x=m+1}^{n} b(x, n, p) < \frac{b(m, n, p)(n-m)p}{m-np} \quad \text{for} \quad m > np, \tag{4}$$

which is obtained from (3) for $d = m - kr = m - np$.

Finally we shall indicate two simple proofs of Bernoulli's law of large numbers.

For the right tail probability, we have

$$\sum_{1}^{k} f_i > k f_k \quad \text{and} \quad \sum_{k+1}^{ks} f_i < (s-1) p f_k,$$

according to (3). The ratio of the two sums is therefore larger than $k/(s-1)p$, which in turn must be larger than or equal to c to satisfy Bernoulli's requirement. It follows that $k \geqslant c(s-1)p \vee c(r-1)q$. The crude evaluation of the two sums thus leads to a value of n proportional to c instead of $\ln c$, as in Bernoulli's proof.

Another proof is based on the Bienaymé–Chebychev inequality, which applied to the binomial yields

$$\Pr\left\{ |h_n - p| \leqslant \frac{1}{t} \right\} \geqslant 1 - t^2 \frac{pq}{kt} > \frac{c}{c+1},$$

so that $k \geqslant tpq(c+1)$. This proof depends only on the mean and the variance of the binomial distribution. At the time of the Bernoullis, however, the variance was an unknown concept.

After completion of the present chapter, Uspensky's 1913 translation of Part 4 of *Ars Conjectandi* has been reprinted in *J. Bernoulli: On the Law of Large Numbers* (In Russian), edited by Yu. V. Prohorov (1986). In addition to the translation, this book contains many notes on *Ars Conjectandi*, a paper by Sheynin on "J. Bernoulli and the beginnings of probability theory," a paper by Prohorov on "The law of large numbers and the evaluation of the probability of large deviations," and a "Biography of J. Bernoulli" by Youshkevitch. Prohorov's paper contains a detailed discussion of Bernoulli's proof and some remarks on later developments; for example, he points out that Pearson's result (3) was found by Chebychev in 1846, see also Maistrov (1974, p. 196).

16.5 A SHARPENING OF BERNOULLI'S THEOREM

We shall show that we can sharpen the results of the Bernoullis by using only the elementary mathematics known at their time. Like Bernoulli, we shall assume that np is an integer, but otherwise we shall make no restrictions. This assumption makes the notation simpler without essentially affecting the result. Bernoulli disregarded the central term of the binomial in his proof; we shall include it taking one-half to each side. We shall only give the theorem and the proof for the right-hand side, the results for the left-hand side are obtained by interchange of p and q. Hence, we consider the inequality

$$R = \frac{\sum\limits_{0}^{d} f_i - \frac{1}{2} f_0}{\sum\limits_{d+1}^{nq} f_i} > c, \qquad d = 1, 2, \ldots, nq - 1.$$

Like Bernoulli, we introduce f_0/f_d and write

$$R = \frac{f_0}{f_d} \frac{\sum\limits_{0}^{d} (f_i/f_0) - \frac{1}{2}}{\sum\limits_{d+1}^{nq} (f_i/f_d)}.$$

The problem is to find lower bounds for the two factors in the numerator and an upper bound for the denominator. We shall denote the "generalized variance" by

$$s^2 = npq - dp.$$

Theorem. $R > AB/C$, where

$$A = \left(1 + \frac{d}{np}\right)^{np+d} \left(1 - \frac{d}{nq+1}\right)^{nq+1-d} \left(1 + \frac{1}{nq}\right)^{-d},$$

$$B = s\sqrt{2\pi}\left\{\Phi\left(\frac{d + \frac{3}{2}}{s}\right) - \frac{1}{2}\right\} - 1,$$

and $C = s^2/d$.

Proof. As previously noted, the fundamental quantity is

$$v_i = \ln\frac{f_i}{f_{i+1}} = \ln\frac{q(np + i + 1)}{p(nq - i)} = \ln\frac{1 + (i + 1)/np}{1 - i/nq}.$$

Considering the slope $v_i - v_{i-1}$, it is easy to prove that v_i is positive and increasing and that v_i is first concave and then convex if $p < \frac{1}{2}$, whereas v_i is convex if $p \geq \frac{1}{2}$. Setting

$$-v_i = \ln(1 - t_i) > \frac{-t_i}{1 - t_i} \qquad \text{for} \quad 0 < t_i < 1,$$

it follows that

$$v_i < \frac{i+1-p}{npq(1-i/nq)}.\tag{1}$$

The first problem is to find a lower bound for

$$\ln\frac{f_0}{f_d} = \sum_0^{d-1} v_i.$$

From the properties of v_i, it follows that

$$\sum_0^{d-1} v_i > \int_{-1}^{d-1} v_x\,dx = \int_0^d \ln\frac{q(np+x)}{p(nq-x+1)}\,dx$$

$$= (np+d)\ln\left(1+\frac{d}{np}\right) + (nq+1-d)\ln\left(1-\frac{d}{nq+1}\right)$$

$$-\,d\ln\left(1+\frac{1}{nq}\right),$$

which equals $\ln A$. This result is due to Kiefer (1961).

A sketch of the relations between the various bounds and approximations has been given in Fig. 16.5.1 for $p > \frac{1}{2}$ and $d = 12$. The sum of the 12 v's is

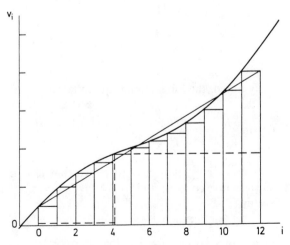

Fig. 16.5.1. Lower bounds and approximation to $v_0 + \cdots + v_{11}$.

represented by the areas of the corresponding rectangles. James Bernoulli's lower bound is given by the area below the dashed horizontal line through v_4; Nicholas Bernoulli's approximation by the area below the straight line joining v_0 and v_{11}; the lower bound A by the area below the continuous curve from the abscissa -1 to 11.

To find a lower bound for $\sum_0^d f_i/f_0$, we first consider

$$\ln \frac{f_i}{f_0} = \sum_{j=0}^{i-1} (-v_j) > -\sum_{j=0}^{i-1} \frac{j+1}{npq(1-(i-1)/nq)}$$

$$= -\frac{i(i+1)}{2npq(1-(i-1)/nq)},$$

where we have used (1) in the evaluation of the lower bound. It follows that

$$\frac{f_i}{f_0} > \exp\left(-\frac{i(i+1)}{2[npq-(i-1)p]}\right).$$

Since this function is decreasing, we have

$$\sum_0^d \frac{f_i}{f_0} > \int_0^{d+1} \exp\left(-\frac{x(x+1)}{2s^2}\right) dx.$$

Setting $x = us - \frac{1}{2}$, we get

$$\sum_0^d \frac{f_i}{f_0} > s \exp\left(\frac{1}{8s^2}\right) \int_{1/2s}^{(d+3/2)/s} \exp\left(-\frac{u^2}{2}\right) du.$$

Replacing the factor $\exp(\frac{1}{8}s^2)$ by 1 and introducing the standard notation for the normal distribution, the lower bound B follows.

The upper bound C follows directly from Pearson's inequality (4.4). This completes the proof.

The three factors are rather easy to calculate. It is, however, not possible to give an explicit solution of the equation $AB/C = c$ with respect to n, so n has to be found numerically. For Bernoulli's example, we find $n = 6660$ compared with Bernoulli's 25,550 and de Moivre's 6500.

The normal distribution is not introduced as the limit of the binomial for large n but occurs quite naturally in the derivation of the lower bound. Of course, the Bernoullis did not know the normal distribution function, but they could calculate the integral involved by numerical integration according to Newton's method.

It should be noted that the lower bound has been derived without any restrictions on d in relation to n. For this reason the "standard deviation" s depends on both n and d. It is interesting to analyze the lower bound in the three cases with $d = n\epsilon$ (Bernoulli's case), $d = u\sqrt{npq}n^a$, where u and a are constants, $u > 0$, and $0 < a < \frac{1}{2}$, (the case of large deviations), and $d = u\sqrt{npq}$ (de Moivre's case, the normal distribution). We shall leave this to the reader, who should also try to improve the three bounds by using more advanced mathematical methods.

CHAPTER 17

Tests of Significance Based on the Sex Ratio at Birth and the Binomial Distribution, 1712–1713

This Equality of [adult] Males and Females is not the Effect of Chance but Divine Providence.

—JOHN ARBUTHNOTT, 1712

The point is to find out how much one ought to bet against one that what has happened at London would not have happend if the births had depended on chance.

—G. J. 'sGRAVESANDE, 1712

There is no reason to be surprised at the fact that the number of infants of each sex do not differ more [than observed].

—NICHOLAS BERNOULLI, 1713

17.1 ARBUTHNOTT'S STATISTICAL ARGUMENT FOR DIVINE PROVIDENCE

The data and ideas in Graunt's *Observations* (1662) inspired mathematicians to construct probabilistic models to describe and explain the phenomena recorded. We have previously discussed the probabilistic interpretation of Graunt's life table in Chapter 8. Just as Huygens likened the chances of death derived from Graunt's life table to the chances in a lottery, so Arbuthnott compared the yearly number of male and female births in London with the

275

outcome of a number of throws with a two-sided die. This binomial model led to the first tests of significance.

In his youth, John Arbuthnott (1667–1735) earned his living as a teacher of mathematics in London. At that time he translated Huygens' treatise into English under the title *On the Laws of Chance* (1692), see § 12.2. He then studied medicine and received his degree in 1696. He became a man of wide-ranging learning, a fashionable physician, a scientist, a literary wit, and a political satirist. He was a collaborator and friend of Jonathan Swift, Alexander Pope, and John Gay. In 1710, when he read his paper to the Royal Society, he gave his title as "Physician in Ordinary to Her Majesty, and Fellow of the College of Physicians and the Royal Society." Short biographies have been written by Allan Birnbaum (1967) and Pearson (1978).

The title of his paper is *An Argument for Divine Providence, taken from the constant Regularity observed in the Births of both Sexes*, read 1710, published 1712. Graunt (1662) had published the yearly number of christenings of males and females in London for the period 1629–1660 and noted that there were about 14 males born for 13 females (see § 7.6). Arbuthnott publishes the observations for the period 1629–1710 without reference to Graunt.

Arbuthnott's Data on the Yearly Number of Male and Female Christenings in London from 1629 to 1710

YEAR	MALE	FEMALE	YEAR	MALE	FEMALE
1629	5218	4683	1646	3768	3395
1630	4858	4457	1647	3796	3536
1631	4422	4102	1648	3363	3181
1632	4994	4590	1649	3079	2746
1633	5158	4839	1650	2890	2722
1634	5035	4820	1651	3231	2840
1635	5106	4928	1652	3220	2908
1636	4917	4605	1653	3196	2959
1637	4703	4457	1654	3441	3179
1638	5359	4952	1655	3655	3349
1639	5366	4784	1656	3668	3382
1640	5518	5332	1657	3396	3289
1641	5470	5200	1658	3157	3013
1642	5460	4910	1659	3209	2781
1643	4793	4617	1660	3724	3247
1644	4107	3997	1661	4748	4107
1645	4047	3919	1662	5216	4803

(*Continued*)

YEAR	MALE	FEMALE	YEAR	MALE	FEMALE
1663	5411	4881	1687	7737	7214
1664	6041	5681	1688	7487	7101
1665	5114	4858	1689	7604	7167
1666	4678	4319	1690	7909	7302
1667	5616	5322	1691	7662	7392
1668	6073	5560	1692	7602	7316
1669	6506	5829	1693	7676	7483
1670	6278	5719	1694	6985	6647
1671	6449	6061	1695	7263	6713
1672	6443	6120	1696	7632	7229
1673	6073	5822	1697	8062	7767
1674	6113	5738	1698	8426	7626
1675	6058	5717	1699	7911	7452
1676	6552	5847	1700	7578	7061
1677	6423	6203	1701	8102	7514
1678	6568	6033	1702	8031	7656
1679	6247	6041	1703	7765	7683
1680	6548	6299	1704	6113	5738
1681	6822	6533	1705	8366	7779
1682	6909	6744	1706	7952	7417
1683	7577	7158	1707	8379	7687
1684	7575	7127	1708	8239	7623
1685	7484	7246	1709	7840	7380
1686	7575	7119	1710	7640	7288

Source: Phil. Trans., 1712, Vol. 27, 189–190.

Arbuthnott does not comment on the quality of the data and considers the number of christenings to be proportional to the number of births. He neither gives the totals of the numbers of males and females nor the ratio of these numbers; he merely notes that the number of males exceeds the number of females,

> in almost a constant proportion for each of the 82 years. [He says that] to judge of the wisdom of the Contrivance, we must observe that the external Accidents to which Males are subject (who must seek their Food with danger) do make a great havock of them, and that this loss exceeds far that of the other Sex, occasioned by Diseases incident to it, as Experience convinces us. To repair that Loss, provident Nature, by the Disposal of its wise Creator, brings forth more Males than Females, and that in almost a constant proportion.

So much for the data and his theological explanation. There is really no reason to go further with the statistical analysis; anyone who looks at the data will be convinced, like Graunt and Arbuthnott, that the chance of a male birth is greater than that of a female birth. Nevertheless, Arbuthnott continues with a probabilistic "proof" of this assertion, and herein consists the originality of his paper.

He begins by stating that the number of chances of the different outcomes by throwing n two-sided dice is given by the coefficients of the binomial expansion of $(M + F)^n$. For an even value of n he finds that the probability of getting as many M's as F's tends to zero as n tends to infinity; however, what we have to investigate is not "Mathematical but Physical" equality of males and females, which means that we have to include some terms next to the middle one. If chance governs, i.e., if the probability of a male birth equals $\frac{1}{2}$, then in a given year the probability of more males than females is smaller than or equal to $\frac{1}{2}$. He then refers to the data and writes, "Now, to reduce the Whole to a Calculation, I propose this *Problem*. A lays against B, that every year there shall be born more Males than Females: To find A's Lot, or the Value of his Expectation."

To get an upper bound for A's probability of winning, he sets the probability that in each year there shall be born more males than females equal to $\frac{1}{2}$. It follows that the probability of more males than females every year for 82 years becomes $(\frac{1}{2})^{82}$, which equals 1 divided by 4.836×10^{24}. Not being satisfied with this small probability he continues:

> But if A wager with B, not only that the Number of Males shall exceed that of Females, every Year, but that this Excess shall happen in a constant Proportion, and the Difference lye within fixed limits, and this not only for 82 Years, but for Ages and Ages, and not only at *London*, but all over the World,... then A's Chance will be near an infinitely small Quantity, at least less than any assignable Fraction. From whence it follows that it is Art, not Chance, that governs.

In modern terminology Arbuthnott wants to test the hypothesis that the probability, p, say, of a male birth equals $\frac{1}{2}$ against the alternative that $p > \frac{1}{2}$. By means of the binomial distribution he first proves that for any number of births $\Pr\{M > F | p = \frac{1}{2}\} \leqslant \frac{1}{2}$, and next he uses this result to transform the original hypothesis and its alternative to the hypothesis that $\Pr\{M > F\} \leqslant \frac{1}{2}$ and the alternative $\Pr\{M > F\} > \frac{1}{2}$ for the yearly number of births. In that way he avoids the difficulties stemming from the varying numbers of births and reduces the comparison of the 82 binomial distributions to a simple sign test.

Hacking (1965, pp. 75–81) has discussed Arbuthnott's test in relation to modern testing theory.

Despite the brevity and simplicity of Arbuthnott's paper, it greatly influenced contemporary statisticians and theologians. The dispute among statisticians will be discussed in the following sections.

Pearson (1978) has described the influence of the works of Graunt and Arbuthnott on the clergyman and scientist William Derham (1657–1735), who collected and analyzed a great amount of population statistics in his *Physico-Theology: A Demonstration of the Being and Attributes of God from His Works of Creation* (1713). Derham and other theologians considered the regularity of statistical ratios as a proof of Divine Design.

The following discussion is to a large extent similar to the one given by Shoesmith (1985, 1987); however, Shoesmith offers more details on the theological background, whereas we present a statistical analysis based on a reconstruction of Nicholas Bernoulli's calculations.

17.2 'sGRAVESANDE'S TEST OF SIGNIFICANCE

The Dutch scientist, W. J. 'sGravesande (1688–1742), who later became professor of mathematics, astronomy, and philosophy at Leiden, presented an improvement of Arbuthnott's test in 1712 in a paper *Démonstration Mathématique du soin que Dieu prend de diriger ce qui se passe dans ce monde, tirée du nombre des Garcons et des Filles qui naissent journellement* (A mathematical demonstration of the care, which God takes in governing that which happens in this world, drawn from the daily numbers of male and female births), which he circulated among his friends. The main results of his analysis were published in 1715 by B. Nieuwentyt in *Het regt gebruik der Wereldbeschouwingen*, translated into English as *The Religious Philosopher: Or, the Right Use of Contemplating the Works of the Creator*, see Pearson (1978) for a detailed description of this work. 'sGravesande's paper is published in his *Oeuvres* (1774, Vol. 2, pp. 221–236), which also contains a short exposition of elementary probability theory (pp. 82–97).

In 1712 Nicholas Bernoulli met 'sGravesande in The Hague on his tour to the Netherlands, England, and France. They discussed Arbuthnott's paper, and a correspondence ensued which is published in the *Oeuvres* (1774, Vol. 2, pp. 236–248).

In his test Arbuthnott uses only the fact that for each of the 82 years, the number of males is greater than the number of females. 'sGravesande further makes use of the fact that the relative number of male births varies between $7765/15,448 = 0.5027$ in 1703 and $4748/8855 = 0.5362$ in 1661. Because of the different yearly number of births, the numbers of males are not directly comparable, and 'sGravesande therefore transforms the observations above by multiplying the relative frequencies by the average number of births for the 82 years, which he finds to be 11,429. This gives him a fictitious minimum and maximum number of male births: 5745 and 6128. He then considers the data as 82 observations from the same binomial distribution with $n = 11,429$ and all the observations contained in the interval [5745, 6128].

To find the probability of this event under Arbuthnott's hypothesis, he calculates the terms of the binomial for $p = \frac{1}{2}$ and $n = 11,429$ and sums the 384 terms corresponding to the interval in question. Actually, he uses the recursion

$$\binom{n}{x+1} = \binom{n}{x}\frac{n-x}{x+1},$$

and tabulates

$$10^5 \binom{n}{x} \bigg/ \binom{n}{5715}$$

from $x = 5715$ to 5973, after which the tabular values become smaller than $\frac{1}{2}$. He finds that

$$\Pr\{5745 \leqslant x \leqslant 6128 \,|\, p = \tfrac{1}{2}\} = 3,849,150/13,196,800,$$

and remarks that he has added a small amount to the numerator to make sure that the probability is not undervalued because terms in the tail smaller than $\frac{1}{2} \times 10^{-5}$ have been neglected. ('sGravesande probability equals 0.292, and the normal approximation gives 0.287.)

Under the hypothesis, the probability of the observed event becomes the 82nd power of the probability above, which gives 1 divided by 7.5598×10^{43} ('sGravesande gives all 44 digits), which is only a small fraction of the probability found by Arbuthnott.

17.3 NICHOLAS BERNOULLI'S COMPARISON OF THE OBSERVED DISTRIBUTION WITH THE BINOMIAL

During his stay in London in 1712, Nicholas Bernoulli discussed Arbuthnott's paper with other Fellows of the Royal Society; he explained his views in a letter to the mathematician Burnet and sent a copy to 'sGravesande. This letter contains the proof of Nicholas Bernoulli's theorem (see §16.3) but was not published until 1774 in 'sGravesande's *Oeuvres*. However, Bernoulli gave the proof in essentially the same form in his letter to Montmort (1713, pp. 388–393). He derived his theorem to obtain a general formula for the tail probability of the binomial and thus avoid the heavy computational work which had been necessary for 'sGravesande.

Bernoulli does not seem much interested in the theological debate; his attitude is pragmatic, like that of a modern statistician. The main points of his letters are (1) to estimate the probability of a male birth from the observations; (2) to compare the distribution of the observations with the binomial distribution to determine whether the observed variation can be explained by this model; and (3) to provide the mathematical tool of this comparison by finding an approximation to the binomial for large n.

Let m_i and f_i, $i = 1, \ldots, 82$, denote the yearly number of male and female births, respectively, and set $n_i = m_i + f_i$ and $h_i = m_i/n_i$. The relative frequency of male births for the whole period is

$$\bar{h} = \frac{\sum m_i}{\sum n_i} = \frac{484,382}{938,223} = 0.5163,$$

and the average yearly number of births is 11,442.

From the observed values of h_i, Bernoulli calculates the number of male births under the assumption that the yearly number of births is constant and equal to 14,000. He wants to show that the 82 values of $x_i = 14,000h_i$ can be considered observations from a binomial distribution. He finds $\bar{x} = 7237$, $x_{min} = 7037$, and $x_{max} = 7507$, and using (16.3.3) with $kr = 7237$, $ks = 6763$, and $d = 7237 - 7037 = 200$, he gets

$$P_d = \Pr\{7037 \leqslant x \leqslant 7437\} > \frac{303}{304} \quad \text{for} \quad p = \frac{7237}{14,000} = 0.5169.$$

He does not comment on the four observations larger than 7437. Noting that $(303/304)^{100} \cong 1000/1389$, he concludes that the variation observed may be explained by the binomial distribution if only the correct value is chosen for the probability of a male birth. These results are to be found in his letter to Burnet and 'sGravesande.

In his reply, 'sGravesande acknowledges Bernoulli's proof but says that Bernoulli has misunderstood Arbuthnott, who only intends to prove that p is larger than $\frac{1}{2}$. Bernoulli replies that Arbuthnott's paper also implies that the variation in the number of male births is smaller than could be expected, and it is only this assertion that he has tried to refute. It is difficult to see that Arbuthnott asserts what Bernoulli postulates.

In his letter to Montmort, Bernoulli chooses another value of p, presumably to obtain agreement with the assumptions of his theorem, which require that $p = r/(r + s)$ and $n = k(r + s)$, where r, s, and k are integers. Since $14,000 = 2^4 \times 5^3 \times 7$, it seems reasonable to choose $r + s = 35$, and since $35 \times 0.5169 = 18.1$, we are led to $r = 18$, so that $p = 18/35 = 0.5143$. Bernoulli does not explain his choice of $p = 18/35$; he merely notes that the observed ratio of males to females is a little larger than $18/17$. He also does not explain the choice of $n = 14,000$ rather than a value in the neighborhood of the average. Instead of Arbuthnott's two-sided die, he explains his model by referring to a die with 35 sides, 18 being white and 17 black.

Bernoulli divides his investigation into two parts. First, he calculates the probability of an observation within a rather large interval covering the main part of the observed distribution; next, he finds the probability of the 11 "outliers" among the 82 observations.

For $n = 14,000$ and $p = 18/35$, he finds that the expected number of male births is 7200, the distance to x_{min} is $d = 7200 - 7037 = 163$, and

$$P_d = \Pr\{7037 \leqslant x \leqslant 7363\} > 0.9776,$$

as shown in §16.3.

Outside this interval, i.e., larger than 7363, there are 11 observations. (The corresponding number in his letter to 'sGravesande is only four because there he considers an interval about the mean, 7237.) Using $1 - P_d \leqslant 0.0224$, Bernoulli states that the probability of obtaining 2 outliers at most among the 82 observations is larger than $\frac{1}{2}$ and the probability of 10 outliers at most is larger than $226/227 = 0.9956$. He does not explain how he has calculated these probabilities. It is easy to check that the probabilities in question become 0.72 and 0.999997 if we use a binomial distribution with $n = 82$ and $p = 0.0224$.

Bernoulli concludes that "there is a large probability that the numbers of males and females fall within limits which are even closer than those observed," see Montmort (1713, p. 388). According to our standards this would lead to rejection of the hypothesis because the observed variation is larger than expected; Bernoulli, however, seems to be satisfied with his model because it explains the greater part of the variation.

One reason that Bernoulli does not note the discrepancy between the data and the model may have been that his formula gives too large a value of the tail probability. As noted in §16.3, $1 - P_d = 0.0058$, whereas he presumably uses $1 - P_d = 0.0224$. Another and more important reason is that he stresses only the large probability of getting an observation between the chosen limits without also considering the corresponding small probability for observations outside the limits. From his own calculations it follows that the expected number of observations outside the limits is at most $82 \times 0.0224 = 1.8$, so that the probability of getting 11 (or more) outliers is extremely small.

Last, let us compare Bernoulli's results with a modern analysis. Bernoulli notes that one may make more comparisons of the data and the model by using other values of d. To do so in a comprehensive manner we have found the standard deviation of Bernoulli's observations, $x_i = 14,000h_i$. For the variance we have

$$V\{x_i\} = (14,000)^2 \frac{pq}{n_i} \cong 14,000pq\left(\frac{14,000}{11,442}\right),$$

which gives a standard deviation of 65. We have then computed

$$\frac{x_i - 7200}{65}, \qquad \frac{x_i - 7237}{65}, \qquad u_i = \sqrt{n_i}\,\frac{h_i - \bar{h}}{\sqrt{\bar{h}(1 - \bar{h})}}, \qquad \bar{h} = 0.5163.$$

Under the hypothesis, the two last-mentioned quantities should be nearly normally distributed with mean 0 and standard deviation 1. The grouped distributions of the three standardized deviations are shown in the following table:

Distribution of Standardized Deviations from Arbuthnott's data

Standardized deviation	Deviations from			Expected from normal distribution
	7200/14,000	7237/14,000	0.5163	
	Number of observations			
(−5)–(−4)	0	0	0	0.0
(−4)–(−3)	0	1	1	0.1
(−3)–(−2)	1	3	2	1.8
(−2)–(−1)	14	18	18	11.1
(−1)–0	17	23	23	28.0
0–1	20	19	19	28.0
1–2	16	8	9	11.1
2–3	7	6	8	1.8
3–4	4	2	2	0.1
4–5	3	2	0	0.0
Total	82	82	82	82.0

Under the binomial hypothesis, $\sum u_i^2$ should be distributed as χ^2 with 81 degrees of freedom. We find, however, that $\sum u_i^2 = 169.7$, which shows that the observed variation is significantly larger than the binomial. This is also obvious from a comparison with the expected numbers according to the normal distribution and from Fig. 17.3.1. The distribution of Bernoulli's observations about the mean, 7237, gives nearly the same result as the distribution of the u_i's, whereas the distribution about 7200 results in too

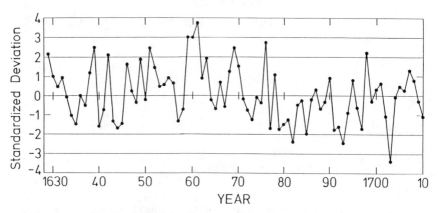

Fig. 17.3.1. Graph of the standardized deviations u_i from Arbuthnott's data.

many positive deviations (the 11 outliers) because Bernoulli's choice of $p = 18/35$ leads to an expected number that is too small, with about half a standard deviation.

To reject the hypothesis of a binomial distribution of the number of male *births* based on the above analysis of the number of male *christenings* presupposes that the yearly number of male and female christenings have the same proportion to the yearly number of male and female births and that there are no recording errors. The number of christenings recorded in 1704 is conspicuously small; Struyck (*Oeuvres*, p. 185) points out that parish records generally are unreliable and as an example he states that the correct numbers for 1704 are 8153 and 7742 instead of Arbuthnott's 6113 and 5738. This means that the relative number of males becomes 0.5129 instead of 0.5158. Graunt points out that the number of christenings in relation to the number of births decreases drastically after 1642 (see §7.4). From Arbuthnott's data it will be seen that the yearly number of christenings before 1642 is about 10,000. In the 1650s it decreases to about 6000 and then steadily increases to about 15,000 in 1700. The turning point is about the Restoration, and it is worth noting that the three largest positive values of u occur in 1659–1661, see Fig. 17.3.1. The smallest negative value occurs in 1703. These four observations contribute 43.6 to the χ^2 value of 169.7. It seems safe to conclude that the variability of the sex ratio at birth is considerably smaller than that of the sex ratio of the christenings.

Bernoulli's results were not appreciated by his contemporaries, perhaps because his mathematical and statistical arguments were new and difficult to follow (the *Ars Conjectandi* had not yet been published), and perhaps because he found Arbuthnott's statistical analysis incomplete and thus indirectly criticized the religious argument. It was not yet understood that Bernoulli's analysis should be considered an extension of Arbuthnott's. De Moivre (1756, pp. 252–253) replied to Bernoulli's criticism of Arbuthnott in much the same way as 'sGravesande and concluded that "if we were shewn a number of Dice, each with 18 white and 17 black faces, which is Mr. Bernoulli's supposition, we should not doubt but that those Dice had been made by some Artist; and that their form was not owing to *Chance*, but was adapted to the particular purpose he had in View." It may have been the authority of de Moivre that led Todhunter (pp. 130–131, 197–198) to underrate the contributions of Nicholas Bernoulli.

17.4   A NOTE ON THEOLOGY AND POLITICAL ARITHMETIC

As related in §11.1, many natural scientists in the latter half of the 17th century and the beginning of the 18th interpreted the newly found natural laws as evidence of Divine Design and the existence of God. They were now joined by social scientists, in many cases represented by theologians, who used the regularities observed in population and vital statistics to the same end. These authors contributed to statistics in the form of political arithmetic by collecting, tabulating, and comparing data from different countries, stressing the similarities and glossing over the discrepancies.

Pearson (1978, Chap. 9) has given a lively account of the life and works of Derham, Nieuwentyt, Süssmilch, and Struyck. Westergaard (1932) discusses the same period in his Chap. 7 and writes (p. 72), "But even though quite naturally we are induced to take notice of deviations from the average, and to ask how these irregularities can be explained, it is just as natural that our predecessors first of all were struck by the regularity and cared less for the deviations."

This line of thought culminated with the publication in 1761 of the second edition of *Die göttliche Ordnung in den Veränderungen des menschlichen Geschlechts aus der Geburt, dem Tode und der Fortpflanzung desselben erwiesen* (The divine order in the changes of the human race demonstrated from the birth, the death, and the reproduction of the same) by the German theologian J. P. Süssmilch (the first edition had been published in 1741).

CHAPTER 18

Montmort and the *Essay d'Analyse sur les Jeux de Hazard*, 1708 and 1713

I very willingly acknowledge his [Montmort's] Solution to be extreamly good, and own that he has in this, as well as in a great many other things, shewn himself entirely master of the doctrine of Combinations, which he has employed with very great Industry and Sagacity.

— *De MOIVRE, 1718*

18.1 MONTMORT AND THE BACKGROUND FOR HIS *ESSAY*

Pierre Rémond de Montmort (1678–1719) was born into a wealthy family of the French nobility. As a young man he traveled in England, the Netherlands, and Germany. Shortly after his return to Paris in 1699 his father died and left him a large fortune. He studied Cartesian philosophy under Malebranche and studied the calculus on his own. For some years he was a canon of Notre Dame. In 1704 he bought the estate of Montmort, and in 1706 he resigned his canonry and settled at Montmort, where he stayed for the rest of his life, interrupted only by visits to Paris and London. He seems to have lived a happy life as a country gentleman, occupied by his interests in natural philosophy and mathematics. On a visit to Paris he contracted smallpox and died at the age of 40.

The main source for an account of the life of Montmort is the eulogy by Fontenelle (1721), see also David (1962).

Montmort corresponded with Leibniz whom he greatly admired. He was also on good terms with Newton whom he visited in London. In 1709 he printed 100 copies of Newton's *De Quadratura* at his own expense. As noted earlier, through John Bernoulli, he also offered to print *Ars Conjectandi*. He was on friendly terms with Nicholas Bernoulli and Brook Taylor. He kept his balance in the priority dispute between Newton and Leibniz and their followers.

Besides his work on probability theory he worked on the summation of infinite series and was working on a history of mathematics when he died.

The *Essay d'Analyse sur les Jeux de Hazard* (Analytical Essay on Games of Chance) was published in 1708. It consists of a preface and three parts on card games, dice games, and some further problems of games of chance, respectively. A second edition was published in 1713. Neither edition contains the author's name on the title page nor in the text. In the second edition the name Montmort occurs only as a place name in letterheads.

The prefaces to the two editions are identical apart from the section in which the contents are briefly mentioned. An *Avertissement* of 18 pages was added to the second edition. The section on combinatorics in the first edition is much expanded and was made into the first part of the second edition; then follow the three parts from the first edition with additions and generalization and, finally, as the fifth part, 132 pages of letters between Montmort and John and Nicholas Bernoulli. A survey of the contents is given below.

Table of Contents of the *Essay* by Montmort, 1708 and 1713

	PAGES	
TOPIC	1708 EDITION	1713 EDITION
Preface	III–XXIV	III–XXIV
Avertissement	—	XXV–XLII
On combinations	79–100	1–72
On card games	1–108	73–172
Pharaon, lansquenet, treize, bassete, piquet, triomphe, l'ombre, brelan, imperiale, quinze		
On dice games	109–155	173–215
Quinquenove, hazard, esperance, trictrac, trois dez, passe-dix, raffle, jeu des noyaux		
Solutions of various problems on games of chance	156–185	216–277
Huygens' five problems, de Méré's problem, the problem of points, Robartes' problem, the lottery of Lorraine, the duration of play		
Four problems for solution	185–189	278–282
Letters between Montmort, John Bernoulli, and Nicholas Bernoulli	—	283–414

Montmort was inspired by what he had read about James Bernoulli's work. The most precise description of the contents of *Ars Conjectandi* known in 1708 is the one given by Saurin (1706), from which we quote:

> The author there determines and reduces to a calculus the different degrees of certainty or probability of conjectures which one may make on matters dependent on chance, and he extends the same to civil life and private affairs.... It is in the fourth part that the author extends his method and his reasonings, as we have already said, to matters regarding civil life and private affairs. The fundament of this part is an important problem, which he first solves, and which he considers more important than the quadrature of the circle. It is the question to determine whether an increasing number of observations of an event at the same time leads to a corresponding increase of the degree of probability of finding the true ratio of the number of cases where the event may occur and the number where it may not occur; in this way one can finally reach a degree of probability larger than any given degree, that is, veritable certainty.

No indication of Bernoulli's method of proof is given.

We shall let Montmort himself relate the background story of his book by giving a free translation, somewhat condensed, of the most important points of his preface to the first edition.

> Mathematics has been of great importance for the development of the natural sciences. It would be even more glorious for mathematics if it could be used to set up rules for the judgments and actions of man in practical matters of life. Bernoulli has done so in his *Ars Conjectandi*, but his premature death prevented him from completing his work. Fontenelle and Saurin have both given short descriptions of his work in their eulogies of 1705 and 1706, respectively. In the first three parts of his book Bernoulli gives the solutions of several problems of games of chance, using infinite series and combinatorics, and he also gives the solution of Huygens' five problems. In the fourth part he applies the methods developed in the first three to the solution of moral, political, and civil problems. We do not know what problems in games of chance and what political and moral problems Bernoulli has discussed but we are confident that he has solved the problems perfectly.

> Several of my friends have for a long time asked me to try to determine the advantage of the banker in the play Pharaon by means of Algebra. I should not have dared to embark on this project had it not been for the success of the late Bernoulli, which I heard about. For some years I have worked on this problem and I have been more fortunate than I could hope. Not alone have I solved the problem of Pharaon but I have also discovered some general principles for solving similar and more difficult problems. I realized that it is possible to proceed rather far into this country where nobody yet has been; I hoped that one could there make an ample harvest of new and interesting results, and this induced me to

seek to the bottom of this matter in a desire to compensate the public for the loss of the excellent work of Bernoulli. Various reflections have also confirmed me in this plan.

Players are superstitious in many ways, in particular they believe (wrongly) that the outcomes of previous games influence the future outcomes. One could say nearly the same about the conduct of people in all the situations of life in which chance plays a part; it is the same kind of prejudices that govern their actions. I therefore think that it will be useful not alone for players but also for people in general to learn about the rules for events depending on chance. For players it will be useful to be able to judge whether a game is fair or not. We do not know the future but we may know that some events are more probable than others. It seems that before now it has not been apprehended that it is possible to give infallible rules for calculating the differences which exist between various probabilities. We attempt to give in this book an exposition of this new art, applying Analysis to matters that until now have been considered obscure and unfit for such a treatment.

Montmort then gives a short account of the contents of his book for which we refer to the previous table. He continues,

Contrary to Bernoulli I shall not add a fourth part but limit myself to applications of the methods developed to games of chance. What has kept me back is the difficulty that I have found in making the hypotheses which could lead and support me in this research, and since I am not satisfied I think it will be better to leave this work to another occasion or leave it to another person more able than me. I shall, however, first make some brief remarks on the relationship between games of chance and these other matters and on the necessary points of view for reaching a solution.

Precisely speaking, nothing depends on chance. Everything is determined by the Creator by exact laws, but if the causes are unknown to us we may regard the events as dependent on chance. According to this definition one may say that human life is a game ruled by chance. Like in games of chance one should in practical problems count the number of cases for and against the happening of an event, or to speak in mathematical terms, one should determine the possible advantage of a decision by multiplying what we hope to obtain by the degree of probability of obtaining it and subtract the stake. Continuing this comparison we note two reasons for our ignorance of future events. First, our uncertainty about the actions to choose, and second, that the number of causes and comparisons to make is so large that it is impossible for us to keep track of them. The prudence of man is insufficient to penetrate into the future.

After a long discussion in the same vein on the similarities of games of chance and problems in civil life, Montmort concludes with the following two rules for delineating the problems from real life that may be treated by methods from games of chance.

> To end this analogy between the problems of games of chance and the questions that may be proposed on economical, political and moral matters one should observe that in both cases there is a type of problem which may be solved by observing the following two rules: (1) limit the question, which one takes up, to a small number of suppositions established on facts that are certain, (2) disregard all circumstances in which freedom (free will) of man, that eternal obstacle to our understanding, might have a share. One would believe, that Bernoulli has taken such rules into regard, and it is certain that with these two restrictions one will be able to treat several political and moral matters with the exactness of mathematics.

As an "admirable example" Montmort then discusses Halley's paper (1694) and also mentions Petty's work on political arithmetic.

Finally, Montmort gives a short account of the history related to de Méré, Pascal, Fermat, and Huygens.

It will be seen that Montmort is motivated partly by what he has read about Bernoulli's book and partly by problems on games of chance raised by his friends. He expresses views on determinism and chance similar to Bernoulli's. However, most practical problems are so complicated that it is impossible to evaluate the number of chances for and against the happening of an event and thus calculate its probability, and he therefore limits himself to problems on games of chance. Nevertheless, he tries (not very successfully) to formulate restrictions on those problems amenable to probabilistic analysis. In that respect he is more sensible than Bernoulli.

The *Essay* (1708) is the first published comprehensive text on probability theory, and it represents a considerable advance compared with the treatises of Huygens (1657) and Pascal (1665). Montmort continues in a masterly way the work of Pascal on combinatorics and its application to the solution of problems on games of chance. He also makes effective use of the methods of recursion and analysis to solve much more difficult problems than those discussed by Huygens. Finally, he uses the method of infinite series, as indicated by Bernoulli (1690).

When the *Essay* was written, card games had become very fashionable, and this is reflected in the prominent place occupied by the probabilistic analysis of such games. Also with respect to dice games Montmort's analysis is much more extensive and general than previous discussions.

Montmort's pioneering work had three features: (1) It demonstrated the four methods of solution mentioned above; (2) it gave the solutions of many important new problems; and (3) it inspired Nicholas Bernoulli and de Moivre to generalize these problems and develop new methods of proof.

Pioneering works are often difficult reading, and the *Essay* (1708) is no exception. In many places it is also tedious because of numerous details. These difficulties are enhanced by Montmort's decision to present the solution of the most interesting problems by numerical examples only and state the general solution without proof. However, in the *Avertissement* of the second edition, Montmort (1713) explains that he omitted some proofs in the first edition to stimulate the curiosity of the reader—he certainly succeeded with respect to Nicholas Bernoulli and de Moivre. He adds that he has now included these proofs at the request of some friends.

In the *Avertissement* Montmort mentions the publication of *De Mensura Sortis* by de Moivre in 1712 and quotes from the preface:

Huygens was the first that I know who presented rules for the solution of this sort of problems, which a French author has very recently well illustrated with various examples; but these distinguished gentlemen do not seem to have employed that simplicity and generality which the nature of the matter demands; moreover, while they take up many unknown quantities, to represent the various conditions of gamesters, they make their calculation too complex; and while they suppose that the skill of the gamesters is always equal, they confine this doctrine of games within limits too narrow.

It is of course a great injustice to characterize Montmort's work by the remark that he has well illustrated the rules developed by Huygens with various examples. Montmort vigourously refutes the critical remarks by de Moivre, and on pp. 361–370 he gives a "review" of *De Mensura Sortis* and a comparison with his own work (we shall return to this in §22.2). De Moivre regretted his remarks, and in the preface of the *Doctrine of Chances* (1718) he wrote,

However, had I allowed my self a little more time to consider it, I had certainly done the Justice to its Author, to have owned that he had not only illustrated *Huygens*'s Method by a great variety of well chosen Examples, but that he had added to it several curious things of his own Invention. ⋯ Since the printing of my Specimen [*De Mensura Sortis*], Mr. *de Monmort*, Author of the *Analyse des jeux de Hazard*, Published a Second Edition of that Book, in which he has particularly given many proofs of his singular Genius, and extraordinary Capacity; which Testimony I give both to Truth, and to the Friendship with which he is pleased to Honour me.

The remaining part of the *Avertissement* is taken up by a history of probability theory before 1713, which we have discussed in §12.2.

The second edition of the *Essay* (1713) is a greatly improved version of the first, not only because Montmort collected the combinatorial theorems in an introductory part, but also because he added various proofs and generalizations and his correspondence with the Bernoullis. There is no doubt that Montmort was much influenced by Nicholas Bernoulli and that the improvements owe much to collaboration with him.

In publishing his correspondence with the Bernoullis, Montmort carefully left questions of priority to his readers.

The correspondence with Nicholas Bernoulli is fascinating reading. It has the form of a series of friendly challenges of increasing difficulty. It usually begins with a problem set by Montmort, who illustrates the solution by a numerical example without disclosing his proof; Nicholas Bernoulli then provides a proof not only of the original problem but also of a generalization.

We have discussed some of Montmort's results in §§14.1 and 14.3. Today he is mostly known for his solution of the problem of coincidences, which we shall discuss in Chapter 19. He also formulated the problem of the duration of play and gave the solution for some special cases, as related in §§20.2 and 20.4. In the remainder of the present chapter we shall discuss the other problems treated in the *Essay*.

Judging from the number of references to the three books, Montmort's *Essay* never came to enjoy the popularity of Bernoulli's *Ars Conjectandi* and de Moivre's *Doctrine of Chances*. One reason for this is that the second edition of the *Essay* is a combination of a textbook and a scientific correspondence on the further development of the most important problems in the text, so it is rather difficult for the reader to follow the solution of a given problem because it is treated in many different places of the book. Another reason may be that Bernoulli and de Moivre were renown mathematicians so that other mathematicians would naturally turn to their works instead of to the work of an "amateur." The contributions of Montmort (and of Nicholas Bernoulli) have therefore been overlooked or undervalued by many authors. There exist, however, two comprehensive expositions of Montmort's work, namely, by Todhunter (Chapter 8) and by Henny (1975), on which we shall draw in the following.

18.2 MONTMORT'S COMBINATORIAL ANALYSIS AND THE OCCUPANCY DISTRIBUTION

Montmort writes that Pascal (1665) has given the best and most comprehensive exposition of combinatorics. As related in §5.2, Pascal's work consists of separate sections on the binomial coefficients, the figurate numbers, the binomial expansion, combinations, and the sums of powers of integers, respectively, which entail many repetitions. Montmort gives an integrated theory of these topics, much like Bernoulli's, but for some results with simpler proofs following Pascal. We shall not repeat these results, which may be found in §5.2 and §15.4. The reader who wants to compare the expositions of Bernoulli and Montmort should note that Montmort defines the figurate numbers slightly differently than Bernoulli, namely, as

$$f_n^m = \sum_{i=1}^{n+m-2} a_{i,m-1} = \binom{n+m-2}{m-1},$$

compare with (15.4.1). It is, however, remarkable how closely the choice of topics, proofs, and examples agree.

Montmort does not arrive at Bernoulli's formula for the sums of powers of integers. On the other hand, Bernoulli does not go as far in the analysis of the occupancy problem as Montmort.

Montmort introduces the symbol

$$\overset{n}{\underset{m}{\square}}$$

for the binomial coefficient; however, we shall retain the modern symbol introduced by Euler.

Let us first consider Montmort's results for the multivariate hypergeometric distribution and the multinomial distribution. Let N elements be divided into f classes, with N_i elements in the ith class, $N_1 + \cdots + N_f = N$. The number of ways in which n elements may be selected without replacement among the N elements such that n_i elements are selected from the ith class, $n = n_1 + \cdots + n_f$, equals

$$\binom{N_1}{n_1} \cdots \binom{N_f}{n_f}. \tag{1}$$

This is the number of chances of getting the specified outcome. Dividing by the total number of chances $\binom{N}{n}$ results in what today is called *the multivariate hypergeometric distribution*.

Consider next n independent trials with f equally likely outcomes in each trial, for example, n throws of a die having f faces. The number of ways of getting n_i outcomes of a specified kind, for example, n_i throws each with face number i, $n_1 + \cdots + n_f = n$, equals

$$\binom{n}{n_1}\binom{n-n_1}{n_2} \cdots \binom{n-n_1-\cdots-n_{f-1}}{n_f} = \frac{n!}{n_1! \cdots n_f!}. \tag{2}$$

Dividing by f^n, we get *the multinomial distribution*.

of (1) and (2) are done by straightforward combinatorial
ng successive conditionings.

presents three interpretations of the multinomial coefficient (2),
namely, as the number of combinations, as the number of permutations, and
as a coefficient in the multinomial expansion.

In the subsequent discussion Montmort uses three examples:

1. Drawings without replacement from a pack of cards (random sampling
   from a finite population).
2. Throwing a number of dice each having f faces (random sampling
   from an infinite population or sampling with replacement from a finite
   population).
3. The coefficients in the multinomial expansion $(a_1 + \cdots + a_f)^n$.

Today, many more examples of this kind have been analyzed, e.g., the
sampling of populations classified into several categories; the distribution of
the number of accidents over the days of the week; and the distribution of
a number of particles into compartments, usually discussed in terms of a
random distribution of balls into cells, for which reason these problems are
called *occupancy problems*.

Consider a two-dimensional array of elements (a_{ij}), say, for example, a
pack of cards, where $i = 1, \ldots, f$ denotes the face value and $j = 1, \ldots, s$ denotes
the suit number, so that the total number of elements equals $N = fs$.

Suppose that n elements are drawn without replacement from the N
elements and that we count the number of elements in the sample for each
face value, disregarding the suit numbers. This leads to the occupancy
distribution, $n_1, \ldots, n_f$, where n_i denotes the number of elements with face
value equal to $i, n_1 + \cdots + n_f = n, n_i = 0, 1, \ldots, s$, and $0 \leqslant n_i \leqslant n$. According to
(1), the corresponding probability distribution is

$$\binom{s}{n_1} \cdots \binom{s}{n_f} \bigg/ \binom{sf}{n}. \tag{3}$$

To find the distribution of the n_i's according to size, we have to count the
number of face values for which the occupancy numbers equal $0, 1, \ldots, s$,
respectively. Let the resulting distribution be

Value of n_i	k_1	k_2	$\cdots$	k_c
No. of face values	r_1	r_2	$\cdots$	r_c

where

$$0 \leqslant k_1 < k_2 < \cdots < k_c \leqslant s, \quad \sum_1^c r_j = f, \quad \text{and} \quad \sum_1^c k_j r_j = n.$$

For the pack of cards, this distribution gives the number of face values missing in a hand, the number of singletons, the number of doubles, etc. In the example of the balls, the distribution gives the number of empty cells, the number of cells singly occupied, doubly occupied, etc.

With this notation (3) may be written

$$\binom{s}{k_1}^{r_1} \cdots \binom{s}{k_c}^{r_c}.$$

However, this expression does not give the number of chances of getting the distribution in question because we have not taken into account the fact that the face values are to be disregarded. We are interested only in the number of triples, say, occurring in the sample, not in their face values. The number of chances above must therefore be multiplied by the number of ways in which the given number of face values may be chosen among the f face values; that is, by

$$\binom{f}{r_1}\binom{f-r_1}{r_2} \cdots \binom{f-r_1-\cdots-r_{c-1}}{r_c} = \frac{f!}{r_1! \cdots r_c!},$$

so that the final result becomes

$$\binom{s}{k_1}^{r_1} \cdots \binom{s}{k_c}^{r_c} \frac{f!}{r_1! \cdots r_c!}. \tag{4}$$

This is Montmort's result given in 1708, Proposition 14, without proof, and in 1713, Proposition 15, with a proof similar to the one above.

Let us next consider the throw of n dice, each having f faces numbered from 1 to f. This setup is analogous to the previous one, with the modification that the s suits have been replaced by n dice and that the outcome for each die is independent of the outcomes of the other dice, corresponding to drawings with replacement instead of without replacement. Disregarding the numbering of the dice, the distribution of the n outcomes according to face value is given by $(n_1, \ldots, n_f)$, $n_i = 0, 1, \ldots, n$, and $n_1 + \cdots + n_f = n$. The corresponding probability distribution is obtained by dividing the multinomial coefficient (2) by the total number of outcomes f^n. It should be noted that for drawings without replacement, we consider the $\binom{fs}{n}$ samples equally likely, whereas for drawings with replacement, we have f^n equally likely samples.

As before we now disregard the face numbers and consider the distribution of the n_i's according to size. Using the same notation, except that $k_c \leqslant n$ instead of $k_c \leqslant s$, the number of chances for the given occupancy distribution becomes

$$\frac{n!}{(k_1!)^{r_1} \cdots (k_c!)^{r_c}} \frac{f!}{r_1! \cdots r_c!}, \tag{5}$$

where the multinomial coefficient (2) has been written in terms of the k's and r's and then multiplied by the number of ways in which the given number of face values may be chosen among the f face values. This is Montmort's result given in 1708, Proposition 30, without proof, and with a printing error in the formula, but with the correct formula used in the numerical examples, and in 1713, Proposition 15, with the proof as given above. For the use of dice players, Montmort provides a table of the values of (2) and (5) for $f = 6$ and $n = 2, \ldots, 9$, see 1708, pp. 138–140, and 1713, pp. 200–202. Montmort returns to a discussion of (5) and its applications (1713, pp. 353–355) and claims priority for this formula.

A similar discussion of Montmort's results has been given by Henny (1975).

For a throw with n dice Montmort notes that the total number of different results when order is taken into account is f^n, whereas the number of different results is only

$$\binom{f + n - 1}{n}$$

when order is ignored (see 1708, Proposition 32; 1713, Propositions 10 and 11). He also proves that the number of chances of getting k aces, say, is

$$\binom{n}{k}(f - 1)^{n-k}. \tag{6}$$

Finally, he finds the distribution of the total number of points by throwing n dice, as discussed in §14.3.

We have discussed Montmort's formula for the summation of a series having constant differences of a certain order in connection with (15.4.13).

In further work on infinite series he proved that

$$\sum_1^\infty u_n x^n = \sum_1^\infty \left(\frac{x}{1-x} \right)^n \Delta^{n-1} u_1, \qquad (7)$$

if the series converges for $|x| < 1$, see Chrystal (1900, Vol. 2, 405–408).

There is no sharp distinction between combinatorial and probability problems in Montmort's book; many problems treated later could have been included in Part 1, and some of the problems in Part 1 could have been treated as card or dice games.

Montmort's combinatorial analysis is deeper and more comprehensive than Bernoulli's and requires more perseverance of the reader because of its greater difficulty. Furthermore, because Montmort does not write in the elaborate pedagogical style of Bernoulli, his text did not become as popular as Bernoulli's.

18.3   MONTMORT ON GAMES OF CHANCE

To demonstrate Montmort's methods of proof we shall discuss two card games, Pharaon and Lansquenet, and a game of bowls, Robartes' problem, in which he uses combinatorics, recursion, conditional probabilities, and summation of figurate number to find the expectations of the players. The excitement that followed Montmort's work can be seen in the discussion inspired by his derivation of the banker's expectation in Pharaon, which led to contributions from John and Nicholas Bernoulli, Struyck, de Moivre, Daniel Bernoulli, and Euler.

Montmort begins by defining the expectation of a player as $e = ps$, where p is the probability of winning and s is the total stake. Considering a game with two players, he defines the game as fair if the ratio of the stakes of the players equals the ratio of their expectations. He defines the *advantage* of a player as his expectation minus his stake. Assuming that the probability of winning is proportional to the number of chances favorable to the player, he proves that the player's expectation equals $e = (ms + n \times 0)/(m + n)$ if there are m cases for getting s and n cases for getting 0. He thus finds the probability of winning as the number of favorable cases divided by the total number of possible cases. He does not state explicitly that all the cases have to be equally likely, but it is clear from the context that this is what he has in mind.

Pharaon

Pharaon is played with an ordinary pack of 52 cards. At the beginning of the game, the banker's adversary, the punter, stakes the amount 1 on one of the 13 face values. The banker begins by drawing two cards in succession. If the first card drawn has the same face value as the punter's, the banker wins the punter's stake. If the second card drawn has the same face value as the punter's, the banker has to pay 1 to the punter. If both cards have the same face value as the punter's, the banker wins half the punter's stake. If both cards differ from the punter's, the game continues with the remaining 50 cards, the punter choosing a face value, the banker drawing two cards, and so on. These general rules are amended by a rule to the advantage of the banker, namely, that the *second* card in the *last* drawing does not count.

Montmort analyzes a case in which the stock consists of p cards, p being even, and the punter's card occurs q times in the stock, $1 \leqslant q \leqslant p$. We shall denote the cards by $a_1, \ldots, a_q; b_1, \ldots, b_{p-q}$. Montmort considers the $p!$ equally likely permutations and notes for each whether the banker gets the amount 2, 0, or $\frac{3}{2}$, or equivalently, whether his gain is 1, -1, or $\frac{1}{2}$. The corresponding averages give the banker's expectation $e_p(q)$ and the banker's advantage $u_p(q) = e_p(q) - 1$. The formula for u_p is simpler than that for e_p because the cases with $+1$ and -1 balance each other.

Montmort begins with a detailed discussion of the permutations for $p = 2$, 4, and 6 and calculates the corresponding values of $e_p(q)$. He notes that when a permutation begins with two b's, e_p may then be expressed by means of e_{p-2}, so that the problem may be solved by recursion. Turning to the general case he derives a recursion formula, as indicated in the following table.

DERIVATION OF RECURSION FORMULA

Outcome	ab	ba	aa	bb
No. of chances	$q(p-q)$	$(p-q)q$	$q(q-1)$	$(p-q)(p-q-1)$
Prize	2	0	$\frac{3}{2}$	$e_{p-2}(q)$
Gain	1	-1	$\frac{1}{2}$	$u_{p-2}(q)$

The total number of chances is $p(p-1)$. For $q > 2$ it follows that

$$e_p(q) = \frac{2q(p-q) + \frac{3}{2}q(q-1) + e_{p-2}(q)(p-q)(p-q-1)}{p(p-1)}, \tag{1}$$

$$u_p(q) = \frac{\frac{1}{2}q(q-1) + u_{p-2}(q)(p-q)(p-q-1)}{p(p-1)}. \tag{2}$$

For $q = 1$ Montmort considers the $p!$ permutations of $a_1, b_1, \ldots, b_{p-1}$. If a_1 occurs at an odd place, the banker gets 2; if a_1 occurs at an even place, the banker gets 0, except for the case where a_1 occurs last, which gives 1.

Hence,

$$e_p(1) = \left\{ 2\left(\frac{p}{2}\right) + 1 \right\} \frac{(p-1)!}{p!} = 1 + \frac{1}{p}, \tag{3}$$

so that $u_p(1) = 1/p$.

For $q = 2$ the only contribution to u_p comes from the $p/2$ cases, where the banker draws a_1 and a_2 together. The $p/2 - 1$ cases give the banker $\frac{1}{2}$, and the last case gives 1. Hence,

$$u_p(2) = \left\{ \frac{1}{2}\left(\frac{p}{2} - 1\right) + 1 \right\} \frac{2!(p-2)!}{p!} = \frac{p+2}{2p(p-1)}. \tag{4}$$

Montmort does not derive $u_p(2)$ directly as we have done above, but he gets the same result by a modification of the formula for $u_p(q)$ for $q > 2$, which we are going to prove.

By successive substitutions in (2) Montmort finds that

$$u_p(q) = \frac{q^{(2)}}{2p^{(2)}} \sum_{i=0}^{h} \frac{(p-q)^{(2i)}}{(p-2)^{(2i)}}, \qquad h = \begin{cases} \dfrac{p-q}{2} & \text{for } q \text{ even,} \\[2mm] \dfrac{p-q-1}{2} & \text{for } q \text{ odd.} \end{cases} \tag{5}$$

Multiplying the numerator and denominator by $(p - 2 - 2i)^{(q-2-2i)}$, he obtains

$$u_p(q) = \frac{q^{(2)}}{2p^{(2)}} \sum_{i=0}^{h} \frac{(p - 2 - 2i)^{(q-2)}}{(p-2)^{(q-2)}} \tag{6}$$

$$= \frac{q^{(2)}}{2p^{(2)}} \sum_{i=0}^{h} \left. \binom{p - 2 - 2i}{q - 2} \right/ \binom{p-2}{q-2} \tag{7}$$

$$= \sum_{i=0}^{h} \left. \binom{p - 2 - 2i}{q - 2} \right/ 2\binom{p}{q}. \tag{8}$$

Noting that $u_p(q)$ depends on the sum of alternate figurate numbers of order $q - 1$, Montmort states that such sums may be found by means of his summation formula (15.4.13) but that he will use a more direct method "based on a curious property of the figurate numbers."

Montmort defines the kth figurate number of order m by the recursion

$$f_k^m = \sum_{j=1}^{k} f_j^{m-1}, \quad f_k^1 = 1, \quad m = 2, 3, \ldots, \tag{9}$$

so that

$$f_k^m = \binom{m+k-2}{m-1}. \tag{10}$$

Denoting the sum of alternate figurate numbers by F_k^m, we have

$$F_k^m = f_k^m + f_{k-2}^m + \cdots, \tag{11}$$

the last term being $f_2^m = m$ for k even and $f_1^m = 1$ for k odd.
From the definitions it follows that

$$f_k^m - f_{k-1}^m = f_k^{m-1},$$

$$f_k^m + f_{k-1}^m + f_{k-2}^m + f_{k-3}^m + \cdots = f_k^{m+1},$$

$$f_k^m - f_{k-1}^m + f_{k-2}^m - f_{k-3}^m + \cdots = F_k^{m-1}.$$

Adding the last two equations and dividing by 2, Montmort gets the recursion

$$F_k^m = \tfrac{1}{2} f_k^{m+1} + \tfrac{1}{2} F_k^{m-1},$$

which leads to

$$F_k^m = \frac{1}{2} f_k^{m+1} + \frac{1}{4} f_k^m + \frac{1}{8} f_k^{m-1} + \cdots + \frac{1}{2^m} (f_k^2 + c_k), \tag{12}$$

where $c_k = 0$ for k even, and $c_k = 1$ for k odd.
For later use Montmort also proves that

$$F_k^m = \tfrac{1}{2} f_{k+1}^{m+1} - \tfrac{1}{2} F_{k+1}^{m-1},$$

which leads to

$$F_k^m = \frac{1}{2} f_{k+1}^{m+1} - \frac{1}{4} f_{k+2}^m + \frac{1}{8} f_{k+3}^{m-1} - \cdots - \frac{(-1)^m}{2^m} (f_{k+m}^2 + d_{k+m}), \tag{13}$$

where $d_{k+m} = 0$ for $k + m$ even, and $d_{k+m} = -1$ for $k + m$ odd.

Writing (8) in the form

$$u_p(q) = F_{p-q+1}^{q-1} \Big/ 2\binom{p}{q},$$

it follows from (12) that

$$u_p(q) = \left\{ \frac{1}{2}\binom{p-1}{q-1} + \frac{1}{4}\binom{p-2}{q-2} + \cdots \right\} \Big/ 2\binom{p}{q}$$

$$= \sum_{i=1}^{q-1} \frac{1}{2^{i+1}} \frac{q^{(i)}}{p^{(i)}} + \frac{1}{2^q} \frac{q^{(q)}}{p^{(q)}} g_q, \tag{14}$$

where $g_q = 1$ for q even, and $g_q = 0$ for q odd. By means of this formula, Montmort tabulates $u_p(q)$ for $q = 3, 4, \ldots, 8$; the first two results are

$$u_p(3) = \frac{3}{4(p-1)} \quad \text{and} \quad u_p(4) = \frac{2p-5}{2(p^2-4p+3)}. \tag{15}$$

By means of (13), Montmort obtains the alternative form

$$u_p(q) = \sum_{i=1}^{q-1} (-1)^{i+1} \frac{1}{2^i} \frac{q^{(i)}}{(p-q+i)^{(i)}}. \tag{16}$$

Of the results given above the following appear in the first edition (1708): the recursion (1) for $e_p(q)$; the general solution (8) in terms of figurate numbers is given without proof on p. 23; the lemma about the sum of alternate figurate numbers is indicated without proof on pp. 24–25; and the values of $u_p(q)$ for $q = 1, \ldots, 8$ are given on pp. 24–25.

Having obtained $u_p(q)$ in the form (8) by a combination of combinatorial reasoning and recursion, it is curious that Montmort did not derive this result afterward by a direct combinatorial argument. However, this was done by John Bernoulli in his letter to Montmort (1713, pp. 284–287). Bernoulli points out that the proof becomes simpler if one considers only the $\binom{p}{q}$ permutations of q a's and $p-q$ b's rather than the $p!$ permutations of p different elements. He demonstrates this for $p = 4$ and 6 and then states without proof that

$$u_p(q) = \left\{ 1 + \binom{q}{2} + \binom{q+2}{4} + \cdots + \binom{p-2}{p-q} \right\} \Big/ 2\binom{p}{q}$$

$$= \frac{(q-2)^{(q-2)} + q^{(q-2)} + \cdots + (p-2)^{(q-2)}}{p^{(q-2)}} \frac{q^{(2)}}{2p^{(2)}} \tag{17}$$

for q even, and gives a similar formula for q odd. Bernoulli writes that Montmort will find that this formula gives the same particular results, i.e., $u_p(q)$ for $q = 1, \ldots, 8$, as stated in his book; Bernoulli seems (like Todhunter) to overlook the fact that Montmort has given the general formula (8) and that (17) is only slightly different in form. The interesting question is what kind of combinatorial reasoning John Bernoulli used. It is tempting to suggest that John had his result from James, who used combinatorics to solve a very similar problem in the card game Bassette; see §15.5.

In his reply, Montmort (1713, pp. 303–304) points out that he has for a long time known of formula (5), which differs only slightly in its form from Bernoulli's.

Formula (16) is given without proof by Nicholas Bernoulli in a letter to Montmort (1713, p. 299) and proved by Montmort on p. 99 by means of (13). Presumably, Nicholas understood Montmort's indication of the proof of (12) and then derived (13) for use in his own proof.

Referring to Montmort's *Essay* (1713), Struyck (1716, pp. 104–107) gives a purely combinatorial proof. He notes that any permutation begining with a may be decomposed into two beginning with ab and aa, respectively, and that the corresponding probabilities are

$$\frac{q}{p} = \frac{q(p-q) + q(q-1)}{p(p-1)}.$$

Comparison with permutations beginning with ba, which have a probability of $(p-q)q/p(p-1)$, will show that the banker's advantage may be expressed in terms of the difference of these probabilities.

Ignoring the rule that the last card does not count, Struyck states that the winner of the game is the one who first draws an a from a bowl containing $q\,a$'s and $p-q\,b$'s, the banker and the punter drawing alternately. This problem is, however, the same as Huygens' second problem generalized to drawings without replacement; see §14.2. Hence, the banker's probability of winning is

$$P_1 = \frac{q}{p} + \frac{(p-q)^{(2)}}{p^{(2)}} \frac{q}{p-2} + \frac{(p-q)^{(4)}}{p^{(4)}} \frac{q}{p-4} + \cdots$$

$$= \left\{ \binom{p-1}{q-1} + \binom{p-3}{q-1} + \cdots \right\} \bigg/ \binom{p}{q},$$

and the punter's probability is similarly,

$$P_2 = \left\{ \binom{p-2}{q-1} + \binom{p-4}{q-1} + \cdots \right\} \bigg/ \binom{p}{q}.$$

The banker's advantage, depending on his priority, is thus $\frac{1}{2}(P_1 - P_2)$. Since,

$$\binom{p-x}{q-1} - \binom{p-x-1}{q-1} = \binom{p-x-1}{q-2},$$

the banker's advantage becomes

$$\left\{\binom{p-2}{q-2} + \binom{p-4}{q-2} + \cdots\right\}\bigg/ 2\binom{p}{q},$$

in agreement with (8).

De Moivre (1718, pp. 40–44, 55–58) derives $u_p(q)$ for $q = 1, \ldots, 4$ by a combinatorial argument similar to that used by Struyck. In his preface (1718, p. XI), de Moivre writes that "I own that some great Mathematicians have before me taken the pains of calculating the Advantage of the Banker... But still the curiosity of the Inquisitive remained unsatisfied; The Chief Question, and by much the most difficult, concerning *Pharaon* or *Bassette*, being what it is that the Banker gets *per Cent* of all the Money adventured." He then goes on to make q a random variable. He imagines that m times in succession the punter puts his stake on a face value chosen at random, his first choice being made before the first drawing. The average advantage of the banker becomes

$$\bar{u} = \frac{1}{m} \sum_{i=0}^{m-1} \sum_{q=0}^{4} u_{p-2i}(q) \binom{p-2i}{q}\binom{2i}{4-q}\bigg/ \binom{p}{4}, \qquad (18)$$

where the factor to $u_{p-2i}(q)$ equals the probability that q of the four cards remains in the stock. De Moivre reduces this expression to a polynomial in p and m divided by $p^{(4)}$. For $p = 52$ and $m = 23$, he finds $\bar{u} = 0.0299$, which means that the banker's average advantage is about 3% of the stake. In his preface de Moivre is very enthusiastic about this solution. As noted by Todhunter (p. 152), de Moivre's idea is rather unrealistic from a gambler's point of view. Today the idea is interesting mainly as an early example of a randomized procedure.

According to Henny (1975, p. 486), the game of Pharaon has also been discussed by Daniel Bernoulli in 1724 and by Euler in 1764 without essential new results.

Montmort's discussion on the card game *Bassette* is similar that on Pharaon, and we have discussed Bassette in §15.5.

Lansquenet

Lansquenet is a card game with n players placed at random at a round table. From an ordinary pack of cards the banker deals a card to each of the $n - 1$

players and then to himself. Each player stakes the amount 1, except for the banker, who stakes $n - 1$. Turning the remaining cards successively, the banker wins if the card has the same value as the player's, and he loses his own stake if the card has the same value as his own. The game stops when the banker has won all the stakes or lost his own. Although these general rules seem to make the game a fair one, all participants having the expectation zero, there are some supplementary rules that depend on the occupancy numbers, i.e., the number of singletons, doublets, etc., that give the banker an advantage. Further, the rule of continuation, also depending on the occupancy numbers, gives the banker a certain probability for continuing as banker. The banker's total advantage thus depends on his expected gain, g say, in a single game, and his probability of continuing as banker, p, say, so that the total advantage becomes

$$u = g + pg + p^2g + p^3g + \cdots = \frac{g}{1 - p}.$$

The problem is to determine g and p.

Montmort (1708, pp. 30–53; 1713, pp. 105–129) calculates these quantities for $n = 3, 4, \ldots, 7$. Even the patient Todhunter (p. 91) says of Montmort's discussion that "It does not appear to present any point of interest, and it would be useless labour to verify the complex arithmetical calculations which it involves." It seems that Todhunter has overlooked the fact that Montmort, after having derived the results for $n = 3$ and 4 by simple enumeration, points out for $n = 5$ that the general solution is obtained by means of the occupancy distribution, which he then goes on to apply. We shall indicate Montmort's general method and use the case $n = 4$ for illustration.

The occupancy distribution of n cards [see (2.4)] is

$$\binom{4}{0}^{r_0} \binom{4}{1}^{r_1} \binom{4}{2}^{r_2} \cdots \frac{13!}{r_0!, r_1!, r_2!, \cdots} \bigg/ \binom{52}{n},$$

where $\sum r_i = 13$, $\sum i r_i = n$, r_0 is the number of face values missing among the n cards, r_1 the number of singletons, r_2 the number of doublets, etc. We shall denote the number of chances given by the numerator by N_i corresponding to vector number i in a given ordering of the vectors $(r_0, r_1, r_2, \ldots)$.

To find the banker's expected gain in a single game, Montmort considers the occupancy distribution for the $n - 1$ cards, and for each outcome he calculates the expected gain according to the special rules of the game. Finally he obtains g as the average of the conditional gains, as shown in the following table.

Montmort's calculation of the expected gain in a single game with four players

i	EXAMPLE[a]	r_0	r_1	r_2	r_3	$N_i/52$	EXPECTED GAIN, g_i
1	abc	10	3	0	0	352	63/245
2	aab	11	1	1	0	72	197/245
3	aaa	12	0	0	1	1	275/245
Total:						425	

[a] a, b, and c denote different face values.

The total number of chances equals $\binom{52}{3} = 22{,}100$; dividing by 52 we get 425, as shown in the table. It follows that

$$g = \frac{\sum N_i g_i}{\sum N_i} = \frac{7327}{20{,}825}.$$

   To find the probability of continuation, Montmort considers the occupancy distribution for n cards, and for each outcome he calculates the probability of continuation according to the special rules of the game. Finally he obtains p as the average of the conditional probabilities as shown in the following table.

Montmort's calculation of the probability of continuation in a game with four players

i	EXAMPLE[a]	r_0	r_1	r_2	r_3	r_4	$N_i/13$	Probability p_i
1	abcd	9	4	0	0	0	14,080	20/80
2	aabc	10	2	1	0	0	6336	29/80
3	aabb	11	0	2	0	0	216	40/80
4	aaab	11	1	0	1	0	192	50/80
5	aaaa	12	0	0	0	1	1	80/80
							20,825	

The banker's total expected gain thus becomes

$$\frac{g}{1-p} = \frac{36,635}{73,896}.$$

Montmort ends his analysis by giving examples of the calculation of the expected loss of each player using the same method as above.

In his letter to Montmort (1713, pp. 287–289) John Bernoulli points out that a more realistic analysis should be based on the assumption that the players successively become banker, so that when the banker loses his stake, the player on his right takes over as banker. Bernoulli analyzes this problem for two players. Let the outcome be aa with probability p and ab with probability $q = 1 - p$. In the first case the banker gets the amount 2, in the second case he gets nothing. Hence his expected gain in a single game is $g = 2p - 1 = p - q$, so that his total expected gain according to Montmort's assumption becomes $g/(1 - p) = (p - q)/q$. Let u denote the banker's total expected gain under Bernoulli's assumption. Bernoulli then solves the problem by means of the equation

$$u = p(1 + u) + q(-1 - u),$$

which reflects the fact that the two players change roles with probabilities p and q. The solution is $u = (p - q)/2q$, just half of the previous result.

It will be seen that John Bernoulli's result is a special case of the result given by James in his Problem 19, see §15.5.

In Nicholas Bernoulli's brief remarks to Montmort (1713, pp. 299–301), he generalizes to n players and a finite number of games, m, say. Let the players $A_1, \ldots, A_n$ have the expected gains $g_1, \ldots, g_n, \sum g_i = 0$, in a single game in which A_1 is banker, and let p be the probability of continuing as banker for each player. Player A_2 sits to the left of A_1, A_3 to the left of A_2, and so on, until A_n, who sits to the right of A_1. When A_1 loses the game, A_n becomes banker, so that A_1 now has the same position in relation to the banker as A_2 had before, which means that his expected gain now is g_2. Since the game is "circular," we have $g_i = g_{n+i} = g_{2n+i} = \cdots$ if more than n games are involved. Without proof Nicholas Bernoulli gives the expected gain of A_1 as

$$u_1 = g_1 + (pg_1 + qg_2) + (p^2 g_1 + 2pqg_2 + q^2 g_3) + \cdots,$$

the series containing m terms. Analogous expressions hold for the other players. The proof is straightforward. In the first game, A_1's gain is g_1. If he wins he continues as banker and gets g_1; if he loses, A_n becomes banker, and A_1's gain is g_2, and so on. Without proof Bernoulli then gives another form of this expression. We shall indicate one proof; another is given by Todhunter (pp. 116–119).

We have

$$u_1 = \sum_{k=0}^{m-1} \sum_{i=0}^{k} \binom{k}{i} p^{k-i} q^i g_{1+i} = \sum_{i=0}^{m-1} g_{1+i} q^i \sum_{k=0}^{m-i-1} \binom{k+i}{k} p^k,$$

after changing the order of summation.

Note that

$$\binom{k+i}{k} q^{i+1} p^k$$

is the probability of getting the $(i+1)$st failure at the last trial in $k+i+1$ trials, i.e., the binomial waiting time distribution (14.1.3). To reduce the sum of such terms to a sum of binomial probabilities we note that

$$\sum_{x=0}^{c} \binom{n}{x} q^x p^{n-x} = \sum_{y=n+1}^{\infty} \binom{y-1}{c} q^{c+1} p^{y-c-1},$$

because the left-hand side gives the probability of at most c failures in n trials, which is the same as the probability that the $(c+1)$st failure occurs at the $(n+1)$st trial or later, as given by the right-hand side. It follows that

$$1 - B(c, n, q) = q^{c+1} \sum_{y=0}^{n-c-1} \binom{y+c}{y} p^y.$$

By substitution we obtain

$$u_1 = q^{-1} \sum_{i=0}^{m-1} g_{1+i}[1 - B(i, m, q)],$$

which is Bernoulli's result.

As explained in §14.1, Montmort knew the relation between the binomial and the negative binomial distributions in 1713 but presumably not in 1708. Whether Nicholas Bernoulli made use of this relation we do not know.

Finally Bernoulli gives the solution for $m \to \infty$ as

$$u_1 = \frac{(n-1)g_1 + (n-2)g_2 + \cdots + g_{n-1}}{nq}.$$

We shall leave the proof to the reader. Besides Todhunter, Henny (1975) has discussed Bernoulli's results.

Other Games

The card game *Treize* is discussed in Chapter 19.

After having discussed the games Pharaon, Lansquenet, Treize, and Bassette, which are pure games of chance, Montmort turns to some other card games that also depend on the knowledge and decisions of the players. He states that it is impossible to give a complete discussion of such games and proceeds to discuss some special situations that may be solved by simple combinatorial methods.

The next part of the *Essay* is on dice games. We have discussed the game *Quinquenove* (Cinq et neuf) in §15.5. The most important result for dice games is Montmort's formula for the number of chances of getting s points by throwing n dice each having f faces and the corresponding table, which we have discussed in §14.3.

In the following part Montmort solves Huygens' five problems, (see §14.2) and the problem of points (see §14.1).

On pp. 248–257 and p. 366 of the second edition, there follows an analysis of the problem of points in a game of bowls. According to de Moivre (Preface, 1712 and 1718), this problem was posed to him by Francis Robartes, and de Moivre gave the solution of a special case in *De Mensura Sortis* (1712, Problems 16 and 17). *Robartes' problem* was generalized by Montmort as follows:

Players A and B play a game of bowls, A with m bowls and B with n. The skill of A is to the skill of B as r to s. In each game the winner gets a number of points equal to the number of his bowls that are nearer to the jack than any of the loser's. If the play is interrupted when A needs a points and B b points in winning, how should the stake be divided equitably between them?

Montmort explicitly defines "skill" by referring to a game with one bowl for each player. Player A's skill, $r/(r+s)$, is then A's probability of getting nearer to the jack than B. He also points out that for $m = n = 1$, we have the classical problem of points.

Montmort assumes that the total stake is 1 so that A's expectation, $e(a, b)$, say, equals his probability of winning the stake.

To solve the problem Montmort introduces an urn with mr white chips and ns black chips, representing A's and B's chances of winning. The total number of chips is $t = mr + ns$.

Let P_i be A's probability of getting at least i points, i.e., the probability of getting a run of i white chips by drawings without replacements from the urn. It follows that

$$P_i = \frac{mr}{t} \frac{mr - r}{t - r} \cdots \frac{mr - (i-1)r}{t - (i-1)r}, \qquad i = 1, 2, \ldots, m.$$

Let p_i be A's probability of getting exactly i points, i.e., the probability of getting a run of i white chips followed by a black. Hence,

$$p_i = \frac{mr}{t} \frac{mr-r}{t-r} \cdots \frac{mr-(i-1)r}{t-(i-1)r} \frac{ns}{t-ir}, \qquad i = 1, 2, \ldots, m.$$

P_i and p_i are defined as zero otherwise.

The corresponding probabilities for B will be denoted by Q_i and q_i, $i = 1, 2, \ldots, n$, and they are obtained from P_i and p_i by interchanging (m, r) and (n, s).

Note that $p_i = P_i - P_{i+1}$ and that

$$\sum_1^m p_i + \sum_1^n q_i = P_1 + Q_1 = 1.$$

Montmort gives the solution as the recursion

$$e(a, b) = P_a + \sum_{i=1}^{a-1} p_{a-i} e(i, b) + \sum_{i=1}^{b-1} q_i e(a, b-i), \qquad \begin{cases} a = 1, 2, \ldots, \\ b = 1, 2, \ldots, \end{cases}$$

and $e(a, 0) = e(0, b) = 1$. The proof follows directly from the addition and multiplication theorems.

If the players have only one bowl each, then $P_1 = p_1 = r/(r+s)$, $Q_1 = q_1 = s/(r+s)$, and

$$e(a, b) = p_1 e(a-1, b) + q_1 e(a, b-1),$$

which is the recursion for the classical problem of points.

Montmort states that similar results hold for any number of players and gives a numerical example for three players.

In his formulation and discussion of Robartes' problem, de Moivre (1712) assumes that the players are of equal skill and have the same number of bowls. He then derives the formulae for $e(2, 1)$ and $e(3, 1)$ and states that the general formula may be found by the same method. In the *Doctrine of Chances* (1718, Problems 27 and 28), de Moivre acknowledges Montmort's general solution and derives $e(a, b)$ for $m = n$.

In modern terminology, Robartes' problem may be described as a random walk in two dimensions, the horizontal steps being of length 1 or 2, ..., or m, and the vertical steps being of length 1 or 2, ..., or n. Using the same graph as in Fig. 5.3.1, A wins if the random walk crosses the vertical line through $(a, 0)$ before crossing the horizontal line through $(0, b)$.

In 1711 Montmort published a paper in *Journal des Sçavans* containing a problem on the *Loterie de Loraine*, giving his own solution in the form of an anagram. He argued that this was another example of the usefulness of mathematics in civil affairs and urged the magistrates to consult mathematicians before making decisions on such matters. The rules of the lottery are much to the disadvantage of the owner. This lettter is reprinted together with a discussion of the solution in the *Essay* (1713, pp. 257–260, 313, 326, 346). In another example, Montmort considers the problem of choosing between two candidates for an office by majority voting (1713, pp. 260–261). These problems are not particularly difficult, and we have used them as problems for the reader in §18.7.

18.4   THE CORRESPONDENCE OF MONTMORT WITH JOHN AND NICHOLAS BERNOULLI

Montmort sent a copy of his *Essay* to John Bernoulli who responded with a letter of 17 March 1710. Bernoulli writes that the *Essay* contains many beautiful, interesting, and useful results and goes on with many detailed remarks on the games Pharaon, Lansquenet, and Treize; on Huygens' problems; on the multinomial coefficient; the problem of points; and the duration of play. Except for the statement on the solution of the problem of points (see §14.1), there are no essential contributions in the letter; we have discussed most of the remarks in connection with the topics in question.

Bernoulli ends by praising Montmort's deep insight and indefatigable patience in carrying out long and laborious calculations. He expresses his hope that Montmort will continue his work and produce a more ample and rich book, and he points out that there are many more problems to investigate, in particular moral and political matters, which his brother James had begun to discuss in a book that, apparently, will never be published.

It must have been greatly encouraging for Montmort, who was rather unknown, to receive such a letter from a famous mathematician. In his reply Montmort thanks Bernoulli for the honor he has shown him by reading his book and for his many learned and judicious remarks. He replies to the various remarks and adds two new results on the probability of getting a given sum of points by throwing n dice (see §14.3) and on the duration of play (see §20.4), respectively.

John Bernoulli lent his copy of the *Essay* to his nephew Nicholas, who wrote some important remarks which were sent to Montmort together with John's letter. A correspondence ensued of which the *Essay* contains seven letters from Nicholas and six from Montmort. In one of the letters, Montmort (1713, p. 322) asks the Bernoullis' permission to include their letters in the new edition of the *Essay* which he is planning.

The correspondence between Montmort and Nicholas Bernoulli is a paragon of a friendly scientific correspondence, showing the creativity and ingenuity of each and how they inspired one another to formulate and solve ever more difficult problems and thus develop probability theory from an elementary level to a discipline on a par with other branches of mathematics.

The results obtained are too extensive and important to be discussed in a single section; the more important topics are therefore treated separately: Pharaon, Lansquenet, and Robartes' problem in §18.3; the game of tennis in §18.5; the strategic game Her in §18.6; the problem of coincidences (Treize) in §19.3; the duration of play in §20.3; and Waldegrave's problem in §21.2. We have discussed Nicholas Bernoulli's approximation to the binomial in §16.3 and his analysis of the ratio of male to female births in §17.3. An account of Montmort's and Bernoulli's discussion of de Moivre's *De Mensura Sortis* is given in §22.2.

On 26 February 1711, Nicholas Bernoulli mentions that he has submitted Montmorts' offer to print the manuscript of *Ars Conjectandi* to his cousin, James Bernoulli's son, who is in posession of the manuscript, and that he has also written to Hermann, asking him to see that it is printed quickly but that so far he has not had any reply. He also writes that it is desirable that someone complete the last part of the book and that he knows of no one more capable of succeeding than Montmort. Montmort, however, briefly outlines the difficulties of such a project and declares himself insufficient to this task.

In his last letter Bernoulli writes that in return for the many problems posed by Montmort he shall now pose five problems for Montmort to solve. We have used the first three as problems for the reader in §18.7. The fourth problem is as follows: Player B throws successively with an ordinary die and gets x crowns from A if a 6 occurs for the first time at the xth throw, $x = 1, 2, \ldots$; find B's expectation. Obviously, B's expectation is six crowns; we shall leave the proof to the reader. [Todhunter's solution (p. 134) is wrong].

The fifth problem, known today as the *Petersburg problem*, is obtained from the fourth by letting the prize increase in geometric progression, for example, as 2^{x-1}. This leads to an infinite expectation. Reviews of the history of this problem have been given by Jorland (1987) and Dutka (1988).

Montmort replies that he has not yet solved the two first problems, that he has generalized and solved the third, and that the two last problems do not present any difficulty, since the solution is obtained as the sum of an infinite series. He does not mention that the second sum is infinitely large.

The last of the published letters is from Montmort to Bernoulli and dated

15 November 1713. Montmort writes that his friend, Mr. Waldegrave, is taking care of the printing of the new edition of the *Essay* and that it is nearly finished.

18.5 MONTMORT AND NICHOLAS BERNOULLI ON THE GAME OF TENNIS

The problem of points for the game of tennis is discussed in the correspondence between Nicholas Bernoulli and Montmort (1713, pp. 333–334, 349–350, 352–353, 371). The background for their discussion is the solution of the classical problem of points for two players, which we have given in §14.1, and James Bernoulli's discussion on tennis (see §15.6).

Nicholas Bernoulli, who knew the content of James Bernoulli's *Lettre* before it was published in 1713, writes in a letter of 10 November 1711 to Montmort that James has solved many interesting and useful problems on the game of tennis. He quotes four of the problems without giving the solutions or indicating James's method of solution and asks Montmort to solve the problems for comparison with James's solutions. In his reply of 1 March 1712, Montmort does live up to the challenge; he gives an explicit formula for A's probability of winning when A lacks a points and B lacks b points to win.

Montmort considers the problem as a generalization of the classical problem of points. He first refers to the solution of this problem in terms of the binomial distribution, see (14.1.1), and without further comment he notes that the corresponding formula for the problem of points under the rules valid for tennis becomes

$$e_t(a,b) = \sum_{i=a}^{a+b-2} \binom{a+b-2}{i} p^i q^{a+b-2-i} + \binom{a+b-2}{a-1} p^{a-1} q^{b-1} \frac{p^2}{p^2+q^2}.$$

In the ordinary game of tennis, $a = 4 - i$ and $b = 4 - j$, but the formula holds for any other number than 4.

It will be seen that Montmort's elegant result comprises all the formulae which James Bernoulli laboriously derived by recursion, since

$$g(i,j) = e_t(4-i, 4-j).$$

Montmort says that $e_t(a,b)$ is obtained from $e(a,b)$, replacing $a+b-1$ by $a+b-2$ and multiplying the last term by $p^2/(p^2+q^2)$. This is also the proof of the formula, since the first term gives A's probability of getting a

points before B gets $b - 1$ points, and the last term gives the probability of winning after a deuce.

Noting that the first term of $e_t(a, b)$ equals $e(a, b - 1)$, Montmort's formula may also be written in the form

$$e_t(a, b) = p^a \sum_{i=0}^{b-2} \binom{a - 1 + i}{i} q^i + \binom{a + b - 2}{a - 1} p^{a-1} q^{b-1} \frac{p^2}{p^2 + q^2}.$$

For example,

$$e_t(4, 4) = p^4(1 + 4q + 10q^2) + \frac{20p^5 q^3}{p^2 + q^2}.$$

Montmort does not mention this alternative form of his formula in the published part of his letter, but according to Henny (1975), the corresponding form of $e(a, b)$ is given in the letter. Presumably Montmort left it out of the letter and transferred it to his general discussion of the problem of points (1713, pp. 245–246) for systematic reasons.

To solve the problem of handicaps for a given relative strength of the players, Montmort solves the equation $1 - e_t(a, b) = \frac{1}{2}$ with respect to $m = a + b - 2$ for given values of a and p/q. He considers the example with $a = 4$ and $p/q = 2$. First, he solves the corresponding equation for the problem of points; as pointed out by Todhunter (p. 125), the equation given by Montmort (1713, p. 342) is wrong, but the root is correct so that he must have had the correct equation. For the game of tennis Montmort just gives the solution that B's handicap should be $j = 2\frac{11}{244}$, without giving the equation, which obviously is

$$1 + 2\binom{m}{1} + 4\binom{m}{2} + \frac{8}{5}\binom{m}{3} = \frac{3^m}{2},$$

or

$$8m^3 + 36m^2 + 16m + 30 = 5 \times 3^{m+1},$$

with the approximate solution $m = b + 2 = 3\frac{213}{224}$.

As a check Montmort inserts $a = 4$ and $b = 1\frac{213}{224}$ in the equation $e_t(a, b) = \frac{1}{2}$, which leads to a homogeneous algebraic equation of the sixth degree in p and q from which he finds that $p/q = 2$.

Since the handicap will normally lie between two integers, Montmort states that the solution requires randomization to be carried out in practice.

In the example above a chip is drawn from a bag with 213 white and 11 black chips, and if the chip drawn is white B gets a handicap of 2 points, if black only 1 point.

In his reply Nicholas Bernoulli acknowledges Montmort' solution and makes a further generalization. He assumes that A's probability of winning a point in odd- and even-numbered games equals p_1 and p_2, respectively, thereby taking into account the fact that A's probability of winning depends on whether he is serving or not. Let $a + b - 1 = m + n, m = n$ if $m + n$ is even, and $m = n + 1$ if $m + n$ is odd. Bernoulli then gives A's probability of winning for the problem of points in the form

$$\sum_{i=0}^{b-1} \sum_{j=0}^{i} \binom{m}{i-j} p_1^{m-i-j} q_1^{i-j} \binom{n}{j} p_2^{n-j} q_2^{j}.$$

He adds that for the game of tennis, $m + n$ should be replaced by $m + n - 1$ and that the term for $i = b - 1$ should be multiplied by $p_1 p_2 / (p_1 p_2 + q_1 q_2)$.

18.6 THE DISCUSSION OF THE STRATEGIC GAME
HER AND THE MINIMAX SOLUTION

For two players, A and B, the card game called Her may be described as follows. Player B holds a pack of cards consisting of four suits, the cards being numbered from 1 to 13. Player B draws a card at random, which he gives to A, and afterward he draws a card for himself; the main object of the game is to obtain the higher card. However, if A is not content with the card received, he may compel B to exchange cards with him, unless B has a 13. If B is not content with the card first obtained or with the card that A has compelled him to take, he may change it by drawing a card at random among the 50 cards remaining in the pack; but if he draws a 13, he is not allowed to change his card. The players then compare cards, and the player with the higher card wins; if they have cards of the same value, B wins.

The game is described for four players and used as a problem for the reader by Montmort (1708, pp. 185–187). The description is repeated for three players by Montmort (1713, pp. 278–279) and again posed as a problem. It is discussed and solved for two players in the correspondence with Nicholas Bernoulli, see Montmort (1713, pp. 321, 334, 338–340, 348–349, 361–362, 376–378, 400, 403–406, 409–412, 413).

It will be seen that the game consists of three steps. The first is the drawing of two cards at random, and the outcome thus depends on chance only. The second step depends on A's decision whether to use his right to exchange cards with B or not. The third step depends on B's decision whether to use his right to exchange his card with a card from the pack or not. Since A's probability of winning depends not only on chance but also on the decisions made by the players, the game is called a strategic game in today's terminology. In the first letter in which Montmort mentions this game he writes that "The difficulties of this problem are of a singular nature."

Dresher (1961, pp. 6–7, 59–60) has used Her as an example of a strategic game but without historical comments; Henny (1975) has given a complete account of the history of Her and elucidated the solution by means of modern game theory. We shall make use of both these expositions.

It is intuitively clear that a good strategy consists in changing cards of low value and retaining cards of high value and, furthermore, that the line of demarcation between low and high must be about 7. Without discussion Montmort and Bernoulli concur in the opinion that A always should change cards of value less than 7 and retain cards of value higher than 7; his decision with regard to 7 depends on the strategy chosen by B. Further, they agree that a similar rule holds for B but with 7 replaced by 8. The evaluation of the effect of these rules on A's probability of winning depends on calculations which are published by Montmort as an appendix to his last letter "to spare the reader for the trouble to do it himself." We shall begin by explaining these calculations; similar explanations have been given by Trembley (1804), Todhunter (pp. 106–110), Fisher (1934), and Dresher (1961).

In each game there are at most three cards involved, their values being i, j, and k, say, such that A gets i, B gets j, and if the game ends with B drawing a card from the pack, he gets k. Player A's probability of winning may be written as

$$p(i, j; S) = \frac{p(i, j)c(i, j; S)}{50},$$

where $p(i, j)$ is the probability of getting the values (i, j) of the two cards drawn, and $c(i, j; S)$ is the number of cards among the remaining 50 cards which makes A win, given that the players use the strategy S. Assuming that the stake is unity, A's expectation equals his probability of winning.

Obviously, $p(i, j)$ equals $(4/52)(4/51)$ for $i \neq j$ and $(4/52)(3/51)$ for $i = j$. To illustrate the calculation of $c(i, j; S)$, we have (like Dresher) considered the case where A retains cards of value 7 and above and changes cards of lower value, and B retains cards of value 9 and above and changes cards of lower value. The results are shown in the accompanying table.

TABLE OF $c(i, j; S)$

$i \backslash j$	B changes								B retains					$c(i; S)$
	1	2	3	4	5	6	7	8	9	10	11	12	13	
A changes														
1	0	7	11	15	19	23	27	31	35	39	43	47	0	297
2	0	0	11	15	19	23	27	31	35	39	43	47	0	290
3	0	0	0	15	19	23	27	31	35	39	43	47	0	279
4	0	0	0	0	19	23	27	31	35	39	43	47	0	264
5	0	0	0	0	0	23	27	31	35	39	43	47	0	245
6	0	0	0	0	0	0	27	31	35	39	43	47	0	222
A retains														
7	27	27	27	27	27	27	24	24	0	0	0	0	0	204
8	31	31	31	31	31	31	31	28	0	0	0	0	0	238
9	35	35	35	35	35	35	35	35	0	0	0	0	0	280
10	39	39	39	39	39	39	39	39	50	0	0	0	0	362
11	43	43	43	43	43	43	43	43	50	50	0	0	0	444
12	47	47	47	47	47	47	47	47	50	50	50	0	0	526
13	50	50	50	50	50	50	50	50	50	50	50	50	0	600
Total														4251

The combinations of i and j for which A loses, $c = 0$, and A wins, $c = 50$, follow immediately from the rules of the game. Consider next one of the combinations which has $1 \leqslant i \leqslant 6$ and $i < j < 13$. For $i \leqslant 6$, A compels B to exchange cards, whereafter B has a card of value i, and he knows that A has a card of value j larger than i. Player B therefore draws a card from the pack and gets a card of value k. Player A wins if $k < j$ or $k = 13$, and the number of cards satisfying this condition is

$$c(i, j; S) = 3 + 4(j - 1) = 4j - 1.$$

For $7 \leqslant i \leqslant 12$ and $1 \leqslant j \leqslant 6$, a similar argument gives $c(i, j; S) = 4i - 1$. We shall leave it to the reader to check the remaining values.

For a given value of i A's probability of winning becomes

$$p(i; S) = \sum_j p(i, j; S) = \frac{2c(i; S)}{13 \times 51 \times 25},$$

where

$$c(i; S) = \sum_{j \neq i} c(i, j; S) + \frac{3}{4} c(i, i; S).$$

The total probability of winning for A then becomes

$$p(S) = \sum_i p(i; S) = \frac{2}{13 \times 51 \times 25} \sum_i c(i; S).$$

For the strategy defined in the table above,

$$p(S) = \frac{2 \times 4251}{13 \times 51 \times 25} = \frac{2834}{5525} = 0.5129.$$

Montmort does not explain these rather simple calculations; he just provides a table with the values of $2c(i; S)$ for the four strategies in question.

MONTMORT'S TABLE OF $2c(i; S)$ AND A'S PROBABILITY OF WINNING

	A changes 7 and under		A retains 7 and over	
Value of A's card	B changes 8 and under	B retains 8 and over	B retains 8 and over	B changes 8 and under
1	594	594	594	594
2	580	580	580	580
3	558	558	558	558
4	528	528	528	528
5	490	490	490	490
6	444	444	444	444
7	390	390	360	408
8	476	434	434	476
9	560	590	590	560
10	724	746	746	724
11	888	902	902	888
12	1052	1058	1058	1052
13	1200	1200	1200	1200
$p(S)$	2828/5525 = 0.5119	2838/5525 = 0.5137	2828/5525 = 0.5119	2834/5525 = 0.5129
Strategy	CC	CR	RR	RC

Source: Montmort (1713), p. 413. (C) change; (R) retain.

We have added the decimal fractions for $p(S)$ and labeled the strategies in the last line of the table C and R, denoting change and retain, respectively. Only these strategies are discussed in the correspondence.

It will be seen that the numbers in the last column of Montmort's table are obtained as $2c(i; S)$ from the previous table.

Returning to the correspondence, the first letter from Montmort on Her contains the remark that A's advantage lies between 1/85 and 1/84. Since

$$\frac{2828}{5525} - \frac{1}{2} = \frac{131}{11,050} = \frac{1}{84.4},$$

this indicates that Montmort has found that A's probability of winning is (at least) 2828/5525.

In his reply Bernoulli states that if one supposes that each player chooses the strategy which is most advantageous for him, then when A chooses C, B also chooses C, and when A chooses R, B also chooses R, and in both cases A's probability of winning becomes 2828/5525. Nevertheless, he adds, it is more advantageous for A to use strategy C than R, and "this is an enigma that I leave for you to expound." Obviously, Bernoulli's remark refers to the fact that strategy C gives either 2828/5525 or 2838/5525, whereas strategy R gives only 2828/5525 or 2834/5525.

Montmort replies that Bernoulli's solution agrees with his own. However, before asking for Bernoulli's opinion he had for some time discussed the problem with two of his friends, M. l'Abbé d'Orbais (also called Monsoury) and an English gentleman, Mr. Waldegrave, and they are of another opinion; M. l'Abbé d'Orbais holds that it is impossible to determine the expectation of either A or B because one cannot find A's optimal strategy without knowing B's strategy and vice versa, and this leads to a circular argument. We have illustrated this in the following table, assuming that A begins by choosing strategy C.

TABLE OF A'S EXPECTATION ACCORDING TO THE
FOUR STRATEGIES

A's strategy	B's strategy		
	C		R
C	2828/5525	←	2838/5525
	↓		↑
R	2834/5525	→	2828/5525

Montmort had shown Bernoulli's letter to his friends, and in response Waldegrave writes a letter to Montmort stating their objections. Although using more sophisticated reasoning than shown above, they arrive at the circular argument and maintain that whether the players use strategy C or R makes no difference. They conclude that Bernoulli's solution is false because he has "contented himself with considering the fractions which express the different expectations of A and B without taking notice of the probability of what the other player will do."

It is no wonder that Bernoulli does not understand the implication of this remark, since the writers themselves have not grasped the full implication of their point of view. Bernoulli is not convinced; he now gives the arguments for his previous solution. If A has chosen a strategy, C or R, then B chooses the same strategy, because this choice minimizes A's expectation and thus maximizes his own; for both these combinations of strategies A's expectation is 2828/5525. However, Bernoulli states that if it is impossible for the players to decide which strategy to use, they may leave the decision to chance and each choose a strategy with a probability of $\frac{1}{2}$, which makes A's expectation equal to 2832/5525. He also states that this procedure does not lead to a unique solution because A's expectation is 2833/5525 if he always uses strategy C and B leaves the decision to chance. (Bernoulli uses slightly different numbers in his reasoning.)

In the last letter on Her, Montmort writes that he is now convinced that the original considerations by himself and Bernoulli do not represent a solution because they presuppose that B knows the strategy of A. He adopts the idea of choosing a strategy at random and gives A's expectation as

$$\bar{p} = \frac{2828ac + 2834bc + 2838ad + 2828bd}{5525(a + b)(c + d)},$$

where $a/(a + b)$ and $c/(c + d)$ are A's and B's probabilities of choosing strategy C, respectively. Montmort is unable to find the optimal values of these probabilities and concludes that it is impossible to solve the problem. However, before he has finished his letter he receives one from Waldegrave with the solution, which he includes in the letter to Bernoulli.

Setting $a = 3$ and $b = 5$, Waldegrave finds that $\bar{p} = 2831.75/5525 = 0.5125$ whichever strategy B chooses and thus, also, whichever values of c and d he chooses. He gives a clear discussion of this solution, showing that if A chooses any other strategy, there exists a strategy for B which makes A's expectation smaller. Hence, $a = 3$ and $b = 5$ is the optimal strategy for A. Furthermore, Waldegrave observes that B is able to limit A's expectation to 2831.75/5525 by choosing $c = 5$ and $d = 3$, and this is the optimal strategy for B. The introduction of a chance device thus resolves the problem of the circular argument and gives a uniquely determined value of A's expectation. Waldegrave does not explain how he has found this solution.

According to Henny (1975), Bernoulli acknowledged Waldegrave's solution in a letter to Montmort in 1714. Neither Waldegrave nor Bernoulli generalized their analysis. A mathematical theory of games of strategy was not developed until the 1920s with the works of Borel and von Neumann and in the fundamental book by von Neumann and Morgenstern (1944).

Like Henny (1975) we shall comment briefly on the relation between Waldegrave's solution and modern theory. Let a_{ij}, $i = 1, 2$ and $j = 1, 2$, denote the values of A's expectation in the 2×2 table above, i and j denoting the strategies of A and B, respectively. Under the assumptions made by Montmort and Bernoulli, A gets at least

$$\max_i \min_j a_{ij} = \frac{2828}{5525},$$

and at most

$$\min_j \max_i a_{ij} = \frac{2834}{5525}.$$

Since

$$\max_i \min_j a_{ij} < \min_j \max_i a_{ij},$$

there is no unique solution by means of pure strategies.

Introducing mixed strategies and setting $x = a/(a + b)$ and $y = c/(c + d)$, Montmort's formula for A's expectation becomes

$$\bar{p}(x, y) = a_{22} + (a_{12} - a_{22})x + (a_{21} - a_{22})y - (a_{21} - a_{22} + a_{12} - a_{11})xy.$$

Just as in the example, we shall assume that the coefficients of x and y are positive and the coefficient of xy negative.

Waldegrave determines the optimal value of x, x_0, say, by the condition that $\bar{p}(x_0, y)$ has to be independent of y. Setting the coefficient of y equal to zero, we obtain

$$x_0 = \frac{a_{21} - a_{22}}{a_{21} - a_{22} + a_{12} - a_{11}} = \frac{3}{8},$$

$$\bar{p}(x_0, y) = \frac{a_{22} + (a_{12} - a_{22})(a_{21} - a_{22})}{a_{21} - a_{22} + a_{12} - a_{11}} = \frac{2831.75}{5525}.$$

Further, Waldegrave states that if A chooses a strategy different from x_0, then there exists a strategy y for B such that $\bar{p}(x, y) < \bar{p}(x_0, y)$ for $x \neq x_0$. To prove this, consider the difference

$$\bar{p}(x_0, y) - \bar{p}(x, y) = (x - x_0)\{-(a_{12} - a_{22}) + (a_{21} - a_{22} + a_{12} - a_{11})y\}.$$

If $x > x_0$, then B chooses $y = 1$ to make the difference as large as possible and positive. If $x < x_0$, then B chooses $y = 0$, which again makes the difference as large as possible and positive. Hence,

$$\bar{p}(x_0, y) = \max_x \min_y \bar{p}(x, y).$$

Analogously it is found that the optimal value of y equals

$$y_0 = \frac{a_{12} - a_{22}}{a_{21} - a_{22} + a_{12} - a_{11}} = \frac{5}{8},$$

and that $\bar{p}(x, y_0) = \bar{p}(x_0, y)$, which means that

$$\max_x \min_y \bar{p}(x, y) = \min_y \max_x \bar{p}(x, y).$$

This is the fundamental minimax theorem, which shows that the use of mixed strategies leads to a uniquely determined value of the game.

Returning to Montmort's considerations it is interesting to see that he has fully grasped the importance of this type of game. He writes that "These questions are rather simple, but I think impossible to solve; if this is so it is a great pity because this difficulty occurs in many instances of civil life: for example, when two persons do business each of them will adjust his behaviour after the other; it also takes place in several games" (1713, p. 406).

Trembley (1804) does not understand Waldegrave's reasoning leading to the minimax solution. Referring to Nicholas Bernoulli he says that if each player is ignorant of the strategy chosen by the other player, then each of them should choose a strategy with a probability of $\frac{1}{2}$; he does not notice Bernoulli's reservation about this solution. Presumably the choice of a uniform a priori distribution seemed natural for Trembley in view of the fact that Laplace in the meantime had introduced the principle of insufficient reason as a general solution of such problems. Trembley extends his analysis to the case of three players.

Todhunter's discussion (pp. 106–110) is incomplete; he has overlooked Montmort's randomized strategy and Waldegrave's minimax solution. Fisher

(1934) discusses the problem based on Todhunter's exposition; he derives the table of $c(i, j; S)$ given above, and by randomization he finds Waldegrave's solution. It seems that he has not read Montmort's book and, misled by Todhunter, he consider his solution to be new; his exposition of the randomization principle is of course clearer than that given by Waldegrave.

18.7   PROBLEMS FROM MONTMORT'S *ESSAY*

1. Drawing seven cards at random from a pack of 52, show that the probability of getting three doubles and a single is 16,632/900,473 (1708, p. 99).

2. Drawing 13 cards at random from two packs (104 cards), find the probability of getting two quadruples, two doubles, and a single (1708, p. 99).

3. Suppose that a pack of cards consists of five kings, four queens, two jacks, two tens, and one ace. Drawing four cards at random among the 14, show that the number of chances for getting two doubles is 93; (1713, p. 31). Montmort derives this result by a generalization of (2.4). Find the general formula.

4. Throwing nine ordinary dice, show that the number of different outcomes without regard to order is 2002 (1713, p. 36).

5. Throwing nine ordinary dice, show that the number of chances for getting exactly three aces is 1,312,500 (1713, p. 39).

6. Throwing nine ordinary dice, show that the number of chances for getting a quadruple, a double, and three singles is 907,200 (1708, p. 140).

7. For the multinomial $(a + b + c + d + e + f)^9$, consider all the terms in which any letter occurs in the fourth power, any combination of two letters in the second power, and any letter in the first power. Show that the sum of the coefficients of these terms equals 680,400 (1713, pp. 45–46).

8. Show that

$$\sum_{i=0}^{n-1} \binom{2+i}{2}^2 = \frac{3n^5 + 15n^4 + 25n^3 + 15n^2 + 2n}{60};$$

(1713, p. 64, 144).

9. The Lottery of Lorraine consists of 1,000,000 tickets at a price of $\frac{1}{2}$ livre each, and the prizes amount to 425,000 livres distributed over 20,000 drawings with replacement. Any player buying 50 tickets without getting a prize will get his 25 livres refunded. Supposing that 20,000 players each buy 50 tickets, show that the expected loss of the Lottery amounts to $184,064 - 75,000 = 109,064$ livres (1713, pp. 257–260, 313, 327, 346). Montmort's result is not quite correct because of rounding errors; find the correct result.

10. Suppose that 12 persons by a majority vote are to decide whether A or B should hold a certain office. It is known that A has three votes and B two and that A and B have the same chance for getting each of the remaining seven votes. However, before the voting, three of the voters fall ill; it is assumed that they have been chosen at random and that it is not known how they would have voted. Show that A's probability of being elected is 109/176 (1713, pp. 260–262).

11. Consider a play consisting of at most three games, each game being for two points. Of the two players, A and B, the one who first wins two games is the winner. To equalize the chances of winning, the stronger player A gives B one point in the first game and one in the third game, if it is to be played. Assuming that A's probability of winning a point is p and B's probability q, show that A's probability of winning the play may be expressed as a polynomial of the seventh degree in p (1713, pp. 343, 350,-352). According to Montmort, p/q is approximately equal to 1.77.

12. Players A and B play alternately with a four-sided die having the face values 0, 1, 2, and 3. Player A pays a certain number of crowns to a common stock and begins the play. Each player gets as many crowns from the stock as the number of points thrown, with the following exceptions for B. If B gets 0 points, he pays a crown to A, and if he gets a larger number of points than the number of crowns remaining in the stock, he gets nothing but has to pay to the stock as many crowns as his number of points exceeds the number of crowns in the stock. They continue in this way until the stock is exhausted. How many crowns should A pay to the stock at the beginning of play for A's and B's expectations to be equal? Solve the problem also under the assumption that B pays a crown to the stock when he throws 0 instead of paying a crown to A (1713, p. 401). These are the two first problems posed by Nicholas Bernoulli to Montmort.

13. Players A and B play alternately with an ordinary die; A pays s crowns to the stock at the beginning of play. Player A gets a and player B gets b crowns from the stock if he throws an even number, if not A has to pay c, and B has to pay d crowns to the stock. Player B has the first throw. The problem is to find (a) the expectations of A and B, respectively; (b) the value of s for given values of a, b, c, and d so that the expectations are equal; (c) the probability of a given duration of play (1713, pp. 401–402, 407–408). This is Montmort's generalization of the third problem posed by Bernoulli.

14. Player A has a number of counters in his hand and asks B to guess whether the number is even or odd. If B's guess is correct, he gets four crowns when the number is even and one crown when the number is odd. If B's guess is wrong, he gets nothing. Find the optimal strategies of the two players (1713, p. 406). The problem is due to Montmort.

15. *Jeu de la Ferme*. This game is played with an ordinary pack of cards after removal of the four 6's. Each player pays one crown to the stock before play begins. The players bid for the position of banker, and the highest bidder becomes banker, paying his bid to the stock. The banker deals two cards to each player, and the sum of each player's points is compared with 16. If all the players have more than 16 points, the banker continues, and each player pays a crown to the stock; if not, the following rules apply. If the number of points exceeds 16, the player has to pay as many crowns to the stock as indicated by the difference. If the number of points is less than 16, the player may either keep his cards or he may exchange them with two new ones from the banker. If the number of points equals 16, the player wins the stock and takes over as banker. If nobody has 16 points, the player who is nearest below 16 takes over as banker. If two or more players have the same number of points, the player sitting farthest to the right of the banker wins. For the banker two special rules apply: (a) if he is closer to 16 than any other player by one point, he loses the position of banker; (b) if he and another player both have 16 points, the play starts all over again with a new round of bidding. The problem is to find how much a player should bid so that the game is a fair one.

This is the third problem posed by Montmort (1708, pp. 187–188) for the reader to solve. Montmort does not indicate the solution.

16. *Jeu des Tas*. The game is played with a pack of 40 cards distributed at random into 10 piles of 4 cards each. The player turns up the top card of each pile and removes cards of the same face value in pairs. He next

turns up the cards immediately below the cards removed and continues the pairing and removal of pairs of equal value. He wins if he succeeds in removing all the cards. Find the best strategy of pairing and the probability of winning. Generalize to a pack of $N = sf$ cards; s suits, each of f face values.

This is the fourth problem posed by Montmort (1708, pp. 188–189) for the reader to solve. For $N = 2n$, Montmort (1713, p. 321) gives the probability of winning as $(n - 1)/(2n - 1)$; a proof has been given by Todhunter (pp. 111–113).

CHAPTER 19

The Problem of Coincidences and the Compound Probability Theorem

In the 24th and 25th Problems, I explain a new sort of Algebra, whereby some Questions relating to combinations [of events] are solved by so easy a Process, that their solution is made in some measure an immediate consequence of the Method of Notation. I will not pretend to say that this new Algebra is absolutely necessary to the Solving of those Questions which I make to depend on it, since it appears by Mr. De Montmort's Book, that both he and Mr. Nicholas Bernoully have solved, by another Method, many of the cases therein proposed: But I hope I shall not be thought guilty of too much Confidence, if I assure the Reader, that the Method I have followed has a degree of Simplicity, not to say of Generality, which will be hardly attained by any other Steps than by those I have taken.

—De MOIVRE, 1718

19.1 INTRODUCTION

In its simplest form the problem of coincidences (matches, recontres) may be formulated as follows: Let there be n objects numbered from 1 to n, and let them be ordered at random, assuming that the $n!$ permutations are equally

326

probable. A coincidence occurs if object number i is found at the ith place. The problem is to find the number of permutations with at least one coincidence or, equivalently, the probability of at least one coincidence.

The problem may be generalized in many ways. Consider for example s sets of n objects each, the objects within each set being numbered from 1 to n, and draw consecutively n objects at random among the ns objects. What is the probability of getting at least one coincidence?

It is practical to introduce two parallel sets of symbols, the one giving the number of permutations and the other the corresponding probabilities. As usual we have two types of distributions corresponding to the *number* of coincidences and the *waiting time* for the first coincidence, respectively.

Considering the simple problem described above, we denote the number of permutations with exactly k coincidences by $c_n(k)$,

$$c_n(0) + \cdots + c_n(n) = n!. \tag{1}$$

The probability of exactly k coincidences then becomes $p_n(k) = c_n(k)/n!$, and the probability of at least k coincidences equals

$$P_n(k) = p_n(k) + \cdots + p_n(n), \qquad P_n(0) = 1. \tag{2}$$

For the probability of at least one coincidence, we introduce the abbreviation P_n.

Let $d_n(i)$ denote the number of permutations with the *first* coincidence occurring at place number i, that is, the waiting time, or duration, equals i. The number of permutations with at least one coincidence then equals

$$d_n = d_n(1) + \cdots + d_n(n). \tag{3}$$

Obviously,

$$d_n + c_n(0) = n!, \tag{4}$$

or, equivalently,

$$P_n + p_n(0) = 1, \tag{5}$$

since $P_n = d_n/n!$.

The notation is easily generalized to $P_{n,s}$, and so on.

There exists a large literature on the problem of coincidences. The two recent papers by Henny (1975) and Takács (1980) cover almost the same ground as the present chapter.

19.2 MONTMORT'S FORMULA FOR THE PROBABILITY OF AT LEAST ONE COINCIDENCE, 1708

The first probabilistic discussion on the problem of coincidences is given by Montmort (1708, pp. 54–64) in connection with the card game *Thirteen* (*le Jeu du Treize*). The rules of the play are as follows: From a pack of 52 cards, 4 suits of 13 cards each, each player draws a card to decide who should be the banker. Any number of players may participate. After shuffling the pack, the banker turns the cards over successively, simultaneously calling out the names of the cards in order of rank, i.e., ace, 2,..., king. If no coincidence occurs among the 13 cards, the banker has lost and pays to each player what that player has staked. The player sitting to the right of the banker then takes over as banker. If, however, a coincidence occurs, for example, the third card turned is a 3, but the first is not an ace, and the second is not a 2, then the banker has won and collects all the stakes. He continues as banker and begins a new round, turning over the next card in the pack and calling out ace, 2, etc. If the banker, after having won several games, runs short of cards in a game because the whole pack has been used, he simply reshuffles the pack and continues the game, calling out the remaining names of the suit.

Montmort says that since it is difficult to find the banker's advantage, he will just give the solution of two simpler problems involving only one suit of 13 cards. This may be of some help in solving the general problem which Montmort uses as a problem for the reader on p. 185.

Suppose then that the banker has n different cards in random order. What is the probability of at least one coincidence when the banker turns over the n cards successively?

Montmort begins by giving the solution for $n = 2,...,5$, in each case following the same principle which obviously is the basis for his proof. For $n = 5$, his argument is as follows: The number of permutations of 5 cards is $5! = 120$. Among these there are 24 in which 1 is in first place, 18 in which 2 is in second place without 1 being first, 14 in which 3 is in third place without 1 being first or 2 being second, 11 in which 4 is in fourth place without 1 being first, 2 being second or 3 being third, and, finally, 9 in which 5 is in fifth place, the other four being out of their places. The probability of at least one coincidence is therefore

$$\frac{24 + 18 + 14 + 11 + 9}{120} = \frac{76}{120} = \frac{19}{30}.$$

He remarks that it will take up too much space to give the general proof and then states the general solution in two forms: as a recursion formula and as an explicit solution in the form of a series.

Denoting the probability of at least one coincidence by P_n, Montmort's recursion formula may be written as

$$P_n = \frac{(n-1)P_{n-1} + P_{n-2}}{n}, \quad n \geqslant 2, \quad P_0 = 0, \quad \text{and} \quad P_1 = 1. \quad (1)$$

By means of this formula Montmort tabulates P_n for $n = 2, \ldots, 13$. He finds

$$P_{13} = \frac{109,339,663}{172,972,800} = 0.632,120,558.$$

Montmort states that the solution may also be expressed by means of the numbers in the arithmetic triangle arranged in an alternating series. The ith term of the series is $\binom{n}{i}/n^{(i)} = 1/i!$, so that Montmort's formula for the probability of at least one coincidence becomes

$$P_n = 1 - \frac{1}{2!} + \frac{1}{3!} + \cdots + \frac{(-1)^{n-1}}{n!}, \quad n \geqslant 1. \quad (2)$$

Referring to a paper by Leibniz he proves that

$$\lim_{n \to \infty} P_n = 1 - e^{-1} = 0.632,120,558.$$

(Montmort does not give the decimal fraction here and above and does not discuss the rate of convergence to the limit. As we have seen above, $P_5 = 19/30 = 0.633$.)

Montmort does not discuss the relationship between the two formulae. Presumably he has noted that (1) may be written in the form

$$P_n - P_{n-1} = -\frac{P_{n-1} - P_{n-2}}{n}, \quad (3)$$

and that (2) gives

$$P_n - P_{n-1} = \frac{(-1)^{n-1}}{n!}, \quad (4)$$

from which it is easy to see how (2) may be derived from (1), and vice versa.

A clue to Montmort's proof may be found from his examples. He divides the set of permutations with at least one coincidence into n disjoint sets defined by the place where the first coincidence occurs. To find $d_n(i)$, the number of permutations with a coincidence at the ith place and no coincidence before, we first note that $d_n(1) = (n-1)!$, since 1 has to be at the first place and the remaining $n-1$ numbers may be permuted in all possible ways. For $d_n(2)$, we first fix 2 at the second place, and from the resulting $(n-1)!$ permutations we have to deduct the number of permutations with 1 at the first place, which number equals $(n-2)!$, so that

$$d_n(2) = (n-1)! - (n-2)! = d_n(1) - d_{n-1}(1).$$

In general, we have

$$d_n(i+1) = d_n(i) - d_{n-1}(i), \qquad n \geqslant 2, \quad i = 1, \ldots, n-1. \tag{5}$$

Presumably Montmort knew this recursion formula and used it for calculating $d_n(i)$ for $i = 2, \ldots, 5$ in his examples. In the *Essay* (1713, p. 137), he uses it to calculate $d_n(i)$ up to $n = 8$.

The recursion formula is a simple example of the application of the method of inclusion and exclusion. To prove the formula we first note that $d_n(i)$ may be interpreted as the number of permutations with a coincidence at the last of i consecutive places and no coincidence at the $i-1$ places before. Considering places $2, \ldots, (i+1)$, $d_n(i)$ thus gives the number of permutations with a coincidence at place $(i+1)$; no coincidence at place $2, \ldots, i$; and no restriction on place 1. To find $d_n(i+1)$, we have to deduct the number of permutations with a coincidence at place 1, i.e., the number of permutations with a 1 on place 1, no coincidence at the following places, and a coincidence at place $(i+1)$, which number equals $d_{n-1}(i)$. This proof is due to Todhunter (p. 92).

In the problem above, Montmort considers only a single game. He next gives a short discussion of the banker's advantage under a rule of continuation but does not reach a general solution.

19.3 THE RESULTS OF MONTMORT AND NICHOLAS BERNOULLI, 1710–1713

Montmort continued his research on the problem of coincidences, and in a letter of November 1710 to John Bernoulli he wrote that he had found "many curious results on this matter," and as an example he gave the banker's advantage for the case with 13 cards drawn from an ordinary pack of 52 cards, which means that he had found the probability of at least one coincidence, $P_{13,4}$. However, he did not give a formula but only the numerical result, see Montmort (1713, p. 304).

In his correspondence with Nicholas Bernoulli from 1710 to 1712, the problem of coincidences is frequently discussed, see Montmort (1713, pp. 301–302, 308–309, 315, 317–318, 323–324, 327–328, 344). Furthermore, Montmort gives the results from 1708 supplemented with some new results in the *Essay* (1713, pp. 130–143). Montmort does not provide proofs for his results from 1708; he writes in the preface (p. XXV) that the proofs may be found in a letter from Nicholas Bernoulli (pp. 301–302) and that "I would not have been able to do it better myself."

Bernoulli's Proof of Montmort's Formula for P_n

Bernoulli first states that $d_n(1) = (n-1)!$, and

$$d_n(2) = (n-1)! - (n-2)!,$$
$$d_n(3) = (n-1)! - 2(n-2)! + (n-3)!,$$
$$d_n(4) = (n-1)! - 3(n-2)! + 3(n-3)! - (n-4)!,$$

and in general,

$$d_n(i) = \sum_{v=0}^{i-1} (-1)^v \binom{i-1}{v} (n-1-v)!. \tag{1}$$

Bernoulli does not attempt to give a rigorous proof of this result; the proof may be completed by inserting (1) into the recursion formula (2.5). We shall leave that to the reader.

Bernoulli next finds the probability that the first coincidence occurs at the ith place as $d_n(i)/n!$ and the probability of a coincidence at the mth place or before as

$$\frac{1}{n!} \sum_{i=1}^{m} d_n(i) = \sum_{v=0}^{m-1} (-1)^v \frac{(n-1-v)!}{n!} \sum_{i=v+1}^{m} \binom{i-1}{v}$$

$$= \sum_{v=0}^{m-1} (-1)^v \binom{m}{v+1} \frac{(n-1-v)!}{n!},$$

without mentioning the (obvious) intermediate step. Setting $m = n$, Bernoulli obtains

$$P_n = \sum_{i=1}^{n} \frac{d_n(i)}{n!} = \sum_{v=0}^{n-1} \frac{(-1)^v}{(v+1)!},$$

in agreement with Montmort's formula (2.2).

Bernoulli's Proof of Montmort's Recursion Formula for P_n

To find the probability of at least one coincidence, $P_n = \Pr\{C\}$, say, Bernoulli uses the theorem

$$\Pr\{C\} = \Pr\{A\} \Pr\{C|A\} + \Pr\{\bar{A}\} \Pr\{C|\bar{A}\},$$

where A denotes the event that object 1 of n objects is at place 1. Hence $\Pr\{A\} = 1/n$ and $\Pr\{C|A\} = 1$, so that

$$\Pr\{C\} = \frac{1}{n} + \frac{n-1}{n} \Pr\{C|\bar{A}\}.$$

Bernoulli then proves that

$$\Pr\{C|\bar{A}\} = \frac{(n-1)!P_{n-1} - (n-2)! + (n-2)!P_{n-2}}{(n-1)!}$$

$$= \frac{(n-1)P_{n-1} - 1 + P_{n-2}}{n-1},$$

which inserted in the result above gives Montmort's recursion formula (2.1).

The permutations corresponding to the event $\{C|\bar{A}\}$ are of the form $(j, i_1, \ldots, i_{n-1}), j \neq 1$, with at least one coincidence at the last $(n-1)$ places. Bernoulli states that among the $(n-1)!$ permutations of the i's there will be less than d_{n-1} permutations with at least one coincidence, since one of the i's is a 1. The transfer of j to place 1 has the effect that permutations with a coincidence only at place j will be changed to permutations without coincidences, the number of such permutations among $(n-2)$ elements being $(n-2)! - d_{n-2}$. The resulting number of permutations with at least one coincidence thus becomes

$$d_{n-1} - (n-2)! + d_{n-2},$$

as was to be proved.

Finally, Bernoulli derives (2.4) and thus shows how the series for P_n may be derived from the recursion formula.

Some Further Results by Montmort

As mentioned in the previous section, Montmort uses the recursion (2.5) to calculate $d_n(i)$ for $n = 1, \ldots, 8$, as shown in the following table.

MONTMORT'S TABLE OF $d_n(i)$ AND d_n

				i					
n	1	2	3	4	5	6	7	8	d_n
1	1								1
2	1	0							1
3	2	1	1						4
4	6	4	3	2					15
5	24	18	14	11	9				76
6	120	96	78	64	53	44			455
7	720	600	504	426	362	309	265		3186
8	5040	4320	3720	3216	2790	2428	2119	1854	25487

Source: Montmort (1713), p. 137.

From this table it is easy to find $P_n = d_n/n!$ and $c_0(n) = n! - d_n$. Furthermore, Montmort gives the recursion formula

$$d_n = nd_{n-1} + (-1)^{n-1}, \qquad n \geqslant 1, \quad d_0 = 0, \tag{2}$$

which follows immediately from the formula for $n! P_n$. For given n the probability that the first coincidence occurs at the ith place equals

$$\frac{d_n(i)}{n!} = \frac{d_n(i)}{d_{n+1}(1)}.$$

Finally, Montmort (1713, p. 138) states that

$$d_n(n) = c_{n-1}(0) = (n-1)! - d_{n-1}, \tag{3}$$

which is intuitively obvious (see also the table above).

Montmort also derives the distribution of the number of coincidences. He states that the number of permutations with exactly k coincidences equals

$$c_n(k) = \binom{n}{k} c_{n-k}(0), \tag{4}$$

because the places with the k coincidences may be chosen in $\binom{n}{k}$ ways, and there should be no coincidences at the remaining $n - k$ places.
From

$$c_n(0) = n! - n! P_n,$$

he obtains

$$c_n(0) = (-1)^n \{ 1 - n + n^{(2)} - n^{(3)} + \cdots + (-1)^{n-2} n^{(n-2)} \}. \tag{5}$$

Expressing these results in terms of probabilities, we get

$$p_n(0) = 1 - P_n = \sum_{i=0}^{n} \frac{(-1)^i}{i!}, \tag{6}$$

$$p_n(k) = \frac{p_{n-k}(0)}{k!} = \frac{1}{k!} \sum_{i=0}^{n-k} \frac{(-1)^i}{i!}. \tag{7}$$

By means of these formulae Montmort calculates the distribution of the number of coincidences for $n = 13$ and finds the banker's expectation $1 - 2p_n(0)$. A summary of his results, expressed as decimal fractions, follows:

$k =$	0	1	2	3	4	5	6	Total
$p_{13}(k) =$	0.368	0.368	0.184	0.061	0.015	0.003	0.001	1.000

Furthermore, $p_{13}(13) = 1/6,227,020,800$.

Montmort also states that the expected number of coincidences equals 1. His proof is equivalent to the following. Let ϵ_i be the characteristic random variable for a coincidence at place i so that $\epsilon_i = 1$ if a coincidence occurs at place i and $\epsilon_i = 0$ otherwise. Then,

$$E\{k\} = E\{\epsilon_1 + \cdots + \epsilon_n\} = nE\{\epsilon_i\} = n\left(\frac{1}{n}\right) = 1.$$

The Montmort–Bernoulli Formula for $P_{n,s}$

Montmort's result for $P_{13,4}$ occasioned Bernoulli to send a letter with a formula for $P_{n,s}$ and a slight correction of Montmort's numerical result. In his reply Montmort points out that the formula is wrong because of a writing error and gives the correct formula, which is stated again in a slightly more convenient form by Bernoulli in his next letter.

Setting $N = ns$, the Montmort–Bernoulli formula becomes

$$P_{n,s} = \sum_{i=1}^{n} \frac{(-1)^{i-1} s^i n^{(i)}}{i! N^{(i)}}. \tag{8}$$

They do not supply a proof, but Bernoulli states that the method of proof is the same as for P_n. The proof was carried out by Struyck (1716, pp. 102–104).

Montmort proposes the *Jeu du Treize* as a problem for solution (1708, p. 185; 1713, p. 278). With regard to the second edition, Todhunter (p. 105) writes, "It is not obvious why this problem is repeated, for Montmort stated the results on his pages 130–143, and demonstrations by Nicolas Bernoulli are given on pages 301, 302." The explanation is, however, that Montmort writes about the Jeu du Treize as it is played in practice with a rule of continuation and not about the simplified version, which we have called the problem of coincidences relating to one game only.

Three Corollaries to Montmort's Results

Corollary 1. For d_n and $c_n(0)$, the following recursions hold:

$$d_n = (n-1)(d_{n-1} + d_{n-2}), \qquad d_0 = 0, \quad d_1 = 1, \tag{9}$$

$$c_n(0) = (n-1)[c_{n-1}(0) + c_{n-2}(0)], \qquad c_0(0) = 1, \quad c_1(0) = 0. \tag{10}$$

These results follow immediately from Montmort's recursion formula for P_n.

Corollary 2. For $c_n(0)$ the following recursions hold:

$$c_n(0) = nc_{n-1}(0) + (-1)^n, \qquad c_0(0) = 1, \tag{11}$$

$$\sum_{k=0}^{n} \binom{n}{k} c_{n-k}(0) = n!, \qquad c_0(0) = 1. \tag{12}$$

The first result follows immediately from (2), and the second is obtained by inserting (4) into (1.1). A recursion formula for $c_n(0)$ is obtained by solving (12) for $c_n(0)$. An explicit formula for $c_n(0)$ is given by (5).

Corollary 3. For $n \to \infty$ and k fixed, we have

$$p_n(k) \to \frac{e^{-1}}{k!}. \tag{13}$$

This result follows from (7). It means that the distribution of the number of coincidences for large n may be approximated by the Poisson distribution with unity as mean.

Montmort does not himself formulate these corollaries. We include them for later reference.

19.4 DE MOIVRE'S DERIVATION OF THE PROBABILITY OF COMPOUND EVENTS, 1718

Reading Montmort's *Essay* (1713), de Moivre realized that all the results on the problem of coincidences may be derived from a general theorem on the probability of compound events. He derived this theorem by means of the method of inclusion and exclusion, a method that was used by Montmort and Bernoulli in deriving the formula for $d_n(i)$ and also used by Montmort on several other occasions. However, they did not, as did de Moivre, give their results as special cases of a general theorem. That de Moivre was aware of the importance of the theorem is evident from the quotation used as epigraph.

For two events it was well known that

$$\Pr\{A_1 + A_2\} = \Pr\{A_1\} + \Pr\{A_2\} - \Pr\{A_1 A_2\}.$$

Without comment this result had been used by de Moivre himself for the solution of Problem 26 in *De Mensura Sortis* (1712). A related method had been used by Halley (1694) for finding the probability of survivorships for two and three independent lives, see §9.3.

Let $A_1, \ldots, A_n$ denote n events, and let the probabilities of the simultaneous occurrence of any number of them be known. It is assumed that the events are exchangeable or symmetric such that the probabilities satisfy the relation

$$P(A_{i_1} \cdots A_{i_r}) = P(A_1 \cdots A_r), \tag{1}$$

for any set of r different integers $(i_1, \ldots, i_r)$ selected from among the integers $(1, \ldots, n)$. The complement of the event A_i is denoted by $\bar{A}_i$. The problem solved by de Moivre is to express the probability of the simultaneous occurrence of a set of A's and $\bar{A}$'s in terms of the probabilities (1).

De Moivre's proof is by incomplete induction and is based on the identity

$$P(\bar{A}_1 A_2) = P(A_2) - P(A_1 A_2) = P(A_1) - P(A_1 A_2). \tag{2}$$

Expressed in the "new sort of algebra" used by de Moivre, the equation

$$P(A_2) = P(A_1 A_2) + P(\bar{A}_1 A_2),$$

takes the form

$$(+A_2) = (+A_1 + A_2) + (-A_1 + A_2).$$

This is an ingenious notation for the present purpose but not easy to use in other connections. We have therefore translated de Moivre's equations to the standard notation used today.

Using (2) repeatedly, de Moivre finds

$$P(\bar{A}_1 A_2 A_3) = P(A_2 A_3) - P(A_1 A_2 A_3) = P(A_1 A_2) - P(A_1 A_2 A_3),$$

$$P(\bar{A}_1 \bar{A}_2 A_3) = P(\bar{A}_1 A_3) - P(\bar{A}_1 A_2 A_3) = P(A_1) - 2P(A_1 A_2) + P(A_1 A_2 A_3),$$

$$P(\bar{A}_1 \bar{A}_2 \bar{A}_3 A_4) = P(\bar{A}_1 \bar{A}_2 A_4) - P(\bar{A}_1 \bar{A}_2 A_3 A_4); \tag{3}$$

$$P(\bar{A}_1 \bar{A}_2 A_3 A_4) = P(\bar{A}_1 A_3 A_4) - P(\bar{A}_1 A_2 A_3 A_4)$$

$$= P(A_1 A_2) - 2P(A_1 A_2 A_3) + P(A_1 A_2 A_3 A_4). \tag{4}$$

From (3) and (4) it follows that

$$P(\bar{A}_1 \bar{A}_2 \bar{A}_3 A_4) = P(A_1) - 3P(A_1 A_2) + 3P(A_1 A_2 A_3) - P(A_1 A_2 A_3 A_4). \tag{5}$$

Generalizing these results de Moivre states without proof that

$$P(A_1 \cdots A_k \bar{A}_{k+1} \cdots \bar{A}_{k+r}) = \sum_{i=0}^{r} (-1)^i \binom{r}{i} P(A_1 \cdots A_{k+i}), \tag{6}$$

for $k \geq 1$, $r \geq 1$, $k + r \leq n$, and that

$$P(\bar{A}_1 \cdots \bar{A}_r) = 1 + \sum_{i=1}^{r} (-1)^i \binom{r}{i} P(A_1 \cdots A_i), \quad 1 \leq r \leq n \tag{7}$$

It is straightforward to complete the proof by induction.

The probability of the simultaneous occurrence of exactly k of the n events equals

$$p_n(k) = \binom{n}{k} P(A_1 \cdots A_k \bar{A}_{k+1} \cdots \bar{A}_n). \tag{8}$$

De Moivre finds the probability of the occurrence of at least k events, $P_n(k)$, by incomplete induction. We shall prove his result by summation of (8). From (6) and (8) we get

$$
\begin{aligned}
P_n(k) &= \sum_{x=k}^{n} \binom{n}{x} \sum_{i=x}^{n} (-1)^{i-x} \binom{n-x}{i-x} P(A_1 \cdots A_i) \\
&= \sum_{i=k}^{n} \binom{n}{i} P(A_1 \cdots A_i) \sum_{x=k}^{i} (-1)^{i-x} \binom{i}{i-x} \\
&= \sum_{i=k}^{n} (-1)^{i-k} \binom{i-1}{i-k} \binom{n}{i} P(A_1 \cdots A_i) \\
&= \binom{n}{k} \sum_{i=0}^{n-k} (-1)^{i} \frac{k}{k+i} \binom{n-k}{i} P(A_1 \cdots A_{k+i}).
\end{aligned}
\tag{9}
$$

In the proof we have used the relation

$$
(-1)^{x} \binom{n}{x} = \Delta_x (-1)^{x-1} \binom{n-1}{x-1},
$$

which leads to

$$
\sum_{x=k}^{i} (-1)^{i-x} \binom{i}{i-x} = (-1)^{i-k} \binom{i-1}{i-k}.
$$

De Moivre's formulae (6)–(9) are very important and of great generality, as he mentions in his preface (1718, p. XI). The results are to be found in the *Doctrine of Chances* (1718, pp. 59–66; 1738, pp. 95–103; 1756, pp. 109–117). In the text, however, all the results are couched in the language of the problem of coincidences and are intermingled with the solution of special cases of this problem. We have separated the general results from the applications, which are discussed in the following section.

19.5   DE MOIVRE'S SOLUTION OF THE PROBLEM OF COINCIDENCES

The problem is formulated by de Moivre in Problem 25 (1718): "Any given number of Letters a, b, c, d, e, f etc. all of them different, being taken promiscuously, as it Happens: To find the Probability that some of them shall be found in their places, according to the rank they obtain in the Alphabet; and that others of them shall at the same time be found out of their places."

It will be seen that de Moivre has generalized the problem posed by Montmort.

Let there be n letters, and let A_i denote the event that a coincidence occurs at place i. Then (4.6) gives the probability that k letters are at the right place, that r letters are out of place, and that no restrictions are put on the remaining $n - k - r$ letters. The solution is obtained by inserting

$$P(A_1 \cdots A_i) = \frac{1}{n^{(i)}}.$$

The probability of at least one coincidence is obtained from (4.7) as

$$P_n = 1 - P(\bar{A}_1 \cdots \bar{A}_n). \tag{1}$$

In this way de Moivre derives all Montmort's results as special cases of (4.6)–(4.8). Furthermore, he uses (4.9) to find

$$P_n(k) = \frac{1}{(k-1)!} \sum_{i=0}^{n-k} \frac{(-1)^i}{i!(k+i)}. \tag{2}$$

De Moivre next considers the problem of coincidences for n drawings from a pack of N cards consisting of s suits with n cards in each. He notes that

$$P(A_1 \cdots A_i) = \frac{s^i}{N^{(i)}},$$

which inserted into (1) and (4.7) leads to the Montmort–Bernoulli formula (3.8) for at least one coincidence.

From (4.9) we find

$$P_{n,s}(k) = \frac{1}{k!} \sum_{i=0}^{n-k} \frac{(-1)^i}{i!} \frac{k}{k+i} \frac{s^{k+i} n^{(k+i)}}{N^{(k+i)}}. \tag{3}$$

De Moivre does not give the general formula; he finds $P_{n,s}(k)$ for $k = 1, \ldots, 4$ and states that "The Law of the continuation of these Series being manifest, it will be easy to reduce them all to one general Series."

Finally, de Moivre considers a generalization to m packs of cards, each pack containing n different cards. A coincidence is defined as the occurrence of the same card in the same place for all the packs. Since the one pack may be laid out in natural order and the remaining $m - 1$ packs in random order, we have

$$P(A_1 \cdots A_i) = \frac{1}{(n^{(i)})^{m-1}}.$$

De Moivre gives the probability of at least one coincidence as

$$\sum_{i=1}^{n} \frac{(-1)^{i-1}}{i!(n^{(i)})^{m-2}}, \tag{4}$$

which is an immediate consequence of (4.7).

19.6 SOME NOTES ON LATER DEVELOPMENTS

The problem of coincidences is an example of a problem occurring in many different contexts and therefore solved by many different authors, many of them unaware of the fundamental contributions by Montmort, Nicholas Bernoulli, and de Moivre. Even after the publication of the Montmort–Bernoulli formula for $P_{n,s}$ by Laplace (1812), independent derivations of Montmort's formulae by essentially the same method occurred steadily. The following notes should not be taken for a complete history of the problem but only as a guide for further study.

Euler (1753, 1811) wrote two papers on the problem, the latter read in 1779 and published posthumously in 1811. In the first paper he begins with some numerical examples and later derives the recursion

$$d_n(i) = (n-1)! - \sum_{j=1}^{i-1} d_{n-1}(j),$$

which he uses to tabulate $d_n(i)$ up to $n = 10$. In his discussion of the tabulation he notes that the simpler recursion (2.5) holds, which Montmort used to construct his table. He then uses this formula in much the same manner as Bernoulli to find P_n; he tabulates P_n up to $n = 15$ and shows the agreement with the limiting value $1 - e^{-1}$. In the second paper he derives the recursion

$$d_{n+1}(n+1) = (n-1)[d_n(n) + d_{n-1}(n-1)].$$

However, since $d_{n+1}(n+1) = c_n(0)$, this relation is identical to (3.10). He also derives (3.11). Compared with the results in Montmort's *Essay* (1713), there are essentially no new results or methods of proof in Euler's papers. He does not refer to any of his predecessors.

Inspired by Euler's first paper, Lambert (1773) takes up the problem in order to discuss the reliability of predictions of the weather and other events to be found in the popular almanacs in Germany. He says that variations in the weather from the one day to the next are produced by an infinite number of unknown causes, just like the outcome of a game of chance. This implies that the reliability of predictions may be tested by comparing the number of coincidences of actual weather and predicted weather for a number of days with the expected number of coincidences. He then derives (3.11) and (3.12), which gives the expected number of coincidences. He does not, however, provide any observations as a basis for carrying out the test.

Laplace (1812) discusses the problem of coincidences for drawing n balls from an urn containing $N = ns$ balls, the balls being numbered from 1 to n, each number being repeated s times. Using the method of inclusion and exclusion, Laplace derives $P_{n,s}$ by the same method as Bernoulli and Struyck and furthermore derives de Moivre's result $P_{n,s}(k)$ and formula (5.4). He also proves that $p_{n,s}(0) \to e^{-1}$ for $n \to \infty$. He does not refer to Montmort and de Moivre, but in the introductory remarks to his book there is a general reference to their works. Laplace's results are reproduced in Book 2, Chap. 2, §9 in all editions of his book.

Kendall (1968) has pointed out that the theory of coincidences has been discussed by the English physician and scientist Thomas Young, who also did some of the first successful work in deciphering Egyptian hieroglyphic inscriptions. In a paper on the probability of errors in physical observations and on the density of the earth by Young (1819), it is surprising to find a section on the application of the theory of coincidences to linguistic and historical problems. He writes,

There are cases in which some little assistance may be derived from the doctrine of chances with respect to matters of literature and history: but even here it would be extremely easy to pervert this application in such a manner, as to make it subservient to the purpose of clothing fallacious reasoning in the garb of demonstrative evidence. Thus if we were investigating the relations of two languages to each other, with a view of determining how far they indicated a common origin from an older language, or an occasional intercourse between two nations speaking them, it would be important to inquire, upon the supposition that the possible varieties of monosyllabic or very simple words must be limited by the extent of the alphabet to a certain number; and that these names were to be given promiscuously to the same number of things, what would be the chance that 1, 2, 3 or more of the names would be applied to the same things in two independent instances.

Like Lambert, he finds the recursion formula (3.12) for $c_n(k)$, which he uses to tabulate $p_n(k)$ for $n = 10$. He compares $p_{10}(k)$ with the Poisson approximation $e^{-1}/k!$. Since he does not have an explicit expression for $p_n(k)$ such as (3.7), he cannot find the limiting value directly. Instead he derives the Poisson distribution by noting that the probability of no coincidence equals $[(n-1)/n]^n$, which is approximately e^{-1} for large values of n. He continues his "proof" by stating that "if n is increased by 1, each of these cases of no coincidence will afford 1 of a single coincidence; if by two, each will afford one of a double coincidence, but half of them will be duplicates; and if by three, the same number must be divided by 6, since all the combinations of three would be found six times repeated." From these numerical results Young concludes that

> It appears therefore that nothing whatever could be inferred with respect to the relation of two languages from the coincidence of the sense of any single word in both of them; and that the odds would only be 3 to 1 against the agreement of two words; but if three words appeared to be identical, it would be more than 10 to 1 that they must be derived in both cases from some parent language, or introduced in some other manner; six words would give near 1700 chances to 1, and 8 near 100,000; so that in these last cases the evidence would be little short of absolute certainty.

It will be seen that Young uses the tail probabilities of the Poisson distribution as a test of significance of the null hypothesis that there is no relation between the two languages.

Young's paper does not contain any reference to previous results on the theory of coincidences.

In his textbook on combinatorics, Oettinger (1837, pp. 100–111) generalizes the problem by considering the number of coincidences obtained by drawing only r elements, $1 \leqslant r \leqslant n$, of the n or $N = ns$ elements. Setting $r = n$, he later finds all the previously mentioned formulae except (5.4). In 1837 he gives the solution in terms of the number of permutations in question, and in his textbook on probability (1852, pp. 37–44) he gives the solution in terms of probabilities. He uses the method of inclusion and exclusion in his proofs, and he refers to the previous literature. Using the same symbols as before but with r added as a superscript, Oettinger's results may be written as

$$P_n^r = \sum_{i=1}^{r} \frac{(-1)^{i-1} r^{(i)}}{i! n^{(i)}}, \tag{1}$$

$$p_n^r(k) = \frac{r^{(k)}}{k! n^{(k)}} \sum_{i=0}^{r-k} \frac{(-1)^i (r-k)^{(i)}}{i! (n-k)^{(i)}}. \tag{2}$$

Analogous formulae hold for the other probabilities.

According to Takács (1980), the formula for $p_n^r(k)$ has also been found by Catalan (1837).

Schneider (1974) has given a detailed discussion of one of the first applications of probability theory in physics, namely, Clausius' (1849) use of the Poisson distribution with mean unity as the limit of the distribution of the number of coincidences in a mathematical model of light radiation in the atmosphere.

It seems that all the authors mentioned above have overlooked de Moivre's elegant general solution by means of the compound probability theorem.

Using the method of inclusion and exclusion, M. C. Jordan (1867) proved the compound probability theorem without de Moivre's restriction to symmetric events. The fundamental quantities in the theorem are the sums

$$S_k = \sum P(A_{i_1} A_{i_2} \cdots A_{i_k}), \qquad 1 \leqslant i_1 \leqslant i_2 \cdots \leqslant i_k \leqslant n, \quad k = 1, \ldots, n, \qquad (3)$$

which have $\binom{n}{k}$ terms. Using that

$$\binom{n}{k}\binom{n-k}{i} = \binom{k+i}{i}\binom{n}{k+i},$$

and under the assumption of symmetry that

$$S_{k+i} = \binom{n}{k+i} P(A_1 \cdots A_{k+i}),$$

de Moivre's results (4.8) and (4.9) may be written as

$$p_n(k) = \sum_{i=0}^{n-k} (-1)^i \binom{k+i}{i} S_{k+i}, \qquad (4)$$

$$P_n(k) = \sum_{i=0}^{n-k} (-1)^i \binom{k+i-1}{i} S_{k+i}, \qquad (5)$$

Jordan proved that these formulae are also valid for the nonsymmetrical case, with the definition of S_k given in (3); he does not refer to de Moivre. The proof may be carried out by the same method used by de Moivre; by means of indicator functions, as done by Loéve in Parzen (1960); or by symbolic operations, as done by King (1902), K. Jordan (1956, 1972), and Riordan (1958).

Using King's symbolic representation, the results may be written in the form

$$p_n(k) = S^k(1 + S)^{-k-1}, \tag{6}$$

$$P_n(k) = S^k(1 + S)^{-k}, \tag{7}$$

where S^i after expansion in series should be replaced by S_i and $S_i = 0$ for $i > n$.

Takács (1967) has discussed the history of the compound probability theorem and generalized the theorem.

Using the method of inclusion and exclusion, Whitworth (1878) proved the following theorem (Whitworth, 1901, p. 70):

If there be N events or operations and if (out of r possible conditions $\alpha, \beta, \gamma, \dots$) every one condition (such as α) be fulfilled in N_1 of the events; and every combination of two simultaneous conditions (such as α, β) be fulfilled in N_2 of the events; and every combination of three simultaneous conditions (such as α, β, γ) be fulfilled in N_3 of the events; and so on; and finally all the r conditions be simultaneous fulfilled in N_r of the events; then the number of events free from all these conditions is

$$N - \frac{r}{1}N_1 + \frac{r(r-1)}{1 \cdot 2}N_2 - \frac{r(r-1)(r-2)}{1 \cdot 2 \cdot 3}N_3 + \cdots \pm N_r.$$

Obviously, Whitworth's result is just a description of the formula for $p_n(0)$. He noted that the formula may be derived by a symbolic method. He used his result to prove some theorems on "derangements." Irwin (1955) discussed Whitworth's theorem and used it to derive some important statistical distributions. Whitworth gives no references.

The compound probability theorem was used by de Moivre (1725) and Simpson (1742) to find the probabilities of survivorships for several lives; it has been used in life insurance mathematics ever since. King (1902) has given a comprehensive account of such applications.

The distribution of the number of coincidences has been used for various statistical purposes, mainly in connection with testing guessing and classification abilities, for example, in psychological, graphological, and tasting experiments, see Vernon (1936) and some critical comments by Irwin (1955). There are also examples of applications to genetic problems of random mating.

An exposition of the theory of coincidences within the framework of combinatorics and with some historical comments has been given by Netto (1901). A more advanced treatment is due to MacMahon (1915).

The problem has been generalized to multiple coincidences of several packs of cards with suits of unequal numbers of cards and with many different definitions of a multiple coincidence. There exist many papers on these problems; the later development of the mathematical theory involved has been discussed by Barton (1958).

A monograph on the compound probability theorem and its applications to the problem of coincidences has been written by Fréchet (1940, 1943).

Textbook expositions have been given by K. Jordan (1956, 1972), Riordan (1958), and David and Barton (1962).

The history of a closely related problem about the seating arrangements of $2n$ couples at a round table, the *problème des ménages*, has been discussed by Takács (1981).

19.7 PROBLEMS

1. Prove that

$$P_n = \sum_{i=1}^{n} \frac{1}{i!} - \sum_{i=1}^{n} \frac{P_{n-i}}{i!}.$$

This result is due to John Bernoulli, see Montmort (1713, p. 290).

2. Prove the recursion formula (3.10) directly by combinatorial arguments.

3. Tabulate $p_{10}(k)$ and the Poisson approximation $e^{-1}/k!$. Verify Young's (1819) statement about the tail probabilities, see §19.6. Find limits for the difference $p_n(k) - e^{-1}/k!$.

4. Prove Oettinger's formula for $p_n^r(k)$ [see (6.1)].

5. Show that $P_{13,4} = 0.643$. This result is due to Montmort (1713, p. 324) and N. Bernoulli. Find the limiting value of $P_{n,s}$ for $n \to \infty$.

6. De Moivre (1718, p. 66) writes that "the Odds that two or more like Cards in two different Packs will not obtain the same Position, are very nearly as 736 to 264 or 14 to 5." Check this statement.

7. Solve the problem of the Jeu du Treize in its original formulation with a rule of continuation for the banker, see §19.2.

8. Use de Moivre's method of proof for proving Jordan's generalization of the compound probability theorem.

9. Use the compound probability theorem to compare $P_n(1) = 1 - p_n(0)$ when the n events are (a) independent, (b) symmetric, (c) independent and symmetric, see Jordan (1972, p. 198).

10. Prove Montmort's result (14.3.1) on the number of chances of getting s points by throwing n dice by means of the compound probability theorem.

11. Let $p_n(k)$ be the probability that exactly k lives of n lives will survive t years. Find $p_n(k)$ in terms of the probabilities of the joint lives surviving t years assuming independence and the same life table for all lives, see King (1902, p. 14).

12. Prove that the number of ways in which an ordered series of n objects may be "deranged" so that no object is followed by the object that originally followed it equals

$$n! - \binom{n-1}{1}(n-1)! + \binom{n-1}{2}(n-2)! - \cdots (-1)^{n-1}\binom{n-1}{n-1},$$

and that this expression may be reduced to

$$c_n(0) + c_{n-1}(0) = \frac{c_{n+1}(0)}{n},$$

which is the integer nearest to $(n+1)!/ne$. This result is due to Whitworth (1901, p. 103).

13. Suppose that n married couples are seated at a round table such that the women first take alternate seats and the men choose the remaining seats at random. Find the probability that exactly k husbands are sitting next to their wives. *Hint*: Use the compound probability theorem and prove that

$$S_k = \frac{2n}{2n-k}\binom{2n-k}{k}(n-k)!.$$

Takács (1981) has generalized this problem by considering the distribution of the number of husbands sitting next to their wives' right and left sides, respectively.

CHAPTER 20

The Problem of the Duration of Play, 1708–1718

20.1 FORMULATION OF THE PROBLEM

The problem of the duration of play, also called the ruin problem, may be formulated as follows: consider two players, A and B, having a and b counters, respectively. In each game, A has probability p and B has probability $q = 1 - p$ of winning, and the winner gets a counter from the loser. The play continues until one of the players has lost all his counters. What is the probability that the play ends at the nth game or before?

A complete probabilistic description of the play is given by the function $u_n(x, a, b)$, $-a < x < b$, denoting the probability of the compound event that neither player is ruined in n games and that A after n games has won x counters from B. The probability that B is ruined at the nth game, which is the same as the probability that A wins the play at the nth game, thus becomes

$$r_n(a, b) = p u_{n-1}(b - 1, a, b),$$

an analogous formula being valid for the probability $r_n^*(a, b)$ that A is ruined. Hence, the probability of a duration of exactly n games equals

$$d_n = r_n + r_n^*.$$

Since the parameters a and b usually are fixed in a given problem, we shall often suppress them as in the formula above.

Cumulative sums of $\{r_n\}$ and $\{d_n\}$ are denoted by the corresponding capital letters, so that

$$D_n = \sum_{i=0}^{n} d_i = R_n + R_n^*$$

denotes the probability of a duration of at most n games. The probability of a duration of more than n games, $1 - D_n$, may also be expressed as

$$U_n(a, b) = \sum_{x = -a + 1}^{b - 1} u_n(x, a, b),$$

so that

$$D_n + U_n = R_n + R_n^* + U_n = 1.$$

If A has infinitely many counters, we shall write $r_n(b)$ instead of $r_n(\infty, b)$ for B's probability of being ruined at the nth game.

The first solutions of the problem of the duration of play were given by Montmort, Nicholas Bernoulli, and de Moivre between 1708 and 1718, mostly without proofs or with incomplete proofs. In the present chapter we shall discuss these results and indicate how they may have been proved.

In Chapter 23 we shall return to the problem and discuss the solutions derived by de Moivre, Lagrange, and Laplace by the method of difference equations.

Another solution using generating functions was given by Laplace in 1812. Since then the problem of the duration of play has become a standard topic in advanced textbooks on probability theory. Like so many other of the classical problems of games of chance, it has steadily been given new interpretations and applications.

In today's terminology the play may be described as a random walk (t, g_t), $t = 0, 1, \ldots$, with absorbing barriers at $-a$ and b, g_t denoting the gain of A at the end of t games. The path moves one step up when A wins and one step down when A loses, and the play ends when the path for the first time hits one of the barriers.

This description lends itself to generalizations by varying the time interval between jumps and the size of the jumps and, further, by making these quantities random instead of deterministic. In this way a theory of stochastic processes with many applications has been developed.

Letting the number of steps per time unit tend to infinity and the size of the jumps tend to zero, the random walk model tends to a diffusion model. Hence, the random walk model has been used as a first approximation to describe the random motion of particles in statistical mechanics and kinetic theory. In statistical quality control and later on in the theory of testing statistical hypotheses, a generalized version of the random walk model has led to Wald's sequential analysis. In insurance mathematics a generalized model has been used to find the probability of ruin by comparing the accumulated claims with the company's capital. These applications and many more are today discussed in specialized treatises; their relations to the classical problem have been indicated by Thatcher (1957), Takács (1969), and Feller (1970).

20.2 MONTMORT'S DISCUSSION OF THE DURATION OF PLAY IN 1708

In continuation of his discussion on the problem of points, Montmort stresses an essential difference between this problem and the ruin problem. In the problem of points the play will end after at most $a + b - 1$ games because the winner of each game gets one point and the loser gets zero; in the ruin problem, however, the winner of each game gets one point and the loser loses one point, so that the play may continue indefinitely without any of the players being ruined. Montmort (1708, p. 178) recounts an experience with two players having six counters each at the beginning of the play, none of them being ruined in the course of 30 games when they decide to stop. Montmort proposes to divide the total stake between them in the ratio of the number of counters they have, assuming that the players are of equal skill; however, he only discusses numerical examples. In his letter to Montmort (1713, pp. 295–296), John Bernoulli states that the ratio of the probabilities of winning is as $n + x : n - x$ when A has $n + x$ counters and B has $n - x$, $x = 0, 1, \ldots, n - 1$; a proof is given by Nicholas Bernoulli in a letter to Montmort (1713, p. 311). The problem corresponds to Huygens' fifth problem with the modification that the ratio of the probabilities of winning a single game is as $1 : 1$ instead of as $9 : 5$. The reader may prove the result as a special case of the three proofs given in §14.2.

Montmort (1708, pp. 184–185) then generalizes Huygens' fifth problem by considering the probability that either one or the other of the players will be ruined in at most a given number of games, the problem of the duration of play. As an example he considers two players of equal skill having three counters each. He states that the probability of a duration of at most $2n + 1$ games equals

$$\sum_{i=1}^{n} \frac{3^{i-1}}{4^i}, \qquad n = 1, 2, \ldots . \tag{1}$$

He gives no proof. Presumably he has followed his usual procedure of considering some examples and then has derived the general formula by incomplete induction. Rashly he adds that "Without much difficulty one may find similar formulae for other cases which will lead one to a rather interesting research." As we shall see in the next section he continued his investigations, and contrary to his expectations the research proved to be rather difficult.

20.3 NICHOLAS BERNOULLI'S FORMULA FOR THE RUIN PROBABILITY, 1713

On 15 November 1710, Montmort (1713, pp. 306–307) answered John Bernoulli's letter and returned to the problem of the duration of play, noting that he now had the general solution for players of equal skill and with the same number of counters. He returns to the example of two players, each having six counters, and states that the probability of a duration of at most 26 games is $16,607,955/33,554,432$, which is a little less than $\frac{1}{2}$, whereas the probability of a duration of at most 28 games is $70,970,250/134,217,728$, which is a little larger than $\frac{1}{2}$. We have given the correct denominator of the second probability; as pointed out by Nicholas Bernoulli there is an error in the result as originally stated by Montmort. Montmort writes that he has found these results nearly without calculation but does not disclose his formula.

From his numerical results and later remarks it seems reasonable to assume that he had found the probability of a duration of at most n games in the form

$$
\begin{aligned}
D_n(b,b) &= \sum_{k=0}^{\infty} \left[2 \sum_{i=0}^{m-2kb-1} \binom{n}{i} + \binom{n}{m-2kb} \right] \Big/ 2^{n-1} \\
&\quad - \sum_{k=0}^{\infty} \left[2 \sum_{i=0}^{m-2kb-b-1} \binom{n}{i} + \binom{n}{m-2kb-b} \right] \Big/ 2^{n-1} \\
&= \sum_{k=0}^{\infty} \sum_{i=m-2kb-b+1}^{m-2kb} \binom{n+1}{i} \Big/ 2^{n-1}, \qquad n = b + 2m, \quad p = \tfrac{1}{2}. \qquad (1)
\end{aligned}
$$

The reduction from the first to the second expression is obtained by means of the formula

$$
2 \sum_{i=0}^{m-1} \binom{n}{i} + \binom{n}{m} = \sum_{i=0}^{m} \binom{n+1}{i}.
$$

The two probabilities mentioned above equal

$$
\sum_{i=5}^{10} \binom{27}{i} \Big/ 2^{25} \qquad \text{and} \qquad \sum_{i=6}^{11} \binom{29}{i} \Big/ 2^{27},
$$

respectively.

Let us digress a little from the course of the historical events and summarize Montmort's numerical results. In the correspondence with the Bernoullis he gives several examples of determining the *median duration* of play, i.e., he solve the equation $D_n(b,b) = \frac{1}{2}$ for n,

$$D_{26}(6,6) = 0.495 < \tfrac{1}{2} < D_{28}(6,6) = 0.529,$$

$$D_{35}(7,7) = 0.485 < \tfrac{1}{2} < D_{37}(7,7) = 0.511,$$

$$D_{61}(9,9) = 0.505.$$

He also mentions that he believes that $D_{122}(12,12) < \frac{1}{2} < D_{124}(12,12)$, but as pointed out by Nicholas Bernoulli the correct solution is

$$D_{108}(12,12) = 0.499 < \tfrac{1}{2} < D_{110}(12,12) = 0.507.$$

We have given the probabilities as decimal fractions, whereas Montmort naturally uses the ratio of two integers.

Finally, Montmort (1713, p. 276) states that the equation $D_n(b,b) \geqslant \frac{1}{2}$ for b odd has the approximate solution

$$n \geqslant \tfrac{3}{4}b^2 + \tfrac{1}{4}. \tag{2}$$

He does not indicate how he has obtained this result and adds that he has not been able to find a similar formula for b even. Using (2) to find n for the four examples above, we get 27.25, 37.00, 61.00, and 108.25. Montmort does not give these result but mentions as an example that the median duration for $b = 19$ will be $n = 271$. A formula similar to (2) was proved by de Moivre in 1738, see (23.2.10).

Returning to the letter to John Bernoulli, Montmort writes that he will send his formula if Bernoulli is interested. At the time John Bernoulli was 43 years old and, except for Leibniz, the most renowed mathematician on the Continent. He was busy with his own research at Basel, and Montmort could not expect Bernoulli to spend more time on his problems. Montmort therefore indicated that he would be glad if John would pass on the problem to his nephew Nicholas, "who seems to me to be capable of solving the most dificult problems and, being young, perhaps he has the leisure to search for the solution which certainly is worthy for him." Nicholas was 23 years old, had just received his degree in 1709, and was preparing himself for his Grand Tour to England, the Netherlands, and France. He did live up Montmort's expectations. On 26 February 1711 he sent the complete solution to Montmort (1713, pp. 309–311). It was really an extraordinary achievement on the part of Nicholas Bernoulli to solve this complicated problem that had until then proved too difficult for the more experienced probabilists Montmort and de Moivre. He gave no proof of his formula.

In his reply Montmort expresses his admiration for Bernoulli's solution but admits that he is not quite able to understand it and asks for a numerical example. Bernoulli answers that Montmort's difficulties are understandable in view of the fact that he has used ambiguous notation which he now corrects. He then derives formula (1) as a special case of his formula and gives two numerical examples. Montmort answers that he now understands Bernoulli's formula and that the special case agrees with his own previous result. The correspondence regarding the ruin problem can be found in Montmort (1713, pp. 315–316, 324–326, 344–345, 368–369, 375, 380).

In the formula Bernoulli assumes that $n - b$ is even because B can be ruined only at games number $b, b + 2, b + 4, \ldots$; it follows that

$$R_{b + 2m + 1}(a, b) = R_{b + 2m}(a, b), \qquad m = 0, 1, \ldots .$$

Nicholas Bernoulli's Formula for the Ruin Probability $R_n(a, b)$

Setting $c = a + b$, we have

$$R_n(a, b) = p^b \sum_{k=0}^{\infty} (qp)^{kc} \sum_i \binom{n}{i} (p^{n-b-2kc-i} q^i + q^{n-b-2kc-i} p^i)$$

$$- p^b \sum_{k=0}^{\infty} (qp)^{a+kc} \sum_i \binom{n}{i} (p^{n-b-2kc-2a-i} q^i + q^{n-b-2kc-2a-i} p^i), \qquad (3)$$

where $0 \leqslant 2i \leqslant n - b - 2kc$ in the first sum; $0 \leqslant 2i \leqslant n - b - 2kc - 2a$ in the second sum; and where only one of the two identical members of the last term of the sums over i should be included when the upper limit for $2i$ is even.

As a corollary Bernoulli gives the complete solution of Huygens' fifth problem, which is obtained for $n \to \infty$. Setting $\lim R_n = R$, his solution becomes

$$R(a, b) = \frac{p^c - p^b q^a}{p^c - q^c}, \qquad a \neq b, \quad p \neq q,$$

$$R(a, a) = \frac{p^a}{p^a + q^a},$$

$$R(a, b) = \frac{a}{a + b}, \qquad p = \frac{1}{2},$$

(see the discussion in §14.2). The result is given without proof; we shall leave the proof to the reader.

We have already mentioned that Bernoulli also derived formula (1) for $D_n(b,b), p = \frac{1}{2}$.

It is of course easy be means of Bernoulli's formula to derive the other probabilities defined in §20.1, except for the function $u_n(x,a,b)$. Interchanging the roles of A and B we get $R_n^*(a,b)$ and thus $D_n(a,b)$, and by differencing we get r_n, r_n^*, and d_n. Bernoulli does not give these results.

We shall finally indicate how Bernoulli may have derived his formula.

A Proof of Bernoulli's Formula

Takács (1969) has suggested that Bernoulli derived his formula using the method of inclusion and exclusion and the method of reflection. We believe that Takács is right, and in the following we shall report Takács' proof with some comments in relation to Bernoulli's formula.

Since the sum of the exponents of p and q in (3) equals n, and the coefficient equals $\binom{n}{i}$, it is clear that Bernoulli, like Pascal and Fermat in their combinatorial solution of the problem of points, has analyzed a series of n games where the last ones may be fictitious because one of the players has been ruined before the series is completed. We shall therefore consider series of n games in which A wins i games and B wins $n - i$ games, $i = 0, 1, \ldots, n$; the probability of any such series is $p^i q^{n-i}$. Consequently we rewrite Bernoulli's formula in the form

$$R_n(a,b) = \sum_{k=0}^{\infty} \left[\sum_{2i-n \geq b} \binom{n}{i+kc} p^i q^{n-i} + \sum_{2i-n < b} \binom{n}{i-b-kc} p^i q^{n-i} \right]$$
$$- \sum_{k=0}^{\infty} \left[\sum_{2i-n \geq b} \binom{n}{i+kc+a} p^i q^{n-i} + \sum_{2i-n < b} \binom{n}{i-c-kc} p^i q^{n-i} \right], \quad (4)$$

which is obtained by introducing the power of p in each of the sums of (3) as a new variable.

The formula is an alternating series, indicating that Bernoulli has used the method of inclusion and exclusion, which is also affirmed by Montmort and de Moivre who explicitly use this method in their comments on Bernoulli's result. The series is finite because the binomial coefficients become zero for k sufficiently large.

We shall now consider various subsets of the set of series of games in which A wins i games and B wins $n - i$ games to find the set, R, say, in which A ruins B. The set R is defined by the property that A's gain reaches b *before* B's gain reaches a. To find R consider the set C_1 in which A wins b counters from B at least once in n games. Compared with R, the set C_1 is obviously too large because it includes the set in which A's gain reaches $-a$ before it reaches b. We therefore deduct the set of series of games, C_2, say, in which A's gain at least once passes from $-a$ to b. This is the beginning of an alternating series in which the terms are defined as follows: $C_{2k}, k = 1, 2, \ldots$, is the set of series of games in which A's gain at least k times passes from $-a$ to b, and $C_{2k+1}, k = 0, 1, \ldots$, is the set of series of games in which A's gain at least once reaches b and afterward at least k times passes from $-a$ to b; clearly C_{2k+1} is a subset of C_{2k}.

The definition of these sets is indicated by the form of (3), since the factor p^b means that A's gain at least once reaches b, and the factor $(qp)^{kc}$ indicates that A's gain at least k times goes from $-a$ to b.

Using the method of inclusion and exclusion, the set of series of games leading to the ruin of B is found as

$$R = \sum_{k=1}^{\infty} (-1)^{k-1} C_k. \tag{5}$$

This result may also be seen by writing R in the form

$$R = C_1 - \sum_{k=1}^{\infty} (C_{2k} - C_{2k+1}),$$

since C_1 is the set of series of games in which A's gain at least once reaches b, and $\{C_{2k} - C_{2k+1}; k = 1, 2, \ldots\}$ is the set of series not belonging to R because A's gain reaches $-a$ before reaching b.

To find $\Pr\{R\}$, we thus have to find

$$\Pr\{C_k\} = N_k(i) p^i q^{n-i},$$

where $N_k(i)$ denotes the number of series of games of length n belonging to C_k. Formula (4) indicates that we have to distinguish between series in which A's gain, $g_n = 2i - n$, is larger and smaller than b, respectively.

The total number of different series of games in which A wins i games and B wins $n - i$ games equals $\binom{n}{i}$. Consider now $N_1(i)$. All the series in which $2i - n \geq b$ are members of C_1, since A's gain necessarily reaches b at least once, and we thus have

$$N_1(i) = \binom{n}{i} \quad \text{for} \quad 2i - n \geq b.$$

All the series in which $2i - n < 2b - n$ are not members of C_1, since $i < b$, so that

$$N_1(i) = 0 \qquad \text{for} \quad 2i - n < 2b - n.$$

In the intermediate case, where $2b - n \leqslant 2i - n < b$, we have to find the number of series of games in which A's gain at least once reaches b and ends with being smaller than b. Consider the random walk $(t, g_t), t = 0, 1, \ldots, n$, and reflect the section of the path after it has reached b for the first time in the horizontal line through b. Since $g_n = 2i - n$, the reflected path will lead to a gain of

$$b + (b - g_n) = 2b - (2i - n) = (n + b - i) - (i - b) \geqslant b,$$

so that the reflected path represents a series of n games in which A wins $n + b - i$ games, and B wins $i - b$ games. A graph of the random walk shows that there is a one-to-one correspondence between the original and the reflected path. Hence,

$$N_1(i) = \binom{n}{i - b} \qquad \text{for} \quad 2b - n \leqslant 2i - n < b,$$

because this is the total number of series in which A wins $n + b - i$ games, B wins $i - b$ games, and A's gain reaches b at least once.

Reflection corresponds to changing the outcome of a game from $+1$ to -1 and vice versa, as indicated by the interchange of p and q in the inner sums of (3).

Summing over i we get the probability that A's gain reaches b at least once as

$$\sum_{2i - n \geqslant b} \binom{n}{i} p^i q^{n-i} + \sum_{2i - n < b} \binom{n}{i - b} p^i q^{n-i}, \tag{6}$$

which is the first term of (4). Since we have disregarded the possibility of A being ruined, this probability obviously equals $R_n(b)$.

The proof may be extended to find the four binomial coefficients in (4) by reflecting the path each time A's gain reaches $-a$ or b; we shall leave that to the reader. The complete proof has been given by Takács (1969).

We do not believe that Bernoulli carried out a formal proof such as the one above; presumably he simply worked through some examples, and from the structure of the solution he constructed his formula by incomplete induction. Nevertheless, the solution of this complicated problem requires great combinatorial insight and testifies to his ingenuity.

Besides printing the correspondence with Bernoulli on this problem, Montmort (1713, pp. 268–277) wrote a section in the main part of his book giving Bernoulli's formula with a first attempt at a proof and some numerical examples. From Montmort's arguments one gets the general idea of the proof, but as pointed out by de Moivre, some of the details are wrong. De Moivre (1718, pp. 122–124) does not quote Bernoulli's formula, only the example discussed by Montmort for $n = 15$, $a = 3$, and $b = 2$. He then improves Montmort's argument as follows:

> The first Series of the first Branch expresses the number of Chances there are for A to win b stakes of B, including the number of Chances there are for B, before the expiration of the n games, to be in a circumstance of winning a Stakes of A; which number of Chances may be deduced from óur foregoing Problem [to find $R_n(b)$]. The second Series of the first Branch is a part of the first, and expresses the number of Chances there are, for B to win a Stakes of A, out of the number of Chances there are for A in the first Series, to win b Stakes of B. It is to be observed about this Series, *First*, that the Chances of B expressed by it are not restrained to Happen in any Order, that is, either before or after A has won b Stakes of B.

(We have replaced de Moivre's numbers by n, a, and b, respectively.) De Moivre's argument is correct, but it must certainly have been very difficult to understand. It will be seen that he gives a verbal formulation of (5), the first part of the proof above. With respect to the second part of the proof, he simply refers to formula (6), which he states (without proof) immediately before. In the next section we shall show that he presumably proved (6) and some other results by induction without using the combinatorial methods that Bernoulli and Montmort used.

20.4 DE MOIVRE'S RESULTS IN *DE MENSURA SORTIS*, 1712

Independently of Nicholas Bernoulli and about the same time, de Moivre solved some problems on the duration of play; his results are given as solutions to Problems 20–26 in *De Mensura Sortis*, read to the Royal Society in 1711 and published in 1712. *De Mensura Sortis* has been translated into English by McClintock (1984).

De Moivre gives a general rule for finding the probability of continuation $u_n(x, a, b)$, which we shall call de Moivre's algorithm.

De Moivre's Algorithm. Multiply $p + q$ by itself n times, and after each multiplication reject those terms in which the exponent of p exceeds the exponent of q by b and in which the exponent of q exceeds the exponent of p by a. The remaining terms will then give the probability of continuation after n games.

The algorithm is obviously correct; in principle it solves the problem completely, but in practice it is useful only for small values of n, a, and b. Naturally, de Moivre had hoped to derive a formula for the continuation probability, but he did not succeed.

After having formulated the algorithm he gives an example for $n = 7$, $a = 3$, and $b = 2$, which leads to

$$u_7(1) = 13p^4q^3 \quad \text{and} \quad u_7(-1) = 21p^3q^4;$$

we shall leave the derivation to the reader.

He then turns to the simpler problem of finding U_n for $a = b$. First, he proves that

$$U_{b+2m} = (2pq)^{1+m}, \quad a = b = 2, \quad m = 0, 1, \ldots,$$

$$U_{b+2m} = (3pq)^{1+m}, \quad a = b = 3, \quad m = 0, 1, \ldots.$$

Using the algorithm to find the first few terms, it is straighforward to prove these results by induction.

However, even for $a = b = 4$ the problem becomes so complicated that de Moivre cannot find an explicit solution. Instead he uses the algorithm to tabulate $U_n(4, 4)$ for $n \geq 4$, ending with

$$U_{12}(4, 4) = 560p^7q^5 + 792p^6q^6 + 560p^5q^7,$$

and without proof he gives recursion formulae for the coefficients, so that tabulation can easily be continued.

He solves the symmetric equations $U_4(4, 4) = \frac{1}{2}$ and $U_6(4, 4) = \frac{1}{2}$ with respect to p and finds that $p/q = 5.274$ and 2.576, respectively. He shows how such a reciprocal equation of the nth degree can be transformed to an equation of degree $\frac{1}{2}n$.

Commenting on de Moivre's result, Montmort (1713, p. 275) writes that although the algorithm is simple and ingenious, Bernoulli's and his formula is much easier to use. As an example he mentions that his formula (3.1) gives $D_{61}(9, 9) = 581{,}928 \times 10^{12}/2^{60}$, which he had found in less than an hour; but de Moivre's rule implies the calculation of $(1 + 1)^9$, followed by 26 multiplications with $1 + 2 + 1$, and rejection of the two extreme terms after each multiplication, which requires an immense amount of calculation.

Without referring to Montmort, de Moivre (1718, p. 106) later took up the idea of using the algorithm on $(1 + 1)$ instead of $(p + q)$ and after having found these numbers, to multiply by the adequate powers of p and q; presumably they both considered this procedure a modification of the arithmetic triangle, which is generated by $(1 + 1)^n$, $n = 0, 1, \ldots$.

Setting

$$u_n(x, a, b) = c_n(x, a, b)p^{(n+x)/2}q^{(n-x)/2}, \tag{1}$$

where $c_n(x, a, b)$ denotes the number of ways in which A may win x counters in n games without any of the players being ruined, it is clear that de Moivre's algorithm leads to the recursion

$$c_n(x) = c_{n-1}(x - 1) + c_{n-1}(x + 1), \qquad -a < x < b, \quad n \geq 1, \tag{2}$$

with the initial values

$$c_0(0) = 1 \quad \text{and} \quad c_0(x) = 0 \quad \text{for} \quad x \neq 0,$$

and the boundary conditions

$$c_n(-a) = c_n(b) = 0.$$

This is just a formalization of de Moivre's numerical examples. An example of this procedure has been given in Fig. 20.4.1, corresponding to de Moivre's previously mentioned example for $n = 7$, $a = 3$, and $b = 2$.

The coefficients of the ruin probabilities are listed just outside the boundaries; these coefficients are equal to $c_{n-1}(b - 1)$ and $c_{n-1}(-a + 1)$, respectively, since $r_n = pu_{n-1}(b - 1)$.

Besides the algorithm, the most important result given by de Moivre is *the probability $R_n(b)$ of B being ruined in at most n games when A's capital is unlimited*:

$$R_n(b) = \sum_{i \geq (n+b)/2} \binom{n}{i} p^i q^{n-i} + \sum_{i < (n+b)/2} \binom{n}{i-b} p^i q^{n-i}. \tag{3}$$

He gives no proof, but he has presumably used the algorithm to calculate some values of $R_n(b)$ from which he has inferred the formula, or he may have used the algorithm to derive a few terms of the series r_n and U_n, $n \geq b$, as done by Laplace (1776), and then found the formula by incomplete induction. We shall indicate the complete proof.

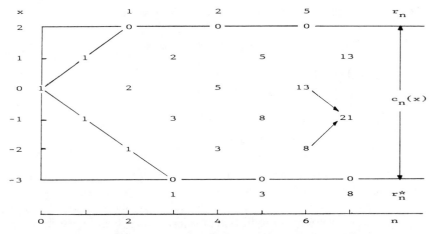

Fig. 20.4.1. Modified arithmetic triangle according to de Moivre's algorithm, showing the calculation of $c_n(x, a, b), r_n(a, b)$, and $r_n^*(a, b)$ for $a = 3$ and $b = 2$.

Proof of (3). Since $r_{b+2m+1} = 0, m = 0, 1, \ldots$, we shall set $n = b + 2m$ and use the algorithm with two steps at a time. From the expansion of

$$(p + q)^b = r_b + U_b,$$

it follows that

$$r_b = p^b \quad \text{and} \quad U_b = \sum_{i=1}^{b} \binom{b}{i} p^{b-i} q^i. \tag{4}$$

Continuing with the expansion of $(p^2 + 2pq + q^2)U_b$, we find r_{b+2} and U_{b+2}, and so on. After a few steps it is easy to see that the general formulae become

$$r_{b+2m} = \left\{ \binom{b + 2m - 2}{m} - \binom{b + 2m - 2}{m - 2} \right\} p^{b+m} q^m \tag{5}$$

$$U_{b+2m} = \sum_{i=1}^{m} \left\{ \binom{b + 2m}{m + i} - \binom{b + 2m}{m - i} \right\} p^{b+m-i} q^{m+i}$$

$$+ \sum_{i=m+1}^{b+m} \binom{b + 2m}{m + i} p^{b+m-i} q^{m+i}, \tag{6}$$

To prove by induction that these formulae hold, we use

$$(p^2 + 2pq + q^2)U_{b+2m} = r_{b+2m+2} + U_{b+2m+2}.$$

The binomial coefficients of order $b + 2m$ on the left-hand side are changed to order $b + 2m + 2$ by means of the formula

$$\binom{n}{i} + 2\binom{n}{i+1} + \binom{n}{i+2} = \binom{n+2}{i+2}, \tag{7}$$

which is obtained by using the recursion formula for the arithmetic triangle two times in succession. In this way the left-hand side may be brought into a form equal to the one obtained from (5) and (6), with $m + 1$ substituted for m, which concludes the induction. Since the formulae hold for $m = 0$, formulae (5) and (6) have been proved.

From U_n we obtain R_n,

$$R_{b+2m} = (p + q)^{b+2m} - U_{b+2m}$$

$$= \sum_{i=0}^{m} \binom{b+2m}{i} p^{b+2m-i}q^i + \sum_{i=m+1}^{2m} \binom{b+2m}{2m-i} p^{b+2m-i}q^i, \tag{8}$$

which proves that (3) holds for $n = b + 2m$.

Since $r_{b+2m+1} = 0$, we have

$$R_{b+2m+1} = (p + q)R_{b+2m},$$

which leads to an expression for R_{b+2m+1} equal to (8), with $2m$ replaced by $2m + 1$ so that (3) holds also for $n = b + 2m + 1$. This concludes the proof.

De Moivre points out that the binomial coefficients in (3) are symmetric with respect to $i = \frac{1}{2}(n + b)$ and that the first sum equals the upper tail of the binomial $(p + q)^n$. This is of course in agreement with the first term of Bernoulli's formula for $R_n(a, b)$, see (3.6).

20.5 DE MOIVRE'S RESULTS IN THE *DOCTRINE OF CHANCES*, 1718

In 1711 de Moivre was 44 years old and a renowned mathematician. It must have been a great surprise for him to learn that a young and unknown mathematician, Nicholas Bernoulli, had solved a problem which de Moivre at the same time had tried solving without finding a satisfactory solution. In the *Doctrine of Chances* (1718, p. XIII), he acknowledged Bernoulli's result as follows:

> Mr. *de Monmort*, and Mr. *Nicholas Bernoully*, have each of them separately given the Solution of my XXXIXth Problem [to find D_n], in a Method differing from mine, as may be seen in Mr. *de Monmort*'s second Edition of his Book. Their Solutions, which in the main agree together, and vary little more than in the form of Expression, are extreamly beautiful; for which reason I thought the Reader would be well pleased to see their Method explained by me, in such a manner as might be apprehended by those who are not so well versed in the nature of Symbols: In which matter I have taken some Pains, thereby to testify to the World the just Value I have for their Performance.

This remark is omitted from the following editions of the *Doctrine* in which he claimed, without giving any evidence, that he had the solution in 1711 but was "preserving this Problem by me in order to be published when I should think it proper" (1738, p. 181). De Moivre never gave a full account of Bernoulli's formula, he only discussed a numerical example; Todhunter followed de Moivre, and for this reason Bernoulli's result was overlooked for a long time.

De Moivre's important new results in the *Doctrine* may be summarized as follows:

1. a formula for $r_n(a, b)$;
2. a recursion formula for $d_n(a, b)$;
3. a method for summing recurring series to find $D_n(a, b)$; and
4. a solution of the difference equation corresponding to the recursion giving a formula for $U_n(b, b)$.

De Moivre did not indicate how he obtained these results, except for the summation formula under (3); he gave only some numerical examples of how to use the formulae. For the generation of mathematicians after de Moivre it became a challenge to prove the recursion formulae. The young Laplace (1774, 1776) formulated de Moivre's algorithm as the partial difference equation

$$u_n(x) = pu_{n-1}(x - 1) + qu_{n-1}(x + 1), \qquad -a < x < b, \quad n \geqslant 1, \qquad (1)$$

with the boundary conditions $u_0(0) = 1$, and

$$u_n(x) = 0, \quad x > (b - 1) \wedge n \qquad \text{and} \qquad x < (-a + 1) \vee (-n), \quad n \geqslant 0,$$

and showed how it could be transformed to a linear difference equation with respect to n only, in this way proving de Moivre's recursion formulae, see §23.4.

Todhunter (pp. 169–172) gives up reconstructing de Moivre's proofs; instead, he uses generating functions due to Laplace (1812) to derive de Moivre's results. Todhunter's defeat leads him to write (p. 193) that "Our obligations to De Moivre would have been still greater if he had not concealed the demonstration of the important results which we have noticed in Art. 306."

It seems, however, that Todhunter has overlooked a clue given by de Moivre (1718, p. XIII):

All the Problems which in my *Specimen* [De Mensura Sortis] related to the Duration of Play, have been kept entire in the following Treatise; but the Method of Solution has received some Improvements by the new Discoveries I have made concerning the Nature of those Series which result from the Consideration of the Subject; however, the Principles of that Method having been laid down in my *Specimen* I had nothing now to do, but to draw the Consequences that were naturally deducible from them.

We interpret this statement to mean that de Moivre derived the recursion formulae from his algorithm, and we shall therefore give a proof based on this method and induction. We further believe that de Moivre found the formula for r_n by differencing Bernoulli's formula, even if he did not refer to Bernoulli. We shall prove the formula by this method.

We shall give page references to the *Doctrine* (1718) only; the reader may find the same results in the second and third edition, as indicated in §22.4.

De Moivre's formula for $r_n(a, b)$

Setting $c = a + b$ and

$$r_{b+2m}(a, b) = p^b (pq)^m \beta_m(a, b), \qquad m = 0, 1, \ldots, \tag{2}$$

we have

$$\beta_m(b) = \frac{b}{b + 2m} \binom{b + 2m}{m}, \tag{3}$$

$$\beta_m(b, b) = \frac{b}{b + 2m} \sum_{k=0}^{[m/b]} (-1)^k (2k + 1) \binom{b + 2m}{m - kb}, \tag{4}$$

$$\beta_m(a, b) = \sum_{k=0}^{[m/c]} \frac{b + 2kc}{b + 2m} \binom{b + 2m}{m - kc} - \sum_{k=0}^{[(m-a)/c]} \frac{a + (2k + 1)c}{b + 2m} \binom{b + 2m}{m - a - kc}. \tag{5}$$

These results are given by de Moivre on pp. 121, 110, and 118–119, respectively.

Proof. Setting

$$S_n(m) = \sum_{i=0}^{m-1} \binom{n}{i}(qp)^i(p^{2m-2i} + q^{2m-2i}) + \binom{n}{m}(qp)^m,$$

Bernoulli's formula (3.3) may be written as

$$R_n(a, b) = p^b \sum_{k=0}^{\infty} (qp)^{kc} S_n(m - kc)$$

$$- p^b \sum_{k=0}^{\infty} (qp)^{a+kc} S_n(m - a - kc), \qquad n = b + 2m. \qquad (6)$$

To find

$$r_{b+2m} = R_{b+2m} - R_{b+2m-2},$$

we therefore have to find $S_n(m) - S_{n-2}(m-1)$. Inserting

$$\binom{n}{i} = \binom{n-2}{i} + 2\binom{n-2}{i-1} + \binom{n-2}{i-2}$$

into $S_n(m)$, multiplying $S_{n-2}(m-1)$ by $p^2 + 2pq + q^2$, and comparing terms of the same form, we get

$$S_n(m) - S_{n-2}(m-1) = (pq)^m \left\{ \binom{n-2}{m} - \binom{n-2}{m-2} \right\}$$

$$= (pq)^m \frac{n-2m}{n} \binom{n}{m}. \qquad (7)$$

Since

$$R_{b+2m}(b) = p^b S_{b+2m}(m),$$

(3) follows. This result agrees with (4.5).

The differences of the S's in the two terms of (6) are

$$(pq)^{m-kc} \frac{n-2m+2kc}{n} \binom{n}{m-kc} = (pq)^{m-kc} \frac{b+2kc}{b+2m} \binom{b+2m}{m-kc},$$

$$(pq)^{m-a-kc} \frac{n-2m+2a+2kc}{n} \binom{n}{m-a-kc}$$

$$= (pq)^{m-a-kc} \frac{a+(2k+1)c}{b+2m} \binom{b+2m}{m-a-kc}.$$

Inserting these results into (6), (4) and (5) are easily found. By analogy we get

$$r^*_{a+2m}(a, b) = q^a(pq)^m \beta^*_m(a, b), \qquad \beta^*_m(a, b) = \beta_m(b, a),$$

which concludes the proof. A similar proof has been given by Hald (1988).

Before proving de Moivre's recursion formulae we shall prove two lemmas, which presumably would have been known to de Moivre. They are easily found by incomplete induction.

De Moivre gave two forms of $R_n(b)$, one is a linear combination of the terms $p^i q^{n-i}$, $i = 0, 1, \ldots, n$, see (4.3) and the other a linear combination of the terms $p^b(pq)^i$, $i = 0, 1, \ldots, m$, see (2) and (3). Hence, the first form corresponds to the usual expansion of the binomial $(p + q)^n$, whereas the second requires an expansion in terms of $(pq)^i$.

Lemma 1. An Expansion of the Binomial.

$$(p + q)^n = p^n + q^n + \sum_{i=1}^{[n/2]} (- 1)^{i-1} \alpha_i(n)(pq)^i(p + q)^{n-2i}, \qquad n \geq 1, \qquad (8)$$

where

$$\alpha_i(n) = \frac{n(n - i - 1)^{(i-1)}}{i!} = \frac{n}{n-i}\binom{n-i}{i}. \qquad (9)$$

Proof. Expanding the binomial in powers of pq we get

$$(p + q)^2 = p^2 + q^2 + 2pq,$$

$$(p + q)^3 = p^3 + q^3 + 3pq(p + q),$$

$$(p + q)^4 = p^4 + q^4 + 4pq(p + q)^2 - 2(pq)^2,$$

and so on. These expansions are of the form (8), and from the pattern of the coefficients it is easy to see that (9) holds.

To give a rigorous proof of (9) we multiply (8) by $p + q$ and eliminate $p^{n-1} + q^{n-1}$ from the right-hand side, in this way obtaining an expansion of $(p + q)^{n+1}$ analogous to (8), but with coefficient $\alpha_i(n) + \alpha_{i-1}(n - 1)$. Replacing n by $n + 1$ in (8) and comparing coefficients in the two expansions, we get the difference equation

$$\alpha_i(n+1) = \alpha_i(n) + \alpha_{i-1}(n-1), \qquad n \geqslant 1, \quad i \geqslant 1,$$

where $\alpha_0(n) = 1$ for $n \geqslant 0$. It is easy to prove that the complete solution is given by (9). The proof is essentially due to Laplace (1776).

By means of the lemma it may be proved that

$$\sum_{i=1}^{k \wedge [n/2]} (-1)^{i-1} \alpha_i(n) \binom{n-2i}{k-i} = \binom{n}{k}, \qquad 0 < k < n, \tag{10}$$

$$1 + \sum_{i=1}^{k \wedge [n/2]} (-1)^{i-1} \binom{n-i}{i} \binom{n-2i}{k-i} = \binom{n}{k}, \qquad 0 < k < n. \tag{11}$$

The proofs are due to Hald and Johansen (1983) and Hald (1988).

In his search for recursion formulae it seems reasonable to suppose that de Moivre began by investigating $r_n(b)$ as the simplest of the ruin probabilities.

Lemma 2. The Recursion Formula for $r_n(b)$.

$$r_n(b) = \sum_{i=1}^{\infty} (-1)^{i-1} \binom{n-1-i}{i} (pq)^i r_{n-2i}(b), \qquad n > b. \tag{12}$$

Proof. Since $r_{b+2m+1}(b) = 0$, we need only consider the case $n = b + 2m$. The proof is based on (4.5) for r_{b+2m}, starting from $r_b = p^b$. We have

$$r_{b+2} - bpqr_b = 0,$$

and from the difference $r_{b+4} - (b+2)pqr_{b+2}$ combined with (4.5), we get

$$r_{b+4} = \binom{b+2}{1} pqr_{b+2} - \binom{b+1}{2} (pq)^2 r_b,$$

and so on. This leads to (12) by incomplete induction. The series contains only m terms since $r_n = 0$ for $n < b$.

To give a rigorous proof we insert (4.5) in (12) and use the fact that

$$\binom{n-2}{m} - \binom{n-2}{m-2} = \binom{n-1}{m} - \binom{n-1}{m-1},$$

so that (12) is reduced to the equation

$$\binom{n-1}{m} - \binom{n-1}{m-1} = \sum_{i=1}^{m} (-1)^{i-1} \binom{n-1-i}{i} \left[\binom{n-1-2i}{m-i} - \binom{n-1-2i}{m-i-1} \right],$$

which holds according to (11).

De Moivre's Recursion Formulae

As observed by Montmort, de Moivre's algorithm requires too many additions to be of practical use for n large. Bernoulli's formula remedies this deficiency by replacing additions by binomial coefficients. However, $R_n(a, b)$ is also difficult to calculate, being a complicated polynomial of the nth degree. In his search for a better solution, de Moivre got the ingenious idea of transforming the algorithm into a recursion formula with respect to n, and by summing the recurring series, he found a solution that is easier to calculate.

Writing $D_n(a, b)$ in the form

$$D_n(a, b) = R_n(a, b) + R_n^*(a, b)$$

$$= \sum_{i=0}^{[(n-b)/2]} r_{b+2i}(a, b) + \sum_{i=0}^{[(n-a)/2]} r_{a+2i}^*(a, b), \tag{13}$$

where r and r^* are defined in (2), de Moivre's recursion formula for the coefficients in the two series becomes

$$\beta_m(a, b) = \sum_{i=1}^{[(a+b-1)/2]} (-1)^{i-1} \binom{a+b-1-i}{i} \beta_{m-i}(a, b), \qquad m \geqslant a, \tag{14}$$

where $\beta_m^*(a, b)$ satisfies the same recursion for $m \geqslant b$.

For $a = b$ the recursion may be reduced to

$$\beta_m(b, b) = \sum_{i=1}^{m \wedge [b/2]} (-1)^{i-1} \frac{b}{b-i} \binom{b-i}{i} \beta_{m-i}(b, b), \qquad m \geqslant 1. \tag{15}$$

Both (14) and (15) are given by de Moivre on pp. 116 and 109, respectively.

We shall prove (14) only; the same method may be used to prove (15). One may of course wonder why de Moivre did not give the recursion for $a = b$ as a special case of the general formula; the reason is that (15) contains fewer terms and is computationally simpler than (14) for $a = b$. A proof of the reduction of (14) to (15) has been given by Hald and Johansen (1983), see Problems 11 and 12 in §20.6.

Proof of (14). De Moivre states that the beginning of the series for D_n is

$$\sum_{i=0}^{a-1} r_{b+2i}(b) + \sum_{i=0}^{b-1} r^*_{a+2i}(a).$$

This follows from the fact that for small values of n, only one of the boundaries can lead to ruin. For example, if $n < b + 2a$, as in the first sum, and A wins at least b games more than B, then it is impossible for B to ruin A regardless of the size of a. Hence,

$$d_n(a, b) = r_n(b) + r^*_n(a), \qquad 0 \leqslant n < \{(a + 2b) \wedge (b + 2a)\}. \tag{16}$$

Since a and b are positive integers, the equation holds for at least $n \leqslant a + b$. Using Lemma 2, we thus have

$$d_n(a, b) = \sum_{i=1}^{\infty} (-1)^{i-1} \binom{n-1-i}{i} (pq)^i d_{n-2i}(a, b), \qquad (a \wedge b) < n \leqslant a + b. \tag{17}$$

To find a recursion for larger values of n, we need the auxiliary formula

$$u_n(x) = \sum_{i=1}^{[(a+b-1)/2]} \alpha_i u_{n-2i}(x), \qquad n \geqslant a + b - 1, \qquad -a < x < b, \tag{18}$$

where the coefficients $\{\alpha_i\}$ are independent of n. Supposing that this formula holds, summation over x shows that the same recursion holds for U_n, and since $D_n = 1 - U_n$, we get

$$D_n(a, b) = \sum \alpha_i D_{n-2i}(a, b) + C, \qquad n \geqslant a + b - 1,$$

C being a constant. We have here, as in the following, left out the limits of summation, which are the same as in (18). From $d_n = D_n - D_{n-1}$ we obtain

$$d_n(a, b) = \sum \alpha_i d_{n-2i}(a, b), \qquad n \geqslant a + b. \tag{19}$$

Comparing (17) and (19) for $n = a + b$, we see that (19) holds for

$$\alpha_i = (-1)^{i-1} \binom{a+b-1-i}{i} (pq)^i, \qquad i = 1, 2, \ldots, [\tfrac{1}{2}(a+b-1)]. \tag{20}$$

Inserting

$$d_n(a, b) = r_n(a, b) + r^*_n(a, b),$$

and the expressions for r and r^* in terms of β and β^*, the recursion formula (14) follows.

We note that (16) and (17) show that other recursions exist if a or b is larger than unity. However, these recursions are of he same form as (19) and (20) but with $a + b$ replaced by a larger number, so that (20) gives the optimum solution because it has the smallest number of terms, the smallest coefficients, and is valid for the largest n interval.

We shall finally prove (18) by means of de Moivre's algorithm as given by (1).

Starting from $x = b - 1$ and using $u_n(b) = 0$, we get

$$u_n(b - 1) = pu_{n-1}(b - 2), \qquad n \geqslant 1.$$

Continuing with smaller values of x and eliminating $u_{n-1}(x + 1)$, we obtain

$$u_n(b - 2) = pqu_{n-2}(b - 2) + pu_{n-1}(b - 3), \qquad\qquad n \geqslant 2,$$

$$u_n(b - 3) = 2pqu_{n-2}(b - 3) + p[u_{n-1}(b - 4) - pqu_{n-3}(b - 4)], \qquad n \geqslant 3,$$

$$u_n(b - 4) = 3pqu_{n-2}(b - 4) - (pq)^2 u_{n-4}(b - 4)$$
$$+ p[u_{n-1}(b - 5) - 2pqu_{n-3}(b - 5)], \qquad\qquad n \geqslant 4,$$

and so on. It follows that $u_n(x)$ may be written as

$$u_n(x) = \sum_{i=1} \alpha_i(x)u_{n-2i}(x) + \sum_{i=1} \beta_i(x)u_{n-2i+1}(x - 1), \qquad n \geqslant b - x, \quad (21)$$

the total number of terms on the right-hand side being $b - x$. The coefficients α_i are of the form $(-1)^{i-1}k_i(pq)^i$, $i = 1, 2, \ldots$, where the k's are constants. If $b - x$ is even there are $\frac{1}{2}(b - x)$ α's and β's, and if $b - x$ is odd, there are $\frac{1}{2}(b - x - 1)$ α's and $\frac{1}{2}(b - x + 1)$ β's. The summations are from $i = 1$ to the appropriate upper limits.

Setting $x = -a + 1$, and using $u_n(-a) = 0$, we get

$$u_n(-a + 1) = \sum_{i=1}^{[(a+b-1)/2]} \alpha_i(-a + 1)u_{n-2i}(-a + 1), \qquad n \geqslant a + b - 1, \quad (22)$$

which is a recursion formula of the form (18). Since

$$qu_n(-a + 1) = r^*_{n+1}(a, b),$$

It follows that r^*_n satisfies the same recursion for $n \geqslant a + b$.

The idea of the proof of (22) is due to Laplace, and in §23.4 we shall see how Laplace found $\alpha_i(x)$ and $\beta_i(x)$.

In the following we shall write α_i for $\alpha_i(-a+1)$. The number of terms in the recursion is $[\frac{1}{2}(a+b-1)]$, since $b-x=b+a-1$ for $x=-a+1$.

From Fig. 20.4.1 it will be seen that every second value of $u_n(x)$, $n=0,1,\ldots$, for a given value of x equals zero, so that the recursion (18) is trivially true if $u_n(x)=0$. We shall therefore assume that $u_{a+b-1}(-a+1)>0$, which implies that $u_{a+b-1}(x)>0$ for $x=-a+3, -a+5,\ldots$, and that the intermediate values equal zero. It follows that $u_{a+b}(-a+1)=0$, $u_{a+b}(-a+2)>0$, and so on. The reader should draw a diagram like Fig. 20.4.1 to see these and the following results.

From (1) we have

$$qu_n(x+1)=u_{n+1}(x)-pu_n(x-1), \qquad n\geqslant 0, \quad -a<x<b.$$

If the recursion (18) holds for $u_n(x-1)$ and $u_{n+1}(x)$, then it also holds for $u_n(x+1)$ because the coefficients α_i are independent of n. Equation (22) shows that the recursion holds for $u_n(-a+1)$, $n\geqslant a+b-1$. Furthermore,

$$qu_n(-a+2)=u_{n+1}(-a+1),$$

so that the recursion also holds for $u_n(-a+2)$, $n\geqslant a+b$. Hence it holds for $u_n(-a+3)$, $n\geqslant a+b-1$, for $u_n(-a+4)$, $n\geqslant a+b$, for $u_n(-a+5)$, $n\geqslant a+b-1$, and so on. This completes the proof of (18).

The proof of the recursion for $d_n(a,b)$ given above is a modified version of a proof by Hald (1988). We note that formula (5.15) in that paper holds for $n=a+b-1$ as shown above and not for $n\leqslant a+b-1$ as stated in the paper; however, this mistake does not influence the remainder of the proof.

Using $c_n(x)$ instead of $u_n(x)$, the proof becomes simpler because p and q disappear. We do not believe that de Moivre carried out a formal proof of the recursion formula, presumably he looked at a few examples of the modified arithmetic triangle (see Fig. 20.4.1) from which the number pattern is easily found. The reader may be convinced by calculating $c_n(x)$ for $(a,b)=(3,5)$, $(4,5)$, $(4,6)$, say.

Inserting $r_{b+2m}(b)$ in terms of β_m into (12) we get the recursion

$$\beta_m(b)=\sum_{i=1}^{m}(-1)^{i-1}\binom{b+2m-1-i}{i}\beta_{m-i}(b), \qquad m\geqslant 1, \quad \beta_0(b)=1. \quad (23)$$

De Moivre (p. 116) gives the final result in the form

$$D_n(a, b) = p^b \left[\sum_{i=0}^{a-1} \frac{b}{b+2i} \binom{b+2i}{i} (pq)^i + \sum_{i=a}^{[(n-b)/2]} \beta_i(a, b)(pq)^i \right]$$

$$+ q^a \left[\sum_{i=0}^{b-1} \frac{a}{a+2i} \binom{a+2i}{i} (pq)^i + \sum_{i=b}^{[(n-a)/2]} \beta_i(a, b)(pq)^i \right], \quad (24)$$

where $\beta_a(a, b)$ and the following coefficients are found from the preceding ones by the recursion (14). De Moivre does not notice that the first a terms might have been expressed recursively by means of (23).

As an example, for $a = 3$ and $b = 2$ de Moivre finds

$$D_n(3, 2) = p^2 [1 + 2pq + 5(pq)^2 + 13(pq)^3$$
$$+ 34(pq)^4 + 89(pq)^5 + 233(pq)^6 + \cdots]$$
$$+ q^3 [1 + 3pq + 8(pq)^2 + 21(pq)^3$$
$$+ 55(pq)^4 + 144(pq)^5 + 377(pq)^6 + \cdots].$$

The recursion (14) gives $\beta_i = 3\beta_{i-1} - \beta_{i-2}$, which is used for $i \geqslant 3$ in the first series and for $i \geqslant 2$ in the second. The number of terms given corresponds to $n = 15$, which is the example used by Montmort to illustrate the application of Bernoulli's formula. Of course, both Bernoulli's and de Moivre's formulae are polynomials in p of the nth degree, but de Moivre shows that the organization of his formula as a power series in pq combined with the recursion formula lead to simpler calculations than Bernoulli's formula.

For $a = b$, de Moivre's formula becomes

$$D_{b+2m}(b, b) = (p^b + q^b) \sum_{i=0}^{m} \beta_i(b, b)(pq)^i, \qquad \beta_0(b, b) = 1, \quad (25)$$

where $\beta_i(b, b)$ satisfies the recursion (15). This may be proved by means of the algorithm combined with Lemma 1, see Hald (1988).

On the Summation of Recurring Series

To find D_n without calculating each term of the series it is necessary to have a formula for finding the sum of a recurring series.

De Moivre defines a recurring sequence $\{r_n\}$, $n = 0, 1, \ldots$, of order k by means of two sets of k numbers, $r_0, r_1, \ldots, r_{k-1}$ and $a_1, a_2, \ldots, a_k$, where $a_k \neq 0$, and the recurrence relation

$$r_n = \sum_{i=1}^{k} (-1)^{i-1} a_i r_{n-i}, \qquad n \geqslant k.$$

The coefficients $\{(-1)^{i-1} a_i\}$ are called the scale of relation, and the polynomial

$$A(x) = \sum_{i=0}^{k} (-1)^i a_i x^i, \qquad a_0 = 1,$$

is called the differential scale.

A recurring series is defined as the power series

$$R(x) = \sum_{n=0}^{\infty} r_n x^n.$$

In the *Doctrine* (1718, pp. 128–130), de Moivre proves that

$$R(x) = \frac{1}{A(x)} \sum_{n=0}^{k-1} x^n \sum_{i=0}^{n} (-1)^i a_i r_{n-i}, \tag{26}$$

which means that $R(x)$ may be found as the ratio of two polynomials of degree $k-1$ and k, respectively, the coefficients depending only on the $2k$ numbers defining the recursion. De Moivre adds that the terms of $R(x)$ should be "continually decreasing" for the sum to exist. This is the main result published by de Moivre in 1718; he later proved many other results on recurring series which we shall discuss in §23.1.

The sum of the first m terms of $R(x)$ may be found as the difference between $R(x)$ and the power series beginning with the $(m+1)$st term, so that

$$A(x) \sum_{n=0}^{m-1} r_n x^n = \sum_{n=0}^{k-1} x^n \sum_{i=0}^{n} (-1)^i a_i r_{n-i} - x^m \sum_{n=0}^{k-1} x^n \sum_{i=0}^{n} (-1)^i a_i r_{m+n-i}. \tag{27}$$

In 1718 de Moivre expressed the sum by means of the k values immediately before r_m instead of the k values used here, which he introduced in 1738.

From (14) and (15) it follows that $\{\beta_m(b, b)\}$ and $\{\beta_m(a, b)\}$ are recurring sequences with scales defined by

$$a_i = \frac{b}{b-i}\binom{b-i}{i} \qquad \text{and} \qquad a_i = \binom{a+b-1-i}{i},$$

respectively. Ignoring the factors p^b and q^a, the sums entering the formulae for D_n are thus partial sums of recurring series and may therefore be evaluated by means of (27), using (4) and (5) to calculate the first $[\frac{1}{2}b]$ and $[\frac{1}{2}(a + b - 1)]$ values, respectively, and the same number of terms following the last term of the sum. The result will be a ratio of two polynomials in pq.

The development leading to this end result may be summarized as follows. First, de Moivre shows that the ruin probability $r_{b+2m}(a, b)$ may be written as $p^b(pq)^m\beta_m(a, b)$, where the value of β_m may be found by means of an algorithm analogous to the one leading to the binomial coefficients. Second, he gives a formula for β_m which is too complicated to be used for tabulating β_m. Third, he gives a recursion formula for β_m so that each value is a linear function of a fixed number of previous values, which makes the calculation easy. Fourth, he derives a formula for the summation of recurring series so that only $2a$ values of r_{b+2m} have to be calculated to find the sum R_{b+2m}, $m \geq 2a$, regardless of the value of m.

De Moivre's Trigonometric Formula for $U_n(b, b)$

In a casual remark at the end of his discussion of the problem of the duration of play, de Moivre (pp. 149–150) gives a completely new solution. Without proof he states that the probability of continuation may be found as

$$U_n(b, b) = \sum_{j=1}^{[b/2]} c_j t_j^{n/2},$$ (28)

where

$$t_j = 2pq\left[1 + \cos\left\{\frac{(2j - 1)\pi}{b}\right\}\right],$$

$$c_j = \frac{\prod_{i \neq j}(1 - t_i)}{\prod_{i \neq j}(t_j - t_i)};$$

for b odd, n should be replaced by $n - 1$ on the right-hand side of (28).

He writes that he takes this solution to be "as expeditious as the nature of the Problem can admit of." Obviously, the new solution offers great theoretical and computational advantages compared with the previous ones both because of its form and its small number of terms; but the form of the solution and the lack of explanation must have baffled the reader. However, de Moivre knew that Montmort and Nicholas Bernoulli were working on the same problem, and he wanted to secure his priority by publishing the solution. He later developed the theory of recurring series and the trigonometric solution of the ruin problem in more detail; we shall give an account of his theory in §§23.1 and 23.2.

In 1718 he limited himself to present two numerical examples. In the first he showed that $U_{40}(4, 4) = 0.05085$; in the second he solved the equation $D_n(4, 4) = 100/101$ and found that n is about 60.

In addition to those references already mentioned, we have used in particular the papers by Fieller (1931), Schneider (1968), and Kohli (1975b).

20.6 PROBLEMS

1. Derive the solution of Huygens' fifth problem from Nicholas Bernoulli's formula for $R_n(a, b)$.

2. Derive Montmort's formula for $D_n(b, b)$, $p = \frac{1}{2}$, from Nicholas Bernoulli's formula for $R_n(a, b)$.

3. Use Montmort's formula for $D_n(b, b)$, $p = \frac{1}{2}$, to prove that

$$D_n(2, 2) = 1 - (\tfrac{1}{2})^{n/2}, \qquad D_n(3, 3) = 1 - (\tfrac{3}{4})^{(n-1)/2},$$

as stated by Montmort (1713, pp. 275–276) (see also Todhunter, pp. 103–104). Compare with de Moivre's proof.

4. Check Nicholas Bernoulli's result that

$$D_{108}(12, 12) < \tfrac{1}{2} < D_{110}(12, 12), \qquad p = \tfrac{1}{2}.$$

Use de Moivre's trigonometric formula to solve the same problem.

5. Find an approximate solution of the equation

$$D_n(4, 4) = \tfrac{1}{101}, \qquad p = \tfrac{1}{2},$$

by means of de Moivre's trigonometric formula (De Moivre, 1718, p. 151).

6. Calculate $D_{15}(3, 2)$ for $p = \frac{1}{2}$. Montmort (1713, p. 270) finds the result 26,606/32,768, whereas de Moivre (1718, p. 117) gets 31,171/32,768.

7. "To Find what Probability there is, that in a given number of Games, A may be winner of a certain number q of Stakes; and at some other time, B may likewise be winner of the number p of stakes, so that both circumstances may Happen" (De Moivre, 1718, Problem 41; 1712, Problem 26).

8. "To Find what Probability there is, that in a given number of Games, A may win the number q of Stakes; with this farther condition, that B, during the whole number of Games, may never have been winner of the number p of Stakes" (De Moivre, 1718, Problem 42).

9. "Supposing A and B, whose proportion of Skill is as a to b, to Play together, till A either wins the number q of Stakes, or loses the number p of them; and that B Sets at every Game the sum G to the sum L: It is required to find the Advantage, or Disadvantage of A" (De Moivre, 1718, Problem 43).

10. "If A and B, whose proportion of Skill is supposed equal, play together till Four Stakes be won or lost on either side; and that C and D, whose proportion of Skill is also supposed equal, play likewise together till Five Stakes be won or lost on either side: What is the Probability that the Play between A and B will be Ended in fewer Games than the Play between C and D?" (De Moivre, 1718, Problem 46).

11. Suppose that A has infinitely many counters and that B has b counters. Prove by recursion that the number of ruin permutations ending with a gain for A of $g_n = 2i - n < b$ equals $\binom{n}{i-b}$.
   Hint: Set $i = b + m$, $m = 0, 1, \ldots, [\frac{1}{2}(n - b)]$, and let $N(b, m, n)$ denote the number of permutations leading to ruin. Prove that

$$N(b, m, n) = \binom{n - b}{m} + \sum_{k=1}^{b-1} N(k + 1, m - 1, n - b + k - 1)$$

$$+ \sum_{k=1}^{m} N(b + k - 1, m - k, n - k - 1)$$

and $N(b, 0, n) = 1$.

12. Prove that

$$\sum_{i=0}^{k \wedge [b/2]} \alpha_i(b) \binom{b - 1 - k + i}{k - i} = \binom{2b - 1 - k}{k}, \qquad 0 \leqslant k < b.$$

Hint: Use the generating functions $(p^b - q^b)/(p - q)$ and $p^b + q^b$. A combinatorial proof has been given by Hald and Johansen (1983).

13. Compare de Moivre's recursion formula for $d_n(b, b)$ with that obtained from $d_n(a, b)$ by setting $a = b$. Use the result in Problem 12 to prove that the two formulae give the same result.

CHAPTER 21

Nicholas Bernoulli

*All this [Nicholas Bernoulli's theorem on the binomial distribution
and his solution of Waldegrave's problem] is indeed very difficult
and a great work. You are a terrific man; being ahead of you I
thought that I would not have been catched up so soon, but I now
see that I am mistaken; I am at present well behind you and forced
to use my whole aspiration to follow you at a distance.*
    —MONTMORT TO NICHOLAS BERNOULLI IN A LETTER OF
                     20 AUGUST 1713

21.1 *DE USU ARTIS CONJECTANDI IN JURE,* 1709

Nicholas Bernoulli obtained his master's degree in mathematics in 1704 and
his doctor's degree in jurisprudence in 1709 at the age of 21. His thesis is
designated as *Dissertatio Inauguralis Mathematico-Juridica* and has the title
On the Use of the Art of Conjecturing in Law. It may be considered an attempt
to carry out part of the program outlined by James Bernoulli in the last part
of *Ars Conjectandi.* As pointed out by Kohli (1975d), the thesis is greatly
influenced by the work of James Bernoulli in the *Meditationes* and *Ars
Conjectandi* and contains many statements that may be traced to these works.
Nevertheless, it also shows its author's ability to complete the analyses
initiated by James and his originality in taking up and solving new problems.
The contents is shown in the following table.

    We have used the translation by Thomas Drucker (1976, unpublished).
We shall survey the problems discussed by Nicholas Bernoulli and present
some of his solutions. A similar commentary has been given by Kohli (1975d).

    The thesis abounds with references to civil and canon law and to the
works of many jurists. In many cases Nicholas Bernoulli severely criticizes

375

Table of contents of *De Usu Artis Conjectandi in Jure*, 1709

the arguments and results of his juridical predecessors because they have not taken into account the fact that the frequency of occurrence of uncertain events may be estimated from observation, so that the value of contracts concerning such events may be assessed by means of the calculus of probability. We shall not relate the juridical discussion but keep to Bernoulli's probabilistic arguments.

The first chapter contains the definitions of the concepts of probability and expectation in accordance with the works of Huygens and James Bernoulli. Nicholas stresses the analogy between the expectation, the arithmetic mean, and the center of gravity.

The main part of the thesis is found in Chapters 2–5, which contain a discussion of the probabilities of survival for one and two persons, the median and the expected lifetime, and the calculation of life annuities based on Graunt's life table. (We have previously discussed these chapters in §§8.2, 9.1, and 9.2.)

In Chapter 6 Bernoulli solves three problems of marine and life insurance by simple calculation of expectations from known probabilities and known values of the insured.

Bernoulli defines a lottery as fair if the expected value of the prizes equals the price of a ticket; however, a small surplus must be allowed to cover expenses. If lotteries are held to provide money for public works or for charity, a larger surplus may be considered reasonable.

In Chapter 7 he takes the *Genoese lottery* as an example of an unfair lottery. It is held each year in Genoa just before the election of five members to the city council from among the 100 senators. Each player names five candidates and gets a prize of 1, 10, 300, 1500, and 10,000 crowns if it turns out that he has correctly guessed the names of one, two, three, four, and five candidates, respectively. The price for participating is 1 crown. In his analysis Bernoulli assumes that each player chooses the five names at random among the 100, so that the probability of guessing i names correctly is

$$p_i = \binom{5}{i}\binom{95}{5-i} \Big/ \binom{100}{5}, \qquad i = 0, 1, \ldots, 5.$$

Setting the corresponding prize equal to x_i, $x_0 = 0$, he further assumes that the prizes should vary inversely as the probabilities of winning, which means that $x_1 p_1 = \cdots = x_5 p_5 = c$, say, so that the equation $\sum x_i p_i = 1$ gives $c = \frac{1}{5}$. He thus finds the following prizes: 0.95; 10.9; 337; 31,700; and 15,057,504. The actual prizes are thus too small, except for the first, which is a little too large. Bernoulli also finds that the expected value of the actual prizes equals 0.58 crowns, so that the lottery has a surplus of 0.42 crowns per ticket; he therefore concludes that the Magistracy should not permit wagers of this sort. (We have expressed Bernoulli results in decimals; he naturally gives the exact fractions.)

As an example of a fair lottery, Bernoulli mentions a lottery from the Netherlands, issued in 1709, where the prizes are life annuities that may be converted to capitals equal to 11.75 times the yearly payment if the winner of the prize so desires.

In Chapter 8 a question of inheritance is discussed. Suppose that a man dies and leaves a son and a pregnant wife; what part of the inheritance can the son request? According to civil law the son will get one-fourth, since the Roman jurists reasoned that the pregnant wife may give birth to a number of children between one and five, the mean being three. Bernoulli says that it is unreasonable to use this mean; one should instead use the arithmetic mean and also take into account the possibility of a miscarriage, which means that on the average, only one child will be born. If the division of the inheritance cannot wait until the birth of the child, then it should be divided into two equal parts.

In Chapter 9 Bernoulli says that the trustworthiness of a witness should be measured empirically by the relative frequency of the number of times in which he has been found to speak the truth.

To find the probability of the innocence of an accused person, Bernoulli states that one should find the probability that any piece of evidence presented against the accused indicates his innocence and thus the probability of his innocence relative to the total evidence. For example, if each piece of evidence indicates his guilt with a probability of $\frac{1}{3}$ and his innocence with a probability of $\frac{2}{3}$, then the probability of his innocence relative to n pieces of evidence will be $(\frac{2}{3})^n$.

In view of the background for the work of James and Nicholas Bernoulli in applied probability, they obtained important and original results, however, as noted earlier they seem to be unaware of the pathbreaking paper by Halley (1694) and of the other English contributions mentioned in Chapter 12, and on that basis, there is nearly nothing new.

Todhunter's discussion of Nicholas Bernoulli's thesis is based on a summary published in the *Acta Eruditorum*, 1711, pp. 159–170, and it is therefore somewhat incomplete.

21.2 SOLUTIONS OF WALDEGRAVE'S PROBLEM BY NICHOLAS BERNOULLI, MONTMORT, AND DE MOIVRE

Let there be $n + 1$ players, $A_1, \ldots, A_{n+1}$, of equal skill. Players A_1 and A_2 play a game, and the loser pays a crown to a common stock and does not enter the play again until all the other players have played; the winner plays against A_3, and the loser pays a crown to the stock, and so on. If the winner of the first game beats all the rest, the play is finished; if not, the play goes on, each player coming in again in turn until one player has beaten in succession all the other players, and he then receives all the money in the stock.

The problem is to determine

1. the probability of each player winning the stock;
2. the expectation of each player; and
3. the probability of a given duration of the play.

In a letter to Bernoulli of 10 April 1711, Montmort writes that the problem has been proposed to him and also solved by Waldegrave for three players. Independently, de Moivre formulated and solved the problem for three players in *De Mensura Sortis* (1712). Todhunter has proposed the name *Waldegrave's problem*.

Montmort does not state how Waldegrave solved the problem. He offers a solution without proof and also gives part of the solution for four players but leaves it unfinished because it is too difficult.

In this answer Bernoulli presents the general solution of all three problems and uses his formulae to calculate the solutions for three and four players. He writes that the proof is quite long, that it is not based on the use of infinite series as is Montmort's formula, and that "I shall leave to you the pleasure of proving it yourself."

Montmort does not succeed, and in a letter of 30 December 1712, Bernoulli finally sends him the detailed proof which we are going to discuss. Bernoulli states that he values this proof more than anything else he so far has contributed to probability theory.

Bernoulli sent a Latin version of his proof to de Moivre, who writes in the preface of the *Doctrine* (1718),

> The 32d Problem [Waldegrave's problem] having in it a Mixture of the two Methods of Combinations and Infinite Series, may be proposed for a pattern of Solution, in some of the most difficult cases that may occur in the Subject of Chance, and on this occasion I must do that Justice to Mr. *Nicholas Bernoully*, the Worthy Professor of Mathematics at *Padua*, to own he had sent me the Solution of this Problem before mine was Published; which I had no sooner received, but I communicated it to the *Royal Society*, and represented it as a Performance highly to be commended: Whereupon the Society ordered that his Solution should be printed; which was accordingly done some time after in the *Philosophical Transactions*, Numb. 341, where mine was also inserted.

The paper alluded to by de Moivre is his solution for four players, which was later given in English in the *Doctrine* (1718).

Struyck (1716) bases his discussion of Waldegrave's problem on Bernoulli's solution and elaborates two of Bernoulli's formulae.

The references to the discussion on Waldegrave's problem as outlined above are as follows: Montmort (1713, pp. 318–320, 328–331, 345–346, 350–351, 366, 375, 380–387); N. Bernoulli *Phil. Trans.*, 1717, pp. 133–144; de Moivre (1712, Problem 15; *Phil. Trans.*, 1717, pp. 145–158; 1718, Problems 31–32); Struyck (1716, pp. 116–118, 206–210).

For three players it is easy to list the sequence of games for any duration of play and to find the corresponding values of the probabilities and the stock and thus to obtain the solution by summation of three infinite series. This is the method used by Montmort and de Moivre. Bernoulli writes to Montmort that it is difficult to obtain the general solution by this method, and he has therefore solved the problem by Huygens' analytical method. During his visit to London in 1712, Bernoulli discussed the problem with de Moivre, telling about his own results and pointing out the difficulties with the generalization of de Moivre's method. De Moivre reacted by publishing the solution for four players in the *Phil. Trans.* and in the *Doctrine*.

Nicholas Bernoulli's Solution

Theorem 1. Let p_i denote A_i's probability of winning the stock. The p's may be found from the equations

$$p_1 = p_2; \quad p_{i+1} = cp_i, \quad i = 2, \ldots, n, \quad c = \frac{2^n}{1 + 2^n}; \quad \sum_{i=1}^{n+1} p_i = 1. \quad (1)$$

Proof. In each game a new player enters and plays against the winner of the preceding game who may have won 0 or 1 or $\cdots n - 1$ games in succession. Assuming that he has won i games in succession, we shall denote the entering player's (conditional) probability of winning the stock by r_i, $i = 0, 1, \ldots, n - 1$, $r_0 = p_1 = p_2$.

In each game one of the players leaves the game as loser, but, except for the last game, he has the possibility of coming back and winning the stock. Assuming that the winner of the game has won i games in succession, we shall denote the loser's (conditional) probability of winning the stock by s_i, $i = 1, \ldots, n - 1$, $s_n = 0$.

These are the fundamental concepts introduced by Bernoulli who then derives three sets of linear equations among the p's, r's, and s's, eliminates r and s, and thus obtains the set of equations between the p's given in (1).

Since A_i enters the play at the $(i - 1)$st game and plays against an adversary who has won either 1 or 2 or $\cdots i - 2$ games, we have $p_3 = r_1$ and

$$p_i = \tfrac{1}{2}r_1 + (\tfrac{1}{2})^2 r_2 + \cdots + (\tfrac{1}{2})^{i-3}r_{i-3} + (\tfrac{1}{2})^{i-3}r_{i-2}, \quad i = 4, \ldots, n+1, \quad (2)$$

where the last term is the sum of two equally large probabilities because the adversary may be either A_1 or A_2.

Considering a player entering a game against an adversary, who has won i games in succession, the entering player may either lose and thus leave the adversary with $i + 1$ victories, or he may win either 1 or 2 or $\cdots n - 1$ games and thus leave the game with an adversary having just one victory, or he may win n games and thus the stock. This reasoning leads to the equation

$$r_i = \tfrac{1}{2}s_{i+1} + [(\tfrac{1}{2})^2 + \cdots + (\tfrac{1}{2})^n]s_1 + (\tfrac{1}{2})^n, \quad i = 0, 1, \ldots, n-1. \quad (3)$$

Consider next the case where a player leaves the game as loser and the winner has won $n - 1$ games in succession. The condition for the loser to enter the play again is that the winner loses to the next player, and in that case the player in question is situated as A_{n+1} at the beginning of the play; his probability of winning the stock therefore becomes

$$s_{n-1} = \tfrac{1}{2}p_{n+1}.$$

Continuing this reasoning we obtain

$$s_i = (\tfrac{1}{2})^{n-i}p_{i+2} + (\tfrac{1}{2})^{n-i-1}p_{i+3} + \cdots + \tfrac{1}{2}p_{n+1}, \qquad i = 1, \ldots, n-1. \qquad (4)$$

To solve the three sets of equations, Bernoulli first uses (3) and (4) to find that

$$r_i - r_{i+1} = \tfrac{1}{2}(s_{i+1} - s_{i+2}) = (\tfrac{1}{2})^{n-i}p_{i+3}. \qquad (5)$$

From (2) it follows that

$$2^i(p_{i+2} - p_{i+3}) = r_i - r_{i+1},$$

which inserted into (5) gives

$$p_{i+3} = \frac{2^n}{1+2^n}p_{i+2}, \qquad i = 0, 1, \ldots, n-2,$$

and thus Theorem 1. (We have only deviated from Bernoulli's method of solving the equations by amalgamating two simple steps into one.)

Summing the geometric series in (1) Struyck gives the explicit solution

$$p_1 = p_2 = \frac{1-c}{(2-c-c^n)}, \qquad (6)$$

$$p_i = p_2 c^{i-2}, \qquad i = 3, \ldots, n+1. \qquad (7)$$

Theorem 2. Let P_i denote A_i's expectation. The P's may be found from the equations: $P_1 = P_2$,

$$P_{i+1} = \frac{2^n(P_i + p_i) - np_{i+1}}{1+2^n}, \qquad i = 2, \ldots, n; \qquad (8)$$

and $\sum_{i=1}^{n+1} P_i = 0$.

Proof. Following Bernoulli, we shall denote the expectations of the player in the various states considered by the capital letters used for the probabilities in the proof above, i.e., by P, R, and S. The proof follows the same lines as the proof of Theorem 1.

From the definitions we have $P_1 = P_2 = R_0$ and $P_3 = R_1$.

Suppose that A_i in the $(i-1)$st game plays against an adversary who has won just one game. Player A_i's conditional expectation is R_1, and the stock contains $(i-2)$ crowns, which is $i-3$ crowns more than corresponding to R_1. The expected value of this surplus is $(i-3)r_1$, so that the total conditional expectation becomes $R_1 + (i-3)r_1$. Hence,

$$P_i = \tfrac{1}{2}[R_1 + (i-3)r_1] + (\tfrac{1}{2})^2[R_2 + (i-4)r_2] + \cdots$$
$$+ (\tfrac{1}{2})^{i-3}(R_{i-3} + r_{i-3}) + (\tfrac{1}{2})^{i-3}R_{i-2}, \qquad i = 4, \ldots, n+1. \qquad (9)$$

In analogy to (3) and (4), Bernoulli obtains

$$R_i = \tfrac{1}{2}(S_{i+1} - 1) + (\tfrac{1}{2})^2[S_1 - 1 + (i+1)s_1] + \cdots$$
$$+ (\tfrac{1}{2})^n[S_1 - 1 + (n+i-1)s_1] + (\tfrac{1}{2})^n(n+i), \qquad i = 0, 1, \ldots, n-1, \quad (10)$$

$$S_i = (\tfrac{1}{2})^{n-i}[P_{i+2} + (n-1)p_{i+2}] + (\tfrac{1}{2})^{n-i-1}[P_{i+3} + (n-2)p_{i+3}] + \cdots$$
$$+ \tfrac{1}{2}(P_{n+1} + ip_{n+1}), \qquad i = 1, \ldots, n-1. \qquad (11)$$

To solve these equations with respect to P_i, Bernoulli uses (10) and (3) to find

$$R_i - R_{i+1} = \tfrac{1}{2}(S_{i+1} - S_{i+2}) - p_1 + \tfrac{1}{2}s_1. \qquad (12)$$

From (11) and (4) he gets

$$\tfrac{1}{2}(S_{i+1} - S_{i+2}) = (\tfrac{1}{2})^{n-i}(P_{i+3} + np_{i+3}) - \tfrac{1}{2}s_{i+1},$$

which inserted into (12) gives

$$R_i - R_{i+1} = (\tfrac{1}{2})^{n-i}(P_{i+3} + np_{i+3}) - \tfrac{1}{2}s_{i+1} + \tfrac{1}{2}s_1 - p_1. \qquad (13)$$

From (9) it follows that

$$2^i(P_{i+2} - P_{i+3} + p_{i+2}) = R_i - R_{i+1} + r_i,$$

which inserted into (13) gives

$$2^i(P_{i+2} - P_{i+3} + p_{i+2}) - r_i = (\tfrac{1}{2})^{n-i}(P_{i+3} + np_{i+3}) - \tfrac{1}{2}s_{i+1} + \tfrac{1}{2}s_1 - p_1.$$

Using $p_1 = r_0$, it follows from (3) that

$$r_i - \tfrac{1}{2}s_{i+1} + \tfrac{1}{2}s_1 - p_1 = 0,$$

so that

$$P_{i+2} - P_{i+3} + p_{i+2} = (\tfrac{1}{2})^n (P_{i+3} + n p_{i+3}),$$

which proves (8) and thus Theorem 2.

Struyck derives an explicit formula for P_i from Bernoulli's Theorem 2. Setting

$$a = 2^n, \qquad b = 1 + 2^n, \qquad c = \frac{a}{b}, \qquad e = \left(1 - \frac{n}{b}\right) c p_2,$$

Struyck finds from (7) and (8) that

$$P_{i+3} = c P_{i+2} + c^i e,$$

which leads to

$$P_{i+3} = c^{i+1} P_2 + (i+1) c^i e, \qquad i = 0, 1, \ldots, n-2. \tag{14}$$

Using the fact that the sum of the $n + 1$ P's equals zero, it follows that

$$P_2 = \frac{p_2 e(n a^{n-1} + a^n - b^n)}{b^{n-2}}. \tag{15}$$

Inserting this result and the expression (6) for p_2 into (14), Struyck finds P_{i+3} in terms of a, b, and n.

Theorem 3. Let the probability of a duration of play of exactly x games and at most x games be denoted by z_x and Z_x, respectively. Then,

$$z_1 = \cdots = z_{n-1} = 0, \qquad z_n = (\tfrac{1}{2})^{n-1},$$
$$z_x = \tfrac{1}{2} z_{x-1} + (\tfrac{1}{2})^2 z_{x-2} + \cdots + (\tfrac{1}{2})^{n-1} z_{x-n+1}, \qquad x = n+1, n+2, \ldots. \tag{16}$$

Furthermore,

$$z_{n+m-1} = \left(\frac{1}{2}\right)^n + \sum_{i=1}^{[(m-1)/n]} \frac{(-1)^i}{i!} (m - in + i - 2)^{(i-1)}$$
$$\times (m - in + 2i - 1) \left(\frac{1}{2}\right)^{(i+1)n}, \qquad m \geqslant 2, \tag{17}$$

$$Z_{n+m-1} = \sum_{i=0}^{[(m-1)/n]} \frac{(-1)^i}{(i+1)!} (m - in + i - 1)^{(i)}$$

$$\times (m - in + 2i + 1)\left(\frac{1}{2}\right)^{(i+1)n}, \qquad m \geqslant 2. \tag{18}$$

Proof. Obviously, the probability of winning the stock at the nth game equals $2(\frac{1}{2})^n$, since the first or the second player may win the stock in n games. Bernoulli shows that the following values of z_x may be found by the recursion (16).

A duration of x games requires that the winner of the stock, W, say, enters the play just after the $(x - n)$th game and wins the next n games.

The adversary of W, that is, the winner of the $(x - n)$th game, may have won 1 or 2 or $\cdots n - 1$ games, but no more, because otherwise the play would have been finished. Bernoulli proves (16) by conditioning on each of these events. A duration of $x - 1$ games requires the winner of the $(x - n)$th game to have won only one game and thereafter to win $n - 1$ games in succession. *After* having won the $(x - n + 1)$st game, W's probability of winning the stock is the same as the probability that an adversary having won one game wins the stock. Hence, *before* W has won the $(x - n + 1)$st game, his probability of winning the stock is $\frac{1}{2}z_{x-1}$. Similarly, a duration of $x - 2$ games requires the winner of the $(x - n)$th game to have won exactly two games and, thereafter to win $n - 2$ games in succession. *After* having won the $(x - n + 1)$st game and the next, W's probability of winning the stock is the same as the probability that an adversary having won two games wins the stock. Hence, *before* entering the $(x - n + 1)$st game, W's probability of winning the stock is $(\frac{1}{2})^2 z_{x-2}$. Obviously, the following terms may be derived in the same way.

Bernoulli uses the recursion to calculate the probability distribution of the duration of play for five players. It is remarkable, however, that he also derives an explicit solution of the difference equation. He does not give a complete proof, only an essential hint of the solution in these words: "One can easily find the proof of these formulae [(17) and (18)] supposing that the numerator of each term of the series $z_n, z_{n+1}, \ldots$, equals the sum of all the preceding ones minus those that have been taken too much so that only $n - 1$ terms are included." Following this advice we shall reconstruct Bernoulli's proof.

Setting

$$z_x = (\tfrac{1}{2})^{x-1} a_x, \qquad x = 1, 2, \ldots, \tag{19}$$

(16) is transformed to the simpler equation

$$a_x = a_{x-1} + a_{x-2} + \cdots + a_{x-n+1}, \qquad x = n+1, n+2, \ldots, \tag{20}$$

with the starting values $a_1 = \cdots = a_{n-1} = 0$ and $a_n = 1$.

Writing (20) as

$$a_x = \sum_{i=1}^{x-1} a_i - \sum_{i=1}^{x-n} a_i,$$

and setting $x = rn + k; r = 1, 2, \ldots, k = 0, 1, \ldots, n-1$, we obtain

$$a_{rn+k} = 2a_{rn+k-1} - a_{(r-1)n+k}, \tag{21}$$

with the exception that for $r = 1$, the formula does not hold for $k = 0$.

If follows that

$$a_{n+k} = 2a_{n+k-1}, \qquad k = 2, \ldots, n-1,$$

so that

$$a_{n+k} = 2^{k-1}, \qquad k = 1, \ldots, n-1.$$

Similarly, from

$$a_{2n+k} = 2a_{2n+k-1} - a_{n+k},$$

we get

$$a_{2n+k} = 2^{n+k-1} - 2^{k-1}(k+2).$$

From these results and from (21) we see that a_{3n+k} is obtained from a_{2n+k} by substituting $n+k$ for k and adding a term dependent on k only, so that

$$a_{3n+k} = 2^{2n+k-1} - 2^{n+k-1}(n+k+2) + 2^{k-1}b_k,$$

where b_k satisfies the difference equation

$$b_k = b_{k-1} + (k+2).$$

Hence,

$$b_k = \frac{(k+1)(k+4)}{2!}.$$

Proceeding in this way we obtain

$$a_{rn+k}(\tfrac{1}{2})^{k-1} = 2^{(r-1)n} - [(r-2)n + k + 2]2^{(r-2)n}$$

$$+ \frac{1}{2!}[\{(r-3)n + k + 1\}\{(r-3)n + k + 4\}]2^{(r-3)n}$$

$$- \cdots$$

$$+ \frac{(-1)^{r-1}}{(r-1)!}(k+1)(k+2)\cdots(k+r-2)(k+2r-2),$$

which inserted into (19) gives z_{rn+k}. To turn this result into the form used by Bernoulli, we set $rn + k = n + m - 1$, which leads to (17). The probability of a duration of at most $n + m - 1$ games is found by summation.

De Moivre's Solution

De Moivre assumes that each player pays a crown to the common stock before the play begins and that he pays a fine of f crowns to the stock each time he loses a game. This means that all the results of de Moivre involving the size of the stock and the fines may be transformed to the results of Montmort and Bernoulli by a change of origin and scale.

De Moivre does not attempt to find a formula for the probability of a given duration of play.

He denotes the players by A, B, C, and D, and we shall use this notation in addition to the previous one.

In all of de Moivre's proofs he assumes that B beats A in the first game. We shall denote A_i's probability of winning the stock under this assumption

Table 21.2.1. Outcomes and Probabilities for a Sequence of Games with Three Players

Duration	Winner	Probability	Stock	Outcome of the sequence of games for the given duration
1	B		$3 + f$	BA[a]
2	B	$\frac{1}{2}$	$3 + 2f$	BA, BC
3	C	$(\frac{1}{2})^2$	$3 + 3f$	BA, CB, CA
4	A	$(\frac{1}{2})^3$	$3 + 4f$	BA, CB, AC, AB
5	B	$(\frac{1}{2})^4$	$3 + 5f$	BA, CB, AC, BA, BC
6	C	$(\frac{1}{2})^5$	$3 + 6f$	BA, CB, AC, BA, CB, CA
7	A	$(\frac{1}{2})^6$	$3 + 7f$	BA, CB, AC, BA, CB, AC, AB

[a]BA denotes that B beats A in a game.

by p_i, so that the unconditional probabilities of winning become

$$\bar{p}_1 = \bar{p}_2 = \tfrac{1}{2}(p_1 + p_2)$$

for A_1 and A_2, whereas the following probabilities are unchanged, since the probability of winning for these players does not depend on whether A_1 or A_2 wins the first game. (The reader should note the change of notation compared with the notation used in the proof of Bernoulli's theorems.)

De Moivre's procedure for finding the probability of winning the stock in a given number of games has been summarized in Table 21.2.1.

Because of the rules for reentering the play, each player has the possibility of winning at every third game, for example, for B at games $2, 5, 8, \ldots$, and the probabilities of winning constitute a geometric series.

Summing the probabilities of winning, de Moivre obtains

$$p_2 = \tfrac{4}{7}, \qquad p_3 = \tfrac{2}{7}, \qquad p_1 = \tfrac{1}{7},$$

so that the unconditional probabilities become $\bar{p}_1 = \bar{p}_2 = 5/14$.

The expectation of B is found as

$$e_2 = \sum_{i=0}^{\infty} \left(\frac{1}{2}\right)^{1+3i} [3 + (2 + 3i)f] = \frac{1}{2} \sum_{i=0}^{\infty} \left(\frac{1}{8}\right)^{i} (3 + 2f + 3fi)$$

$$= \frac{12}{7} + \frac{68}{49} f.$$

Similarly,

$$e_3 = \frac{6}{7} + \frac{48}{49} f, \qquad e_1 = \frac{3}{7} + \frac{31}{49} f,$$

so that the unconditional expectations are $\bar{e}_1 = \bar{e}_2 = 15/14 + (99/98)f$.

From these expectations, de Moivre deducts the expected values of the fines. Looking at Table 21.2.1, it will be seen that B has a probability of $\frac{1}{2}$ of being fined (losing) in the second game, a probability of 0 of being fined in the third game, a probability of $(\frac{1}{2})^3$ of being in the fourth game, and so on. These probabilities form a geometric series with every third term missing, so that the total probability becomes

$$\sum_{i=1}^{\infty} \left(\frac{1}{2}\right)^{i} - \frac{1}{4} \sum_{i=0}^{\infty} \left(\frac{1}{8}\right)^{i} = \frac{5}{7}.$$

Hence, B's expected gain is

$$g_2 = e_2 - \frac{5}{7}f - 1 = \frac{5}{7} + \frac{33}{49}f.$$

Similar calculations for C and A give

$$g_3 = -\frac{1}{7} + \frac{6}{49}f, \qquad g_1 = -\frac{4}{7} - \frac{39}{49}f,$$

so that the unconditional expected gains are

$$\bar{g}_1 = \bar{g}_2 = \frac{1}{14} - \frac{3}{49}f.$$

De Moivre states that the same reasoning will lead to the solution if the skills of the players are in a given proportion, $a:b:c$, say. In the *Doctrine* (1718) he indicates the proof. From Table 21.2.1 it will be seen that the sequence CB, AC, BA occurs at the fifth, sixth and seventh game, so that the probability of winning at these games equals the probability of winning at the second, third, and fourth games times the probability that the sequence in question occurs. Since this property is repeated for each of the following sets of three games, B's probability of winning becomes of sum of the terms of a geometric series with the ratio

$$\Pr\{CB\}\Pr\{AC\}\Pr\{BA\} = \frac{c}{c+b}\frac{a}{a+c}\frac{b}{b+a} = m,$$

say, so that

$$p_2 = \frac{b}{b+c}(1 + m + m^2 + \cdots) = \frac{b}{b+c}\frac{1}{1-m},$$

which shows how to modify the previous solution.

To discuss de Moivre's solution for four players, we define $u_{r,x}$ as the probability that A_r wins the stock at the xth game assuming that A_2 beats A_1 in the first game, $r = 1,\ldots,4, x = 1,2,\ldots, u_{r,1} = u_{r,2} = 0$. The probability that A_r wins the stock then becomes

$$p_r = \sum_{x=1}^{\infty} u_{r,x}.$$

De Moivre's procedure has been summarized in Table 21.2.2.

Table 21.2.2. Outcomes and Probabilities for a Sequence of Games with Four Players

x	$u_{2,x}$	$u_{3,x}$	$u_{4,x}$	$u_{1,x}$	Stock	Sequence of outcomes
3	1/4	0	0	0	$4 + 3f$	BA, BC, BD
4	0	1/8	0	0	$4 + 4f$	BA, CB, CD, CA
5	0	0	2/16	0	$4 + 5f$	BA, CB, DC, DA, DB
						BA, BC, DB, DA, DC
6	0	0	0	3/32	$4 + 6f$	BA, CB, DC, AD, AB, AC
						BA, CB, CD, AC, AB, AD
						BA, BC, DB, AD, AC, AB
7	3/64	2/64	0	0	$4 + 7f$	To be filled in by the reader
8	3/128	3/128	2/128	0	$4 + 8f$	
9	0	3/256	6/256	4/256	$4 + 9f$	
10	4/512	2/512	6/512	9/512	$4 + 10f$	

De Moivre continues the table of $u_{r,x}$ to $x = 12$, and by studying the pattern of numbers he derives the following set of recursion formulae for $x \geqslant 3$:

$$u_{2,x} = \tfrac{1}{2}u_{1,x-1} + \tfrac{1}{4}u_{1,x-2}, \qquad u_{3,x} = \tfrac{1}{2}u_{2,x-1} + \tfrac{1}{4}u_{4,x-2},$$
$$u_{4,x} = \tfrac{1}{2}u_{3,x-1} + \tfrac{1}{4}u_{2,x-2}, \qquad u_{1,x} = \tfrac{1}{2}u_{4,x-1} + \tfrac{1}{4}u_{3,x-2}. \tag{22}$$

These equations are the key to the solution; the remaining part of the proof consists of summations leading to four linear equations between the probabilities of winning and the expected durations of play, respectively, which are then solved numerically.

De Moivre relates a simple proof of the recursion formulae communicated to him by Brook Taylor.

Summing (22) over x, de Moivre finds the four equations

$$p_2 = \tfrac{1}{4} + \tfrac{3}{4}p_1, \qquad p_3 = \tfrac{1}{2}p_2 + \tfrac{1}{4}p_4,$$
$$p_4 = \tfrac{1}{2}p_3 + \tfrac{1}{4}p_2, \qquad p_1 = \tfrac{1}{2}p_4 + \tfrac{1}{4}p_3,$$

which have the solution

$$p_2 = \frac{56}{149}, \qquad p_3 = \frac{36}{149}, \qquad p_4 = \frac{32}{149}, \qquad p_1 = \frac{25}{149}.$$

The expected value of the stock for A_r is

$$e_r = \sum_{x=1}^{\infty} u_{r,x}(4 + fx) = 4p_r + fv_r,$$

say, where $v_r = \sum u_{r,x} x$ is the expected duration of play. Inserting (22) in v_r, four linear equations between the v's are obtained, which de Moivre solves numerically.

De Moivre analyzes the probability of being fined in the same manner as in Table 21.2.2 and finds that the same recursion formulae hold as for the u's, but with other starting values. Using the same procedure as above, he then calculates the expected fine for each player.

Finally, he finds the expected gain as $e_r - 1$ minus the expected fine. For A_1 and A_2 the average gain is calculated just as for three players. The results are

$$\bar{g}_1 = \bar{g}_2 = \frac{13}{149} - \frac{2700}{22{,}201} f, \qquad g_3 = -\frac{5}{149} + \frac{1176}{22{,}2201} f,$$

$$g_4 = -\frac{21}{149} + \frac{4224}{22{,}201} f.$$

De Moivre states that this method of proof may be used for any number of players, and he gives the recursion formulae for six players. The method is theoretically simple but in practice very cumbersome for more than three players. Furthermore, a new set of equations has to be derived and solved for each n so that the method does not show how the solution depends on the number of players; it is therefore unsatisfactory as a general solution of the problem.

Bernoulli's solution is not mentioned in the text, only in the preface as quoted at the beginning of the present section.

In the second edition of the *Doctrine* (1738) de Moivre reprints the proof for three players from the first edition and gives a slightly shortened version of the proof for four players.

Presumably, the unsatisfactory solution kept nagging him, for in the third edition of the *Doctrine* (1756, p. 151), after having reprinted the exposition in the second edition, he adds a *Remark* of nine pages in which he makes use of Bernoulli's Theorem 1 to obtain a simplification of his own solution, and at the same time he extends the problem as indicated by the following introductory remark: "As the Application of the Doctrine contained in these Solutions and Corollaries may appear difficult when the Gamesters are many, and when it is required to put an end to the play by a fair distribution of the money in the *Poule* [stock]; which I look upon as the most useful Question concerning this Game: I shall explain this Subject a little more particularly."

This means that he takes up the problem of an equitable division of the stock when the play is stopped before a player has won n games in succession. This problem is of course considerably more complicated than the original one, and its solution requires sophisticated reasoning with conditional probabilities and expectations. As in the solution of the original problem, de Moivre uses recursion, which leads to a new set of linear equations for each value of n. He tabulates the solution for three, four, five, and six players.

The only new method employed is to be found in his proof of Bernoulli's first theorem, which runs as follows.

The probability distribution $\bar{p}_1, \bar{p}_2, p_3, \ldots, p_{n+1}$, whose sum is 1, gives the "value" of the place for each of the $n + 1$ players before the play begins. The greater value of the place of A_1 (or A_2) compared to the value of the other places depends on the chance that A_1 has of beating all the other players in the first round. This chance equals $(\frac{1}{2})^n$, and the other players should therefore pay A_1 the amount

$$(\tfrac{1}{2})^n(p_3 + \cdots + p_{n+1})$$

to give up his priority and move to the last place. Since A_1 and A_2 have the same chance of winning, A_2 should not pay anything. If A_1 is moved to the last place, all the other players move up one place. Since A_{i+1} has paid $(\frac{1}{2})^n p_{i+1}$ for this change, we have

$$\bar{p}_2 = p_3 + (\tfrac{1}{2})^n p_3, \qquad p_i = p_{i+1} + (\tfrac{1}{2})^n p_{i+1}, \qquad i = 3, \ldots, n,$$

from which Bernoulli's Theorem 1 follows. Using equilibrium considerations, de Moivre thus obtains a simpler proof than the one given by Bernoulli.

In an Appendix to the *Doctrine* (1756, p. 332), de Moivre mentions Bernoulli's result on the duration of play. It is odd that he does not discuss this aspect of the problem already in 1718 when he used so much space to discuss the duration of play in the ruin problem.

Laplace's Solution

Laplace (1812, Book 2, §11) begins by deriving the difference equation (16) in nearly the same way as Bernoulli and then uses his own method of generating functions to find the solution. The generating function of z_x becomes

$$g(t) = \sum_{x=1}^{\infty} z_x t^x = \frac{(\tfrac{1}{2})^n t^n (2 - t)}{(1 - t)(1 - t - (\tfrac{1}{2})^n t^n)},$$

and the generating function of Z_x is $g(t)/(1 - t)$. The two probabilities z_x and Z_x are then found as the coefficients of t^x in the expansions of the generating functions. In this way he derives Bernoulli's results (17) and (18). His proof is neither shorter nor simpler than Bernoulli's, but it has the advantage of being based on a general method for solving difference equations and thus not requiring ad hoc methods. (Note that Laplace's formula for z_x contains an error; the numerator of the second term should be $x - 2n + 2$.)

De Morgan (1837, §52) complains of the incompleteness of Laplace's derivation of the generating function of Z_x and also provides some details for the derivation of the coefficient of t^x. The account given by Todhunter (pp. 535–539) is based on de Morgan's comments; Todhunter does not discuss the solutions of Bernoulli and de Moivre.

Next, generalizing de Moivre's formula (22) to $n + 1$ players, Laplace derives a recursion formula for $u_{r,x}$ which he transforms to the partial difference equation

$$u_{r,x} - u_{r-1,x-1} + (\tfrac{1}{2})^n u_{r,x-n} = 0.$$

He derives the generating function of $u_{r,x}$, which turns out to be rather complicated, and presumably for that reason he does not give an explicit formula for $u_{r,x}$. He then turns to the simpler problems of the probability of A_r winning the stock and of A_r's expectation and solves these problems by means of the generating function for $u_{r,x}$. His solution agrees with the one found by Bernoulli and Struyck, but his proof is more complicated.

It is understandable that Nicholas Bernoulli rated his own proof of Waldegrave's problem highly. The problem is complicated, and its solution requires sophisticated reasoning with conditional probabilities, proving the first two theorems by means of Huygens' analytical method and the third by solving a difference equation. Bernoulli's proof is not only the first general proof, it is also mathematically simpler than the other proofs mentioned above.

21.3   A SURVEY OF NICHOLAS BERNOULLI'S CONTRIBUTIONS

Since Bernoulli's contributions are scattered throughout Montmort's *Essay* and in the present book as well, we shall now list in chronological order his most important contributions.

In his thesis in 1709 he explained how to use Graunt's life table to solve many legal problems hinging on the lifetime of one or more persons, such as problems of inheritance and life insurance. He derived the expectation of the largest observation among n observations from a uniformly distributed population. He also demonstrated the importance of using probability theory for the evaluation of lotteries and testimony. These problems have been discussed in §§8.2, 9.1, 9.2, and 21.1.

In 1710 he proved Montmort's two formulae for the probability of at least one coincidence among n objects, and together with Montmort he later generalized this result to the case of s groups of n objects each; see §19.3.

In 1711 he solved the ruin problem by giving the probability of a player being ruined in at most n games, each game being a Bernoulli trial with a gain of one unit or a loss of one unit, and as a special case he gave the solution of Huygens' fifth problem on the Gambler's Ruin; see §20.3.

In 1711 Bernoulli also gave the solution of Waldegrave's problem on the probability of winning a tournament, the expected gain, and the probability of a given duration of the tournament. He communicated the proof to Montmort in 1712; see §21.2.

In 1712 he contributed to the discussion on the problem of points in the game of tennis, see §18.5, and to the discussion on the game *Her*, see §18.6.

In 1713 Nicholas improved James Bernoulli's proof of the law of large numbers and derived a simple approximation to the binomial distribution, which he used for testing the significance of the variations of the sex ratio at birth; see §§16.3 and 17.3.

In 1713 he also formulated five problems for solution, among them the Petersburg problem; see §18.4.

In 1713 Montmort openly admitted that Bernoulli was far ahead of him in probability theory; a comparison of the *Essay* (1713) and the *De Mensura Sortis* (1712) shows that Bernoulli was ahead also of de Moivre. Why, then, did Bernoulli at the age of 26 suddenly disappear from the scene of probability theory?

There may be several reasons. First, he obviously missed the inspiration from Montmort, who had become absorbed in a project of writing a history of mathematics. Bernoulli tried to keep up a correspondence on the Petersburg problem, but Montmort was no longer interested, as is evident from their letters from 1714 to 1716, published by Spiess (1975). Second, Bernoulli did not have the abundance of time that he had during his period of study and travel from 1709 to 1713. Returing to Basle in 1713, he had to look for a Job, and in 1716 he became professor of mathematics in Padua. His stay there does not seem to have been satisfactory, for he returned to Basle and became professor of logic in 1719, later becoming professor of law. These jobs may naturally have taken much of his energy, and even though he kept up a correspondence on mathematics, he never again succeeded in making essential contributions to probability theory.

Hence, after the death of Montmort at an early age in 1719 and the engagement of Nicholas Bernoulli in other matters, there was left only de Moivre of the three pioneers to carry on research in probability theory, which he did by producing three editions of the *Doctrine of Chances*.

21.4   A NOTE ON NICOLAAS STRUYCK

Nicolaas Struyck (1687–1769) has born into a burgher family in Amsterdam. He got a good education in classical subjects as well as in mathematics and the natural sciences. Throughout his adult life he earned his living as a teacher of mathematics, cosmography, astronomy, and accountancy and as an author of many books.

The many references to books and papers in foreign journals in his works show that he read profusely and that he carried on a vast correspondence with colleagues in many countries. He was elected a member of the Royal Society of London and the Paris Academy of Sciences.

He wrote about probability theory, life annuities, population statistics, astronomy, geography, and accountancy. Nearly all his publications were written in Dutch, and for that reason his work did not get the influence it deserved. His treatises about probability theory, life annuities, and population statistics have been translated into French by J. A. Vollgraff and published by La Société Générale Néerlandaise d'Assurances sur la Vie et de Rentes Viagères in 1912.

His first book was entitled *Calculation of the Chances in Play, by means of Arithmetic and Algebra, together with a Treatise on Lotteries and Interest* (1716). It shows that he masters probability theory as given in the works of Huygens, Montmort, James and Nicholas Bernoulli and de Moivre to whom he refers. The problems treated are formulated as games of chance and solved by means of combinatorics, the method of inclusion and exclusion, infinite series, and difference equations. For his solution of Huygens' five problems we refer to §14.2, for the problem of coincidences to §19.3, and for Waldegrave's problem to §21.2. It will be seen that he gives slight generalizations and improvements of previous proofs and results. He also mentions de Moivre's Poisson approximation, and he used Montmort's idea of randomization of the number of trials as shown in §14.4. Furthermore, he discusses the binomial and the multinomial distributions and uses them to solve many problems including the problem of points. He stresses the important distinction between drawings with and without replacements, illustrating this by his solution of Huygens' second problem and his discussion of the Lottery of Lorraine. He solves the occupancy problem in the same way as de Moivre. Without proof he gives Montmort's formula for the distribution of the sum of points at dicing and shows that this formula agrees with the one given without proof by de Moivre, see the two forms of formula (14.3.1); like Montmort he tabulates the distribution for 2, 3, ..., 9 ordinary dice. Finally he discusses the duration of play using Nicholas Bernoulli's formula for the ruin probability and de Moivre's algorithm. He illustrates all the theorems by many examples. He does not mention James and Nicholas

Bernoulli's theorems on the binomial distribution, nor does he take up problems outside the field of games of chance.

Struyck's book in interesting from a historical point of view because it shows that a person with a good mathematical education in 1716 was able to fully understand, apply, and slightly improve the fundamental results obtained by the four pioneers between 1708 and 1713. It is the only mathematical work written by Struyck. His book has not been given much attention outside Holland, neither by his contemporaries nor in the historical literature; it is not mentioned by Todhunter.

Struyck's treatises on life annuities and population statistics are to be found in two of his books published in 1740 and 1753, which also treat of many other topics.

In his treatises on annuities he refers to de Witt and Halley but not to de Moivre (1725). He points out that the value of annuities should be calculated from life tables based on observations (as done by Halley) and not from hypotheses (as done by de Witt). However, he considers the construction of Halley's table as unsatisfactory because Halley had access to the number of deaths only and not to the corresponding number of living and because it rests on population data. His first problem was, therefore, to provide a reliable life table for annuitants. He realizes, as Kersseboom and Deparcieux about the same time and de Witt and Hudde before them, that such a table may be constructed from the registers of annuitants. His observations comprise 794 male and 876 female annuitants who bought their annuities in Amsterdam in 1672–1674 and 1686–1689, most of them being dead before 1738. For each five-year group, 0–4 years, 5–9 years, etc., he tabulates the number of annuitants entering at a give age and the number of survivors at any later age. Assuming that mortality at a given age does not change over time, he sums the numbers exposed to risk and the numbers of deaths for each age group and calculates the rates of mortality from which he derives the life table l_x for $x = 5, 6, \ldots, 94$. He does not explain how he has graduated the raw mortality rates and how he has interpolated to l_x for each year of age. He stresses that the mortality of females is smaller than that of males and contends that he is the first to calculate life tables for males and females separately.

Using Halley's formula and his own life table Struyck tabulates the value of annuities for every fifth year of age for an interest rate of $2\frac{1}{2}\%$ and finds that there is good argeement with the average values found from his observations. Like Halley, he points out that the government is selling annuities too cheaply. He calculates the value of annuities based on Halley's table and finds that these values for ages between 15 and 55 years are a little larger than the average value for males and females found from his own tables. He also gives examples of increasing, decreasing, and deferred annuities. These results are given in his 1740 treatise.

In 1753 he supplements the life table from 1740 with values of l_x for $x = \frac{1}{2}, 1, 1\frac{1}{2}, 2, \ldots, 5$, which he has obtained by following 850 newborn boys and 822 newborn girls in the village Broek-in Waterland from 1657 to 1738. He shows that the corresponding values of annuities are increasing with age because of the large infant mortality.

Finally he discusses Deparcieux's results for French annuitants compared with his own, and he investigates some tontines and widow pension funds.

With regard to Struyck's works on population statistics we refer to the *Memoir* published by Algemeene Maatschappij van Levensverzekering (1898, pp. 85–99) and to K. Pearson (1978). We shall, however, briefly mention his method of analysis of the sex ratio at birth.

He supplements the data from London with data from many other regions as demanded by Graunt. For London he finds 783,145 male and 740,113 female births in the period 1629–1744. He uses the ratio of these numbers as a standard which he applies to observations from other regions to find the expected number of female births from the observed number of male births and thus the deviation between the expected and the observed number of female births. For example, in Amsterdam for the years 1700–1739 he finds 14,655 male births, which gives an expected number of female births of 13,850 compared with the observed number 13,844. He finds similar small deviations for several other cities and regions, the only exception being five Dutch villages with approximately the same number of male and female births. He does not give a probabilistic evaluation of the deviations as done by Nicholas Bernoulli.

Despite the larger number of male births Struyck finds in most places a larger number of females than males in the whole population. He mentions that this fact has been explained by Arbuthnott, Nieuwentyt, and Derham by the larger mortality of males due to their more dangerous occupations and participation in wars. Struyck argues, however, that this explanation is incomplete because the number of females surpasses the number of males already from the age of 10.

Among political arithmeticians Struyck stands out as critical and sober-minded. He has a skeptical attitude about the accuracy of data collected by others, and to provide reliable data he arranges for a census of several Dutch villages under his own supervision. He criticizes the existing birth and death registers in several countries and recommends that regular and accurate censuses be carried out. He warns against uncritical use of statistical ratios and stresses that the crude birth rate depends on the composition of the population according to age, sex, and marital status. In most cases he tries to explain facts without involving divine providence. He does not use the binomial or other distributions in his analysis of population data; probability theory and political arithmetic seem to be two different worlds.

CHAPTER 22

De Moivre and the *Doctrine of Chances,* 1718, 1738, and 1756

22.1 THE LIFE OF DE MOIVRE

The following account in based mainly on the fundamental paper by Schneider (1968), who has used the biography by Maty (1755) and the letters, papers, and books by de Moivre.

Abraham de Moivre (1667–1754) was born into a Protestant family in Vitry, France. His father was a surgeon, and the family did not belong to the French nobility; their name was Moivre, the "de" was added to the name by the young Moivre himself when he emigrated from France to England. Moivre received a good education in the humanities in the colleges of Sedan and Saumur and also studied some mathematics on his own; it is stated that he read Huygens' treatise at the age of 16. From 1684 onward he studied mathematics in Paris. When the persecution of the French Protestants (Huguenots) was intensified after the revocation of the Edict of Nantes in 1685, the 18-year-old Moivre was interned in a priory in an attempt to persuade him to change his religion. After three years he was released and immediately left France to seek asylum in England, like thousands of other Huguenots before him. It is not known what became of his family. In London presumably he was first supported by other refugees until he established himself as a visiting tutor of mathematics to the sons of wealthy citizens.

During these first years of hard work assimilating the customs of a foreign country, it must have been exciting for him to read the newly published *Principia* (1687) by Newton, containing not only a new natural philosophy but also the beginnings of a new mathematics. He succeeded in mastering Newton's work, and in 1695 he published his first paper on the method of fluxions and some of its applications in the *Phil. Trans.,* communicated by

Halley, whom he had gotten to know in 1692. Through Halley he was introduced to the circle of scientists around Newton, and in 1697 de Moivre was elected a Fellow of the Royal Society.

He published his papers, 15 in all, in the *Phil. Trans.* between 1695 and 1746. We have previously mentioned the two papers on probability theory, the *De Mensura Sortis* and the paper on Waldegrave's problem. Another paper is on insurance mathematics, and the remaining papers are on pure mathematics or mathematics applied to physical and astronomical problems.

Like other mathematicians of the time, de Moivre worked on the solution of algebraic equations and on infinite series. In particular he showed that the n roots of the equation $z^n = 1$ are equally spaced on the unit circle, and he found the fundamental formula which today bears his name,

$$(\cos x + i \sin x)^n = \cos nx + i \sin nx,$$

although he never wrote it in this form. Another contribution to the theory of equations is his method of reduction of reciprocal equations.

In 1698 he published a paper, "A Method of Raising an infinite Multinomial to any given Power, or Extracting any given Root of the same", in which he gave the rule for finding the coefficient of z^{m+i} in the expansion of $(az + bz^2 + cz^3 \cdots)^m$. The following year he wrote "A Method of extracting the Root of an Infinite Equation", in which he solved the equation

$$az + bz^2 + cz^3 + \cdots = gy + hy^2 + iy^3 + \cdots,$$

expressing z as a power series in y. To find the coefficients of y he wrote z as a power series in y with undetermined coefficients, and inserting this series in the given equation he determined the coefficients successively by comparing the coefficients of the same power of y.

Besides the papers he wrote several outstanding books. In 1718 he published *The Doctrine of Chances: or, A Method of Calculating the Probability of Events in Play*, which he considerably improved in the second (1738) and third (1756) editions. In 1725 he published the *Annuities upon Lives*, which appeared in expanded editions in 1743, 1750, and 1752; in 1756 it was incorporated into the *Doctrine of Chances*. These books contain many original contributions to probability theory and insurance mathematics, and they were for many years the best textbooks in these fields.

In 1730 he published the *Miscellanea Analytica de Seriebus et Quadraturis* summarizing much of his previous work on mathematics with improved versions of some proofs and also giving important new results. Among the new contributions are the proofs of his theorems on recurring series, his approximation to $\log m!$ (Stirling's formula), and his proof by means of a generating function of the probability of throwing a given number of points with any given number of dice. In 1733 he derived the normal approximation to the binomial distribution. All these results were incorporated in the second edition of the *Doctrine*. We shall return to these topics in the following sections.

De Moivre's work on probability theory and insurance mathematics are purely deductive; he did not collect or analyze observations to demonstrate the applicability of his theories.

Schneider (1968) has given an annotated list of de Moivre's letters; his extensive correspondence with John Bernoulli has been published with commentaries by Wollenschläger (1933).

In 1703 the physician and mathematician G. Cheyne published a book on the inverse method of fluxions (integration) in which he gave a rather unsatisfactory exposition of Newton's theory. He also mentioned de Moivre's theorem on the expansion of the multinomial without giving de Moivre credit for this result. In a book published in 1704, de Moivre protested and gave a critical evaluation of Cheyne's book. Cheyne responded in 1705 by publishing a pamphlet with corrections, some of which he ascribed to John Bernoulli, and a severe personal attack on de Moivre. K. Pearson (1978, p. 143) writes,

De Moivre allowed the controversy to drop, for it had passed from the field of mathematics to that of personalities. Somewhere in the twenties (5th Edn 1725) Cheyne published his essay on health, and the effect of a vegetable and milk diet on the truculent medicine man was obvious: "The defense of that book (the *Methodus fluxionum inversa*) against the learned and acute Mr Abraham de Moivre being written in a spirit of levity and resentment, I most sincerely retract, and wish undone, so far as it is personal or peevish, and ask him and the world pardon for it."

The controversy with Cheyne gave de Moivre the opportunity to begin a correspondence with John Bernoulli by sending him a copy of the *Animadversiones* (1704). The correspondence continued until 1714, when the priority dispute between Newton and Leibniz placed de Moivre and Bernoulli in opposite camps.

De Moivre's occupation as a wandering private tutor did not leave him much time for study and research. In 1707, at the age of 40, he had published six papers in the *Phil. Trans.*, and with that background he began to look for a university post. In letters to Bernoulli in 1707 and 1708 he explained his latest research and his working conditions and asked Bernoulli to intercede with Leibniz to secure a university appointment for him somewhere on the Continent, but nothing came of it.

His luck was no better in England. When Newton retired as professor of mathematics at Cambridge he was succeeded by William Whiston, who like Newton was an Arian but unlike Newton spoke and wrote openly about his religious beliefs, with the result that he was expelled from the university in 1710. The chair went to the 29-year-old Nicholas Saunderson in 1711 and not to de Moivre, who was a Huguenot and perhaps still considered to be a foreigner, deeply rooted in French culture as he was.

Saunderson died in 1739 without having published anything essential; his lectures on algebra were published by his friends in 1740. Stigler (1983) has suggested that Saunderson may be the one who discovered Bayes' theorem and communicated it to the English psychologist David Hartley, who quoted a version of the theorem in his book from 1749. However, Edwards (1986) contests that the quotation refers to inverse probability and suggests instead that Hartley's formulation has its root in de Moivre's considerations on the estimation of a probability by means of his normal approximation to the binomial distribution. Gillies (1987) maintains that Hartley's source was Bayes himself.

Schneider (1968) indicates that de Moivre's correspondence with John Bernoulli in 1708 made him realize that he had no future as a pure mathematician, and his efforts in vain to secure a university appointment must have confirmed this conclusion. From 1708 onward, de Moivre therefore turned more and more to applied mathematics in the form of probability theory and insurance mathematics. Besides the income from the sale of his books he also earned money as a consultant to gamblers and insurance brokers, his "office" being Slaughter's Coffe House at the upper end of St. Martin's Lane, where he usually came after his daily teaching work.

Like many other scientists at the time, de Moivre became involved in some priority disputes besides the one with Cheyne. We mentioned the dispute with Montmort earlier in §18.1, and we shall comment further in the next section. In 1742 Thomas Simpson published a treatise on annuities against which de Moivre reacted in the second edition of his treatise (1743), attacking Simpson for plagiarism; we shall return to this matter in §25.1.

De Moivre did Newton many services. He saw the Latin edition of the *Opticks* through the press and helped with the French edition. He functioned as interpreter when Nicholas Bernoulli in 1712 and Montmort in 1715 visited London and were introduced to Newton. He became a member of the Royal Society Committee to investigate the priority of the invention of the calculus, and he translated some of the material in this dispute into French. After the death of Leibniz in 1716, the priority dispute continued for many years with Newton and John Bernoulli as the main opponents. Much of the correspondence went on with de Moivre and the French mathematician Pierre Varignon as mediators; they tried to end the dispute but without much success. It is said that Newton often called at Slaughter's Coffe House and took de Moivre home with him for mathematical and philosophical discussions. De Moivre dedicated the first edition of the *Doctrine of Chances* (1718) to Newton, thanking him for the inspiration received from "your incomparable Works, especially your Method of Series" and adding: "But one Advantage which is more particularly my own, is the Honour I have frequently had of being admitted to your private Conversation; wherein the Doubts I have had upon any Subject relating to *Mathematics*, have been resolved by you with the greatest Humanity and Condescension." After the death of Newton in 1727, de Moivre wrote a memorandum on Newton's life based on these conversations; see Westfall (1980, pp. 385, 403).

De Moivre was rather old when he began his research in mathematics; and not until the age of 41 did he begin his work on probability theory. Nevertheless, he succeeded in becoming the leading probabilist from 1718 until his death, and he found one of his most important results, the normal approximation to the binomal distribution, in 1733 at the age of 66.

He had arrived in London as a penniless refugee the age of 21, and he died there in poverty, 87 years old.

Todhunter (p. 135) writes, "In the long list of men ennobled by genius, virtue, and misfortune, who have found an asylum in England, it would be difficult to name one who has conferred more honour on his adopted country than de Moivre."

Maty's small *Mémoire* (1755, 46 pp.) contains a list of de Moivre's 21 books and papers and a survey of their contents.

For other biographies of de Moivre, we refer to Wollenschläger (1933), Walker (1934), David (1962), and K. Pearson (1978).

22.2   *DE MENSURA SORTIS*, 1712

De Moivre's first paper on probability theory is the *De Mensura Sortis, seu, de Probabilitate Eventuum in Ludis a Casu Fortuito Pendentibus* (On the Measurement of Chance, or, on the Probability of Events in Games Depending Upon Fortuitous Chance), read to the Royal Society in 1711 and published in the *Phil. Trans.* in 1712. It is dedicated to Francis Robartes (c. 1650–1718), later the third Earl of Radnor, politician, musician, member of the Royal Society, and "patron of the Mathematical Sciences." It is a peculiar coincidence that the first important book on statistics, Graunt's *Observations* (1662), is dedicated to Lord Roberts (Robartes), later the first Earl of Radnor, and the first important paper on probability published in England is dedicated to his son.

De Mensura Sortis is important in the history of probability because it is the fourth of the great treatises, the first three being those by Huygens (1657),

Pascal (1665), and Montmort (1708). In the preface, and in the preface to the *Doctrine of Chances* (1718, pp. I–II), de Moivre states that he took up probability theory in 1708 at the exhortation of Robartes, who after having read Montmort's *Essay* proposed "to me some Problems of much greater difficulty than any he had found in that Book," these problems being the problem of points in a game of bowls (Robartes' problem) (see §18.3) and the occupancy problem (see §22.5). That these problems should be more difficult or require other methods for their solution than the problems solved by Montmort is not true.

Furthermore, de Moivre (1718) writes that

> I had not at that time [1708] read any thing concerning this Subject, but Mr. *Huygens's* Book,... As for the French Book, I had run it over but cursorily, by reason I had observed that the Author chiefly insisted on the Method of *Huygens*, which I was absolutely resolved to reject, as not seeming to me to be the genuine and natural way of coming to the solution of Problems of this kind.

It will be seen that de Moivre, like James Bernoulli, has overlooked Pascal's treatise, which for de Moivre is rather odd, since it is mentioned by Montmort, and, contrary to Bernoulli, he repudiates Huygens' methods.

We have previously quoted Montmort's reply to the unjust remarks on the *Essay* and de Moivre's apology, see §18.1. In 1714 Montmort sent a copy of the second edition of the *Essay* to de Moivre, who thanked for the present, and a friendly correspondence later ensued. It is, however, not correct as indicated by Todhunter (p. 187) that a lasting reconciliation was achieved. The publication of the *Doctrine* (1718) upset Montmort because he felt that de Moivre had not given proper credit to him and to Nicholas Bernoulli, although in his preface de Moivre apologized for his previous error and expressed his admiration for the second edition of the *Essay*; see Schneider (1968, pp. 193–195). In the *Miscellanea Analytica* (1730), de Moivre defended himself, and in the following editions of the *Doctrine* he removed the laudatory remarks about Montmort from the preface.

De Moivre wrote his paper for a scientific journal and not as Montmort for a mixed reading circle of gamblers and scientists. He therefore had to be more concise and leave out the many card and dice games treated by Montmort and instead concentrate on the principles, although all the problems naturally were formulated as games of chance.

De Mensura Sortis consists of a preface, a short introduction, and the solution of 26 problems. An English translation has been given by McClintock with a commentary by Hald (1984). Most of the material from *De Mensura Sortis* is included in the *Doctrine*, see §22.4.

In the introduction de Moivre defines the probability of an event as the number of favorable cases divided by the total number of cases in which the event may happen, provided that these cases are equally likely. He defines the expectation of a player as his probability of winning times the prize. In much of the paper the reasoning is expressed in terms of odds or expectations instead of probabilities.

The multiplication theorem for independent events is given in the following form, which also implies the addition theorem:

> If two events have no dependence on each other, so that p is the number of chances by which the first event may happen and q the number of chances by which It may fail, and r is the number of chances by which the second event may happen and s the number of chances by which it may fail: multiply $p + q$ by $r + s$, and the product $pr + qr + ps + qs$ will contain all the chances by which the happenings and failings of the events may be varied amongst one another.

He gives an interpretation of the individual terms and any sum of these and says that this method may be extended to any number of events. As a special case he derives the binomial distribution. He does not discuss the multiplication theorem for dependent events in the introduction, but many of his problems lead to drawings without replacement from a finite population, and in those cases he just uses the multiplication theorem with the adequate conditional probabilities.

We have previously discussed most of de Moivre's solutions, as shown by the following list: the problem of points (§14.1), Huygens' five problems (§14.2), the number of chances of throwing a given number of points with any given number of dice (§14.3), the number of trials giving an even chance of getting at least a given number of successes (§14.4), Robartes' problem (§18.3), Waldegrave's problem (§21.2), and the duration of play for the ruin problem (§20.4). The occupancy problem is discussed in §22.5.

These problems, except for Robartes' problem and Waldegrave's problem, had already been discussed by other authors, but de Moivre does not give specific references to the previous solutions. The most remarkable of de Moivre's contributions are his derivation of the ruin probability in Huygens' fifth problem; his use of the Poisson approximation to solve the binomial equation $B(c, n, p) = \frac{1}{2}$ with respect to n; his solution of the occupancy problem by means of the method of inclusion and exclusion, and the algorithm for the continuation probability in the duration of play for the ruin problem. Furthermore, he gives without proof the probability of getting a given number of points by throwing any given number of dice and the probability of ruin when one of the players has infinitely many counters.

The only contemporary evaluation of these impressive results is the critical review given by Montmort (1713, pp. 362–368) in a letter of 5 September

1712 to Nicholas Bernoulli, about a month after Montmort had received a copy of the paper from de Moivre. In 1712 de Moivre was 45 years old and a renowned mathematician; the 34-year-old Montmort therefore opened de Moivre's treatise with great expectations, hoping to find new problems formulated and solved. Like Huygens before him, Montmort had posed some problems for the reader at the end of his book, and he had hoped that de Moivre had at least tackled these problems, but "you [Bernoulli] will find that his work is limited almost entirely to solve the most simple and easy problems from my book in a more general way than I have done." Montmort adds that there is nearly nothing new in de Moivre's paper since the problems that are not already solved in the *Essay* (1708) have been solved in our letters. This statement is true but unfair to de Moivre, who did not know the contents of these still unpublished letters and therefore in good faith published his solutions as new.

Montmort goes on to a detailed critical commentary on the solution of each of the 26 problems in which he recognizes de Moivre's priority to the Poisson approximation, to Robartes' problem, and to the algorithm for finding the continuation probability in the problem of the duration of play.

22.3 THE PREFACES OF THE *DOCTRINE OF CHANCES*

Planning to write a book on probability in English, de Moivre had to decide what circle of readers to aim at. He was so fortunate to have two excellent models: *Ars Conjectandi* by James Bernoulli and the second edition of the *Essay* by Montmort, both of them published in 1713, the *Ars Conjectandi* being written for mathematicians and the *Essay* for the educated public. One would have expected de Moivre to choose the first solution, but actually he chose the second and wrote the *Doctrine of Chances* as a mixture of *De Mensura Sortis* and the sections on games of chance in the *Essay*, with some extensions of both parts. The reasons for this choice were presumably economic; he hoped in this way to get a larger number of subscribers to finance the printing of his book, and he also wanted to demonstrate his ability to solve all types of problems in popular games of chance and thus to strengthen his business as a consultant.

De Moivre writes in the preface:

I have explain'd in my *Introduction* to the following Treatise, the chief Rules on which the whole Art of Chances depends; I have done it in the plainest manner that I could think of, to the end it might be (as much as possible) of General Use. I flatter my self that those who are acquainted with Arithmetical Operations, will, by the help of the *Introduction* alone, be able to solve a great Variety of Questions depending on *Chance*: I wish, for the Sake of some Gentlemen who have been pleased to subscribe to the printing of my Book, that I could every where have been as plain as in the *Introduction*; but this was hardly practicable, the Invention of the greatest part of the Rules being intirely owing to *Algebra*; yet I have, as much as possible, endeavour'd to deduce from the Algebraical Calculation several practical Rules, the Truth of which may be depended upon, and which may be very useful to those who have contented themselves to learn only common Arithmetick.

This is followed by a paragraph in the first edition only: "It is for the Sake of those Gentlemen that I have enlarged my first Design, which was to have laid down such Precepts only as might be sufficient to deduce the Solution of any difficult Problem relating to my Subject."

De Moivre's decision to write also for readers knowing only common arithmetic had far-reaching consequences for the style of his book. For a mathematician large parts of the *Doctrine* are, like the similar parts of Montmort's *Essay*, elementary, discoursive, and tedious. In his own words (1718, p. 10),

Yet for the sake of those who are not acquainted with Algebraical computation, I shall set down the Method of proceeding in like cases. In order to which, it is necessary to know, that when a Question seems somewhat difficult, it will be useful to solve at first a Question of the like nature, that has a greater degree of simplicity than the case proposed in the Question given; the Solution of which case being obtained, it will be a step to ascend to a case a little more compounded, till at last the case proposed may be attained to. Therefore, to begin with the simplest case, we may suppose that A wants 1 Game of being up, and B 2.

He does not care for his mathematical readers in the same way; in some of the most difficult problems of the duration of play, he gives his results without proof, see §20.5.

The preface falls into four parts: (1) historical remarks; (2) the uses which may be made of the book; (3) a description of the methods used; and (4) a description of the most important problems with some references.

The preface of the first edition is reprinted in the second and third editions with the following footnote: "This Preface was written in 1717." Hence, one gets the impression that the complete preface has been reproduced, but this is not so; we have already noted that the laudatory remarks on Montmort were omitted in the later editions and so is the following remark on the *Ars Conjectandi*:

Before I make and end of this Discourse, I think my self obliged to take Notice, that some Years after my *Specimen* [*De Mensura Sortis*] was printed, there came out a Tract upon the Subject of Chance, being a posthumous Work of Mr. *James*

Bernoulli, wherein the Author has shewn a great deal of Skill and Judgment, and perfectly answered the Character and great Reputation he hath so justly obtained. I wish I were capable of carrying on a Project he had begun, of applying the Doctrine of Chances to *Oeconomical* and *Political* Uses, to which I have been invited, together with Mr. *de Montmort*, by Mr. *Nicholas Bernoully*: I heartily thank that Gentleman for the good opinion he has of me; but I willingly resign my share of that task into better Hands, wishing that either he himself would prosecute that Design, he having formerly published some successful Essays of that Kind, or that his Uncle, Mr. *John Bernoully*, Brother to the Deceased, could be prevailed upon to bestow some of his Thoughts upon it; he being known to be perfectly well qualified in all Respects for such an Undertaking.

It is remarkable that this is the only reference to the *Ars Conjectandi* in the first edition. It is even more remarkable that de Moivre does not mention Bernoulli's theorem or the improved version by Nicholas Bernoulli; only after having found the normal approximation to the binomial does he in the second edition indicate that the two Bernoullis have discussed the same problem. Of course he alludes to Bernoulli's theorem in a loose way, for example, in the preface where he writes, "For, by the Rules of Chance, a time may be computed, in which those cases may as probably happen as not; nay, not only so, but a time may be computed in which there may be any proportion of Odds for their so happening."

We shall let de Moivre recommend his own book by means of the following quotations: "Such a Tract as this may be useful to several ends."

1. For inquisitive persons "to know what foundation they go upon, when they engage in play."

2. "It may serve in Conjunction with the other parts of the Mathematicks, as a fit introduction to the Art of Reasoning."

3. It may "be a help to cure a Kind of Superstition, which has been of long standing in the World, *viz.* that there is in Play such a thing as *Luck*, good or bad."

4. "The same Arguments which explode the Notion of Luck, may, on the other side, be useful in some Cases to establish a due comparison between Chance and Design: We may imagine Chance and Design to be as it were in Competition with each other, for the production of some sorts of Events, and may calculate what Probability there is, that those Events should be rather owing to one than to the other."

5. "One of the Principal Uses to which this *Doctrine of Chances* may be apply'd, is the discovering of some Truths, which cannot fail of pleasing the Mind, by their Generality and Simplicity; the Admirable Connexion of its Consequences will increase the Pleasure of the Discovery; and the seeming Paradoxes wherewith it abounds, will afford very great matter of Surprize and Entertainment to the Inquisitive."

Many experienced gamblers, then as now, have noticed the occurrence of long sequences of successes which they have ascribed to "luck" in contradistinction to chance. For example, Cardano (1564) writes about luck as if it was a quality of certain persons on certain days. At a time when superstition was widespread it is easily understandable that he did not attempt to explain these phenomena by means of his rudimentary calculus of chances. However, the great advance of probability theory led probabilists such as James Bernoulli, Montmort, and de Moivre to deny the existence of luck in play because they could show that all the possible outcomes of a play had positive probability. For example, de Moivre (1718, pp. IV–V) states that the probability of losing 15 games of Piquet equals $(1/2)^{15} = 1/32,768$ "from whence it follows, that it was still possible to come to pass without the Intervention of what they call *Ill Luck*. Besides, This Accident of losing Fifteen times together at Piquet, is no more to be imputed to ill Luck, than the Winning with one single Ticket the Highest Prize, in a Lottery of 32,768 Tickets, is to be imputed to good Luck, since the Chances in both Cases are perfectly equal." Like Arbuthnott and James Bernoulli, de Moivre states that the probability of an equal number of heads and tails in a long series of coin tossings is very small so that "Chance alone by its Nature constitutes the Inequalities of Play, and there is no need to have recourse to Luck to explain them" (1718, p. V).

Inspired by Arbuthnott's discussion of chance versus design, de Moivre makes the comment quoted above under (4) that we "may calculate what Probability there is, that those Events should be rather owing to one than to the other." As an example he considers two new packs of 32 cards each and calculates the probability that the cards are in the same order (the maker's design) or in any other order. He concludes (1718, p. VI) that "there are the Odds of above 26,313,083 Millions of Millions of Millions of Millions to One, that the Cards were designedly set in the Order in which they were found" $(32! = 26,313,083.7 \times 10^{28})$. He does not pursue this important idea in the text; it was not until the third edition of the *Doctrine* (1756, pp. 251–253) that he made a few remarks on Arbuthnott's and Nicholas Bernoulli's analyses of the data on the sex ratio, see §24.5. De Moivre does not discuss James Bernoulli's concept of moral certainty.

Turning to the methods used, de Moivre states that he has first of all used the doctrine of combinations; however, when the play may continue for an infinite number of games with a priority of play for the gamesters, it will be more natural to use the method of infinite series, where "every Term of the Series includes some particular Circumstance wherein the Gamesters may be found."

De Moivre does not mention recursion, and he seems to have an aversion to Huygen's analytic method, although he commends Nicholas Bernoulli's solution of Waldegrave's problem.

In the second edition the modified preface from the first is supplemented with an *Advertisement* describing the many improvements introduced.

In the third (posthumous) edition the *Advertisement* from the second is omitted, and a new is added with the following content:

> The Author of this Work, by the failure of his Eye-sight in extreme old age, was obliged to entrust the Care of a new Edition of it to one of his Friends; to whom he gave a Copy of the former, with some marginal Corrections and Additions, in his own hand writing. To these the Editor has added a few more, where they were thought necessary: and has disposed the whole in better Order; by restoring to their proper places some things that had been accidentally *misplaced*, and by putting all the Problems concerning *Annuities* together; as they stand in the late *improved* Edition of the Treatise on that Subject. An *Appendix* of several useful Articles is likewise subjoined: the whole according to a Plan concerted with the Author, above a year before his death.

22.4   A SURVEY OF THE PROBABILITY PROBLEMS TREATED IN THE *DOCTRINE OF CHANCES*

The *Doctrine* consists of an introduction with definitions and elementary theorems, followed by a series of numbered problems. The first edition contains 53 problems on probability; the second, 75 problems on probability and 15 on insurance mathematics; the third, 74 problems on probability, followed by 33 problems on insurance mathematics, some tables, and an appendix.

The first edition of the *Doctrine* brings the English educated public abreast with the development of probability theory as found in *De Mensura Sortis*, the *Essay*, and *Ars Conjectandi*. The only new results given by de Moivre are the compound probability theorem and some theorems on the duration of play. After the hectic period of progress from 1708 to 1718, there followed a period of nearly stagnation for about 20 years, during which time only de Moivre published essential new results in the *Phil. Trans.* and in his *Miscellanea Analytica*, thus preparing the second edition of the *Doctrine*.

The introduction contains the same definition of probability as in *De Mensura Sortis* but without mentioning the assumption that the total number of cases (chances) has to be equally likely. It is, however, clear from the applications that this is what de Moivre has in mind. He does not mention James Bernoulli's discussion and extension of the concept of probability.

Like Bernoulli he gives a clear definition of independent events and the corresponding multiplication rule for probabilities. He also explicitly defines dependent events and the multiplication rule as follows:

Two Events are dependent, when they are so connected together as that the Probability of either's happening is alter'd by the happening of the other. ... the Probability of the happening of two Events dependent, is the product of the Probability of the happening of one of them, by the Probability which the other will have of happening, when the first shall have been consider'd as having happen'd" (1738, pp. 6–7).

He defines expectation and notes that the expectation of a sum equals the sum of the expectations. Moreover, the introduction contains some elementary examples leading to the binomial distribution and the binomial waiting-time distribution and a discussion of the problem of points.

The three editions of the *Doctrine* show the development of probability theory from 1718 to 1756, each new edition being an enlargement of the previous one by adding new problems and new remarks and corollaries to existing problems. We have exhibited this development in the following list of the problems treated in the *Doctrine* but using todays' terminology to characterize each problem. We have discussed many of these problems in previous sections, and we shall discuss the most important of the remaining problems in the following sections and chapters. Some problems have been used as problems for the reader in §22.7.

A List of the Problems Treated in *De Mensura Sortis* and the Three Editions of the *Doctrine of Chances*, Ordered as in the Third Edition, with Cross References to the Present Book

Problem	*De Mensura Sortis* 1712[a]	*Doctrine of Chances* (edition)[a]			Section[b]
		1718	1738	1756	
Examples of relation between handicap and skill; the problem of points	3, 4	3, 4	1, 2	1, 2	14.1, 18.5
The number of trials giving an even chance of at least c successes; the Poisson approximation	5–7	5–7	3–5	3–5	14.4
The probability of throwing s points with n dice each having f faces	p. 220	p. 17	p. 35	p. 39	14.3

(Continued)

Problem	*De Mensura Sortis* 1712[a]	*Doctrine of Chances* (edition)[a] 1718	1738	1756	Section[b]
The problem of points for any number of players	8	8	6, 69	6	14.1
Huygens' fifth problem; the probability of the Gambler's Ruin	9	9	7	7	14.2
An example of the problem of points for two players	10	10	8	8	14.1
The gambler's expectation in a generalization of Huygens' fifth problem	—	43, 44	9	9	22.7
Huygens' second problem and its generalizations, priority of play and the method of infinite series	11, 12	11, 14	10, 11	10, 11	14.2
The summation of any number of terms of a series having constant differences of a certain order; the Montmort–Bernoulli formula	—	p. 29	p. 52	p. 59	15.4
The expectation of the players in a series of games with priorities of play	—	—	89	12	
Bassette	—	12	12	13	15.5
Pharaon	—	13, 23	13, 32	14, 33	18.3
Elementary theorems on permutations and combinations	—	15–19	14–18	15–19	
Huygens' fourth problem; the hypergeometric distribution	14	20	19	20	14.2

(Continued)

Problem	De Mensura Sortis 1712[a]	Doctrine of Chances (edition)[a]			Section[b]
		1718	1738	1756	
Problems on lotteries solved by combinatorial methods	—	21, 22	20–24	21–25	22.7
An example of the multivariate hypergeometric distribution	—	50	25	26	22.7
Problems on the card game Quadrille solved by combinatorial methods	—	—	26–31	27–32	
The expectation in a play where one has to win the successive games without interruption	—	24	33	34	22.7
The problem of coincidences and the compound probability theorem	—	25, 26	34, 35	35, 36	19.4, 19.5
Robartes' problem; the problem of points in a game of bowls	16, 17	27, 28	36, 37	37, 38	18.3
The probability that m specified faces occur at least once in n throws with a die having f faces; the occupancy problem	18, 19	29, 30	38–41	39–42	22.5
The probability of the occurrence of a given number of events in a given order	—	—	42	43	22.7
Waldegrave's problem	15	31, 32	43, 44	44, 45	21.2
On Hazard, a game of dice	—	47, 53	45, 46	46, 47	
On Raffling	—	49	47, 48	48, 49	
On Whist	—	48	49	50	22.7

(Continued)

Problem	*De Mensura Sortis* 1712[a]	*Doctrine of Chances* (edition)[a] 1718	*Doctrine of Chances* (edition)[a] 1738	*Doctrine of Chances* (edition)[a] 1756	Section[b]
On the card game Piquet	—	51	50–54	51–55	
Simple problems on the expected gain in games with two players	—	—	55, 56	56, 57	
On the duration of play; the ruin problem	20–26	33–46	57–68; 70, 71	58–71	20.4, 20.5; 23.2
An algorithm for the probability of continuation $U_n(b, b)$	20	33	57	58	20.4
A recursion for the duration probability $D_n(b, b)$ and a formula for $d_n(b, b)$	—	34	58	59	20.5
Solution of equations of the form $U_6(4, 4) = \frac{1}{2}$ with respect to p; reciprocal equations	21–23	35–37	59–61	60–62	20.4
An algorithm for the probability of continuation $U_n(a, b)$	24	38	62	63	20.4
A recursion for the duration probability $D_n(a, b)$ and a formula for $d_n(a, b)$	—	39	63	64	20.5
The probability of B being ruined when A's capital is infinite, $R_n(b)$	25	40	64	65	20.4, 20.5
The probability that A wins b games more than B at least once in n games and that B similarly wins a games more than A	26	41	65	66	20.6

(*Continued*)

Problem	De Mensura Sortis 1712[a]	Doctrine of Chances (edition)[a]			Section[b]
		1718	1738	1756	
The probability that A wins b games more than B at least once in n games and that B does not win a game more than A	—	42	66	67	20.6
On the summation of recurring series	—	p. 127	p. 193	p. 220	20.5, 23.1
The probability of continuation $U_n(b, b)$ expressed as a trigonometric series; the median duration of play	—	45	67, 68; p. 202	68, 69	20.5; 23.2
Examples of application of the formulae for $U_n(a, b)$ and $D_n(a, b)$	—	46	70, 71	70, 71	20.5
The mean deviation of the binomial distribution	—	—	86, 87	72, 73	24.2
The normal distribution as approximation to the binomial	—	—	p. 235	p. 243	24.5
The probability of a run of given length	—	—	88	74	22.6

[a] Numbers are problem numbers, unless otherwise noted.
[b] Cross references to section unmbers of the present book.

To help the reader with the summation of a large number of terms of the harmonic series, de Moivre (1756, p. 95) gives the formula

$$\sum_{i=n}^{a-1} i^{-1} = \ln \frac{a}{n} + (2n)^{-1} - (2a)^{-1} + \sum_{k=1}^{\infty} \frac{B_{2k}}{2k}(n^{-2k} - a^{-2k}),$$

a and n being positive integers, $a > n$, where a few terms of the series on the right-hand side will usually give a good approximation.

Two years after the publication of the second edition of the *Doctrine*,

Thomas Simpson (1710–1761), a teacher of mathematics, author of a textbook on the method of fluxions, and in 1743 professor of mathematics at the Royal Military Academy, Woolwich, published an excellent textbook for mathematicians entitled *The Nature and Laws of Chance. The Whole After a new, general, and conspicuous Manner, And illustrated with A great Variety of Examples* (1740). There is, however, nothing "new, general, and conspicous" in Simpson's book; it is simply plagiarism of the mathematical parts of the *Doctrine*. Leaving out the discussions of all the concrete games of chance, such as Bassette, Pharaon, Quadrille, Raffling, and the main part of de Moivre's careful analysis of the duration of play and all the problems on insurance mathematics, Simpson succeeded in writing a much more compact book; in fact, he reduced the number of pages in the text from de Moivre's 256 to 85. It is presumably such a book that de Moivre originally contemplated writing.

Simpson's bad conscience shines through the preface. He praises the works of de Moivre: "I am satisfied, it may be deemed a sort of Presumption to attempt, upon any Account, a Subject like this, after so great a Man as Mr De Moivre." What, then, is the "excuse" for publishing his book? "The *Price* [of the *Doctrine*] must, I am sensible, have put is out of the Power of many to purchase it; and even some, who want no Means to gratify their Desires this way, and who might not be inclinable to subscribe a Guinea for a single Book, however excellent, may not scruple the bestowing of a small Matter on one, that perhaps may serve equally well for their Purpose."

We shall comment on the main points where Simpson deviates from de Moivre. He elaborates a little on the Poisson approximation to the binomial, see §14.4. He does not use de Moivre's martingale reasoning for solving the Gambler's ruin problem but solves instead the difference equation much like Struyck, see §14.2. He does not use de Moivre's generating function to find the probability of getting a given number of points by throwing a given number of dice; instead, he gives a combinatorial proof much like the one by Montmort. He gives a short and very unsatisfactory treatment of the problem of the duration of play without mentioning de Moivre's ingenious solutions; instead, he gives without reference and without proof Nicholas Bernoulli's formula for the probability that A wins in at most n games. He derives the normal approximation to the binomial distribution and states wrongly that de Moivre has omitted the demonstration of this important result. He gives the first proof of de Moivre's result on the probability of getting a run of a given length, see §22.6.

22.5 THE OCCUPANCY PROBLEM

We have previously noted that Montmort in the *Essay* (1708) found the distribution of the size of the occupancy numbers, see §18.2. After having read the *Essay*, Robartes formulated another occupancy problem for de Moivre to solve, which may be found as Problem 18 in *De Mensura Sortis*: Find the probability that r specified faces occur at least once in n throws with a die having f faces.

In the occupancy terminology this means that n balls are distributed at random into f boxes, and that at least one ball is to be found in each of r specified boxes.

De Moivre's derivation of this probability is based on the method of inclusion and exclusion.

The total number of equally likely outcomes is f^n. The number of cases in which the ace, say, does not occur is $(f - 1)^n$, so that the number of cases with at least one ace is $f^n - (f - 1)^n$.

To find the number of cases where the ace and the deuce occur at least once suppose first that the deuce is expunged from the die so that the number of cases for the ace to occur at least once is $(f - 1)^n - (f - 2)^n$. Let now the deuce be restored; the number of cases for the ace to occur without the deuce will be the same as if the deuce were expunged. If from the number of cases for the ace to occur at least once (with or without the deuce), namely, $f^n - (f - 1)^n$, are subtracted the number of cases for the ace to occur without the deuce, namely, $(f - 1)^n - (f - 2)^n$, there will remain the number of cases for the ace and the deuce to occur at least once, which number thus will be

$$f^n - 2(f - 1)^n + (f - 2)^n.$$

Continuing in this manner for three and four specified faces, de Moivre concludes that the general formula is

$$\sum_{i=0}^{r} (-1)^i \binom{r}{i} (f - i)^n,$$

which divided by f^n gives the probability

$$p_{f,r}(n) = \sum_{i=0}^{r} (-1)^i \binom{r}{i} \left(1 - \frac{i}{f}\right)^n.$$

It will be seen that de Moivre's proof is analogous to the proof of the probability of at least one coincidence given independently and at about the same time by Nicholas Bernoulli, see §§19.2 and 19.3.

De Moivre reprinted his proof in the *Doctrine* without noting that is may considered a special case of his compound probability theorem (19.4.7), letting A_i denote the event that face number i does not occur.

If follows directly from de Moivre's proof that the number of favorable cases may be written as $\Delta^r (f - r)^n$ using finite difference notation.

To solve the equation $p_{f,r}(n) = \frac{1}{2}$ with respect to n, de Moivre states that for small values of r/f, an approximation may be obtained by setting

$$\frac{f-i}{f} \cong \left(\frac{f-1}{f}\right)^i, \qquad i = 1, 2, \ldots, r,$$

which leads to

$$p_{f,r}(n) \cong \left\{1 - \left(\frac{f-1}{f}\right)^n\right\}^r,$$

so that

$$\left(\frac{f-1}{f}\right)^n \cong 1 - \left(\frac{1}{2}\right)^{1/r},$$

from which n is easily found.

De Moivre writes in the preface that this artifice of changing an arithmetic progression into a geometric one, when the numbers are large and their intervals small, is due to his friend Halley.

In the *Doctrine* de Moivre generalizes the occupancy problem as follows: Let a die have f_i faces all marked i, $i = 1, \ldots, k$, $\sum f_i = f$, and find the probability that in n throws r specified numbers, $1, \ldots, r$, say, occur at least once. De Moivre gives the number of favorable cases as

$$f^n - \sum_{i=1}^{r} (f - f_i)^n + \sum_{i=1}^{r-1} \sum_{j=i+1}^{r} (f - f_i - f_j)^n - \cdots + (-1)^r (f - f_1 - \cdots - f_r)^n,$$

and the total number of cases as f^n, commenting that the proof is analogous to the previous one.

Laplace (1774a; 1786; 1812, Book 2, §4) generalizes de Moivre's occupancy problem as follows: A lottery consists of f tickets, numbered $1, \ldots, f$, of which m are drawn (without replacement) at each time. After the drawing, the m tickets are replaced and the whole process is repeated n times. Find the probability that after n drawings r specified tickets, $1, \ldots, r$, say, will have occurred at least once.

De Moivre's problem is obtained for $m = 1$.

Laplace solves the problem in the same manner as de Moivre. Let $z_{f,r}$ denote the number of favorable cases. Then, $z_{f,r-1}$ equals the number of cases where $1, \ldots, r-1$ occur together with r and without r, which means that

$$z_{f,r-1} = z_{f,r} + z_{f-1,r-1} \quad \text{or} \quad z_{f,r} = z_{f,r-1} - z_{f-1,r-1} = \Delta z_{f-1,r-1}.$$

In a single drawing the total number of cases is $\binom{f}{m}$, and the number of cases where 1 does not occur is $\binom{f-1}{m}$, so that 1 occurs at least once in

$$\binom{f}{m} - \binom{f-1}{m}$$

cases. In n independent drawings we get, similarly, that

$$z_{f,1} = \left(\frac{f}{m}\right)^n - \left(\frac{f-1}{m}\right)^n = \Delta\left(\frac{f-1}{m}\right)^n.$$

It follows that

$$z_{f,r} = \Delta^r\left(\frac{f-r}{m}\right)^n,$$

and dividing by the total number of cases, the probability is obtained as

$$p_{f,r}(n, m) = \frac{\Delta^r[(f-r)^{(m)}]^n}{(f^{(m)})^n},$$

which is Laplace's result. The probability that all the numbers occur is obtained for $r = f$.

Like de Moivre, Laplace seeks an approximation to

$$p_{f,f}(n, 1) = \frac{\Delta^f 0^n}{f^n};$$

a discussion of his results compared with other approximations has been given by David and Barton (1962, Chap. 16).

Euler (1785) derives the probability that at least $f - i$ specified tickets occur in n drawings. It is in this paper that Euler introduces the symbol $\left[\begin{smallmatrix} n \\ r \end{smallmatrix}\right]$, which later was modified to $\binom{n}{r}$.

Todhunter (pp. 250–256) gives a discussion of the occupancy problem, including a proof of Euler's result.

22.6 THE THEORY OF RUNS

Let the probability of the event A in a single trial be p and the probability of the complementary event $\bar{A}$ be q, $p + q = 1$. A run of length r is defined as an uninterrupted sequence of rA's. In 1738 de Moivre formulated the following problem: What is the probability of getting a run of length r or more in n trials? We shall denote this probability by Z_n, and we shall let z_n be the probability that a run of length r occurs for the first time at the nth trial, so that

$$Z_n = z_1 + z_2 + \cdots + z_n, \qquad n = 1, 2, \ldots, \tag{1}$$

$$z_1 = \cdots = z_{r-1} = 0.$$

Obviously, $Z_r = z_r = p^r$ and $Z_{r+1} = p^r + qp^r$.

De Moivre's Solution

De Moivre gives the solution by means of a generating function without indicating his proof. The probability Z_n may be obtained as the sum of the first $n - r - 1$ terms of the power series expansion of the function

$$f(x) = \frac{p^r}{1 - x - cx^2 - c^2 x^3 - \cdots - c^{r-1} x^r} = \frac{p^r(1 - cx)}{1 - x - cx + c^r x^{r+1}}, \tag{2}$$

where c should be replaced by p/q and x by q in the sum.

As noted by Todhunter (p. 185), there is a mistake in de Moivre's formula as printed in the *Doctrine* (1738, Problem 88); instead of substituting p/q for c, he uses p; however, de Moivre gives several numerical examples in which the formula is used correctly.

The expansion given by de Moivre in the examples is as follows. Setting $m = c + 1$, we get

$$f(x) = \frac{p^r(1 - cx)}{1 - (mx - c^r x^{r+1})} = p^r(1 - cx) \sum_{k=0}^{\infty} (mx - c^r x^{r+1})^k,$$

$$= p^r(1 - cx) \sum_{i=0}^{\infty} (-1)^i (c^r x^{r+1})^i \sum_{j=0}^{\infty} \binom{j+i}{i}(mx)^j. \tag{3}$$

De Moivre notes that the last sum may be obtained by means of a theorem given under his discussion of recurring series, but he does not give the result. We shall carry out the summation following de Moivre's suggestion. From (3) we obtain the first $n - r + 1$ terms of the power series as

$$p^r\left(\sum_{j=0}^{n-r} (mx)^j - c^r x^{r+1} \sum_{j=0}^{n-2r-1} (j+1)(mx)^j\right)$$

$$+ (c^r x^{r+1})^2 \sum_{j=0}^{n-3r-2} \binom{j+2}{2}(mx)^j - \cdots \Bigg)$$

$$- p^r(cx)\left(\sum_{j=0}^{n-r-1}(mx)^j - c^r x^{r+1} \sum_{j=0}^{n-2r-2}(j+1)(mx)^j \right.$$

$$+ (c^r x^{r+1})^2 \sum_{j=0}^{n-3r-3} \binom{j+2}{2}(mx)^j - \cdots \Bigg).$$

Setting $x = q$; $cx = p$; $mx = 1$; and $c^r x^{r+1} = qp^r = k$, say, we get

$$Z_n = p^r\bigg[\{(n-r)q + 1\} - \frac{1}{2!}k(n-2r)\{(n-2r-1)q + 2\}$$

$$+ \frac{1}{3!}k^2(n-3r)^{(2)}\{(n-3r-2)q + 3\} - \cdots\bigg], \tag{4}$$

where the series breaks off when a factor becomes zero or negative. This formula was first given by Simpson (1740) and later by Laplace (1812).

How did de Moivre find his result? It is easy to see that $z_{r+1} = qz_r$,

$$z_{r+2} = qz_{r+1} + pqz_r,$$

and in general that

$$z_n = qz_{n-1} + pqz_{n-2} + p^2qz_{n-3} + \cdots + p^{r-1}qz_{n-r}, \qquad n = r+1, r+2, \ldots. \tag{5}$$

De Moivre may have known this recursion and in some way used it to find (2).

After having given an algorithm for finding Z_n, de Moivre turns to the question of finding "what number of Games are necessary, in all Cases, to make it an equal chance whether or not r Games will be won without intermission." Without proof he gives the following approximation to the solution of the equation $Z_n = \frac{1}{2}$,

$$n \cong \frac{0.7(1 - p^r)^2}{qp^r}.$$

This means that he uses $\exp\{-nqp^r(1 - p^r)^{-2}\}$ as an approximation to Z_n. From (4) it will be seen that for large values of n in relation to r, we have as a first approximation,

$$Z_n \cong 1 - e^{-nk} + p^r e^{-nk},$$

which gives

$$n \cong \frac{0.7}{qp^r} - \frac{1}{q}$$

as a first approximation to n. How de Moivre found the factor $(1 - p^r)^2$ we do not know.

Simpson's Solution

Simpson (1740, Problem 24) is the first to give an explicit expression for Z_n. Setting $p^r = a$ and $qp^r = k$, say, Simpson gives in verbal form the recursion

$$z_n = (1 - Z_{n-r-1})qp^r = (1 - Z_{n-r-1})k, \qquad n = r+2, r+3, \ldots, \tag{6}$$

using the fact that a run of length r occurs for the first time at the nth trial if and only if such a run does not occur in the first $n - r - 1$ trials, that $\bar{A}$ occurs at trial number $n - r$, and that a run of length r occurs in the last r trials.

It follows that

$$z_{r+i} = k(1 - Z_{i-1}) = k, \qquad i = 1, \ldots, r,$$

and from (1) that

$$Z_{r+i} = a + ik, \qquad i = 0, 1, \ldots, r.$$

Moreover,

$$z_{2r+i} = k(1 - Z_{r+i-1}) = k[1 - a - (i-1)k],$$

$$Z_{2r+i} = Z_{2r} + z_{2r+1} + \cdots + z_{2r+i}$$

$$= a(1 - ik) + (r + i)k - \binom{i}{2}k^2, \qquad i = 1, \ldots, r,$$

$$z_{3r+i} = k(1 - Z_{2r+i-1})$$

$$= k\left\{ 1 - a + (i-1)ak - (r - i - 1)k + \binom{i-1}{2}k^2 \right\},$$

$$Z_{3r+i} = Z_{3r} + z_{3r+1} + \cdots + z_{3r+i}$$

$$= a\left\{ 1 - (r + i)k + \binom{i}{2}k^2 \right\} + (2r + i)k - \binom{r+i}{2}k^2 + \binom{i}{3}k^3,$$

$$i = 1, \ldots, r,$$

and so on.

Simpson stops after Z_{2r+i} and writes "Now having proceeded thus far, the Law of Continuation is manifest." Todhunter (p. 208) writes that Simpson's demonstration is imperfect, and we have therefore included one step more to exhibit the law of continuation clearly. Simpson gives the general formula as

$$Z_n = a \left\{ 1 - (n - 2r)k + \binom{n - 3r}{2} k^2 - \binom{n - 4r}{3} k^3 + \cdots \right\}$$
$$+ (n - r)k - \binom{n - 2r}{2} k^2 + \binom{n - 3r}{3} k^3 - \binom{n - 4r}{4} k^4 + \cdots,$$

which obviously is the same formula as the one we have derived from de Moivre's algorithm. Simpson does not mention de Moivre's procedure.

Laplace's Solution

Laplace (1812, Book 2, §12) first derives the recursion (5) and then finds the generating function of z_n as

$$g(t) = \sum_{n=1}^{\infty} z_n t^n = p^r t^r + q(t + pt^2 + \cdots + p^{r-1} t^r) g(t),$$

which leads to

$$g(t) = \frac{p^r t^r (1 - pt)}{1 - t + q p^r t^{r+1}}.$$

Since the generating function of Z_n equals $G(t) = g(t)/(1 - t)$, Laplace finds Z_n as the coefficient of t^n in the power series expansion of $G(t)$, which leads to Simpson's formula.

Todhunter's Solution

Thus far all the proofs have been based on the recursion for z_n. Todhunter (pp. 184–185) gives a soluttion based on the recursion

$$Z_{n+1} = Z_n + (1 - Z_{n-r}) q p^r,$$

from which he finds the generating function $G(t)$.

22.7 PROBLEMS FROM DE MOIVRE'S *DE MENSURA SORTIS* AND THE *DOCTRINE OF CHANCES*

1. The problem of points for three players. Let the number of points needed to win for three players of equal skill be 2, 3, and 5, respectively. Show that the corresponding probabilities of winning are $1433/2187$, $635/2187$, and $119/2187$ (de Moivre, 1756, Problem 6).

2. "Supposing A and B, whose proportion of skill is as a to b, to Play together, till A either wins the number q of Stakes, or loses the number p of them; and that B Sets at every Game the sum G to the sum L: It is required to find the Advantage, or Disadvantage of A" (de Moivre, 1718, Problem 43). *Answer*: $\{(aG - bL)/(a - b)\}\{qa^q(a^p - b^p) - pb^p(a^q - b^q)\}/(a^{p+q} - b^{p+q})$.

3. The two forms of the hypergeometric distribution. Consider an urn containing a white and b black counters, $a + b = n$. Drawing c counters without replacement the probability of getting p white counters is

$$\binom{a}{p}\binom{b}{c-p} \bigg/ \binom{n}{c} = \frac{a^{(p)}b^{(c-p)}c!}{p!(c-p)!n^{(c)}}.$$

Show that this probability may be written as

$$\frac{c^{(p)}(n-c)^{(a-p)}a^{(p)}}{p!n^{(a)}},$$

which is computationally advantageous when n and c are large and a is small (de Moivre, 1718, Problem 20).

4. "In a Lottery consisting of 40,000 Tickets, among which are Three particular Benefits: What is the Probability that taking 8,000 of them, one or more of the Three particular Benefits shall be amongst them?" (de Moivre, 1718, Problem 21). *Answer*: Nearly $61/125$.

5. "To Find how many Tickets ought to be taken in a Lottery consisting of 40,000, among which there are Three particular Benefits, to make it as Probable that one or more of those Three may be taken as not" (de Moivre, 1718, Problem 22). *Answer*: 8252 tickets.

6. "Supposing a Lottery of 100,000 Tickets, where of 90,000 are Blanks, and

10,000 are Benefits, to determine accurately what the odds are of taking or not taking a Benefit, in any number of Tickets assigned" (de Moivre, 1738, Problem 22). *Answer:* Taking six tickets, the probability of getting one prize or more is 0.46857; for seven tickets, it is 0.52172.

Compare the exact (hypergeometric) solution with the approximate solution obtained from the binomial distribution.

7. "Supposing A and B to play together, the Chances they have respectively to win being as a and b, and B obliging himself to Set to A, so long as A wins without interruption: What is the Advantage that A gets by his Hand?" (de Moivre, 1718, Problem 24). *Answer:* $(a - b)/b$.

8. Find the probability of getting all the six faces by throwing a die 12 times (de Moivre, 1712, Problem 18). *Answer:* Nearly 10/23.

9. Find the probability of getting two aces at one time and two sixes at another time in 43 throws with two dice (de Moivre, 1712, Problem 18). *Answer:* Nearly 49/100.

10. "Supposing a regular Prism having a Faces marked 1, b Faces marked 2, c Faces marked 3, d Faces marked 4, etc., what is the Probability that in a certain number of throws n, some of the Faces marked 1 will be thrown, as also some of the Faces marked 2." (de Moivre, 1738, Problem 40). Find the probability of getting 5 points and 6 points at least once in eight throws with a pair of dice. *Answer:* 0.40861.

11. "Any number of Chances being given, to find the Probability of their being produced in a given order without any limitation of the number of times in which they are to be produced." (de Moivre, 1738, Problem 42). Find the probability of throwing with a pair of dice the number of points 4, 5, 6, 8, 9, and 10 before 7. *Answer:* Nearly 0.066.

12. "To find the Probability of taking Four Hearts, Three Diamonds, Two Spades, and One Club in Ten Cards out of a Stock containing Thirty-two" (de Moivre, 1718, Problem 50). *Answer:* 878,080/64,512,240.

13. Find the probability of the dealer and of any other player to get any given number of trumps in Whist (de Moivre, 1756, Problem 50).

14. Find the probability of getting a run of length 4 or more in 21 trials, given that the probability of the event in question in each trial is $\frac{1}{2}$ (de Moivre, 1738, Problem 88). *Answer:* 521,063/1,048,570.

15. Find the number of games necessary to make it an equal chance whether or not six games will be won without intermission, given that the probability of winning a single game is $\frac{1}{3}$ (de Moivre, 1756, Problem 74). *Answer*: Nearly 763 games.

CHAPTER 23

The Problem of the Duration of Play and the Method of Difference Equations

23.1 DE MOIVRE'S THEORY OF RECURRING SERIES

The models employed in mathematical astronomy and physics in the 17th century were continuous, and therefore the infinitesimal calculus was developed as the natural tool for the analysis of such models. In probability theory, however, the models were mostly discrete so that a new type of calculus was needed, the calculus of finite differences.

Finite differences had previously been employed by Harriot and Briggs for the construction of tables, by Gregory and Newton for polynomial interpolation and quadrature, and by James Bernoulli and Montmort for the summation of series. A general theory was, however, missing until Brook Taylor published his *Methodus Incrementorum Directa et Inversa* in 1715.

By his method of increments Taylor attempts to cover both the calculus of finite differences and as a limiting case the infinitesimal calculus. He introduces a notation for increments (differences) and integrals (sums) similar to Newton's dot notation for fluxions and fluents. As the title of his book indicates, his method is based on the fundamental idea that differencing and summation are inverse operations.

He discusses the existence of the general solution of a difference equation of order n and points out that the solution contains n arbitrary constants.

He derives the formula for $\Delta^n(u_x v_x)$, analogous to the Leibniz formula for $D^n(u_x v_x)$, and uses this result to find the formula for summation by parts analogous to the one for integration by parts.

These general results are used to find differences and corresponding sums of elementary functions, such as factorials, inverse factorials, the exponential function, and the exponential times an arbitrary function. He also solves special cases of the difference equation $u_{x+1} = a_x u_x + b_x$ by a method which may be used to find the general solution. All these results are today contained in any textbook on the calculus of finite differences.

For further details on Taylor's work, we refer to Eneström (1879) and Feigenbaum (1985).

Since the time of Pascal and Huygens, recursion had been an important method for solving problems in probability theory. In the previous chapters we have given many examples of both ordinary and partial difference equations and their solution by ad hoc methods; an outstanding example is Nicholas Bernoulli's solution of the homogeneous linear difference equation with constant coefficients for the duration of play in Waldegrave's problem, discussed in §21.2. From about 1712 Montmort, Nicholas Bernoulli, and de Moivre all realized the importance of finding a general method of solution of the linear difference equation. De Moivre (1718) was the first to publish his results, to a large extent without proofs. He writes in the Preface (1718, p. IX): "Those Demonstrations are omitted purposely to give an occasion to the Reader to exercise his own Ingenuity. In the mean Time, I have deposited them with the *Royal Society*, in order to be Published when it shall be thought requisite." In that way de Moivre wanted to secure his priority.

It follows from letters between Montmort and Nicholas Bernoulli during 1718–1719 that they succeeded in finding the general term of a recurrent series of the second order.

De Moivre's proofs, deposited in the Royal Society, are presumably incorporated in two papers communicated to the Royal Society and published in the *Phil. Trans.* in 1724; and improved version is included in the *Miscellanea Analytica* (1730).

Accounts of de Moivre's exposition in the *Miscellanea Analytica* have been given by Kohli (1975b) and Schneider (1968). They have also given excerpts of previously unpublished letters between Montmort, Nicholas Bernoulli, and de Moivre on this topic.

We have previously given some of de Moivre's results on the summation of recurring series in §20.5, based on the *Doctrine* (1718). Here we shall give an account of the complete theory as presented in the second edition of the *Doctrine* (1738, pp. 193–202) and comment on the proofs.

De Moivre's exposition contains nine propositions with some corollaries and in agreement with his pedagogical principles, he gives results for recurring series of order 2, 3, and 4 from which the reader may deduce the general results. We shall give the general results in modern notation.

Propositions 1–3. Here de Moivre derives formula (20.5.26) for the sum of a recurring power series $R(x)$ of order $k = 2, 3, 4$. For a series of order k, the proof is as follows:

$$R(x) - \sum_{n=0}^{k-1} r_n x^n = \sum_{n=k}^{\infty} r_n x^n = \sum_{n=k}^{\infty} \sum_{i=1}^{k} (-1)^{i-1} a_i x^i r_{n-i} x^{n-i}$$

$$= \sum_{i=1}^{k} (-1)^{i-1} a_i x^i \sum_{j=k-i}^{\infty} r_j x^j$$

$$= \sum_{i=1}^{k} (-1)^{i-1} a_i x^i \left\{ R(x) - \sum_{j=0}^{k-i-1} r_j x^j \right\}.$$

Solving for $R(x)$ we find

$$R(x)A(x) = \sum_{n=0}^{k-1} x^n \sum_{i=0}^{n} (-1)^i a_i r_{n-i}. \tag{1}$$

The differential scale $A(x)$, or, as we would say today, the generating function for the coefficients $\{(-1)^i a_i, i = 0, 1, \ldots, k\}$, is a polynomial of degree k, and the right-hand side of (1) is a polynomial of degree $k - 1$ with coefficients that are found as convolutions of $\{(-1)^i a_i\}$ and $\{r_i\}$, so that $R(x)$ is a proper rational function (assuming that x is sufficiently small for the power series to converge).

Corollary. If the scale operates from the beginning of the series so that $r_1 = a_1 r_0$, $r_2 = a_1 r_1 - a_2 r_0$, etc., then $R(x)A(x) = r_0$.

Proposition 4. If $\Delta^k r_n = 0$, $n = 0, 1, \ldots$, then $R(x)$ is a recurring series of order k, and $A(x) = (1 - x)^k$. Moreover, $R(x)$ may be written as

$$R(x) = \sum_{n=0}^{\infty} x^n \sum_{s=1}^{k} c_s \binom{n+s-1}{s-1} = \sum_{s=1}^{k} c_s (1-x)^{-s}, \tag{2}$$

where the k coefficients $\{c_s\}$ are to be found from $r_0, \ldots, r_{k-1}$.

Proof. The hypothesis means that r_n may be represented by a polynomial in n of degree $k - 1$. Writing this as a linear combination of figurate numbers, we have

$$r_n = \sum_{s=1}^{k} c_s \binom{n+s-1}{s-1}.$$

We have to prove that

$$\sum_{i=0}^{k} (-1)^i \binom{k}{i} r_{n-i} = 0, \qquad n = k, k+1, \ldots,$$

or, equivalently, that

$$\sum_{i=0}^{k} (-1)^i \binom{k}{i} \binom{n-i+s-1}{s-1} = 0, \qquad n = k, k+1, \ldots; \quad s = 1, \ldots, k.$$

This result may be obtained by comparison of coefficients of t^n in the expansions of $(1-t)^k (1-t)^{-s}$ and $(1-t)^{k-s}$, respectively.

The last part of the theorem follows from the fact that

$$\sum_{n=0}^{\infty} x^n \binom{n+s-1}{s-1} = (1-x)^{-s}.$$

De Moivre does not give (2) explicitly, but it follows easily from his corollary. The first part of Proposition 4 may be found in the first edition of the *Doctrine*, which also contains Propositions 1–3. De Moivre then adds, "When the Numerical Quantities belonging to the Terms of any Series are restrained to have their last differences equal to Nothing, then may the sums of those Series be also found by the following elegant theorem, which has been communicated to me by *Mr. de Monmort.*" Montmort's formula is

$$\sum_{n=1}^{\infty} r_n x^n = \sum_{i=1}^{k} x^i (1-x)^{-i} \Delta^{i-1} r_1,$$

see (18.2.7). The reference to Montmort is omitted in the following editions, where de Moivre only gives results corresponding to (2).

Proposition 5. The sum of the first m terms of a recurring series of order k times $A(x)$ equals the difference between two polynomials of degree $k-1$ and $k-1+m$, respectively, see (20.5.27). In particular,

$$A(x) \sum_{n=0}^{m-1} \binom{n+k-1}{k-1} x^n = 1 - x^m \sum_{i=0}^{k-1} \binom{m+i-1}{i} (1-x)^i, \qquad A(x) = (1-x)^k.$$

$$(3)$$

Proof. Using (20.5.27), the last result is obtained from the identity

$$\sum_{n=0}^{\infty} \binom{n+k-1+m}{k-1} x^n = \sum_{i=0}^{k-1} \binom{m+i-1}{i}(1-x)^{-(k-i)}.$$

Todhunter (p. 179) proves (3) by induction.

To use Proposition 5, it is necessary to know r_m and the following $k-1$ r's. This raises the question of how to calculate r_m without having to calculate all the previous values.

In the *Miscellanea Analytica*, de Moivre solves this problem by proving that a recurring series of order k under certain conditions may be written as a linear combination of k geometric series. He begins by proving the following theorem for $k = 2$: The sum of two geometric series

$$r_n = r_{1n} + r_{2n}, \quad \text{where} \quad r_{1n} = t_1 r_{1,n-1} \quad \text{and} \quad r_{2n} = t_2 r_{2,n-1},$$

is a recurring series of order 2,

$$r_n = (t_1 + t_2)r_{n-1} - (t_1 t_2)r_{n-2}.$$

Inversely, a recurring series of order 2,

$$r_n = a_1 r_{n-1} - a_2 r_{n-2},$$

may be written as the sum of two geometric series with ratios t_1 and t_2, respectively, if the equation

$$t^2 - a_1 t + a_2 = 0$$

has the two real and different roots t_1 and t_2.

The proof is straightforward. De Moivre gives a similar proof for $k = 3$ and 4.

In the *Doctrine* (1738) these results are given without proof, and the general theorem is formulated in Proposition 6.

Proposition 6. Let $t_1, \ldots, t_k$ be the real and different roots of the characteristic equation

$$f(t) = \sum_{i=0}^{k} (-1)^i a_i t^{k-i} = 0, \tag{4}$$

and let the symmetric function of order i of $(t_1, \ldots, t_k)$ be

$$s_i = \sum t_{v_1} t_{v_2} \cdots t_{v_i}, \qquad 1 \leqslant v_1 < v_2 < \cdots < v_i \leqslant k, \quad i = 1, \ldots, k,$$

and $s_0 = 1$. Let s_i after removal of t_j be denoted by $s_{i(j)}, s_{0(j)} = 1$. Then we have

$$r_n = \sum_{j=1}^{k} c_j t_j^n, \tag{5}$$

$$R(x) = \sum_{j=1}^{k} c_j \sum_{n=0}^{\infty} (t_j x)^n = \sum_{j=1}^{k} \frac{c_j}{1 - t_j x}, \tag{6}$$

where

$$c_j = \frac{\displaystyle\sum_{i=0}^{k-1} (-1)^i s_{i(j)} r_{k-i-1}}{\displaystyle\prod_{\substack{h=1 \\ h \neq j}}^{k} (t_j - t_h)}. \tag{7}$$

Corollary 1. If $R(x) = 1/A(x)$, then

$$c_j = \frac{t_j^{k-1}}{\displaystyle\prod_{\substack{h=1 \\ h \neq j}}^{k} (t_j - t_h)}. \tag{8}$$

Corollary 2. If $r_0 = r_1 = \cdots = r_{k-1} = 1$, then

$$c_j = \frac{\displaystyle\prod_{\substack{h=1 \\ h \neq j}}^{k} (1 - t_h)}{\displaystyle\prod_{\substack{h=1 \\ h \neq j}}^{k} (t_j - t_h)}. \tag{9}$$

Proof. It follows from the definition of the t's that the expression (5) satisfies the recurrence relation, since

$$\sum_{i=0}^{k} (-1)^i a_i r_{n-i} = \sum_{j=1}^{k} c_j t_j^{n-k} \sum_{i=0}^{k} (-1)^i a_i t_j^{k-i} = 0, \qquad n \geqslant k.$$

The remaining part of the proof consists of solving the k linear equations

$$\sum_{j=1}^{k} c_j t_j^n = r_n, \qquad n = 0, 1, \ldots, k-1,$$

with respect to the c's. Presumably, de Moivre simply solved these equations for $k = 2, 3, 4$, and then by incomplete induction formulated the general solution (7). Following Laplace (1771, 1776), we shall solve the equations simply by successive eliminations; perhaps de Moivre also used this method.

The equations are

$$c_1 + c_2 + \cdots + c_k = r_0,$$

$$c_1 t_1 + c_2 t_2 + \cdots + c_k t_k = r_1,$$

$$\vdots$$

$$c_1 t_1^{k-1} + c_2 t_2^{k-1} + \cdots + c_k t_k^{k-1} = r_{k-1}.$$

To find c_1, say, Laplace first eliminates c_k by substracting the first equation multiplied by t_k from the second equation; next, he subtracts the second equation multiplied by t_k from the third equation, and so on. He thus gets $k-1$ linear equations of the form

$$c_1 t_1^i (t_1 - t_k) + c_2 t_2^i (t_2 - t_k) + \cdots$$

$$+ c_{k-1} t_{k-1}^i (t_{k-1} - t_k) = r_{i+1} - t_k r_i, \qquad i = 0, 1, \ldots, k-2.$$

To eliminate c_{k-1} he proceeds in the same way, successively multiplying each equation by t_{k-1} and subtracting the result from the next equation. The final result becomes

$$c_1 (t_1 - t_k)(t_1 - t_{k-1}) \cdots (t_1 - t_2)$$

$$= r_{k-1} - r_{k-2}(t_k + t_{k-1} + \cdots + t_2)$$

$$+ r_{k-3}(t_k t_{k-1} + t_k t_{k-2} + \cdots + t_3 t_2) - \cdots + (-1)^{k-1} r_0 t_k t_{k-1} \cdots t_2,$$

which proves (7) for $j = 1$. The same procedure may be used to solve the equations for c_j. This concludes the proof.

To determine $s_{i(j)}$ recursively, Laplace notes that

$$f(t) = \sum_{i=0}^{k} (-1)^i a_i t^{k-i} = \prod_{i=1}^{k} (t - t_i) = \sum_{i=0}^{k} (-1)^i s_i t^{k-i},$$

$$\prod_{\substack{i=1 \\ i \neq j}}^{k} (t - t_i) = \sum_{i=0}^{k-1} (-1)^i s_{i(j)} t^{k-i-1}.$$

Multiplying the last equation by $t - t_j$, he gets

$$f(t) = t^k + \sum_{i=1}^{k} (-1)^i (s_{i(j)} + t_j s_{i-1,(j)}) t^{k-i},$$

so that

$$s_{i(j)} = a_i - t_j s_{i-1,(j)}, \qquad i = 1, 2, \ldots, k-1.$$

Returning to Proposition 6, it will be seen that formula (6) gives a decomposition of $R(x)$ into partial fractions.

In modern terminology, Proposition 6 gives the solution of the homogeneous linear difference equation with constant coefficients under the assumption that the roots of the characteristic equation are real and different. Proposition 4 gives the solution when all the roots are equal to unity.

Remarks. Noting that $A(x) = x^k f(x^{-1})$, it will be seen that the roots of $A(x) = 0$, x_j, say, equals t_j^{-1}. De Moivre sometimes uses $A(x)$ as a characteristic function, but mostly he uses

$$F(y) = \sum_{i=0}^{k} (-1)^i a_i x^i y^{k-i} = x^k f\left(\frac{y}{x}\right),$$

so that his roots of the equation $F(y) = 0$ becomes $y_j = x t_j$. The reason for this choice is that he considers $r_n x^n$ as terms of a recursive sequence and thus gives the solution as

$$r_n x^n = \sum_{j=1}^{k} c_j y_j^n.$$

In the following propositions it is assumed that the recurring series considered satisfy the conditions stated in Proposition 6.

Proposition 7. Consider the series consisting of every mth term of $R(x)$ beginning with the first, the second, ..., the mth term, respectively. The common differential scale for these series is the polynomial in z obtained by eliminating x from the equations $A(x) = 0$ and $x^m = z$.

This result follows from the representation of r_n given in Proposition 6.

Proposition 8. Let $R(x)$ and $S(x)$ be recurring series with $A(x)$ and $B(x)$ as differential scales. Then $R(x) \pm S(x)$ are recurring series with the differential scale $A(x)B(x)$.

This result follows from Proposition 6.

Proposition 9. Let $R(x)$ and $S(y)$ be recurring series with the differential scales $A(x)$ and $B(y)$. The series with the terms $r_n s_n (xy)^n$, $n = 0, 1, \ldots$, is a recurring series, and the differential scale is the polynomial in z obtained by eliminating x and y from the equations $A(x) = 0$, $B(y) = 0$, and $xy = z$. (An example appears already in the 1718 edition.)

Corollary. If $B(y) = (1 - y)^k$, then the differential scale equals the polynomial in xy obtained as $\{A(xy)\}^k$.

The proof is based on the representation of r_n and s_n given in Proposition 6, see Todhunter (pp. 179–181).

Daniel Bernoulli wrote three papers on recurring series between 1728 and 1731, they were published in 1732 and 1738; a summary has been given by Bouckaert (1982). Bernoulli's results are less comprehensive than de Moivre's; the only new result is

$$\lim_{n \to \infty} \frac{r_{n+1}}{r_n} = \max \{|t_i|, \quad i = 1, \ldots, k\}.$$

The proof follows immediately from (5).

The most essential extension of the theory of recurring series in the second edition of the *Doctrine* compared with the first is Proposition 6, in which de Moivre shows that the general term of a recurring sequence of order k may be represented as a linear combination of k geometric sequences. De Moivre did not formulae this important result as a theorem on the solution of linear difference equations because such terminology did not exist at the time. However, as we shall discuss in §23.3, de Moivre's solution inspired Euler, Lagrange, and Laplace to develop similar solutions of differential and difference equations and to generalize these results.

23.2 DE MOIVRE'S TRIGONOMETRIC FORMULA FOR THE CONTINUATION PROBABILITY

We shall now prove de Moivre's trigonometric formula for $U_n(b, b)$, assuming for convenience that b is even. Setting

$$v_j = \frac{(2j - 1)\pi}{2b}, \qquad j = 1, 2, \ldots, \tfrac{1}{2}b,$$

de Moivre's formula may be written as

$$U_n(b, b) = \frac{p^b + q^b}{b} 2^{(n+2)/2} (pq)^{(n+2-b)/2}$$

$$\times \sum_{j=1}^{b/2} (-1)^{j-1} \frac{\sin 2v_j}{p^2 + q^2 - 2pq \cos 2v_j} (1 + \cos 2v_j)^{n/2}. \qquad (1)$$

De Moivre's proof in *Miscellanea Analytica* is somewhat incomplete; Kohli (1975b) and Schneider (1968) have given complete proofs filling in the gaps; the following proof is based on these expositions.

Proof. De Moivre first notes that U_n is a recurring series of order $\frac{1}{2}b$, with the scale of relation given by

$$a_i = \frac{b}{b-i} \binom{b-i}{i} (pq)^i.$$

The problem is to find the roots of the corresponding characteristic equation

$$f(t) = \sum_{i=0}^{b/2} (-1)^i \frac{b}{b-i} \binom{b-i}{i} (pq)^i t^{(b-2i)/2} = 0.$$

Setting

$$t = 4pq \cos^2 v, \qquad (2)$$

we get

$$f(t) = (pq)^{b/2} \sum_{i=0}^{b/2} (-1)^i \frac{b}{b-i} \binom{b-i}{i} (2 \cos v)^{b-2i}$$

$$= (pq)^{b/2} 2 \cos bv = 0, \qquad (3)$$

according to a well-known relation between $\cos bv$ and $\cos v$. Hence, the roots are

$$t_j = 4pq \cos^2 v_j = 2pq(1 + \cos 2v_j). \qquad (4)$$

Since $U_0 = U_2 = \cdots = U_{b-2} = 1$, the solution follows from Proposition 6 and its Corollary 2 in the previous section.

To find the coefficients c_j we note that the recursion formula (20.5.15) leads to

$$U_b - 1 + \sum_{i=0}^{b/2} (-1)^i \frac{b}{b-i} \binom{b-i}{i} (pq)^i = 0,$$

which means that

$$U_b - 1 + f(1) = 0, \quad \text{or} \quad f(1) = \prod_{i=1}^{b/2} (1 - t_i) = 1 - U_b = p^b + q^b,$$

since

$$f(t) = \prod_{i=1}^{b/2} (t - t_i).$$

Differentiating $\ln f(t)$, we get

$$f'(t) = f(t) \sum_{i=1}^{b/2} (t - t_i)^{-1},$$

so that

$$f'(t_j) = \prod_{i \neq j} (t_j - t_i). \tag{5}$$

From (3) we find

$$f'(t) = (pq)^{b/2}(-2b) \sin bv \left(\frac{dv}{dt} \right),$$

and from (2),

$$\frac{dt}{dv} = -4pq \sin 2v,$$

which leads to

$$f'(t_j) = \frac{b(pq)^{(b-2)/2} \sin bv_j}{2 \sin 2v_j}$$

$$= \frac{(-1)^{j-1} b(pq)^{(b-2)/2}}{2 \sin 2v_j}.$$

Since

$$c_j = \frac{f(1)}{1 - t_j} \frac{1}{f'(t_j)}, \tag{6}$$

we obtain

$$c_j = \frac{(-1)^{j-1}(p^b + q^b)2\sin 2v_j}{b(pq)^{(b-2)/2}(1 - 4pq\cos^2 v_j)}$$

We shall give two forms of U_n, one in terms of v_j and the other in terms of $2v_j$; this just requires using the fact that $2\sin v \cos v = \sin 2v$, and

$$1 - 4pq\cos^2 v = (p+q)^2 - 2pq(1 + \cos 2v) = p^2 + q^2 - 2pq\cos 2v.$$

Inserting the values found for c_j and t_j, we get U_n in the form (1), which also may be written as

$$U_n(b, b) = \frac{p^b + q^b}{b} 2^{n+2}(pq)^{(n+2-b)/2} \sum_{j=1}^{b/2} (-1)^{j-1} \frac{\sin v_j}{1 - 4pq\cos^2 v_j}(\cos v_j)^{n+1}.$$

$$(7)$$

This completes the proof.

In the *Doctrine* (1738), de Moivre presents his results in a rather odd order. In Problem 67 he considers the case with $p = \frac{1}{2}$ and gives the formula

$$U_n(b, b) = \frac{2}{b} \sum_{j=1}^{b/2} (-1)^{j-1} \frac{\cos^{n+1} v_j}{\sin v_j}, \tag{8}$$

which is a special case of (7). In Problem 68 he gives formula (1). Next follows a section entitled "Of the Summation of recurring Series" succeeded by "Some Uses of the foregoing Propositions," which contains formula (1) with a more detailed explanation than in Problem 68, but without proof. (Note that de Moivre uses "versed sine of v" instead of $1 - \cos v$ and "versed cosine of v" instead of $1 - \sin v$.)

In *Miscellanea Analytica* de Moivre indicates that a similar formula may be found for $a \neq b$, but he gives no details; this remark is not included in the *Doctrine*.

As noted by de Moivre, the trigonometric formula gives the simplest solution possible; it represents the recurring series $U_n(b, b)$ of order $\frac{1}{2}b$ by a sum containing $\frac{1}{2}b$ terms. Furthermore, for large values of n, the first few terms will usually give a good approximation to U_n.

To demonstrate the superiority of (8) over Montmort's formula (20.3.1) de Moivre calculates

$$U_{108}(12, 12) = 0.50053 - 0.00008 + \cdots,$$

which shows that the first term alone is a sufficiently good approximation; similarly he finds

$$U_{1519}(45, 45) = 0.50559 - 0.00002 + \cdots.$$

In Problem 67 de Moivre returns to the problem of finding the median duration for two players of equal skill. Assuming that v_1 is small, i.e., that b is large and that n is large also, it follows from (8) that

$$U_n \cong \frac{2}{b} \frac{\cos^{n+1} v_1}{\sin v_1}, \qquad v_1 = \frac{\pi}{2b}. \tag{9}$$

To solve the equation $U_n = \frac{1}{2}$, de Moivre sets $n + 1 \cong n$, $\sin v_1 \cong v_1$, and $\cos v_1 \cong 1 - \frac{1}{2}v_1^2$, which gives

$$n \cong 2\left[\ln 4 - \ln\left(\frac{1}{2}\pi\right)\right]\left(\frac{2b}{\pi}\right)^2 = 0.756b^2. \tag{10}$$

(The constant should have been 0.758.)

De Moivre (1738, p. 189) writes that Montmort's approximation (20.3.2) "Which tho' near the Truth in small numbers, yet is very defective in large ones." He demonstrates this by means of the two examples mentioned above. In the first case Montmort's formula gives $n = 108.3$ and (10) gives $n = 108.9$, whereas in the second case Montmort's formula gives $n = 1519$ and (10) gives 1531. Todhunter (p. 105) notes that "We should differ here with De Moivre, and consider that the results are rather remarkable for their near agreement than for their discrepancy."

Combining Montmort's formula (20.3.1) with de Moivre's formula (8), an approximation to the symmetric binomial is obtained, as shown by de Moivre, see §24.3.

De Moivre rightly considered his theory of recurring series and its use for solving the problem of the duration of play as one of his greatest achievements. He had previously stressed the method of combinations and the method of infinite series as the two most important methods in probability theory, and he could now add the method of difference equations (as it was to be called later) as a third important method.

23.3 METHODS OF SOLUTION OF DIFFERENCE EQUATIONS BY LAGRANGE AND LAPLACE, 1759–1782

De Moivre left three problems for his successors: (1) to prove the recursion formulae for $U_n(b, b)$ and $U_n(a, b)$; (2) to derive the trigonometric formula for $U_n(a, b)$; and (3) to develop a general theory for solving difference equations. These problems were solved by Joseph-Louis Lagrange (1736–1813) and Pierre-Simon Laplace (1749–1827) in the 1770s. Like Taylor they wanted to create a calculus of finite differences analogous to the infinitesimal calculus. An exposition of the calculus of finite differences at about 1800, including a comprehensive list of the early literature, has been given by Lacroix (1819).

In the present section we shall comment on the early history of difference equations, concentrating on contributions by Lagrange and Laplace. In the next section the theory will be applied for solving the problems mentioned above.

In 1743 Euler solved the homogeneous linear differential equation of the mth order with constant coefficients (using the same idea as de Moivre) by guessing at a particular solution of the form $y = ce^{rx}$ and thus deriving the characteristic equation as an algebraic equation of the mth degree. He pointed out that the general solution will be a linear combination of m independent particular solutions. The year before he had proved (incompletely) that a polynomial with real coefficients can be decomposed into linear and quadratic factors with real coefficients, and he could therefore assert that the characteristic equation always has m roots. He found that the particular solution for a single real root has the form $y = ce^{rx}$, for a real root of multiplicity k the form e^{rx} times a polynomial in x of degree k, and for a pair of conjugate complex roots the form

$$y = e^{ax}(c_1 \cos bx + c_2 \sin bx),$$

where c_1 and c_2 are arbitrary constants.

In 1753 Euler solved the nonhomogeneous differential equation of the mth order with constant coefficients by devising a method for reducing the order by unity and thus successively reducing the problem to the solution of a nonhomogeneous differential equation of the first order.

In 1755 Euler introduced the symbol Δ for a finite difference, so that a difference equation could be written in a form analogous to a differential equation.

In the 1760s Lagrange proved that Euler's 1753 theorem also holds for a nonhomogeneous differential equation of the mth order with variable coefficients so that the general solution may be expressed by means of the solution of the homogeneous equation and the solution of an adjoint equation of the first order.

We shall sketch how analogous results on ordinary and partial difference equations were derived by Lagrange and Laplace.

Lagrange's Solution of the Nonhomogeneous Linear Difference Equation with Constant Coefficients, 1759

Lagrange (1759) first considers a difference equation of the first order,

$$\alpha(x)u(x) + \Delta u(x) = \beta(x), \tag{1}$$

where $u(x)$ is unknown, and $\alpha(x)$ and $\beta(x)$ are known functions of the integer-valued variable x. By means of the substitution $u(x) = v(x)z(x)$, he derives two simpler difference equations

$$(\Delta v + \alpha v)z = 0 \qquad \text{and} \qquad (v + \Delta v)\Delta z = \beta,$$

which lead to the solution

$$u(x) = \left\{ \prod \{1 - \alpha(x-1)\} \right\} \left\{ k + \sum \frac{\beta(x)}{\prod \{1 - \alpha(x)\}} \right\}, \tag{2}$$

where $\prod f(x)$ denotes the indefinite continued product of successive values of $f(\cdot)$, beginning with an initial value and ending with $f(x)$, and $\sum f(x)$ denotes the indefinite sum ending with $f(x-1)$; k is an arbitrary (periodic) constant.

Next, he turns to the solution of the nonhomogeneous linear difference equation of the mth order with constant coefficients,

$$u(x) + \sum_{i=1}^{m} \alpha_i \Delta^i u(x) = \beta(x). \tag{3}$$

Referring to d'Alembert's method for solving the corresponding differential equation, he introduces the auxiliary function

$$z(x) = u(x) + \sum_{i=1}^{m-1} (\alpha_i + a_i) \Delta^i u(x), \tag{4}$$

with undetermined constants $a_1, \ldots, a_{m-1}$, which he chooses such that z satisfies a difference equation of the first order.

Writing (3) as

$$u(x) + \sum_{i=1}^{m-1} (\alpha_i + a_i) \Delta^i u(x) - \left\{ \sum_{i=1}^{m-1} a_i \Delta^i u(x) - \alpha_m \Delta^m u(x) \right\} = \beta(x), \tag{5}$$

he determines the a's by requiring that

$$\Delta u + \sum_{i=1}^{m-1} (\alpha_i + a_i)\Delta^{i+1}u = a_1^{-1}\left(\sum_{i=1}^{m-1} a_i \Delta^i u - \alpha_m \Delta^m u\right). \tag{6}$$

Equating coefficients of $\Delta^i u$ from the two sides of this equation leads to the equations

$$\alpha_i + a_i = \frac{a_{i+1}}{a_1}, \qquad i = 1,\dots,m-2; \qquad \alpha_{m-1} + a_{m-1} = -\frac{\alpha_m}{a_1},$$

with the solution

$$a_1^i a_{m-i} = -\sum_{j=0}^{i} \alpha_{m-j}a_1^j, \qquad i = 1,\dots,m-1. \tag{7}$$

The last of these equations is the characteristic equation

$$\sum_{j=0}^{m} \alpha_{m-j}a_1^j = 0, \qquad \alpha_0 = 1, \tag{8}$$

with m roots which each determines a set of the remaining a's by means of (7).

It follows from (5) and (6) that

$$z(x) - a_1\,\Delta z(x) = \beta(x),$$

which has a solution of the form (2).

Each root of (8) gives a solution. Inserting these functions, $z_1(x),\dots,z_m(x)$, say, into (4) and solving the m linear equations, $u(x)$ is obtained as a linear combination of the z's.

Lagrange has thus solved the nonhomogeneous linear difference equation with constant coefficients in terms of the roots of the characteristic equation for the homogeneous case and the solution of a nonhomogeneous equation of the first order.

In accordance with the title of the paper, "On the integration of an equation in finite differences, which contains the theory of recurring sequences," he notes that the difference equation (3) is equivalent to the equation

$$u(x) + \sum_{i=1}^{m} A_i u(x+i) = \beta(x),$$

which may be considered as a generalization of the equation defining a recurring sequence, the ordinary recurring sequence being obtained for $\beta(x) = 0$. He concludes, "Thus the theory of recurring sequences has been reduced to the calculus of differences and in this manner established from direct and natural principles instead of being treated indirectly as formerly done." He does not refer to de Moivre, but he remarks that the main field of application of the theory above is the doctrine of chances and that he intends to return to these applications at another occasion.

Laplace's Method of Reduction of the Order of the Equation by the Method of Undetermined Coefficients, 1771–1776

Lagrange's paper was followed by three closely connected papers by Laplace (1771, 1774, 1776). The first one, "Researches on the integral calculus with infinitely small differences and with finite differences" is written in 1771, and Laplace sent it to the Royal Society of Turin for publication, presumably because Lagrange at the time was a professor of mathematics at Turin and because the first part of the paper contains a new proof of Lagrange's theorem on the integrability of the nonhomogeneous linear differential equation of the mth order with variable coefficients, published in a previous volume of the Society's journal. Laplace's new proof uses the method of undetermined coefficients, which he applies to the solution of both differential and difference equations. The solutions are given in great detail and involve an enormous amount of algebraic manipulations and formulae, written in rather cumbersome notation. It has to be remembered, however, that this paper is the first important mathematical paper written by Laplace, who was only 22 years old at the time. It is a purely mathematical paper; in the introductory remarks he mentions the possibility of applications (of differential equations) to the natural sciences without giving any examples; he does not mention games of chance in connection with his discussion of difference equations, even if he must have known of Lagrange's remark to that effect.

The two other papers, "Memoir on recurro-recurring sequences and on their uses in the theory of chances" and "Researches on the integration of differential equations with finite differences and on their use in the theory of chances," were not published until 1774 and 1776, even though the last one was read to the Paris Academy of Sciences in the beginning of 1773. Stigler (1978) has discussed the chronology of Laplace's early work.

It seems reasonable to suppose that Laplace immediately after finishing the first paper realized that the same method of proof might be used to solve *partial* difference equations.

In the second paper he shows how to solve partial difference equations of the first and second order and demonstrates his method on three examples from games of chance.

In the introduction to the third paper he states that de Moivre was the first to find the general term of a recurring sequence, but that it was Lagrange who as the first realized that this problem depends on the solution of a linear difference equation. He also refers to some works by Condorcet. He mentions that the results of his previous papers have been incorporated into the present one, which is more general in scope.

Hence, the third paper contains the analysis of difference equations from the first paper, rewritten with improved notation and with some new results added; it also contains the analysis of partial difference equations from the second paper with an important generalization; further, he analyzes partial difference equations of higher order, simultaneous difference equations, circular difference equations, and some simple functional equations. Finally, he solves many of the classical problems in probability theory by giving explicit solutions of the corresponding difference equations. The paper is very impressive; it is 125 pages long.

The three papers by Laplace have not been much discussed in the historical literature, perhaps because the methods proposed were superseded by his method of generating functions a few years later. The papers are, however, of historial importance for several reasons. First, they represent a phase in the development of the theory for solving difference equations. Second, they give an analytical method for solving probability problems which previously had been solved by combinatorial methods. Third, they give the first proofs of de Moivre's recursion formulae for the problem of the duration of play, and they also give the first (incomplete) derivation of the trigonometric solution of this problem for the general case.

We shall sketch Laplace's proofs of the two most important theorems on difference equations, using notation resembling that in the third paper. The first theorem and proof is from the 1771 paper, the second from the 1774 paper. We shall also indicate the amendments made in the 1776 paper.

Laplace first solves the difference equation of the first order

$$u(x) = \alpha(x)u(x-1) + \beta(x).$$

By successive substitutions he gets

$$u(x) = \left\{ \prod \alpha(x) \right\} \left\{ k + \sum \frac{\beta(x+1)}{\prod \alpha(x+1)} \right\}, \tag{9}$$

in agreement with Lagrange's result (2).

Next he considers the difference equation of the mth order with variable coefficients

$$u(x) = \sum_{i=1}^{m} \alpha_i(x)u(x - i) + \beta(x). \qquad (10)$$

To reduce this equation to one of the first order, Laplace sets

$$u(x) = a(x)u(x - 1) + b(x), \qquad (11)$$

where $a(x)$ and $b(x)$ are unknown functions to be determined from $\{\alpha_i(x)\}$ and $\beta(x)$. Using the idea from Lagrange's proof, he shows that $b(x)$ may be found as solution of a difference equation of the same form as the difference equation for $u(x)$ but of order $m - 1$ instead of m. Continuing in this manner, the problem is reduced to the solution of a difference equation of the first order.

The function $a(x)$ is found by solving the characteristic equation

$$\sum_{j=0}^{m-1} \alpha_{m-j}(x)a^{(j)}(x - m) - a^{(m)}(x - m) = 0, \qquad (12)$$

where we have introduced the notation

$$a^{(j)}(x - i) = a(x - i + 1)a(x - i + 2)\cdots a(x - i + j),$$
$$i = 1, 2, \ldots, m - 1; \quad j = 1, 2, \ldots, i,$$

and $a^{(0)}(x - i) = 1$. It will be seen that (12) is a generalization of (8); it is difficult to solve without further specialization.

Let $a_1(x)$ be a solution of (12) and set

$$u_1(x) = \prod a_1(x).$$

It then follows from (9) and (11) that

$$u(x) = u_1(x)\left\{k_1 + \sum \frac{b(x + 1)}{u_1(x + 1)}\right\}$$

is a solution of the original difference equation.

Continuing this process, Laplace gives the complete solution in the form

$$u(x) = \sum_{i=1}^{m} u_i(x)\left\{k_i + \sum \frac{\beta(x + 1)}{z_i(x + 1)}\right\}, \qquad (13)$$

where $u_1(x), \ldots, u_m(x)$ are m independent solutions of the homogeneous equation; $z_1(x), \ldots, z_m(x)$ are functions of the u's; and the k's are periodic constants.

After having derived the general solution, Laplace turns to cases that are reducible to equations with constant coefficients. He specifies the coefficients to be

$$\alpha_i(x) = \alpha_i g(x) g(x-1) \cdots g(x-i+1) = \alpha_i g^{(i)}(x-i),$$

where $\{\alpha_i\}$ are given constants, and $g(x)$ a given function. He notes that this case covers all cases discussed in the literature thus far; in particular, he considers $g(x) = 1$ and $g(x) = x$.

To solve the homogeneous equation, Laplace sets $a(x) = tg(x)$, say, where t is an undetermined constant. Since

$$\alpha_{m-j}(x) a^{(j)}(x-m) = \alpha_{m-j} t^j g^{(m)}(x-m),$$

the characteristic equation (12) becomes

$$f(t) = t^m - \sum_{i=1}^{m} \alpha_i t^{m-i} = 0.$$

Suppose first that the characteristic equation has m real and different roots, $t_1, \ldots, t_m$. It then follows from (13) for $\beta(x) = 0$ that

$$u(x) = G(x) \sum_{i=1}^{m} c_i t_i^x, \qquad G(x) = g(1)g(2) \cdots g(x),$$

where the c's have to be determined from m given values of $u(x), u(1), \ldots, u(m)$, say. Let us denote $u(x)/G(x)$ by u_x. The c's then satisfy the equations

$$\sum_{i=1}^{m} c_i t_i^x = u_x, \qquad x = 1, \ldots, m,$$

which Laplace solves by the method indicated in connection with de Moivre's Proposition 6 in §23.1. Laplace's solution is slightly different from de Moivre's because Laplace uses $u_1, \ldots, u_m$, and de Moivre uses $u_0, \ldots, u_{m-1}$ as starting values.

Suppose next that the characteristic equation has a double root, $t_1 = t_2$, say. Laplace sets $t_2 = t_1 + dt_1$ and shows that the two terms in the limit become $t_1^x(A + Bx)$, and continuing in this manner he gets the solution for multiple roots.

In 1771 he does not discuss the case with complex roots, but in 1776 he covers all three cases and presents the complete solution as found in textbooks today.

Finally, he gives the solution of the nonhomogeneous equation as

$$u(x) = G(x) \sum_{i=1}^{m} t_i^{x+m-1} \frac{1}{f'(t_i)} \left\{ k_i + \sum \frac{\beta(x+1)}{G(x+1)t_i^{x+1}} \right\}.$$

As an example, Laplace considers the expansion of $\sin nz$ in terms of $x = \sin z$ and $y = \cos z$. From the formula

$$\sin nz = 2y \sin(n-1)z - \sin(n-2)z, \qquad n = 1, 2, \ldots,$$

it follows by induction that $\sin nz$ may be written as

$$\sin nz = x\{a_n y^{n-1} + b_n y^{n-3} + c_n y^{n-5} + \cdots\}, \qquad n = 2, 3, \ldots.$$

Laplace derives a set of difference equations for the coefficients and solving these he gets

$$\sin nz = x \sum_{i=0}^{[(n-1)/2]} (-1)^i \binom{n-i-1}{i} (2y)^{n-2i-1}. \tag{14}$$

Laplace ends the 1771 paper with a short discussion of the solution of partial differential equations with two independent variables, using the same method as for difference equations above.

Laplace's second paper (1774) contains the beginnings of a theory of partial difference equations. He generalizes the concept of a recurring sequence $\{u(x); x = 1, 2, \ldots\}$ to a recurro-recurring sequence $\{u_n(x); x = 1, 2, \ldots; n = 1, 2, \ldots\}$, which he defines as an infinite double sequence in which each term is a linear combination of a given number of preceding terms, the relation being valid only after an adequate number of initial functions has been specified. Hence, the terms of a recurring sequence satisfies an ordinary difference equation and the terms of a recurro-recurring sequence satisfies a partial difference equation.

At the end of the introduction Laplace writes; "As I have found that such sequences are very useful in the theory of chances and as they, as far as I know, have not been examined by anybody, I believe that it will not be unavailing to develop them here to some extent."

Laplace first considers a partial difference equation of the first order in n and an arbitrary order in x,

$$u_n(x) - \sum_{i=1} \alpha_n(i)u_n(x-i) = \sum_{j=0} \beta_n(j)u_{n-1}(x-j) + \gamma_n, \tag{15}$$

where $\{\alpha_n(i)\}$, $\{\beta_n(j)\}$ and γ_n are given functions, and $u_n(x)$ is the unknown function to be determined from the equation, and where $u_n(0) = 0$. There are finitely many terms in the sums, and we have followed Laplace in not specifying the number of terms to keep the notation simple.

Since the solution depends on an arbitrary function of x, Laplace sets $u_1(x) = f(x)$. Studying $u_2(x)$ and $u_3(x)$ by means of (15), he concludes that the partial difference equation may be reduced to an ordinary difference equation in x, which he writes in the form

$$u_n(x) = \sum_{i=1} a_n(i)u_n(x-i) + b_n(x), \tag{16}$$

where $\{a_n(i)\}$ and $b_n(x)$ are undetermined functions. Inserting (16) for $u_n(x)$ on the left-hand side of (15) and eliminating $u_{n-1}(x-j)$ by means of (16) on the right-hand side, Laplace transforms the two sides of (15) to linear combinations of $\{u_n(x-i)\}$, and equating coefficients he obtains a set of difference equations for the determination of $\{a_n(i)\}$ and $b_n(x)$, which he shows how to solve.

In the 1774 paper, Laplace considers the special case $b_n(x) = b_n$ only; the general case is discussed in the 1776 paper.

Finally, $u_n(x)$ is found by solving (16). The solution has the form

$$u_n(x) = c_{n1}t_{n1}^x + c_{n2}t_{n2}^x + \cdots + k_n(x),$$

where $t_{n1}, t_{n2}, \ldots$ are the roots of the characteristic equation for the ordinary difference equation (16). Inserting this expression into the original partial difference equation (15), the unknown functions $c_{n1}, c_{n2}, \ldots$, and $k_n(x)$ may be found from a set of difference equations.

Laplace extends the proof to a partial difference equation of the second order in n because such equations are of particular importance in the theory of chances (see Problem 7 in §23.5), and he remarks that the same method may be used to reduce the order of any partial difference equation by unity.

The most important general result in Laplace's three papers is the one that a solution always exists and may be obtained by successively reducing the order of the equation, using the method of undetermined coefficients. The price to be paid for each such reduction is, however, that the coefficients in the new equation have to be found as solutions of a set of difference equations, depending on the coefficients of the old equation. In the general case this procedure leads to a formidable set of equations, but in cases where the degree of the original equation is low and the coefficients are constants, the method is manageable. Laplace demonstrates this by solving nine probability problems; we have used some of them as problems for the reader in §23.5.

Lagrange's Solution by Means of the Substitution $u(x, y) = ca^x b^y$, 1777

Lagrange (1777) reacted to Laplace's papers by publishing a large paper entitled "Researches on recurring sequences whose terms vary in several different ways, or, on the integration of linear equations with finite and partial differences, and on the use of these equations in the theory of chances." He refers to his own paper (1759) and to Laplace's papers and writes, "I believe, however, that one may still add something to the work of this illustrious mathematician and treat the same topic in a manner more direct, more simple and especially more general."

In the first section of the paper, Lagrange discusses the linear difference equation with variable coefficients, which he solves by means of the method that today is known as Lagrange's method of variation of parameters.

In the following sections he discusses partial difference equations with constant coefficients and gives a direct method of solution. We shall change our notation from $u_n(x)$ to $u(x, y)$, which fits better into the system of formulae used by Lagrange.

He first considers the double sequence $\{u(x, y); x = 0, 1, \ldots, y = 0, 1, \ldots\}$, where $u(x, y)$ satisfies the equation

$$\alpha_{00}u(x, y) + \alpha_{10}u(x + 1, y) + \alpha_{01}u(x, y + 1) + \alpha_{11}u(x + 1, y + 1) = 0. \quad (17)$$

Setting

$$u(x, y) = ca^x b^y, \quad (18)$$

a, b, and c being undetermined constants, and inserting into (17), he obtains the equation

$$\alpha_{00} + \alpha_{10}a + \alpha_{01}b + \alpha_{11}ab = 0,$$

which gives

$$b = -\frac{\alpha_{00} + \alpha_{10}a}{\alpha_{01} + \alpha_{11}a}.$$

It follows that b^y may be written as

$$b^y = \sum_{i=0}^{\infty} h_i(y)a^{ry-i}, \tag{19}$$

r being an integer depending on the coefficients of (17). Inserting into (18) he finds

$$u(x, y) = c \sum_{i=0}^{\infty} h_i(y)a^{x+ry-i}.$$

Since a and c are arbitrary, and since any linear combination of expressions like that above is a solution of (17), Lagrange concludes that the form of the solution is

$$u(x, y) = \sum_{i=0}^{\infty} h_i(y)g(x + ry - i),$$

where $g(\cdot)$ is an undetermined function.

From $u(x, 0) = ca^x$, it follows that $h_0(0) = 1$ and $h_i(0) = 0$, $i \geqslant 1$, so that $g(x) = u(x, 0)$. The general solution thus becomes

$$u(x, y) = \sum_{i=0}^{\infty} h_i(y)u(x + ry - i, 0), \tag{20}$$

where r and $h_i(y)$ are to be found from the expansion (19). If the series is infinite, the solution will contain values of $u(x, 0)$ for negative arguments; these values have to be found from (20) using the boundary conditions for $u(x, y)$.

Lagrange works out the solutions for $\alpha_{11} = 0$ and $\alpha_{01} = 0$, respectively; it turns out that the series in both cases are finite.

In general, however, the solution (20) will contain infinitely many terms. To derive another solution depending on a finite number of terms only, Lagrange introduces a representation of a and b in terms of the parameter t, say, by the equations

$$a = a_1 t(1 + a_2 t^{-1}) \qquad \text{and} \qquad b = b_1(1 + b_2 t^{-1}),$$

where $a_1 = -1/\alpha_{11}$, $a_2 = \alpha_{01}$, $b_1 = -\alpha_{10}/\alpha_{11}$, $b_2 = \alpha_{01} - \alpha_{00}\alpha_{11}/\alpha_{10}$. From

$$a^x b^y = \sum_{i=0}^{x+y} h_i(x, y)t^{x-i},$$

where

$$h_i(x, y) = a_1^x b_1^y \sum_{j=0}^{i} \binom{x}{j}\binom{y}{i-j} a_2^j b_2^{i-j},$$

it follows by a similar argument as above that the solution has the form

$$u(x, y) = \sum_{i=0}^{x+y} h_i(x, y) f(x - i),$$

where $f(\cdot)$ is an undetermined function.

The value of $f(\cdot)$ may be determined by means of $u(x, 0)$ and $u(0, y)$. From

$$u(x, 0) = a_1^x \sum_{i=0}^{x} \binom{x}{i} a_2^i f(x - i),$$

it follows that

$$\frac{f(x)}{a_2^x} = \sum_{i=0}^{x} (-1)^i \binom{x}{i} \frac{u(x-i, 0)}{a_1^{x-i} a_2^{x-i}}, \qquad x = 0, 1, \ldots,$$

and from

$$u(0, y) = b_1^y \sum_{i=0}^{y} \binom{y}{i} b_2^i f(-i),$$

it follows that

$$b_2^y f(-y) = \sum_{i=0}^{y} (-1)^i \binom{y}{i} \frac{u(0, y-i)}{b_1^{y-i}}, \qquad y = 0, 1, \ldots.$$

The problem has thus been solved in a simple and general manner.

Lagrange extends the analysis to difference equations of order m with two and three independent variables. We shall sketch the solution for the equation

$$\alpha_{00} u(x, y) + \alpha_{10} u(x + 1, y) + \alpha_{20} u(x + 2, y) + \alpha_{01} u(x, y + 1)$$
$$+ \alpha_{11} u(x + 1, y + 1) + \alpha_{02} u(x, y + 2) = 0.$$

The substitution $u(x, y) = c a^x b^y$ leads to the equation

$$\alpha_{00} + \alpha_{10} a + \alpha_{20} a^2 + \alpha_{01} b + \alpha_{11} ab + \alpha_{02} b^2 = 0. \qquad (21)$$

Lagrange writes the expansion of b^y in the form

$$b^y = \sum_{i=0}^{y} h_i(y)a^i + b \sum_{i=0}^{y-1} k_i(y)a^i,$$

where he has used (21) to eliminate powers of b larger than or equal to 2. By the usual argument he concludes that the general solution has the form

$$u(x, y) = \sum_{i=0}^{y} h_i(y)f(x + i) + \sum_{i=0}^{y-1} k_i(y)g(x + i),$$

where f and g are arbitrary functions. If $u(x, 0)$ and $u(x, 1)$ are given, it is easy to show that the solution becomes

$$u(x, y) = \sum_{i=0}^{y} h_i(y)u(x + i, 0) + \sum_{i=0}^{y-1} k_i(y)u(x + i, 1). \tag{22}$$

Lagrange gives several other forms of the expansion of $a^x b^y$ and corresponding forms of the solutions.

Finally, Lagrange solves seven probability problems; we have used some of them as problems for the reader in §23.5.

Laplace's Solution by Means of Generating Functions, 1782

Laplace did not give up; in 1780 he read a paper, published in 1782, to the Paris Academy of Sciences on the use of generating functions for solving difference equations. He introduced the name "generating function" and developed a calculus of generating functions, to a large extent based on symbolic methods, which originally had been introduced by Leibniz and later developed much further by Lagrange (1772). Generating functions had previously been used by de Moivre, Simpson, and Lagrange for finding the distribution of the sum of n identically distributed random variables, see §14.3. We shall illustrate Laplace's method by discussing two important examples.

Let $g(t)$ be the generating function of $u(x)$ such that

$$g(t) = \sum_{x=0}^{\infty} u(x)t^x,$$

and let $u(x)$ satisfy the homogeneous linear difference equation with constant coefficients

$$\sum_{i=0}^{m} \alpha_i u(x + i) = 0.$$

Multiplying this equation by t^x and summing over x, we get

$$\sum_{i=0}^{m} \alpha_i \sum_{x=0}^{\infty} u(x + i)t^x$$

$$= \alpha_0 g(t) + \sum_{i=1}^{m} \alpha_i t^{-i}\{g(t) - u(0) - u(1)t - \cdots - u(i - 1)t^{i-1}\} = 0.$$

Solving for $g(t)$, we obtain

$$g(t) \sum_{i=0}^{m} \alpha_i t^{m-i} = \sum_{i=0}^{m-1} \alpha_{i+1} t^{m-i-1} \sum_{j=0}^{i} u(j)t^j$$

$$= \sum_{j=0}^{m-1} u(j) \sum_{i=0}^{m-1-j} \alpha_{j+i+1} t^{m-i-1}.$$

This is Laplace's solution of the difference equation in terms of the generating function of $u(x)$. The generating function is a proper rational function; to get $u(x)$ one has to find the corresponding power series. This result corresponds to de Moivre's formula (1.1).

It will be seen that the numerator of $g(t)$ is a polynomial of degree $m - 1$ with coefficients depending on $u(0), \ldots, u(m - 1)$. Generally, we may write $g(t) = f(t)/\sum \alpha_i t^{m-i}$, where $f(t)$ is a polynomial of degree $m - 1$, the m coefficients being determined by m given values of $u(x)$.

Examples of Laplace's use of generating functions have previously been given in §21.2 on Waldegrave's problem and in §22.6 on the theory of runs.

Suppose now that $u(x, y)$ satisfies the partial difference equation

$$u(x + 1, y + 1) - au(x, y + 1) - bu(x + 1, y) - cu(x, y) = 0,$$

and define the generating function as

$$g(s, t) = \sum_{x=0}^{\infty} \sum_{y=0}^{\infty} u(x, y)s^x t^y.$$

Multiplying the difference equation by $s^x t^y$ and summing we get

$$\frac{1}{st}\{g(s, t) - g(s, 0) - g(0, t) + u(0, 0)\} - \frac{a}{t}\{g(s, t) - g(s, 0)\}$$

$$- \frac{b}{s}\{g(s, t) - g(0, t)\} - cg(s, t) = 0,$$

where $g(s, 0)$ and $g(0, t)$ are the generating functions of $u(x, 0)$ and $u(0, y)$, respectively. Hence, we have

$$g(s, t)(1 - as - bt - cst) = (1 - as)g(s, 0) + (1 - bt)g(0, t) - u(0, 0).$$

In general, we may write

$$g(s, t) = \frac{h(s) + k(t)}{1 - as - bt - cst},$$

$h(s)$ and $k(t)$ being functions to be determined from the boundary conditions. Considering the partial difference equation

$$\sum_{i=0}^{m} \sum_{j=0}^{n} \alpha_{ij} u(x + i, \, y + j) = 0,$$

Laplace gives the generating function as

$$g(s, t) = \frac{\sum_{i=0}^{n-1} h_i(s)t^i + \sum_{i=0}^{m-1} k_i(t)s^i}{\sum_{i=0}^{m} \sum_{j=0}^{n} \alpha_{ij} s^{m-i} t^{n-j}}, \tag{23}$$

where the arbitrary functions $\{h_i(s)\}$ and $\{k_i(t)\}$ may be found from the generating functions of $u(x, 0)$, $u(x, 1), \ldots, u(x, n - 1)$ and $u(0, y)$, $u(1, y), \ldots, u(m - 1, y)$. The proof may be carried out analogously to that above; Laplace indicates a proof by symbolic methods.

It will be seen that it is easy to find the generating function of u when u satisfies a linear difference equation with constant coefficients, which is the case for the classical problems in probability theory. Laplace had thus reached his goal; the results may be seen in his *Théorie Analytique des Probabilités* (1812), where he gave the solution of many problems by this method.

23.4   SOLUTIONS OF THE PROBLEM OF THE DURATION OF PLAY BY LAPLACE AND LAGRANGE

As mentioned in §20.5, Laplace writes de Moivre's algorithm as the partial difference equation

$$u_n(x) = pu_{n-1}(x - 1) + qu_{n-1}(x + 1), \qquad -a < x < b, \quad n \geqslant 1, \tag{1}$$

with the boundary conditions $u_0(0) = 1$, and

$$u_n(x) = 0, \quad x > (b-1) \wedge n \qquad \text{and} \qquad x < (-a+1) \vee (-n), \quad n \geq 0.$$

Laplace first derives de Moivre's recursion formula by transforming the partial difference equation to an ordinary difference equation. Next, he solves this equation thereby obtaining the trigonometric formula for $D_n(a,b)$.

To reduce the order of (1) with respect to x from two to one, he eliminates $u_{n-1}(x+1)$ by means of the relation

$$qu_{n-1}(x+1) = u_n(x) - pu_{n-1}(x-1), \tag{2}$$

starting from the upper boundary where $u_n(b) = 0$.

As shown in §20.5 he then obtains the expression

$$u_n(x) = \sum_{i=1} \alpha_i(x) u_{n-2i}(x) + \sum_{i=1} \beta_i(x) u_{n-2i+1}(x-1), \qquad n \geq b - x. \tag{3}$$

It will be seen that the cost of the reduction in order from two to one is that the coefficients now depend on x and that the order with respect to n has increased.

To reduce (3) to an ordinary difference equation in n, Laplace uses the boundary condition $u_n(-a) = 0$. Setting $x = -a + 1$, he obtains

$$u_n(-a+1) = \sum_{i=1}^{[(a+b-1)/2]} \alpha_i(-a+1) u_{n-2i}(-a+1), \qquad n \geq a + b - 1, \tag{4}$$

see (20.5.22).

Since $r_n^* = qu_{n-1}(-a+1)$, multiplication by q leads to a recursion formula for the ruin probability. The only problem left is to find $\alpha_i(x)$.

By means of (2) and (3) we obtain

$$qu_{n-1}(x+1) = \sum_{i=1} \alpha_i(x+1)\{u_{n-2i}(x) - pu_{n-2i-1}(x-1)\}$$

$$+ q \sum_{i=1} \beta_i(x+1) u_{n-2i}(x),$$

which inserted into (1) gives

$$u_n(x) = \sum_{i=1} \{\alpha_i(x+1) + q\beta_i(x+1)\} u_{n-2i}(x)$$

$$+ pu_{n-1}(x-1) - p \sum_{i=1} \alpha_i(x+1) u_{n-2i-1}(x-1).$$

Comparing with (3) and equating coefficients, we obtain

$$\alpha_i(x+1) + q\beta_i(x+1) = \alpha_i(x), \qquad i \geqslant 1,$$
$$\beta_1(x) = p, \qquad \beta_i(x) = -p\alpha_{i-1}(x+1), \qquad i \geqslant 2,$$

which gives

$$\alpha_1(x+1) - \alpha_1(x) = -pq, \tag{5}$$

$$\alpha_i(x+1) - \alpha_i(x) = pq\alpha_{i-1}(x+2), \qquad i \geqslant 2. \tag{6}$$

The boundary conditions follow from (3) for $x = b-1$ and $b-2$, which show that

$$\alpha_i(b-1) = 0, \quad i \geqslant 1, \qquad \text{and} \qquad \alpha_i(b-2) = 0, \quad i \geqslant 2.$$

Repeated applications of (6) gives that $\alpha_i(b+1-2i) = \beta_{i+1}(b-2i) = 0, i \geqslant 1$.
The solution of (5) is

$$\alpha_1(x) = (b-1-x)pq,$$

since $\alpha_1(b-1) = 0$. Laplace also finds α_2 and α_3 from (6), and then he writes "and so on." It is easy to check that the general solution of (6) is

$$\alpha_i(x) = (-1)^{i-1}(pq)^i(b-i-x)^{(i)}/i!, \qquad i \geqslant 1, \tag{7}$$

so that

$$\alpha_i(-a+1) = (-1)^{i-1}\binom{a+b-1-i}{i}(pq)^i, \tag{8}$$

which is the coefficient in de Moivre's recursion formula for the ruin and continuation probabilities, see (20.5.14) and (20.5.20).

This is the first proof of de Moivre's recursion formula; it must have been very satisfactory for Laplace (1776) in one of his first papers to prove this theorem which had remained unproved since 1718. He does not, however, refer to de Moivre.

In the 1774 paper Laplace proves the recursion formula for $a = b$, starting from the partial difference equation

$$u_n(x) = 2pqu_{n-2}(x) + p^2u_{n-2}(x-2) + q^2u_{n-2}(x+2), \qquad -b < x < b.$$

His method of proof is the same as above, but because of symmetry he considers only positive values of x, beginning with $x = 0$ and working out to the boundary $x = b$. We shall leave the proof to the reader.

The proof for $a = b$ is also included in the 1776 paper. One may wonder why he did not derive the result as a special case of (8). The reason is that, like de Moivre, he did not know the combinatorial identity leading from the one formula to the other, see §20.6, Problems 12 and 13.

The proof given above is a "polished" version of Laplace's proof, retaining Laplace's idea but using modern notation; it is based on a paper by Hald and Johansen (1983), which also contains an interesting combinatorial proof by Johansen. As the problem is combinatorial in nature, it is surprising that nobody has given a combinatorial proof until 1983.

Laplace (1776) continues his analysis by solving the ordinary difference equation corresponding to (4) but with u_n replaced by $D_n(a, b)$. Setting $c = a + b$, we may write Laplace's equation as

$$\sum_{i=0}^{[(c-1)/2]} (-1)^i \binom{c-1-i}{i} (pq)^i D_{n-2i} = 0, \qquad n \geqslant c - 1.$$

For convenience we shall assume that c is odd so that $\frac{1}{2}(c-1)$ is an integer; a similar proof holds for c even. Setting $D_n \propto t^{(n+c-1)/2}$, we obtain the characteristic equation

$$f(t) = \sum_{i=0}^{(c-1)/2} (-1)^i \binom{c-1-i}{i} (pq)^i t^{(c-1-2i)/2} = 0.$$

To solve this equation Laplace refers to his previous result (3.14), which may be written as

$$\sin cv = \sin v \sum_{i=0}^{(c-1)/2} (-1)^i \binom{c-1-i}{i} (2 \cos v)^{c-1-2i}.$$

By the substitution

$$2 \cos v = \pm \left(\frac{t}{pq} \right)^{1/2},$$

he finds

$$f(t) = (pq)^{(c-1)/2} \frac{\sin cv}{\sin v}.$$

Since

$$\sin cv = 0 \qquad \text{for} \quad v = \frac{j\pi}{c}, \quad j = 0, 1, \ldots,$$

the roots of the characteristic equation become

$$t_j = 4pq \cos^2 v_j, \qquad v_j = \frac{j\pi}{c}, \qquad j = 1, \ldots, \tfrac{1}{2}(c - 1).$$

Finally, Laplace gives the solution as

$$D_n = \sum_{j=1}^{(c-1)/2} c_j t_j^{n/2} = \sum_{j=1}^{(c-1)/2} c_j (4pq)^{n/2} \cos^n v_j.$$

Laplace leaves the solution in this form; he does not attempt to find the coefficients c_j but refers the reader to find them by the method given in his general theory.

Here we have the first, although incomplete, trigonometric expression for $D_n(a, b)$. As we shall see below in connection with Lagrange's solution, the determination of the c's is not completely trivial.

Lagrange's formulation of the problem (1777) is slightly different from the usual one. He considers a player, A, say, playing a series of games in which he wins a point with probability p and loses a point with probability $q = 1 - p$ in each game. After winning i of n games, his score is $g_n = 2i - n$. The play continues as long as the score is between $-a$ and b, a and b being positive integers, and stops as soon as the score becomes $\leqslant -a$ or $\geqslant b$. Lagrange's formulation is thus equivalent to the modern one as a random walk with two absorbing barriers.

Lagrange defines $w_n(x)$ as the probability of stopping, given that the state of the play is such that A has a score of $g = x - a$ and that the play ends after at most n more games. Hence, $w_n(a)$ denotes the probability of a duration of at most n games, i.e., $w_n(a) = D_n(a, b)$ in the previous notation.

The fundamental difference equation is

$$w_n(x) = pw_{n-1}(x + 1) + qw_{n-1}(x - 1), \qquad x = 1, 2, \ldots, a + b - 1;$$
$$n = 1, 2, \ldots,$$

with the boundary conditions

$$w_n(0) = w_n(a + b) = 1, \qquad n = 0, 1, \ldots,$$
$$w_0(x) = 0, \qquad x = 1, 2, \ldots, a + b - 1.$$

It will be seen that the difference equation is of the same type as the one formulated by Laplace; however, the boundary conditions are different which is the advantage of Lagrange's formulation. Besides the new formulation, Lagrange's contribution consists in a new method of transforming the partial difference equation to an ordinary one and the determination of the constants in its solution.

By his usual substitution, $w_n(x) = c\alpha^n\beta^x$, Lagrange finds the equation

$$q - \alpha\beta + p\beta^2 = 0, \tag{9}$$

which has the roots

$$\beta = \frac{\alpha \pm \sqrt{\alpha^2 - 4pq}}{2p} = \{\beta_1, \beta_2\},$$

say, so that

$$\beta^x = \sum_{i=0}^{x} h_i(x)\alpha^i + \beta \sum_{i=0}^{x-1} k_i(x)\alpha^i = A_x(\alpha) + \beta B_x(\alpha), \tag{10}$$

where powers of β higher than one have been eliminated by means of the relation $\beta^2 = (\alpha\beta - q)/p$, and where $A_x(\alpha)$ and $B_x(\alpha)$ denote polynomials in α.

According to (3.22), the solution of the difference equation is

$$w_n(x) = \sum_{i=0}^{x} h_i(x)w_{n+i}(0) + \sum_{i=0}^{x-1} k_i(x)w_{n+i}(1).$$

Using the fact that $w_n(0) = 1$ and the definition of $A_x(\alpha)$ by (10), it follows that

$$w_n(x) = A_x(1) + \sum_{i=0}^{x-1} k_i(x)w_{n+i}(1),$$

which by the substitution

$$w_n(1) = 1 - z_n$$

becomes

$$w_n(x) = A_x(1) + B_x(1) - \sum_{i=0}^{x-1} k_i(x)z_{n+i}.$$

To find $A_x(1) + B_x(1)$ we note that $\beta = 1$ gives $\alpha = 1$ according to (9) so that (10) gives $A_x(1) + B_x(1) = 1$. Hence,

$$w_n(x) = 1 - \sum_{i=0}^{x-1} k_i(x)z_{n+i}, \tag{11}$$

and using the boundary condition $w_n(a + b) = 1$, Lagrange finds that

$$\sum_{i=0}^{a+b-1} k_i(a + b)z_{n+i} = 0,$$

which means that the original partial difference equation has been transformed to an ordinary difference equation of order $a + b - 1$.

The characteristic equation is

$$\sum_{i=0}^{a+b-1} k_i(a + b)t^i = B_{a+b}(t) = 0,$$

so that

$$z_n = \sum_{j=1}^{a+b-1} c_j t_j^n,$$

where $t_1, \ldots, t_{a+b-1}$ are the roots of the characteristic equation. Inserting this expression into (11), the solution becomes

$$w_n(x) = 1 - \sum_{i=1}^{a+b-1} c_i B_x(t_i)t_i^n, \tag{12}$$

where the c's are to be found from the boundary conditions $w_0(x) = 0$, $x = 1, \ldots, a + b - 1$, which give

$$\sum_{i=1}^{a+b-1} c_i B_x(t_i) = 1, \qquad x = 1, \ldots, a + b - 1. \tag{13}$$

It remains to find the t's and the c's.

To find $B_x(\alpha)$, Lagrange solves the equations

$$\beta_1^x = A + \beta_1 B \qquad \text{and} \qquad \beta_2^x = A + \beta_2 B,$$

which give

$$B_x(\alpha) = \frac{\beta_1^x - \beta_2^x}{\beta_1 - \beta_2}$$

$$= p \frac{\{\alpha + \sqrt{\alpha^2 - 4pq}\}^x - \{\alpha - \sqrt{\alpha^2 - 4pq}\}^x}{(2p)^x \sqrt{\alpha^2 - 4pq}}. \qquad (14)$$

To solve the characteristic equation $B_{a+b}(t) = 0$, Lagrange uses the transformation

$$t = 2\sqrt{pq}\cos v,$$

which inserted into (14) leads to

$$B_{a+b}(t) = \left(\frac{q}{p}\right)^{(a+b-1)/2} \frac{\sin(a+b)v}{\sin v} = 0,$$

so that the roots are

$$t_i = 2\sqrt{pq}\cos v_i, \qquad v_i = \frac{i\pi}{a+b}, \qquad i = 1,\ldots,a+b-1.$$

We note that the characteristic equation and the transformation used by Lagrange are analogous to those used by Laplace.

To solve (13) for the c's we note that

$$c_i B_x(t_i) = c_i \left(\frac{q}{p}\right)^{(x-1)/2} \frac{\sin xv_i}{\sin v_i},$$

so that (13) becomes

$$\sum_{i=1}^{a+b-1} c_i \frac{\sin xv_i}{\sin v_i} = \left(\frac{p}{q}\right)^{(x-1)/2}, \qquad x = 1,\ldots,a+b-1. \qquad (15)$$

To solve these equations for the c's, Lagrange uses that

$$\sum_{x=1}^{a+b-1} \sin xv_i \sin xv_j = \tfrac{1}{2}(a+b) \qquad \text{for} \quad i = j \quad \text{and} \quad 0 \text{ otherwise.}$$

Multiplying (15) by $\sin xv_j$ and summing over x, he obtains

$$\frac{\frac{1}{2}(a+b)c_j}{\sin v_j} = \sum_{x=1}^{a+b-1} \left(\frac{p}{q}\right)^{(x-1)/2} \sin xv_j$$

$$= \left(1 \pm \left(\frac{p}{q}\right)^{(a+b)/2}\right) \frac{\sin v_j}{1 - 2\sqrt{p/q}\cos v_j + \frac{p}{q}},$$

the positive sign being valid for j odd, the negative for j even.
Using

$$1 - 2\sqrt{\frac{p}{q}}\cos v + \frac{p}{q} = q^{-1}(1 - 2\sqrt{pq}\cos v),$$

and inserting the values found for c_i and t_i into (12), we get Lagrange's formula

$$1 - w_n(x) = \frac{(4pq)^{(n+1)/2}}{a+b} \sum_{i=1}^{a+b-1} \left\{ \left(\frac{q}{p}\right)^{x/2} + (-1)^{i-1}\left(\frac{p}{q}\right)^{(a+b-x)/2} \right\}$$

$$\times \frac{\sin xv_i \sin v_i \cos^n v_i}{1 - 2\sqrt{pq}\cos v_i}. \tag{16}$$

For $x = a$ we get $U_n(a,b) = 1 - w_n(a)$. Lagrange does not give (16) explicitly, he leaves it to the reader to make the substitutions. [We have corrected a small error in Lagrange's proof; he has forgotten the factor p on the right-hand side of (14). However, it is easy to see that this error does not affect the final result.]

Noting that

$$\sin av_i = \sin(i\pi - bv_i) = (-1)^{i-1}\sin bv_i,$$

we obtain

$$U_n(a,b) = \frac{(4pq)^{(n+1)/2}}{a+b} \sum_{i=1}^{a+b-1} \left\{ \left(\frac{q}{p}\right)^{a/2} \sin av_i + \left(\frac{p}{q}\right)^{b/2} \sin bv_i \right\}$$

$$\times \frac{\sin v_i \cos^n v_i}{1 - 2\sqrt{pq}\cos v_i}. \tag{17}$$

Lagrange indicates another solution of the difference equation, which he obtains by solving (9) with respect to α instead of β. This leads to an expression for U_n as a linear combination of $p^{n-i}q^i$ with a set of equations for determining the coefficients, but he does not give an explicit expression for U_n. He remarks that this result corresponds to that obtained from de Moivre's algorithm, whereas the solution above corresponds to de Moivre's trigonometric solution. He also refers to Montmort and Nicholas Bernoulli.

Lagrange also solves the problem of the duration of play when one of the players has unlimited capital, see Problem 10 in §23.5.

Naturally, Laplace (1812, Book 2, §10) used his method of generating functions as the basis for a comprehensive analysis, which encompasses the results found by de Moivre and Lagrange and also some new asymptotic results. He obtains the correct generating function, even though his proof contains an error, which was corrected only in the Fourth Supplement (1825) to the third edition of his book. He ascribes the revised proof to his son.

Laplace considers the usual case with A and B having a and b counters, respectively. Let $u(x, y)$ denote the probability that A wins, given that the state of the play is such that A has x counters and that he wins after at most y games. The problem is to find $u(a, n) = R_n(a, b)$. From the difference equation Laplace first finds the generating function for $u(x, y)$ and then the generating function for $u(a, y)$ from which $u(a, n)$ may be obtained.

The difference equation is

$$u(x, y) = pu(x + 1, y - 1) + qu(x - 1, y - 1),$$

with the boundary conditions

$$u(0, y) = 0, \quad y = 0, 1, \ldots; \qquad u(x, 0) = 0, \quad x = 1, 2, \ldots, a + b - 1,$$
$$u(a + b, 2i) = 1, \qquad u(a + b, 2i + 1) = 0, \qquad i = 0, 1, \ldots,$$

corresponding to the fact that A can win only for even values of $n - b$.

According to (3.23), the generating function becomes

$$g(s, t) = \frac{f_1(s) + f_2(t) + f_3(t)s}{qts^2 - s + pt},$$

which Laplace writes in the convenient form

$$g(s, t) = \frac{F_1(s)s + F_2(t) + F_3(t)st}{qts^2 - s + pt}.$$

The arbitrary functions are found from the boundary conditions applied to the generating function.

First,

$$g(0, t) = \sum_{y=0}^{\infty} u(0, y)t^y = 0,$$

which shows that $F_2(t) = 0$.

Second,

$$g(s, 0) = \sum_{x=0}^{\infty} u(x, 0)s^x = s^{a+b} + \text{terms of higher powers of } s,$$

and $g(s, 0) = -F_1(s)$ imply that the term $F_1(s)s$ contains the factor s^{a+b+1}. Third, the generating function of $u(a + b, y)$ becomes

$$\sum_{y=0}^{\infty} u(a + b, y)t^y = 1 + t^2 + t^4 + \cdots = (1 - t^2)^{-1}, \tag{18}$$

which is used for determining $F_3(t)$. Consider

$$\frac{F_3(t)st}{qts^2 - s + pt} = \frac{F_3(t)s/p}{\dfrac{q}{p}s^2 - \dfrac{1}{pt}s + 1} \tag{19}$$

and write the denominator as the product $(1 - z_1 s)(1 - z_2 s)$, where (z_1, z_2) obviously are the roots of the equation

$$z^2 - \frac{1}{pt}z + \frac{q}{p} = 0,$$

so that

$$(z_1, z_2) = \frac{1 \pm \sqrt{1 - 4pqt^2}}{2pt}.$$

Hence, (19) may be decomposed into partial fractions

$$\frac{F_3(t)s}{p} \left(\frac{z_1}{1 - z_1 s} - \frac{z_2}{1 - z_2 s} \right) \frac{1}{z_1 - z_2}, \tag{20}$$

and expanding this expression in powers of s, the coefficient of s^{a+b} becomes

$$\frac{F_3(t)(z_1^{a+b} - z_2^{a+b})}{(z_1 - z_2)p}.$$

However, this is also the coefficient of s^{a+b} in the expansion of $g(s,t)$, since $F_1(s)s$ leads to higher exponents and $F_2(t) = 0$. Equating (18) and (20) Laplace finds

$$F_3(t) = \frac{(1 - t^2)^{-1}p(z_1 - z_2)}{z_1^{a+b} - z_2^{a+b}}.$$

The generating function of $u(a, y)$ is the coefficient of s^a in the expansion of $g(s,t)$, and it therefore equals

$$\frac{F_3(t)(z_1^a - z_2^a)}{(z_1 - z_2)p}.$$

Inserting the value of $F_3(t)$, Laplace gets the generating function of $u(a, y)$ in the form

$$\frac{(1 - t^2)^{-1}(z_1^a - z_2^a)}{z_1^{a+b} - z_2^{a+b}}$$

$$= \frac{(2pt)^b}{1 - t^2} \frac{\{1 + \sqrt{1 - 4pqt^2}\}^a - \{1 - \sqrt{1 - 4pqt^2}\}^a}{\{1 + \sqrt{1 - 4pqt^2}\}^{a+b} - \{1 - \sqrt{1 - 4pqt^2}\}^{a+b}}. \tag{21}$$

This is the generating function of $R_n(a, b)$; it follows that the generating function of $r_n(a, b)$ is obtained multiplying by $1 - t^2$.

Laplace derives the trigonometric expression for R_n by means of the transformation $1/t = 2\sqrt{pq} \cos v$. He also indicates how de Moivre's recursion formulae and the formula for r_n may be found from the power series expansion of (21); some details have been provided by Todhunter (pp. 169–175).

Comparing the three proofs, it will be seen (1) that Laplace's 1776 paper contains a derivation of de Moivre's recursion formulae and an incomplete proof of the trigonometric formula, which ought to be amended; (2) that Lagrange's 1777 paper contains a complete proof of the trigonometric formula but no discussion of the recursion formulae; however, some intermediate results could be used for this purpose; (3) that Laplace (1812, 1825) gives a very simple derivation of the generating function for R_n which by routine methods may be used to find all the previous results mentioned.

A simplified version of Lagrange's proof has been given by Ellis (1844), and a modification of Laplace's second proof has been given by Feller (1970).

De Moivre recognized the relation between Montmort and Nicholas Bernoulli's combinatorial expression (20.3.1) for $D_n(b, b)$ and his own trigonometric expression (23.2.8), (see §24.3), but he did not derive the one from the other. This problem has been discussed by Fieller (1931) and Takács (1969).

Apart from some remarks on Laplace's generating function, Todhunter does not discuss the history of difference equations, and he is therefore not able to give a satisfactory discussion of the solution of the problem of the duration of play.

23.5 PROBLEMS

1. The difference equation for Huygens' fifth problem is

$$u(x) = \left(1 + \frac{q}{p}\right)u(x - 1) - \frac{q}{p}u(x - 2),$$

with the boundary conditions $u(0) = 0$ and $u(m + n) = 1$ (see §14.2). Find $u(x)$ by means of de Moivre's theory of recurring series.

2. The problem of even or odd. From an urn containing x counters a random sample of counters is taken. Assuming that all combinations of counters are equally probable, show that the number of favorable cases for an even number, $u(x)$, say, satisfies the difference equation

$$u(x + 1) = 2u(x) + 1,$$

and that the probability of getting an even number in the sample is $(2^{x-1} - 1)/(2^x - 1)$.

Assuming that x is a random variable taking on the values $1, 2, \ldots, n$ with equal probabilities, show that the probability of getting an even number of counters in the sample is $(2^n - n - 1)/(2^{n+1} - n - 2)$. Solved by Laplace (1774, 1776).

Solve the problem by combinatorial methods, see Laplace (1812, Book 2, §5).

3. Consider a die with f faces numbered from 1 to f. Find the probability of getting the f faces in their natural order in x throws. Let $u(x)$ be the number of favorable cases among the f^x total number of cases. Show that

$$u(x) = fu(x - 1) + f^{x-f} - u(x - f).$$

Solve this equation for $f = 2$ and show that the probability in question equals $1 - (x + 1)2^{-x}$. Solved by Laplace (1776).

4. The problem of points for two players. Let the probabilities of winning a point be p and q for the players A and B, respectively, and suppose that A lacks n points and that B lacks $x - n$ points in winning the play. Show that B's probability of winning, $u_n(x)$, say, satisfies the partial difference equation

$$u_n(x) = qu_n(x - 1) + pu_{n-1}(x - 1),$$

and solve this equation. Solved by Laplace (1776).

5. The problem of points for three players. Using an obvious extension of the notation in the previous problem, show that C's probability of winning satisfies the partial difference equation

$$u_{m,n}(x) = ru_{m,n}(x - 1) + qu_{m,n-1}(x - 1) + pu_{m-1,n}(x - 1),$$

and solve this equation. Solved by Laplace (1776).

Laplace (1776) also gives a combinatorial proof much like the one by de Moivre, see §14.1.

6. A generalization of the problem of points for two players. Drawings with replacement are made from an urn containing four counters marked A_1, A_2, B_1, B_2. If A_i is drawn, A gets i points, and if B_i is drawn B gets i points, $i = 1, 2$. Suppose that A needs n points and that B needs $x - n$ points to win. Show that B's probability of winning, $u_n(x)$, say, satisfies the partial difference equation

$$u_n(x) = \tfrac{1}{4}\{u_n(x - 1) + u_n(x - 2) + u_{n-1}(x - 1) + u_{n-2}(x - 2)\},$$

and solve this equation. Solved by Laplace (1776).

7. Because of its importance in the theory of chances Laplace (1776) has given a special analysis of the partial difference equation of the second order in n of the form

$$u_n(x) = \sum_{i=1} \alpha_n(i)u_n(x - i) + \lambda_n + \sum_{i=0} \beta_n(i)u_{n+1}(x - i)$$

$$+ \sum_{i=0} \gamma_n(i)u_{n-1}(x - i).$$

Show that this equation may be reduced to the following one of the first order,

$$u_n(x) = \sum_{i=1} a_n(i)u_n(x-i) + c_n + \sum_{i=0} b_n(i)u_{n+1}(x-i).$$

Setting

$$d_n = 1 - b_{n-1}(0)\gamma_n(0),$$

show that the coefficients a_n, b_n, and c_n satisfy the following difference equations of the first order in n,

$$d_n a_n(k) = \alpha_n(k) + a_{n-1}(k) - \sum_{i+j=k\geqslant 2} \alpha_n(j)a_{n-1}(i)$$

$$+ \sum_{i+j=k\geqslant 1} \gamma_n(j)b_{n-1}(i), \qquad k = 1, 2, \ldots,$$

$$d_n b_n(0) = \beta_n(0),$$

$$d_n b_n(k) = \beta_n(k) - \sum_{i+j=k} \beta_n(j)a_{n-1}(i), \qquad k = 1, 2, \ldots,$$

$$d_n c_n = \lambda_n\left(1 - \sum_{k=1} a_{n-1}(k)\right) + c_{n-1}\sum_{k=0} \gamma_n(k),$$

and discuss the solution of these equations.

8. Let $u_n(x)$ denote the probability of getting at least x successes in n trials. Show that $u_n(x)$ satisfies the difference equation

$$u_n(x) = pu_{n-1}(x-1) + qu_{n-1}(x),$$

with the boundary conditions

$$u_n(0) = 1, \quad n = 0, 1, \ldots, \qquad \text{and} \qquad u_0(x) = 0, \quad x = 1, 2, \ldots.$$

Solve the equations by Lagrange's method, setting $u_n(x) = ca^n b^x$. Lagrange (1777) thus finds the right-hand tail of the binomial distribution expressed by means of the left-hand tail of the negative binomial distribution.

9. Lagrange (1777) formulates the problem of points as follows: Find the probability that b failures occur before a successes. (This means that B

wins in the usual formulation.) Let $u(x, y)$ denote the probability that y failures occur before x successes. Show that

$$u(x, y) = pu(x - 1, y) + qu(x, y - 1),$$

$$u(x, 0) = 1, \quad x = 1, 2, \ldots, \quad \text{and} \quad u(0, y) = 0, \quad y = 1, 2, \ldots,$$

and solve this equation by Lagrange's method.

10. Consider a series of games in which a player wins a point with probability p and loses a point with probability $q = 1 - p$ in each game. The play continues as long as the player's score is less than b and stops as soon as it reaches b. Find the probability of stopping in at most n games. This is Lagrange's formulation of the duration of play in the case with unlimited capital for one of the players.

Let $u_n(x)$ be the probability of stopping when the player needs to win x more points in at most n more games. Show that

$$u_n(x) = pu_{n-1}(x - 1) + qu_{n-1}(x + 1),$$

$$u_0(x) = 0, \quad x = 1, 2, \ldots, \quad \text{and} \quad u_n(0) = 1, \quad n = 0, 1, \ldots.$$

Solve the difference equation by means of the substitution $u_n(x) = c\alpha^n \beta^x$, which leads to the equation

$$p - \alpha\beta + q\beta^2 = 0.$$

The equation may be solved with respect to β or α, and each of these solutions leads to a different form of $u_n(x)$. Compare these forms for $x = b$ with $R_n(b)$ given by de Moivre.

11. Solve the difference equations in the previous problems by the method of generating functions.

12. Find the limiting value of the generating function (4.21) for $a \to \infty$ and use the result to derive de Moivre's formula for $R_n(b)$.

13. Find the value of the generating function (4.21) for $a = b$ and use the result to derive de Moivre's formula for $R_n(b, b)$.

14. Derive de Moivre's trigonometric formula for $U_n(b, b)$ from (4.17).

CHAPTER 24

De Moivre's Normal Approximation to the Binomial Distribution, 1733

In answer to this, I'll take the liberty to say, that this is the hardest Problem that can be proposed on the Subject of Chance, for which reason I have reserved it for the last, but I hope to be forgiven if my Solution is not fitted to the capacity of all Readers; however I shall derive from it some Conclusions that may be of use to every body: in order thereto, I shall here translate a Paper of mine which was printed November 12, 1733, and communicated to some Friends, but never yet made public, reserving to myself the right of enlarging my own Thoughts, as occasion shall require.

A Method of approximating the Sum of the Terms of the Binomial $(a + b)^n$ expanded into a Series, from whence are deduced some practical Rules to estimate the Degree of Assent which is to be given to Experiments.

—DE MOIVRE, 1738

24.1 INTRODUCTION

We shall use the notation

$$b(x) = b(x, n, p) = \binom{n}{x} p^x q^{n-x}, \qquad x = 0, 1, \ldots, n, \quad n = 1, 2, \ldots,$$

and define $b(x)$ as zero otherwise. In §§24.1–24.5, we shall follow de Moivre and assume that np is an integer.

468

The solution of problems of games of chance by combinatorial methods leads to formulae containing binomial coefficients which are cumbersome to calculate for large values of the arguments. For example, Montmort and Nicholas Bernoulli calculated sums of binomial coefficients for n larger than 100 to determine the median duration of play, see §20.3, and 'sGravesande calculated the sum of 259 binomial coefficients for $n = 11,429$ to improve Arbuthnott's test, see §17.2. Hence, there was a great need for simple approximations to the binomial coefficient and the binomial distribution.

The most difficult problems were those requiring the determination of n, the number of trials necessary to obtain a specified probability. De Moivre (1712) derived an approximate solution of the equation $B(c, n, p) = \frac{1}{2}$ with respect to n by using the Poisson approximation to the binomial for small values of p, see §14.4. In 1718 he also found an approximate solution of the equation $U_n(b, b) = \frac{1}{2}$ for the ruin problem by means of his trigonometric formula, which he further improved in 1738, see §20.5 and §23.2. However, in these solutions he did not directly approximate the binomial coefficient.

James and Nicholas Bernoulli (1713) evaluated the sum

$$P_d = \Pr\{|x - np| \leqslant d\} = \sum_{|x - np| \leqslant d} b(x, n, p), \qquad d = 1, 2, \dots.$$

For $p = r/t$, the ratio of two integers, $d = n/t$, and for $P_d > c/(c + 1), c > 0$, James found a lower bound for values of n satisfying these requirements, see §16.2. Nicholas turned the problem around and found a lower bound for P_d for a given value of n without any restrictions on d. Assuming that d is small compared with n, Nicholas also gave an approximation to P_d based on the approximation

$$\frac{b(np)}{b(np + d)} \cong \left(\frac{np + d}{nq - d + 1} \frac{np + 1}{np} \frac{nq}{np} \right)^{d/2}, \tag{1}$$

see §16.3.

In 1721 de Moivre began his investigations of the binomial distribution for $p = \frac{1}{2}$. He first found an approximation to the maximum term and next an approximation to the ratio of the maximum to the term at a distance of d from the maximum. De Moivre's approach thus differed from that of the Bernoullis by seeking an approximation to $b(x, n, p)$ instead of P_d. From 1725 onward James Stirling (1692–1770) worked on the same problem and found that the constant entering de Moivre's formula equals $\sqrt{2\pi}$. After having obtained these results they realized that it would be simpler to begin with an approximation to $\ln n!$, and they both proved Stirling's formula,

$n! \sim \sqrt{2\pi n}\, n^n e^{-n}$. De Moivre's proofs are given in the *Miscellanea Analytica* (MA) (1730), Stirling's in his *Methodus Differentialis* (1730).

It is a remarkable fact that de Moivre's proofs are based on three well-known theorems only, namely, Wallis' (1655) infinite product for $\pi/2$, Newton's infinite series for $\ln\{(1 + x)/(1 - x)\}$ from the 1660s, and Bernoulli's formula (1713) for the sum of powers of integers. Hence, the mathematical tools for finding the expansion of $\ln n!$ were already available to the Bernoullis.

Three years later, de Moivre (1733) simplified his results for $p = \frac{1}{2}$ and showed that the normal density function may be used as an approximation to the binomial. The generalization to an arbitrary value of p is of course very easy, so without proof de Moivre stated that

$$b(np + d, n, p) \sim (2\pi npq)^{-1/2} \exp\left(-\frac{d^2}{2npq} \right), \qquad d = O(\sqrt{n}),$$

and that P_d may be obtained by integration. He also showed how to calculate the standardized normal probability integral and gave the result for one, two, and three times the standard deviation.

24.2   THE MEAN DEVIATION OF THE BINOMIAL DISTRIBUTION

In 1721 Alexander Cuming posed the following problem to de Moivre: A and B have probabilities p and q, respectively, of winning a single game. After n games A shall pay as many counters to a spectator S as he wins games over np, and B as many as he wins games over nq. Find the expectation of S.

Obviously, the expectation of S becomes

$$D_n = \sum_{x=np+1}^{n} (x - np)b(x, n, p) + \sum_{x=0}^{np-1} (n - x - nq)b(x, n, p)$$

$$= \sum_{x=0}^{n} |x - np| b(x, n, p),$$

which today is called the mean deviation of the binomial distribution.

De Moivre (1730, pp. 99–101) proves that

$$D_n = 2npq \binom{n}{np} p^{np} q^{nq}. \tag{1}$$

He carries out the proof for $p = \frac{1}{2}$ only, but his proof is easily generalized as follows. From

$$\sum_{x=0}^{n} (x - np)b(x) = 0,$$

it follows that

$$D_n = 2 \sum_{x=np}^{n} (x - np)b(x) = 2 \sum_{y=0}^{nq} yb(np + y).$$

Evaluating the ratio $b(np + y + 1)/b(np + y)$, we obtain

$$q(np + y + 1)b(np + y + 1) = p(nq - y)b(np + y),$$

which by summation from 0 to nq gives

$$npq \sum b(np + y + 1) + q \sum (y + 1)b(np + y + 1)$$
$$= npq \sum b(np + y) - p \sum yb(np + y),$$

and thus

$$\sum_{y=0}^{nq} yb(np + y) = npqb(np).$$

This completes the proof, which according to de Moivre is from 1721. Johnson (1957) has given a proof without assuming that np is an integer.

To facilitate the calculation of the binomial coefficients, de Moivre (1730, pp. 103–104) tabulated $\log_{10} n!$ to 14 decimal places for $n = 10(10)900$. In 1730 Stirling pointed out some errors in the table, and de Moivre therefore published a corrected table in the *Supplement* (p. 22) to the MA. This table is reproduced in the *Doctrine* (1756, p. 333).

Besides being the spectator's expected gain per game, D_n/n may also be interpreted as a measure of the dispersion of the relative frequency x/n around the probability p. This is the first time that such a measure has been formulated and discussed; however, from a pedagogical point of view, de Moivre's discussion is rather odd. He pretends that he does not know the properties of D_n despite the fact that he immediately after the derivation of (1) shows, for $p = \frac{1}{2}$, that $b(np)$ tends to zero inversely as $\sqrt{n}$ so that D_n tends to infinity as $\sqrt{n}$. In the MA he only mentions that D_n is an increasing function of n. If he had calculated $D_n/\sqrt{n}$ for $p = \frac{1}{2}$, he would have found the following results:

n	100	400	900
$D_n/\sqrt{n}$	0.39795	0.39869	0.39883

This might have inspired him to guess that $D_n/\sqrt{n} \to (2\pi)^{-1/2} = 0.39894$, which would have saved him a great deal of trouble, as we shall see in the following sections.

In the *Doctrine* (1738, Problems 86 and 87; 1756, Problems 72 and 73) de Moivre discusses Cuming's problem and states (1) without proof. He gives a numerical analysis of the properties of D_n and adds, "how to find the middle Terms of those high Powers [of the binomial] will be shown afterwards," but he does not return to a discussion of D_n after he has derived the normal approximation to the binomial, otherwise he would immediately have found that $D_n \sim \sqrt{2npq/\pi}$. He tabulates D_n for $p = \frac{1}{2}$ and $n = 100(100)900$, and he finds that $D_{100}/100 \simeq 1/25$ and $D_{900}/900 \simeq 1/75$. He also finds D_n for $p = \frac{2}{3}$, $n = 6$ and 12, and makes the following comment:

> From this it follows, that if after taking a great number of Experiments, it should be perceived that the happenings and failings have been nearly in a certain proportion, such as of 2 to 1, it may safely be concluded that the Probabilities of happening or failing at any one time assigned will be very near in that proportion, and that the greater the number of Experiments has been, so much nearer the Truth will the conjectures be that are derived from them.

This formulation of the law of large numbers is clearly influenced by James Bernoulli. De Moivre, however, bases his statement on the numerical fact that D_n/n is decreasing. He knew that

$$\frac{D_n}{n} = E\left\{\left|\left(\frac{x}{n}\right) - p\right|\right\} \sim \sqrt{\frac{2pq}{\pi n}},$$

so that the dispersion of x/n tends to zero for $n \to \infty$. He adds that it is possible to improve the statement above by finding the odds against getting a large deviation from the expected value and goes on to derive the normal approximation to the binomial.

24.3   DE MOIVRE'S APPROXIMATIONS TO THE SYMMETRIC BINOMIAL IN *MISCELLANEA ANALYTICA*, 1730

The *Miscellanea Analytica* (MA) may be considered a series of research reports written between 1721 and 1730, supplemented by a letter from Stirling to de Moivre in 1729. Presumably for priority reasons, de Moivre published the reports in their original form so that the reader could judge for himself which contributions were due to de Moivre and Stirling, respectively.

The exposition in the present section and in the next section is based on de Moivre's proofs in MA and its supplement and to a large extent on the comments by Schneider (1968), who quotes extensively from MA. We shall add some new proofs and comments.

De Moivre naturally begins by studying the simplest case, namely, the symmetric binomial. He first finds an approximation to the maximum and next to the ratio of an arbitrary term to the maximum. He states that he obtained these results in 1721; the proofs are in MA, pp. 125–129.

De Moivre's Approximation to the Maximum of the Symmetric Binomial

$$\binom{n}{\frac{1}{2}n}\left(\frac{1}{2}\right)^n \sim 2.168\left(1 - \frac{1}{n}\right)^n \bigg/ \sqrt{n-1}. \tag{1}$$

Proof. For $n = 2m$, we have

$$b(m+d) = \binom{2m}{m+d}\left(\frac{1}{2}\right)^{2m}, \qquad |d| = 0, 1, \ldots, m, \quad m = 1, 2, \ldots.$$

From

$$b(m) = \frac{2m(2m-1)\cdots(m+1)}{1 \times 2 \times \cdots \times m}2^{-2m} = 2^{-2m+1}\prod_{i=1}^{m-1}\frac{m+i}{m-i},$$

we get

$$\ln b(m) = (-2m+1)\ln 2 + \sum_{i=1}^{m-1}\ln\frac{1+i/m}{1-i/m}. \tag{2}$$

By means of Newton's infinite series

$$\ln\frac{1+x}{1-x} = 2\sum_{k=1}^{\infty}\frac{x^{2k-1}}{2k-1}, \tag{3}$$

the sum above may be written as

$$2\sum_{i=1}^{m-1}\sum_{k=1}^{\infty}\frac{1}{2k-1}\left(\frac{i}{m}\right)^{2k-1} = 2\sum_{k=1}^{\infty}\frac{1}{(2k-1)m^{2k-1}}\sum_{i=1}^{m-1}i^{2k-1}. \tag{4}$$

From Bernoulli's formula (15.4.10) for the summation of powers of integers, it follows that

$$\sum_{i=1}^{m-1} i^{2k-1} = \frac{(m-1)^{2k}}{2k} + \frac{1}{2}(m-1)^{2k-1} + \frac{1}{2}(2k-1)B_2(m-1)^{2k-2} + \cdots. \quad (5)$$

Setting $(m-1)/m = t$, say, and inserting (5) into (4) we get

$$2(m-1)\sum_{k=1}^{\infty} \frac{t^{2k-1}}{(2k-1)2k} + \sum_{k=1}^{\infty} \frac{t^{2k-1}}{2k-1} + \frac{B_2}{m}\sum_{k=2}^{\infty} t^{2k-2} + \cdots. \quad (6)$$

To evaluate the first of these sums, de Moivre integrates (3) from $x = 0$ to $x = t$, and dividing the result by t, he obtains

$$2\sum_{k=1}^{\infty} \frac{t^{2k-1}}{(2k-1)2k} = \ln\frac{1+t}{1-t} + t^{-1}\ln(1-t^2)$$

$$= \ln(2m-1) + \frac{m}{m-1}\ln\frac{2m-1}{m^2}.$$

The second sum in (6) is found directly from (3), which gives

$$\sum_{k=1}^{\infty} \frac{t^{2k-1}}{2k-1} = \frac{1}{2}\ln\frac{1+t}{1-t} = \frac{1}{2}\ln(2m-1).$$

The third sum becomes

$$\frac{B_2}{m}\sum_{k=2}^{\infty} t^{2k-2} = \frac{B_2}{m}t^2(1-t^2)^{-1} = \frac{(m-1)^2}{6m(2m-1)},$$

which tends to $1/12$ for $m \to \infty$.

De Moivre also evaluates the next term and finds that it tends to $-1/360$. For the two following terms he only gives the limits.

Inserting these results into (2), de Moivre finds

$$\ln b(m) \sim (2m-\tfrac{1}{2})\ln(2m-1) - 2m\ln(2m) + \ln 2 + \frac{1}{12} - \frac{1}{360} + \frac{1}{1260} - \frac{1}{1680} + \cdots$$

$$\sim (2m-\tfrac{1}{2})\ln(2m-1) - 2m\ln(2m) + 0.7739 + \cdots. \quad (7)$$

Disregarding the remaining terms, he gets the final result

$$b(m) \sim 2.168(2m - 1)^{2m - 1/2}(2m)^{-2m},$$

which is the approximation given in (1).

Noting that

$$2m\{\ln(2m - 1) - \ln(2m)\} = 2m \ln\left(1 - \frac{1}{2m}\right) = -1 - \frac{1}{4m} - \cdots,$$

it follows from (7) that

$$\ln b(m) \sim -\tfrac{1}{2}\ln(2m) - 1 + 0.7739 + \cdots, \tag{8}$$

which gives the approximation

$$b(m) \sim 0.7976/\sqrt{2m}. \tag{9}$$

This result is not given explicitly by de Moivre, but he notes that $(1 - 1/n)^n \sim e^{-1}$, so he only had to calculate $2.168e^{-1} = 0.7976$.

Remarks. The proof is very important because it demonstrates the method which de Moivre uses in his other proofs. By an ingenious combination of rather elementary infinite series, he finds an expansion of $\ln b(m)$. There is only one part of the solution that is not quite satisfactory, and that is the determination of the numerical constant 2.168, based on $\ln 2$ plus the first four terms of the series in (7), beginning with $1/12$. The first four terms are decreasing, and he states that the sum gives a satisfactory approximation; however, one may wonder why he happend to stop exactly at the best place. The explanation may be that he knew that the following terms are increasing. He could hardly fail to see that the complete series equals

$$\sum_{k=1}^{\infty} \frac{B_{2k}}{(2k - 1)2k}.$$

To illustrate his difficulties, we have calculated the first eight terms:

$$0.0833 - 0.0028 + 0.0008 - 0.0006$$
$$+ 0.0008 - 0.0019 + 0.0064 - 0.0296 + \cdots.$$

By publishing only four terms, he avoided the difficult discussion of the properties of the series. Mathematicians at the time did not distinguish clearly between convergent and divergent series. The series continued to trouble him until in 1730, inspired by a result of Stirling, he proved by means of Wallis' formula that $b(m) \sim (\pi m)^{-1/2}$. Later, in the *Supplement* (1730, p. 9), he mentions that the terms after the fourth term do not decrease. In the *Doctrine* (1756, p. 244) he nevertheless writes that the series "converged but slowly."

De Moivre's Approximation to $b(m)/b(m + d)$

$$\ln \frac{b(m)}{b(m + d)} \sim (m + d - \tfrac{1}{2}) \ln (m + d - 1) + (m - d + \tfrac{1}{2}) \ln (m - d + 1)$$

$$- 2m \ln m + \ln \frac{m + d}{m}. \tag{10}$$

Proof. From

$$\frac{b(m)}{b(m + d)} = \frac{m + d}{m} \prod_{i=1}^{d-1} \frac{m + i}{m - i},$$

it follows that the proof depends on the expansion of

$$\sum_{i=1}^{d-1} \ln \frac{1 + i/m}{1 - i/m}.$$

Using the same method as in the previous proof and setting $t = (d - 1)/m$, the first two terms of the expansion become

$$2(d - 1) \sum_{k=1}^{\infty} \frac{t^{2k-1}}{(2k - 1)2k} + \sum_{k=1}^{\infty} \frac{t^{2k-1}}{2k - 1}.$$

Inserting the values of these sums, which have been found above, the theorem is proved.

Comparing the approximative values with the correct ones, obtained from his table of $\log n!$, de Moivre concludes that the approximations are satisfactory. For $n = 900$ and $p = \tfrac{1}{2}$, he finds $b(450) \cong 0.026585$, compared with the correct value 0.026588, and $\log [b(450)/b(480)] \cong 0.86826,62628$, compared with 0.86826,69779. He does not compare with Nicholas

Bernoulli's approximation (1.1), which turns out to be somewhat poorer, since it gives $\log[b(450)/b(480)] \cong 0.86885$. Comparing the values of $b(450)/b(480)$, de Moivre gives the correct value as 7.38358, the two approximations become 7.38357 and 7.39350, respectively.

De Moivre completes his analysis by a comparison of $b(x, n, p)$ and $b(x, n, \frac{1}{2})$, which easily leads to the result

$$b(x, n, p) = (2p)^x (2q)^{n-x} b(\tfrac{1}{2}n + d, n, \tfrac{1}{2}), \qquad d = x - \tfrac{1}{2}n. \tag{11}$$

This means that an approximation to $b(x, n, p)$ may be obtained from the approximation to $b(\frac{1}{2}n + d, n, \frac{1}{2})$. For Cuming's problem this gives

$$D_n = (2p)^{np+1}(2q)^{nq+1}(\tfrac{1}{2}n)b(\tfrac{1}{2}n + d, n, \tfrac{1}{2}), \qquad d = \tfrac{1}{2}n(p - q).$$

For completeness we shall state de Moivre's 1721 approximation to $b(x, n, p)$, obtained by inserting (9) and (10) into (11):

$$b(x, n, p) \sim 0.3988 \frac{n^{n+1/2}}{x(x-1)^{x-1/2}(n-x+1)^{n-x+1/2}} p^x q^{n-x}. \tag{12}$$

De Moivre does not give this formula explicitly. As he proved in 1730, the constant 0.3988 should have been $1/\sqrt{2\pi} = 0.3989$.

De Moivre begins the section "On the binomial $a + b$ raised to high powers" by a summary of the results obtained by James and Nicholas Bernoulli, including their numerical examples, see §§16.2 and 16.3. After having found an approximation to $b(x)$ and solved Cuming's problem, which involves one value of $b(x)$ only, one would have expected him to continue with an analysis of Bernoulli's problem, but he did not succeed in finding a simple expression for the *sum* of the approximate values of $b(x)$. In a way his approximation to $b(m + d)/b(m)$ was too good; like Nicholas Bernoulli, he did not take the decisive step of sacrificing some of the accuracy to bring $b(m + d)/b(m)$ in a form that could be summed by numerical integration without too much work.

The last two pages of the section on the binomial in MA, pp. 109–110, are entitled "On the inflection points of the binomial." Using the language from Newton's paper on interpolation, de Moivre describes how a curve may be fitted to the binomial, but unlike Newton, he does not illustrate his procedure by a graph. If he had, we would have had the first graph of a probability distribution. By interpolation he determines the points of inflection of the symmetric binomial and finds that they are at a distance of $\frac{1}{2}\sqrt{n+2}$, or for large n at a distance of $\frac{1}{2}\sqrt{n}$, from the point of symmetry.

He continues with his first attempt to find P_d. Without proof he states the formula

$$\sum_{x=a-d}^{a+d} f(x) \cong \frac{2d+1}{6d}\{(d+1)[f(a-d)+f(a+d)]+(4d-2)f(a)\}. \quad (13)$$

For $f(x) = b(x, 2m, \frac{1}{2})$ and $a = m$, the formula gives an approximation to P_d; the three values of $b(x)$ may be found by his approximation formulae. De Moivre neither discusses the accuracy of this formula nor does he give a numerical example. Perhaps he realized that the formula is useful only for small values of d because it presumes that the binomial can be approximated by a quadratic. Outside he points of inflection, the approximation therefore becomes very poor.

It is reasonable to assume that he tried out the formula for $n = 900$ and $d = 30$, since he had calculated $b(450)$ and $b(480)$, as quoted above. This would have given the discouraging result

$$P_{30} \cong \frac{61}{180}(62 \times 0.003601 + 118 \times 0.026585) = 1.139.$$

However, Schneider (1968, p. 295) has calculated P_d by means of (13) for $d = \frac{1}{2}\sqrt{n} = 15$ and found the result $P_d \cong 0.708$, which is quite satisfactory in view of the fact that the normal approximation gives 0.699.

We shall indicate how de Moivre may have derived formula (13). Consider the three-term Newton–Cotes formula, today usually called Simpson's formula,

$$\int_{a-d}^{a+d} f(x)\,dx \cong \frac{1}{3}\{d[f(a-d)+f(a+d)]+4df(a)\}. \quad (14)$$

For $d \to \infty$, the sum on the left-hand side of (13) tends to the integral above. It seems therefore reasonable to approximate the sum by a weighted sum of the three ordinates. Since there are $2d + 1$ terms in the sum, we replace the factor $\frac{1}{3}$ by $(2d+1)/6d$, so that

$$\sum_{x=a-d}^{a+d} f(x) \cong \frac{2d+1}{6d}\{(d+c_1)[f(a-d)+f(a+d)]+(4d+c_2)f(a)\}, \quad (15)$$

which asymptotically agrees with (14); the constants c_1 and c_2 are determined from the requirement that the formula should give the correct sum for $d = 1$.

In 1722 de Moivre made a new attempt to find an approximation to P_d based on his results for the duration of play (see MA, pp. 226–229). His "proof" is incomplete and rather difficult; he refers to Problems 25 and 26 in *De Mensura Sortis* instead of to Nicholas Bernoulli's formula for the ruin probability. We shall give a simpler proof, which in essence shows what de Moivre had in mind.

De Moivre had two expressions for the continuation probability U_n for $p = \frac{1}{2}$, one based on Montmort and Nicholas Bernoulli's formulae involving sums of binomial probabilities, and the other based on his own trigonometric formula. For $p = \frac{1}{2}$, n even, and d odd, it follows from Nicholas Bernoulli's formula (20.3.3) that

$$R_n(d, d) = 2^{-n+1} \left\{ \sum_{i=0}^{(n-d-1)/2} \binom{n}{i} - \sum_{i=0}^{(n-3d-1)/2} \binom{n}{i} + \sum_{i=0}^{(n-5d-1)/2} \binom{n}{i} - \cdots \right\}.$$

Since

$$2 \sum_{i=0}^{(n-d-1)/2} \binom{n}{i} \left(\frac{1}{2}\right)^n = 1 - \sum_{i=(n-d+1)/2}^{(n+d-1)/2} \binom{n}{i} \left(\frac{1}{2}\right)^n = 1 - P_{(d-1)/2},$$

we have

$$1 - R_n(d, d) = P_{(d-1)/2} + (1 - P_{(3d-1)/2}) - (1 - P_{(5d-1)/2}) + \cdots,$$

so that

$$P_{(d-1)/2} \cong 1 - R_n(d, d), \qquad \text{if} \quad P_{(3d-1)/2} \cong 1. \tag{16}$$

To introduce U_n we note that

$$2R_n(d, d) = D_n(d, d) = 1 - U_n(d, d),$$

which inserted into (16) gives

$$\sum_{i=(n-d+1)/2}^{(n+d-1)/2} \binom{n}{i} \left(\frac{1}{2}\right)^n \cong \frac{1}{2} + \frac{1}{2} U_n(d, d), \tag{17}$$

which is de Moivre's result.

For $n = 900$ and $d = 45$, he finds

$$\sum_{i=428}^{472} \binom{900}{i} \left(\frac{1}{2}\right)^{900} \cong 0.8662,$$

and states that this is correct to three decimal places. From the normal approximation we find 0.8664. We note that $U_{900}(45, 45)$ according to (23.2.8) equals

$$\frac{2}{45}\left(\frac{\cos^{901} 2°}{\sin 2°} - \frac{\cos^{901} 6°}{\sin 6°} + \frac{\cos^{901} 10°}{\sin 10°} - \cdots\right)$$

$$= \frac{2}{45}(16.5478 - 0.0678 + 0.0000 - \cdots) = 0.7324.$$

Since the series leading to (16) has alternating signs, the absolute value of the error is at most $1 - P_{(3d-1)/2} \cong \frac{1}{2} - \frac{1}{2}U_n(3d, 3d)$.

For comparison with the previous results, we rewrite (17) as

$$P_d = \sum_{i=n/2-d}^{n/2+d} \binom{n}{i}\left(\frac{1}{2}\right)^n \cong \frac{1}{2} + \frac{1}{2}U_n(2d + 1, 2d + 1), \qquad d > \frac{1}{2}\sqrt{n}. \quad (18)$$

We have added the condition $d > \frac{1}{2}\sqrt{n}$, which de Moivre does not mention. It follows from (16) that P_{3d+1} has to be nearly unity for the approximation to hold. Requiring that $3d + 1$ be larger than three times the standard deviation, the condition follows.

For $n = 900$ and $d = 15$, we find

$$P_d \cong \frac{1}{2} + \frac{1}{2} \times 0.4004 = 0.7002,$$

which is as accurate as the normal approximation and slightly better than the result found from the quadrature formula above.

Hence, by 1722 de Moivre had found a numerical solution of the problem of approximating P_d for the symmetric binomial. For $d < \frac{1}{2}\sqrt{n}$ he had the quadrature formula and for $n > \frac{1}{2}\sqrt{n}$ the trigonometric formula. These formulae are of the same accuracy as the normal approximation, and the error of the trigonometric formula is easily evaluated. However, from a theoretical point of view the solution is unsatisfactory, since the two formulae are completely unrelated. Further, he had no solution for $p \neq \frac{1}{2}$. It seems that de Moivre for the moment gave up; he turned to other problems, and in 1725 he published the *Annuities upon Lives*.

24.4 STIRLING'S FORMULA AND DE MOIVRE'S SERIES FOR THE TERMS OF THE SYMMETRIC BINOMIAL, 1730

In 1725 Cuming discussed his problem with Stirling and informed him of de Moivre's approximations. Stirling developed another approximation, which he communicated to de Moivre. When de Moivre prepared MA for publication in 1729, he asked Stirling's permission to include these results, and Stirling set him a letter, published in MA (pp. 170–172), with the following two results without proofs:

$$[b(m)]^{-2} = \pi m \left(1 + \frac{1}{4(m+1)} + \frac{9}{32(m+1)(m+2)} + \cdots \right),$$

$$[b(m)]^2 = \frac{2}{\pi(2m+1)} \left(1 + \frac{1}{4(m+\frac{3}{2})} + \frac{9}{32(m+\frac{3}{2})(m+\frac{5}{2})} + \cdots \right).$$

This is the first time that π occurs explicitly in the approximations to $b(m)$. How did Stirling succeed in getting π into his series? Stirling's derivation of the two series is rather involved [see Stirling (1730) and the summary given by Schneider (1968, pp. 270–273), which we are going to use]. Stirling extended the definition of $m!$ to noninteger values of m by interpolation. Tabulating $\ln m!$ for $m = 6, 7, \ldots, 17$, he determined $\ln(10.5!)$ by interpolation, and using the functional equation $m! = m(m-1)!$, he found $(-\frac{1}{2})!$ to ten decimal places and noted the agreement with the first ten decimal places of $\sqrt{\pi}$. Hence, in the first instance he did not prove that $(-\frac{1}{2})! = \sqrt{\pi}$.

Stirling thus improved de Moivre's approximation in two ways: He found that (3.1) may be written as

$$b\left(\frac{1}{2} n, n, \frac{1}{2} \right) \sim \sqrt{\frac{2}{\pi n}}, \tag{1}$$

and that this approximation is only the first term of a rapidly converging infinite series.

In continuation of Stirling's letter, de Moivre (MA, pp. 173–174) gives the first published proof of (1). The only tool he uses is Wallis' inequality for the determination of $4/\pi$; for simplicity, we shall indicate his proof by means of the modern version of Wallis' formula:

$$\lim_{m \to \infty} \frac{[2 \times 4 \times \cdots \times (2m)]^2}{[1 \times 3 \times \cdots \times (2m-1)]^2} \frac{1}{2m+1} = \frac{\pi}{2}.$$

Since

$$b(m) = \frac{1 \times 3 \times \cdots \times (2m-1)}{2 \times 4 \times \cdots \times (2m)},$$

(1) follows directly from Wallis' formula.

The proof was of great importance for de Moivre because a comparison of (1) and (3.7) showed that his proof of (3.7) would lead to the correct result if he replaced the series

$$1 - \sum_{k=1}^{\infty} \frac{B_{2k}}{(2k-1)2k}.$$

(2)

by $\frac{1}{2}\ln(2\pi)$.

It was only after they had obtained the results for $b(m)$ that Stirling and, later, de Moivre looked for an approximation to $m!$.

Stirling (1730) proved the following Theorem.

Theorem. Let $x + n$, $x + 3n, \ldots, z - n$ be a set of positive numbers in arithmetic progression. Then,

$$\ln\{(x+n)(x+3n)\cdots(z-n)\}$$
$$= \frac{z\ln z}{2n} - \frac{x\ln x}{2n} - \frac{z-x}{2n} - \sum_{k=1}^{\infty} \frac{(2^{2k-1}-1)B_{2k}}{(2k-1)2k} n^{2k-1}(z^{1-2k} - x^{1-2k}).$$

(3)

Stirling's proof is very simple. Denote the right-hand side of (3) by $f(z)$, say. Then,

$$f(z) - f(z-2n) = \ln z - \frac{n}{z} - \frac{1}{2}\left(\frac{n}{z}\right)^2 - \frac{1}{3}\left(\frac{n}{z}\right)^3 - \cdots$$

$$= \ln z + \ln\left(1 - \frac{n}{z}\right) = \ln(z-n),$$

which gives the required result by summation.

In an example, Stirling shows how to obtain the expansion for $\ln m!$ by setting $z - n = m$, $n = \frac{1}{2}$, and $x = \frac{1}{2}$. The resulting series is

$$\ln m! \sim \left(m + \frac{1}{2}\right)\ln\left(m + \frac{1}{2}\right) - \left(m + \frac{1}{2}\right) + \frac{1}{2}\ln(2\pi)$$

$$- \sum_{k=1}^{\infty} \frac{(2^{2k-1}-1)B_{2k}}{(2k-1)2k}(2m+1)^{1-2k}.$$

(4)

Goldstine (1977) has presented a discussion of Stirling's work.

After the publication of first the MA and then Stirling's book, de Moivre felt the need for rewriting and reorganizing his discussion on approximating the binomial, and he therefore published a 22-page *Supplement*. He begins by quoting Stirling's series (4) and his own result (3.7) and then turns to the derivation of a new series for ln *m*! by means of the method used for proving (3.7).

De Moivre's Version of Stirling's Formula

$$\ln m! \sim \left(m + \frac{1}{2} \right) \ln m - m + \frac{1}{2} \ln (2\pi) + \sum_{k=1}^{\infty} \frac{B_{2k}}{(2k-1)2k} m^{1-2k}, \tag{5}$$

or

$$m! \sim \sqrt{2\pi m} m^m \exp\left\{ -m + \frac{1}{12m} - \frac{1}{360m^3} + \cdots \right\}. \tag{6}$$

Proof. From

$$\frac{m^{m-1}}{(m-1)!} = \prod_{i=1}^{m-1} \frac{m}{m-i} = \prod_{i=1}^{m-1} \left(1 - \frac{i}{m} \right)^{-1},$$

we get

$$\ln \frac{m^{m-1}}{(m-1)!} = \sum_{i=1}^{m-1} \sum_{k=1}^{\infty} \frac{1}{k} \left(\frac{i}{m} \right)^k = \sum_{k=1}^{\infty} \frac{1}{km^k} \sum_{i=1}^{m-1} i^k. \tag{7}$$

Using Bernoulli's summation formula and setting $(m-1)/m = t$, say, (7) becomes

$$(m-1) \sum_{k=1}^{\infty} \frac{t^k}{k(k+1)} + \frac{1}{2} \sum_{k=1}^{\infty} \frac{t^k}{k} + \sum_{k=1}^{\infty} \sum_{r=1}^{[k/2]} \frac{1}{k} \binom{k}{2r-1} \frac{B_{2r}}{(2r)m^{2r-1}} t^{k-2r+1}$$

$$= m - 1 - \frac{1}{2} \ln m + \sum_{k=1}^{\infty} \frac{B_{2k}}{(2k-1)2k} (1 - m^{1-2k}),$$

where the summations are carried out by the same method as in the proof of (3.1). Replacing the series (2) by $\frac{1}{2} \ln (2\pi)$ and adding ln *m*, (5) follows.

De Moivre's series (5) is slightly simpler than Stirling's (4). It is known today as Stirling's formula.

By the same method of proof de Moivre finds an expansion of $\ln \prod_{i=1}^{k} (m - id)$, $m > kd$, analogous to (3).

A few years later, Euler and Maclaurin independently derived the formula

$$\sum_{i=0}^{m-1} f(a+i) = \int_a^{a+m} f(x)\,dx - \frac{1}{2}[f(x)]_a^{a+m}$$

$$+ \sum_{k=1}^{\infty} \frac{B_{2k}}{(2k)!}[f^{(2k-1)}(x)]_a^{a+m}, \tag{8}$$

and Maclaurin showed how to obtain Bernoulli's summation formula and Stirling's formula as special cases of this formula. In many applications of the Euler–Maclaurin summation formula, the resulting series will be divergent. However, as for Stirling's formula, a few terms will often give a good approximation to the sum, and the remainder is numerically smaller than the first neglected term and has the same sign if the terms have alternating signs.

One would have expected de Moivre to use his expansion of $\ln m!$ to find an expansion of $\ln b(m)$, but instead he amends his original proof of (3.7) by evaluating the general term of the series and by substituting $\frac{1}{2}\ln(2\pi)$ for the series (2). This leads him to the following improved version of (3.7):

$$\ln b(m) = (2m - \tfrac{1}{2})\ln(2m-1) - 2m\ln(2m) + \ln 2 - \tfrac{1}{2}\ln(2\pi)$$

$$+ 1 - \sum_{k=1}^{\infty} \frac{B_{2k}}{(2k-1)2k}(2m^{1-2k} - (2m-1)^{1-2k}). \tag{9}$$

Finally, he gives a similarly improved version of (3.10):

$$\ln \frac{b(m)}{b(m+d)} = \left(m+d-\frac{1}{2}\right)\ln(m+d-1) + \left(m-d+\frac{1}{2}\right)\ln(m-d+1)$$

$$- 2m\ln m + \ln\frac{m+d}{m}$$

$$+ \sum_{k=1}^{\infty} \frac{B_{2k}}{(2k-1)2k}\{(m+d-1)^{1-2k}$$

$$+ (m-d+1)^{1-2k} - 2m^{1-2k}\}, \tag{10}$$

which follows immediately from Stirling's formula.

Hence in 1730, inspired by Stirling's results, de Moivre succeeded in finding an excellent approximation to $b(m+d, 2m, \frac{1}{2})$ by combining the first few terms of the two infinite series given above.

24.5 DE MOIVRE'S NORMAL APPROXIMATION TO THE BINOMIAL DISTRIBUTION, 1733

In 1733, twenty years after the publication of James and Nicholas Bernoulli's results, de Moivre succeeded in finding a simple and accurate approximation to the binomial distribution. He considered this result so important that he printed a seven-page paper, *Approximatio ad Summam Terminorum Binomii* $(a + b)^n$ *in Seriem expansi*, for private circulation. The Latin text was translated by de Moivre himself, and with some additions it was included in the second and third editions of the *Doctrine of Chances* (1738, pp. 235–243; 1756, pp. 243–254) under the title, "A Method of approximating the Sum of the Terms of the Binomial $(a + b)^n$ expanded into a Series, from whence are deduced some practical Rules to estimate the Degree of Assent which is to be given to Experiments." The last sentence of the title refers to the comments to the original results given in 1738 and 1756. Only six copies of the *Approximatio* have been found (K. Pearson, 1924; Pearson and Daw, 1972); a photographic reproduction is to be found in Archibald (1926). We shall first discuss the *Approximatio* and then de Moivre's comments.

The *Approximatio* is a mathematical paper in which de Moivre continues his research in *Miscellanea Analytica*; he does not mention the improved results in the *Supplement* to the MA, partly because he did not need them and partly because many of his readers had MA without the *Supplement*.

He begins with a reference to the works of James and Nicholas Bernoulli and concludes that "what they have done is not so much an Approximation as the determining of very wide limits, within which they demonstrated that the Sum of the Terms was contained."

Next he refers to MA and gives his approximations to $b(m)$ and $b(m)/b(m + d)$, see (3.1) and (3.10). He mentions that Stirling has expressed the constant by means of π, which "has spread a singular Elegancy on the Solution," which he gives as

$$b\left(\frac{1}{2}n, n, \frac{1}{2}\right) \cong \frac{2}{\sqrt{2\pi n}}.$$

Without proof he states that (3.10) leads to the limiting value

$$\lim_{n \to \infty} \ln \frac{b(\frac{1}{2}n + d, n, \frac{1}{2})}{b(\frac{1}{2}n, n, \frac{1}{2})} = -\frac{2d^2}{n}, \qquad d = O(\sqrt{n}).$$

This is the decisive step that he did not take in the MA. The proof is very simple; one just has to use the series for $\ln(1 + x)$ on each of the terms of

(3.10) and retain the main term. Of course the proof presupposes that d is of the order of $\sqrt{n}$, which we have added above. De Moivre does not mention this condition in the first instance but later writes that the theorem holds, "provided the Ratio of d to n be not a finite Ratio, but such a one as may be conceived between any given number s and $\sqrt{n}$, so that d be expressible by $s\sqrt{n}$."

Finally, de Moivre replaces the sum of the approximate probabilities by the corresponding integral and thus gets an approximation to P_d.

We now summarize his results.

De Moivre's Approximation to the Symmetric Binomial

$$b\left(\frac{1}{2}n + d, n, \frac{1}{2}\right) \cong \frac{2}{\sqrt{2\pi n}} e^{-2d^2/n}, \tag{1}$$

$$P_d = \sum_{|x - n/2| \leq d} b\left(x, n, \frac{1}{2}\right) \cong \frac{4}{\sqrt{2\pi}} \int_0^{d/\sqrt{n}} e^{-2y^2} \, dy. \tag{2}$$

He does not give (2) explicitly but only the equivalent form

$$\frac{2}{\sqrt{2\pi n}} \sum_{x=0}^{d} e^{-2x^2/n} \cong \frac{2}{\sqrt{2\pi n}} \int_0^d e^{-2x^2/n} \, dx.$$

However, he introduces $\sqrt{n}$ as the "*Modulus* by which we are to regulate our Estimation" and shows that P_d depends on $d/\sqrt{n}$ only, so that (2) is his result in modern notation.

To calculate the integral for small values of d, he expands the exponential function and integrates the resulting series, which gives

$$\frac{1}{2} P_d \cong \frac{2}{\sqrt{2\pi n}} \sum_{k=0}^{\infty} \frac{(-1)^k 2^k d^{2k+1}}{k!(2k+1)n^k}.$$

Setting $d = t\sqrt{n}$, he finds

$$\frac{1}{2} P_d \cong \frac{2}{\sqrt{2\pi}} \sum_{k=0}^{\infty} \frac{(-1)^k 2^k t^{2k+1}}{k!(2k+1)}.$$

For $t = \frac{1}{2}$, the sum of seven terms gives 0.341344 so that $P_{\sqrt{n}/2} \cong 0.682688$;

the correct value is 0.682689. He checks the accuracy of this approximation for various values of n and writes, "Still, it is not to be imagined that there is any necessity that the number n should be immensely great; for supposing it not to reach beyond the 900th Power, nay not even beyond the 100th, the Rule here given will be tolerably accurate, which I have had confirmed by Trials." Of course in 1721–1722 he had already calculated such values, as described in §24.3, but he does not refer to these previous attempts to approximate P_d.

To calculate the integral from $\frac{1}{2}\sqrt{n}$ to $\sqrt{n}$ and from $\sqrt{n}$ to $\frac{3}{2}\sqrt{n}$, de Moivre uses numerical integration by means of Newton's three-eighth's rule

$$\int_a^{a+3h} f(x)\,dx \cong \frac{3h}{8}\{f(a) + 3f(a+h) + 3f(a+2h) + f(a+3h)\}.$$

For $f(x) = \exp(-2x^2/n)$, $a = \frac{1}{2}\sqrt{n}$, and $h = \frac{1}{6}\sqrt{n}$, and multiplying by $2/\sqrt{2\pi n}$, he finds the integral in question, which doubled gives 0.27160. Adding this to $P_{\sqrt{n}/2}$ he obtains $P_{\sqrt{n}} \cong 0.95428$; the correct value is 0.95450. He gives the details of these calculations; for the next integral he notes that by the same method it will be found that $P_{(3/2)\sqrt{n}} \cong 0.99874$. Unfortunately, he must have made an error in his calculations; the three-eighth's rule gives

$$\tfrac{1}{8}(0.135335 + 3 \times 0.065729 + 3 \times 0.028566 + 0.011109) = 0.053666,$$

which multiplied by $\sqrt{2/\pi}$ gives 0.042819. Adding this to the previous result we get $P_{(3/2)\sqrt{n}} = 0.99710$ instead of 0.99874; the correct value is 0.99730.

On the method used de Moivre remarks that "the more Ordinates there are, the more exact will the Quadrature be; but here I confine myself to four, as being sufficient for my purpose."

Hence, besides deriving the normal probability integral he showed how to calculate the integral and gave three examples which, using modern terminology, correspond to deviations of one, two, and three times the standard deviation. He also solved the equation $P_d = \frac{1}{2}$ and gave the solution as $\frac{1}{4}\sqrt{2n}$ "very near," i.e., $d/\sqrt{n} = 0.3536$. This means that the probable error, as it was called later, equals 0.7071 times the standard deviation; the correct factor is 0.6745.

De Moivre does not prove that the probability integral from $-\infty$ to $+\infty$ equals 1, but he uses this fact in computing various odds.

On the last page or so of the *Approximatio*, de Moivre generalizes his results to the skew binomial.

De Moivre's Normal Approximation to the Binomial Distribution

$$b(np + d, n, p) \cong \frac{1}{\sqrt{2\pi npq}} e^{-d^2/2npq}, \tag{3}$$

$$P_d = \sum_{|x-np| \leqslant d} b(x, n, p) \cong \frac{4}{\sqrt{2\pi}} \int_0^{d/2\sqrt{npq}} e^{-2y^2} \, dy; \tag{4}$$

where (3) is stated explicitly but (4) is not. However, he writes; "If the Probabilities of happening and failing be in any given Ratio of inequality, the Problems relating to the Sum of the Terms of the Binomial $(a + b)^n$ will be solved with the same facility as those in which the Probabilities of happening and failing are in a Ratio of Equality." Thus ends the *Approximatio*.

It is completely in accordance with de Moivre's style to omit the proof. Because of the extraordinary importance of his result, we shall indicate what his proof may have been, using methods and results only from his previous proofs.

Proof of (3). Using de Moivre's result (4.5) in the form

$$\ln\left(\frac{n^n}{n!}\right) \cong n - \frac{1}{2} \ln n - \frac{1}{2} \ln(2\pi),$$

and writing

$$b(np) = \frac{(np)^{np} (nq)^{nq} \, n!}{(np)! \, (nq)! \, n^n},$$

it follows immediately that

$$b(np) \cong (2\pi npq)^{-1/2}.$$

From

$$\frac{b(np)}{b(np + d)} = \frac{(np + d)!(nq - d)!(nq)^d}{(np)!(nq)!(np)^d} = \left(1 + \frac{d}{np}\right) \prod_{i=1}^{d-1} \frac{1 + i/np}{1 - i/nq},$$

and

$$\ln \frac{1 + i/np}{1 - i/nq} = \frac{i}{npq} + \cdots,$$

it follows that

$$\ln \frac{b(np)}{b(np+d)} = \frac{d(d-1)}{2npq} + \frac{d}{np} + \cdots = \frac{d^2}{2npq} + \cdots ,$$

which completes the proof.

Finally, we shall give the modern formulation of de Moivre's theorem. Let the normal density function be defined as

$$\phi(u) = \frac{1}{\sqrt{2\pi}} \exp\left(-\frac{1}{2}u^2\right), \qquad -\infty < u < \infty,$$

and the normal distribution function as

$$\Phi(u) = \int_{-\infty}^{u} \phi(t)\, dt.$$

For a fixed value of $p, 0 < p < 1$, for $u = (x - np)/\sqrt{npq}$ bounded, and for $n \to \infty$, we have

$$\lim[\sqrt{npq}\, b(x, n, p)] = \phi(u), \qquad (5)$$

$$\lim_{|x-np| \le u\sqrt{npq}} \sum b(x, n, p) = \Phi(u) - \Phi(-u), \qquad 0 < u < \infty, \qquad (6)$$

np and $u\sqrt{npq}$ being positive integers.

De Moivre's Comments in 1738 and 1756

In 1738 de Moivre adds a comment labeled Remark I in 1756, of about one page to the translated *Approximatio*. Nearly all of it is taken up by a discussion of (2), which may be written as

$$P_{t\sqrt{n}} = \Pr\left\{ \left| \frac{x}{n} - \frac{1}{2} \right| \le \frac{t}{\sqrt{n}} \right\} \cong \frac{4}{\sqrt{2\pi}} \int_{0}^{t} e^{-2y^2}\, dy. \qquad (7)$$

He does not state the formula but gives corresponding numerical examples. He has previously found the value of the probability integral for $t = \frac{1}{2}$ to 0.683, so he now states that the odds are a little more than 2 to 1 that the relative frequency does not deviate more than $1/(2\sqrt{n})$ from its true value if n is sufficiently large for the approximation to hold. He illustrates this by choosing $n = 3600$, 14,400, and 1,000,000 and stating the deviations to be $1/120$, $1/240$, and $1/2000$. He adds that

> the Odds would increase at a prodigious rate, if instead of taking such narrow limits on both sides the Term of Equality, as are represented by $\frac{1}{2}\sqrt{n}$ [for the relative frequency $1/(2\sqrt{n})$], we double those Limits or triple them; for in the first case the Odds would become 21 to 1, and in the second 369 [should have been 344] to 1, and still be vastly greater if we were to quadruple them, and at last be infinitely great.

In this way de Moivre makes the meaning and the applicability of his theorem clear for the nonmathematical reader.

He remarks, "What we have said is also applicable to a Ratio of Inequality, as appears from our 9th Corollary." Following up this hint, the mathematical reader is supposed to generalize (7) to the following version of (4):

$$P_{2t\sqrt{npq}} = \Pr\left\{ \left| \frac{x}{n} - p \right| \leqslant 2t\sqrt{\frac{pq}{n}} \right\} \cong \frac{4}{\sqrt{2\pi}} \int_0^t e^{-2y^2} \, dy. \qquad (8)$$

This formula shows the revolutionary result of de Moivre's analysis of the binomial distribution: Measuring the deviation d by the modulus $2\sqrt{npq}$, a quantity t is obtained which determines the probability that

$$np - 2t\sqrt{npq} \leqslant x \leqslant np + 2t\sqrt{npq},$$

or equivalently,

$$p - 2t\sqrt{\frac{pq}{n}} \leqslant \frac{x}{n} \leqslant p + 2t\sqrt{\frac{pq}{n}}. \qquad (9)$$

Inversely, choosing the odds, c to 1, say, or the probability $c/(c + 1)$ for these inequalities to hold, t may be found as a function of c and the corresponding limits for x or x/n determined.

This result is what de Moivre had in mind in 1738 when he added the following sentence to the title of the *Approximatio*: "from whence are deduced some practical Rules to estimate the Degree of Assent which is to be given to Experiments."

He ends his comment with the following formulation of the law of large numbers: "And thus in all cases it will be found, that altho' Chance produces irregularities, still the Odds will be infinitely great, that in process of Time, those Irregularities will bear no proportion to the recurrency of that Order which naturally results from original design."

In 1756 he italicizes the quotation above and adds a three-page Remark II in the same theological vein. He follows the pattern set by Newton and others and uses his limit theorem to support the arguments for divine providence and design that were popular among scientists at the time, see the quotation from Remark II given at the end of §2.3. Similar theological considerations had been proposed by Graunt, based on the observed stability of statistical ratios, see §7.6, and by Arbuthnott and 'sGravesande, supported by both data and theory, see §§17.1 and 17.2. De Moivre refers to N. Bernoulli's discussion of Arbuthnott's data, see §17.3, and concludes that "if we were shown a number of Dice, each with 18 white and 17 black faces, which is Mr. *Bernoulli*'s supposition, we should not doubt but that those Dice had been made by some Artist; and that their form was not owing to *Chance*, but was adapted to the particular purpose he had in View." This is of course correct, but de Moivre misses completely the essential point of Bernoulli's statistical analysis.

Like James Bernoulli, de Moivre was no statistician; they both had the idea that probability theory could be used with advantage in the physical and social sciences, but they themselves never collected nor analyzed observational data, with the exception of de Moivre's primitive analysis of Halley's life table. De Moivre's theological arguments are thus rather weak because he does not provide the empirical documentation necessary to convince the reader that nature behaves as his theory predicts.

De Moivre had previously alluded to the concepts of *Design* and *Chance* in the preface to the *Doctrine* (1718), where he writes;

> From this last Consideration we may learn, in many Cases, how to distinguish the Events which are the effect of Chance, from those which are produced by Design: The very Doctrine that finds Chance where it really is, being able to prove by a gradual Increase of Probability, till it arrive at Demonstration, that where Uniformity, Order and Constancy reside, there also reside Choice and Design.

He does not there give any theological interpretation. It is first in Remark II in 1756 that he observes that "certain Laws according to which Events happen... as well as the original Design and Purpose of their Establishment, must all be *from without*" and hence that "we shall be led, by a short and obvious way, to the acknowledgement of the great MAKER AND GOVERNOUR of all; *Himself all-wise, all-powerful and good.*"

After this discussion of design, de Moivre defines chance as follows: "*Chance*, as we understand it, supposes the *Existence* of things, and their general known *Properties*: that a number of Dice, for instance, being thrown each of them shall settle upon one or other of its Bases. After which, the

Probability of an assigned Chance, that is of some particular disposition of the Dice, becomes as proper a subject of Investigation as any other quantity or Ratio can be." It will be seen that he here uses chance to mean a random event; he speaks of the probability of an assigned chance. This definition is possibly provoked by Bernoulli's discussion of (subjective) probability as a measure of our knowledge in a deterministic world, see §15.7. Schneider (1972, 1975) has pointed out that de Moivre's definition of chance implies the existence of indeterminate (random) events, so that de Moivre considers probability as an objective physical property which can be measured by repeated observations. K. Pearson (1978, pp. 160–162) has discussed de Moivre's remarks from the point of view of a statistician and a freethinker.

Except for these three pages, the *Doctrine* is a mathematical text with an axiomatic foundation.

We shall now comment on some misconceptions of de Moivre's theorem.

It is sometimes stated that de Moivre found the normal approximation for $p = \frac{1}{2}$ only, but as we have seen above this is not so.

De Moivre did not consider the normal distribution as a distribution in its own right (the error distribution) but only as a convenient means for approximating the binomial. The essential property is the one following from (3) that

$$\lim_{n \to \infty} \left[\sqrt{npq}\, b(np + 2t\sqrt{npq}, n, p) \right] = (2\pi)^{-1/2} e^{-t^2},$$

i.e., the limiting distribution depends on t only, and further, that its sum may be evaluated by numerical integration. For de Moivre's purpose it was inessential what form the limiting distribution had if only its sum could be evaluated.

De Moivre's theorem is rightly considered as the first example of the central limit theorem. There is no indication that he himself looked at his result from this point of view; otherwise he would probably have discussed the limiting distribution of other sums of random variables, for example, the sum of the points obtained by throwing several dice, see the remarks in connection with Fig. 14.3.1.

We shall next consider the following three aspects of de Moivre's theorem: (1) calculation of limits for the relative frequency $h_n = x/n$; (2) calculation of limits for p; and (3) the normal approximation to the binomial. The point (3) will be discussed in the following two sections.

To find limits for h_n we have to solve the equation

$$P_{n\epsilon} = \Pr\{|h_n - p| \leqslant \epsilon\} = \frac{c}{c+1}$$

with respect to ϵ. From (8) we get

$$\epsilon \cong 2t_c \sqrt{\frac{pq}{n}}, \tag{10}$$

where t_c is found by setting the probability integral in (8) equal to $c/(c+1)$.

To show the advantages of de Moivre's solution we shall compare with the formulae due to the Bernoullis. According to (16.2.9), James Bernoulli's theorem leads to the equations

$$m = \frac{n\epsilon(p+\epsilon)+q}{1+\epsilon},$$

$$\ln c = m \ln \frac{p+\epsilon}{p} - \ln \frac{q-\epsilon}{\epsilon}, \tag{11}$$

which have to be solved for ϵ by trial and error. According to (16.3.7), Nicholas Bernoulli's theorem gives

$$\ln(c+1) = \tfrac{1}{2}n\epsilon \ln \frac{1+(\epsilon/p)}{1-(\epsilon/q)}. \tag{12}$$

It follows from the derivation of these formulae that the values of ϵ will be too large.

It will be seen that de Moivre's solution directly shows how ϵ depends on each of the three given quantities n, p, and c. The solution is very simple in numerical respects, provided that the probability integral has been tabulated, and the solution is more accurate than the ones obtained from the theorems of the Bernoullis. From the two other solutions it is rather difficult to see how ϵ depends on the three given quantities, and the numerical solution is more involved.

As an example we shall set $n = 14{,}000$ and $p = 18/35 = 0.5143$, as in N. Bernoulli's analysis of Arbuthnott's data. For convenience we choose $c/(c+1) = 0.954$, or $c = 20.7$, corresponding to de Moivre's $t = 1$. The three values of ϵ become 0.0210, 0.0104, and 0.0084, corresponding to formulae (11), (12), and (10), respectively.

The great advantages of de Moivre's theorem are thus obvious; however, neither de Moivre nor anybody else at the time made such comparisons.

Turning to the inverse problem, we observe that neither James Bernoulli nor de Moivre indicates how to find an interval for p from given values of n, h_n, and c, perhaps because they realized that an exact solution is impossible within the given model, where p is an unknown constant and h_n a random variable. Being eminent mathematicians and probabilists with lifelong training in deductive reasoning, it must have been extremely difficult for them to formulate and solve problems of (inductive) statistical inference. Furthermore, in 1738 de Moivre was 71 years old and perhaps not inclined to take up a new and difficult line of research.

A possibility for solving the problem within the deductive framework would have been to determine which values of p were compatible with h_n for a given moral certainty. That de Moivre was not very far from this idea is evident from his introduction to Remark II, where he writes;

> if from numberless Observations we find the Ratio of the Events to converge to a determinate quantity, as to the Ratio of P to Q; then we conclude that this Ratio expresses the determinate Law according to which the Event is to happen. For let that Law be expressed not by the Ratio $P:Q$, but by some other, as $R:S$; then would the Ratio of the Events converge to this last, not to the former: which contradicts our *Hypothesis*.

If he had applied this principle not only to "numberless observations" but also for a finite n, he would have found an interval of "acceptable values of p" corresponding to h_n and the chosen value of the moral certainty, i.e., a confidence interval in modern terminology.

The great statistical challenges at the time of de Moivre were the analyses of Arbuthnott's data and the data on the mortality of annuitants. By means of his new results de Moivre could have improved Nicholas Bernoulli's analysis much as we have done in §17.3, but he did not even indicate the possibility of doing so. It is a curious coincidence that de Moivre uses $n = 14,400$ and $n = 1,000,000$ in his examples when discussing deviations from $p = \frac{1}{2}$, and that Nicholas Bernoulli had previously used $n = 14,000$ in his analysis of Arbuthnott's data and that the total number of observations was 938,223. For $n = 1,000,000$, de Moivre states that the odds are "vastly greater" than 369 to 1 for the deviation of h_n from $p = \frac{1}{2}$ to be at most $4/2000 = 0.002$. Having observed that $h_n = 0.5163$, he could safely have rejected the hypothesis $p = \frac{1}{2}$, and it must have been tempting for him to say that one could be morally certain that p belongs to the interval 0.5163 ± 0.0020, which just contains the value used by Nicholas Bernoulli. All this naturally assumes that the relative frequency of male births varies according to the binomial distribution with the same p throughout the 82 years, which was what Bernoulli had concluded. Generalizing these considerations, one may wonder why de Moivre did not, like Laplace, interchange p and $h = x/n$ in the inequality (9). He would then have found an interval for p of the form $h \pm 2t\sqrt{h(1-h)/n}$, a result that throughout the 19th century was in widespread use and for $2t = 3$, known as the three-sigma rule.

It was first Bayes (1764) and Laplace (1774b) who (independently) realized that to get a probabilistic solution of the problem, the model had to be extended by assuming that p is a random variable with a known (a priori) distribution. They could thus find the conditional probability distribution of p for given h_n and from this derive probability limits for p.

De Moivre and his contemporaries did not use the normal approximation to the binomial in their discussions of demographic data. The first example of such a statistical analysis we have found is due to Daniel Bernoulli in 1771, see §24.8.

24.6 LAPLACE'S EXTENSION OF DE MOIVRE'S THEOREM, 1812

Neither de Moivre nor Laplace comments on the fact that the skew binomial distribution is approximated by the symmetric normal distribution in de Moivre's theorem. Nevertheless, it may have been one of the reasons that Laplace carried de Moivre's approximation one step further to obtain a correction term taking the skewness into account.

In his proof, Laplace (1812, Book 2, §16) uses Stirling's formula and the Euler–Maclaurin formula; however, he gives his own proofs of these formulae.

Laplace does not as Bernoulli and de Moivre assume that np is an integer. He shows that the greatest binomial probability corresponds to $m = [(n + 1)p] = np + z$, say, $-q < z \leqslant p$, and he begins by finding an approximation to

$$b(m + d) = \frac{n!}{(m + d)!(n - m - d)!} p^{m+d} q^{n-m-d},$$

$m + d$ being an integer. Assuming that $d = O(\sqrt{n})$ and disregarding terms of the order of $1/n$, he finds by straightforward expansion of $\ln b(m + d)$ that

$$b(m + d) = \frac{1}{\sqrt{2\pi np'q'}} \left\{ \exp\left(-\frac{d^2}{2np'q'} \right) \right\} \left\{ 1 - \frac{dz}{np'q'} - \frac{d}{2} \left(\frac{1}{np'} - \frac{1}{nq'} \right) \right.$$
$$\left. + \frac{d^3}{6} \left(\left(\frac{1}{np'} \right)^2 - \left(\frac{1}{nq'} \right)^2 \right) + \cdots \right\}, \tag{1}$$

where we have introduced $p' = m/n$ and $q' = (n - m)/n$ to simplify Laplace's formula. This is Laplace's extension of de Moivre's formula (5.3). Setting $z = 0$, it will be seen that the main term coincides with (5.3) and that the next term is of the order of $n^{-1/2}$ times the first. This term gives a correction for skewness; it depends on $q' - p'$ and disappears for $q' = p'$.

Following the tradition from Bernoulli and de Moivre, Laplace seeks the sum of the probabilities symmetrically around $b(m)$, and he therefore introduces

$$f(x) = b(m - x) + b(m + x).$$

From (1) it follows that

$$f(x) \cong \frac{2}{\sqrt{2\pi np'q'}} e^{-x^2/2np'q'}.$$

According to the Euler–Maclaurin formula (4.8), we have

$$\sum_{x=m-d}^{m+d} b(x) = \sum_{x=0}^{d} f(x) - \frac{1}{2}f(0) \cong \int_0^d f(x)\,dx + \frac{1}{2}f(d),$$

the next term being of a smaller order of magnitude, since $f'(x)$ equals $f(x)$ times a factor of the order of $n^{-1/2}$. This result fills out a gap in de Moivre's proof of (5.2). Finally, Laplace gets

$$\sum_{x=m-d}^{m+d} b(x, n, p) = \frac{2}{\sqrt{\pi}} \int_0^t e^{-y^2}\,dy + \frac{1}{\sqrt{2\pi np'q'}} e^{-t^2} + \cdots , \tag{2}$$

where

$$t = \frac{d}{\sqrt{2np'q'}}. \tag{3}$$

This is Laplace's extension of (5.4); a term of the order of $n^{-1/2}$ has been added to de Moivre's formula. Laplace notes that the approximation may be improved by taking more terms of the expansion into account.

Solving the inequality $|x - m| \leqslant d$ in terms of $h_n = x/n$, Laplace finds

$$|h_n - p| \leqslant tc \sqrt{\frac{2}{n}} + \frac{z}{n}, \tag{4}$$

where

$$c^2 = pq + \frac{(q - p)z}{n} - \frac{z^2}{n^2}. \tag{5}$$

He concludes that the probability of (4) being satisfied is given by (2). Like de Moivre, Laplace notes that for $n \to \infty$ and for a fixed value of t, the probability tends to a limit given by the first term of (2) and the limit for $|h_n - p|$ tends to zero. If the deviation d is fixed, t tends to infinity, and the probability integral tends to 1.

As an example, Laplace takes N. Bernoulli's problem of finding $\Pr\{7037 \leqslant x \leqslant 7363\}$ for $p = 18/35$ and $n = 14{,}000$. Using his continued fraction for the probability integral, he finds from (2) that $P = 0.994{,}303$. He does not observe that de Moivre's formula gives 0.994,155; the correct value is 0.994,306.

Laplace continues with the following remark: "If one knows the number of times that the event a has occurred in n trials then the formula (o) [our (2)] will give the probability that p, which is supposed to be unknown, will be included between the limits given.' Solving the inequality (4) with respect to p, he gets the limits for p,

$$h_n \pm t \sqrt{\frac{2h_n(1 - h_n)}{n}},$$

and to find the corresponding probability, he replaces p' by h_n in (2) and (3).

To get an interval estimate of p, he thus boldly solved the inequality (4) and estimated its probability. He did not feel the need to introduce a new term (such as confidence or fiducial probability) for the trustworthiness of this simple procedure, he simply used probability without any qualification or explanation in the hope that his readers would understand.

However, he hastens to add another justification: "One may arrive at these results directly by considering p as a variable between zero and unity and determining the probability of these different values, given the observed event, as will be seen when we treat the probability of causes, deduced from observed events." The derivation of the posterior distribution of p may found in his §26; it was first given in Laplace (1774b).

24.7 THE EDGEWORTH EXPANSION, 1905

As noted by Eggenberger (1894), the second term on the right-hand side of (6.2) may be included into the integral by changing d to $d + \frac{1}{2}$ in the definition of t, which may be seen by a Taylor expansion of the modified integral. This correction has become known as the continuity correction.

Laplace's results are more complicated than necessary because he used an inconvenient standardization of the random variable. He did not realize that instead of using m as location parameter and $\sqrt{m(n-m)/n}$ as scale parameter, he would have obtained simpler results by using np and $\sqrt{npq}$, respectively. The standard deviation, $\sigma = \sqrt{npq}$, had not yet been introduced explicitly. Formally the results for this standardization may be obtained from Laplace's results by setting $m = np$ and $z = 0$.

Taking one more term of the expansion into account and setting $u = d/\sigma$, we obtain

$$\sigma b(np + d, n, p) = \phi(u)\{1 + (u^3 - 3u)(q - p)/6\sigma + [(1 - 4pq)u^6 - 6(2 - 7pq)u^4$$
$$+ 9(3 - 8pq)u^2 - 6(1 - pq)]/72\sigma^2 + \cdots\}. \tag{1}$$

We shall leave the detailed derivation to the reader. Starting from $\ln b(np + d)$ the proof consists of three steps: (1) application of Stirling's formula to the three factorials; (2) expansion of the two logarithms of the form $\ln(1 + x)$; and (3) introduction of the standardized variable $u = (x - np)/\sigma = d/\sqrt{npq}$.

As mentioned above, the first two terms are due to Laplace. The last term was derived by several authors about the end of the 19th century, see for example Czuber (1891, p. 85), K. Pearson (1895, p. 347), and Edgeworth (1905, p. 59).

To find the distribution function we use the Euler–Maclaurin formula in the form

$$\sum_{x=0}^{k} f(x) = \int_{-1/2}^{k+1/2} f(x)\,dx - \frac{1}{24}[f'(x)]_{-1/2}^{k+1/2} + \cdots. \tag{2}$$

Hence the main term obtained by summation of (1) becomes

$$\int_{(-1/2-np)/\sigma}^{(k+1/2-np)/\sigma} \phi(u)\{1 + g(u)\}\,du,$$

where $g(u)$ represents the last two terms in the parenthesis in (1).

To carry out the integration we first note that

$$\Phi(-x) = \int_{-\infty}^{-x} \phi(u)\,du < \frac{\phi(x)}{x}, \qquad x > 0,$$

which may be proved by repeated integrations by parts. This means that the integral from $-\infty$ to $(-\frac{1}{2} - np)/\sigma$ tends to zero faster than $\exp(-\alpha n)$, $\alpha > 0$, and we may therefore change the lower limit of integration to $-\infty$.

We also need the recursion formula

$$\int u^{2m}\phi(u)\,du = -\int u^{2m-1}\phi'(u)\,du$$

$$= -u^{2m-1}\phi(u) + (2m-1)\int u^{2m-3}\phi(u)\,du.$$

Further, we note that the second term of (2) contributes to the sum by

$$-\frac{1}{24}\sigma^2[\phi(u)(-u+\cdots)]_{(-1/2-np)/\sigma}^{(k+1/2-np)/\sigma}.$$

Setting $t = (k + \frac{1}{2} - np)/\sigma$, we finally get the result

$$B(k,n,p) = \Phi(t) - \phi(t)\{(t^2-1)(q-p)/6\sigma$$
$$+ [t^5(1-4pq) - t^3(7-22pq) + t(3-6pq)]/72\sigma^2 + \cdots\}. \tag{3}$$

In most textbooks only the first two terms are given, since they usually will suffice as an approximation to $B(k,n,p)$ for practical purposes. The second term, depending on $(q-p)/\sigma$, is called the correction for skewness.

Returning to the original problem of Bernoulli and de Moivre, we get

$$P_d = [\Phi(t) - \Phi(-t)][1 + O(n^{-1})], \qquad t = \frac{d+\frac{1}{2}}{\sqrt{npq}},$$

np being an integer.

Thus far we have only used the method of proof indicated by de Moivre. It is, of course, possible to continue the expansion, but as the following terms become more and more complicated, the need for a simple and general principle for the formation of the approximating series is obvious. The problem was solved by Edgeworth (1905) who in a more general context derived an approximation to a density function in terms of the normal density and its derivatives. Edgeworth derived the expansion (1) as we have done it above and used this result as the simplest check on his general result. Finally, Cramér (1928, 1937, 1972) gave a rigorous proof of Edgeworth's general asymptotic expansion and its extension to distribution functions.

For the asymptotic expansions of the binomial, we only know the order of the remainder term, but we have no limits for the error. Uspensky (1937, p. 119) writes, "When we use an approximate formula instead of an exact one, there is always this question to consider: How large is the committed error? If, as is usually done, this question is left unanswered, the derivation of Laplace's formula becomes an easy matter. However, to estimate the error comparatively long and detailed investigation is required. Except for its length, this investigation is not very difficult." Uspensky then evaluates the characteristic function of the binomial distribution and finds upper bounds for the absolute value of the remainder term when the two-term approximations are used. For the density, Uspensky's upper bound for the error is

$$[0.15 + 0.25|p - q|]\sigma^{-3} + e^{-3\sigma/2} \qquad \text{for} \quad \sigma \geqslant 5,$$

and for the difference $B(k_2) - B(k_1)$, he finds

$$[0.13 + 0.18|p - q|]\sigma^{-2} + e^{-3\sigma/2} \qquad \text{for} \quad \sigma \geqslant 5.$$

24.8   DANIEL BERNOULLI'S DERIVATION OF THE NORMAL DENSITY FUNCTION, 1770–1771

Daniel Bernoulli (1700–1782) wrote a paper published in two parts in 1770 and 1771, respectively, on the normal density function as approximation to the binomial and its application to the analysis of the variations of the sex ratio at birth. It is a rather curious paper in the sense that it completely ignores all previous works on these topics; it is surprising that Bernoulli either did not know or had forgotten de Moivre's *Miscellanea Analytica* (1730) and the *Doctrine* (1738, 1756). The paper does contain, however, a new method of deriving the normal density function as the limit of the binomial.

The first part of the paper has been discussed by Todhunter (pp. 235–236), who overlooked the second part. The importance of the second part has been pointed out by Sheynin (1970). Both parts have been reprinted in *Die Werke von Daniel Bernoulli*, Band 2, 1982, with a commentary on the first part. We shall first give an account of Bernoulli's theoretical results and then discuss his most important applications.

In the first part Bernoulli discusses the symmetric binomial. For $n = 2m$, he finds the central term

$$b(m) = \frac{1 \times 3 \times \cdots \times (2m - 1)}{2 \times 4 \times \cdots \times (2m)}, \tag{1}$$

and seeks an approximation for $m \to \infty$. From

$$\frac{b(m + 1)}{b(m)} = \frac{2m + 1}{2m + 2},$$

it follows that

$$\frac{\Delta b(m)}{b(m)} = -\frac{1}{2m+2},$$

so that

$$\frac{\Delta b(m-1) + \Delta b(m)}{2b(m)} \cong -\frac{1}{2m+\frac{1}{2}}\Delta m.$$

Replacing the differences by differentials, Bernoulli obtains the differential equation

$$\frac{db(m)}{b(m)} \cong -\frac{dm}{2m+\frac{1}{2}},$$

which has the solution

$$b(m) \cong c(2m+\tfrac{1}{2})^{-1/2} = c\sqrt{2}(4m+1)^{-1/2},$$

where c is a constant of integration. Hence,

$$\frac{b(m)}{b(k)} \cong \left(\frac{4k+1}{4m+1}\right)^{1/2}.$$

For $k = 12$, Bernoulli calculates $b(k)$ by means of (1) and thus finds the approximation

$$b(m) \cong 1.12826(4m+1)^{-1/2}. \tag{2}$$

De Moivre and Stirling had previously found

$$b(m) \cong (\pi m)^{-1/2} = 1.12838(4m)^{-1/2}.$$

To solve the equation

$$P_d = \sum_{x=-d}^{d} b(m+x) = \frac{1}{2}$$

with respect to d (we shall call the solution the probable error; Bernoulli uses "median limits"), we write

$$P_d = b(m) \sum_{x=-d}^{d} \frac{b(m+x)}{b(m)},$$

where

$$\frac{b(m + x)}{b(m)} = \frac{m(m - 1)\cdots(m - x + 1)}{(m + 1)(m + 2)\cdots(m + x)}.$$

Bernoulli tabulates $b(m + x)/b(m)$ to four decimal places for $m = 10{,}000$ and $x = 1(1)50$. By summation he finds that

$$P_{47} \cong 0.4980 \qquad \text{and} \qquad P_{48} \cong 0.5070$$

and concludes that $P_{47.25} \cong \frac{1}{2}$. A similar calculation for $m = 100$ gives $P_4 \cong 0.4753$ and $P_5 \cong 0.5631$, so that $P_{4.3} \cong \frac{1}{2}$. The ratio of the two values of d is thus approximately equal to the square root of the ratio of the m's. Bernoulli concludes that the probable error equals $0.4725\sqrt{m}$. That the probable error is proportional to $\sqrt{m}$ is supported by his results in the second part of the paper. In the first part he states without proof that if $d = O(\sqrt{m})$, then

$$b(m + d) \cong b(m)\, e^{-d^2/m}. \tag{3}$$

He demonstrates the accuracy of this approximation by numerical examples.

In the second part of the paper, Bernoulli proves that

$$b(np + d, n, p) \cong b(np, n, p)\, e^{-d^2/2npq} \tag{4}$$

by a generalization of the previous proof. From

$$-\frac{b(np + x + 1) - b(np + x)}{b(np + x)} = \frac{x + 1 - p}{(np + x + 1)q},$$

he gets the differential equation

$$-\frac{db(np + x)}{b(np + x)} \cong \frac{(x/q) + 1}{np + x + 1}\, dx.$$

Integration from $x = 0$ to $x = d$ gives

$$\ln \frac{b(np)}{b(np + d)} \cong \frac{d}{q} - \frac{np + 1}{q} \ln \frac{np + d + 1}{np + 1} + \ln \frac{np + d + 1}{np + 1}.$$

Assuming that $d = O(\sqrt{n})$ and taking the first three terms of the series for

$\ln\left[1+(d/(np+1))\right]$ into account, Bernoulli finds

$$\ln\frac{b(np)}{b(np+d)} \cong \frac{1}{q}\left(\frac{d^2}{2(np+1)} - \frac{d^3}{3(np+1)^2}\right) + \ln\frac{np+d+1}{np+1},$$

so that

$$\frac{b(np+d)}{b(np)} \cong \frac{np+1}{np+d+1}\exp\left[-\frac{d^2}{2npq+2q}\left(1-\frac{d}{3(np+1)}\right)\right],$$

which leads to (4), and for $p=\frac{1}{2}$ to (3).

To find $b(np, n, p)$ Bernoulli uses that

$$b(x, n, p) = b\left(x, n, \frac{1}{2}\right)\left(\frac{p}{q}\right)^x (2q)^n.$$

Setting $x = np$ and evaluating $b(np, n, \frac{1}{2})/b(\frac{1}{2}n, n, \frac{1}{2})$ by means of (3), he obtains finally,

$$b(np+d, n, p) \cong \frac{0.56413}{\sqrt{\frac{1}{2}n}}\left\{\exp\left(-\frac{1}{2}n(p-q)^2\right)\right\}\left(\frac{p}{q}\right)^{np}(2q)^n\exp\left(-\frac{d^2}{2npq}\right), \quad (5)$$

which is valid only for $p=\frac{1}{2}+O(n^{-1/2})$ and $d = O(n^{1/2})$.

This is Daniel Bernoulli's approximation to the binomial; he did not succeed in simplifying the factor to $\exp(-d^2/2npq)$, which de Moivre previously had evaluated to $(2\pi npq)^{-1/2}$ [see (5.3)]; neither did he attempt to approximate P_d by integration of (5). Furthermore, his formula holds only for values of p in the neighborhood of $p=\frac{1}{2}$ of the order of $n^{-1/2}$. Daniel Bernoulli was one of the great mathematicians at the time; his paper shows that it was difficult for him to find a satisfactory solution to the problem and thus puts de Moivre's result into perspective.

One reason that Bernoulli stops at the unsatisfactory result (5) may have been that he applies his formula only to investigations of the sex ratio. Setting $p/q = 1 + \alpha$ and evaluating (5) for small values of α, he proves that

$$b(np+d, n, p) \cong \frac{0.56413}{\sqrt{\frac{1}{2}n}}e^{-2d^2/n} \quad \text{for} \quad p=\frac{1}{2}+O(n^{-1/2}); \quad (6)$$

this is the formula he uses in his statistical analyses. He also notes that under the same conditions, the probable error equals $0.4725\sqrt{\frac{1}{2}n}$. The correct factor derived from the normal distribution is 0.4769.

Bernoulli discusses the variations of the sex ratio at birth, comparing observations from different years and different European cities and countries taken from the third edition of Süssmilch's book. In most cases he calculates the probability of an observed deviation by means of (6), which leads to rather small probabilities that are impossible to compare. In other cases he uses the probable error to judge whether a hypothetical value of p is reasonable or not. His most interesting analysis concerns the sex ratio in London. He does not mention the previous results by Graunt, Arbuthnott, Nicholas Bernoulli, and Struyck.

The observations consist of the yearly numbers of male and female christenings from 1664 to 1758. The ratio of the total number of male to female christenings is $737,629/698,958 = 1.055$. Looking at the ratios for each decade, Bernoulli observes that the minimum occurs during 1721–1730, for which he gets $92,813/89,217 = 1.040$. To analyze whether this is an essential deviation from 1.055 he calculates the expected number of male christenings for each year of the decade under the two hypotheses

$$p_0 = \frac{1055}{2055} \quad \text{and} \quad p_1 = \frac{1040}{2040}.$$

He studies the deviations between the observed and expected numbers, as shown in the following table.

DANIEL BERNOULLI'S ANALYSIS OF THE SEX RATIO IN LONDON, 1721–1730[a]

Year	x	n	np_0	$np_0 - x$	np_1	$np_1 - x$
1721	9430	18,370	9431	$+1$ NB	9365	-65
1722	9325	18,339	9414	$+89$	9349	$+24$ NB
1723	9811	19,203	9858	$+47$ NB	9790	-21 NB
1724	9902	19,370	9944	$+42$ NB	9875	-27 NB
1725	9661	18,859	9682	$+21$ NB	9614	-47 NB
1726	9605	18,808	9655	$+50$	9588	-17 NB
1727	9241	18,252	9370	$+129$	9305	$+64$
1728	8497	16,652	8548	$+51$	8489	-8 NB
1729	8736	17,060	8758	$+22$ NB	8697	-39 NB
1730	8606	17,118	8788	$+182$	8727	$+121$

[a]The number of male christenings is denoted by x, the total number by n; we have omitted the number of female christenings. Bernoulli has erroneously used $n = 18,759$ instead of 18,859 for the year 1725; we have corrected the whole line accordingly; Bernoulli's two deviations are -31 and -98. Bernoulli indicates deviations less than the probable error by NB.

Bernoulli notes that deviations according to his theory should be symmetrically distributed about zero and that half of the deviations should be numerically less than the probable error. He says that the positive sign is prevalent in the fifth column (nine of ten are positive in his table) and that the negative sign is prevalent in the seventh column (seven of ten are negative). His discussion is very brief, and he does not explicitly draw the conclusion that the probability of a male birth during 1721–1730 is smaller than 1055/2055. Perhaps he would have done so had he not committed the numerical error that led to nine instead of ten positive signs. Nevertheless, the ideas implicit in his statistical analysis are very important; we do not know whether his contribution was noted by his contemporaries, but presumably it was not.

For convenience of the reader who wants to analyze the data, we note that the average yearly number of christenings equals 18,203 and that the corresponding standard deviation equals 67. The reader may carry out both a sign test and a χ^2 test. Remember that the decade 1721–1730 has been selected because it gives the smallest value of p.

Let us summarize here the historical development of the statistical analysis of the sex ratio. Graunt (1662) points out that the ratio of male christenings in London and Romsey is larger than $\frac{1}{2}$, and that the ratio does not vary much over time; he recommends that similar investigations be carried out at other places and times. Arbuthnott (1712) suggests that the variation of the number of male christenings may be binomial; he does not investigate his hypothesis but uses a probability argument, the sign test, to prove that $p > \frac{1}{2}$. Nicholas Bernoulli (1713) compares the data with a binomial distribution for $p = 18/35$ and accepts (wrongly) the hypothesis that the variation in the period considered is binomial. He evaluates the observed deviations from the expected value by means of the tail probability. De Moivre (1738) stresses that deviations should be measured in terms of the modulus $2\sqrt{npq}$ and the corresponding value of the probability integral. Daniel Bernoulli's method is based upon an analysis of the sign and the size of the deviations in relation to the probable error. It is an early example (perhaps the first) of a simple and effective method of analyzing binomial data, which in a more detailed form was to become widespread in the following centuries. Bernoulli's table contains, at least implicitly, a detailed test procedure. First he estimates p from the data, $p_1 = 1040/2040$, and calculates the deviations which he compares with the probable error to find out whether the data are binomially distributed. Next, he repeats the whole procedure with a hypothetical value of $p, p_0 = 1055/2055$, to find out whether or not this value is compatible with the data. Here we have the germ of a method of testing statistical hypotheses.

From 1777 on, the method of comparing observed and expected numbers of deaths found important applications in actuarial science and in vital statistics, see Keiding (1987).

Daniel Bernoulli's method of deriving the normal density from the symmetric binomial seems to have been overlooked or forgotton; however, K. Pearson (1895, pp. 53–54) independently of Bernoulli gave the same proof. He writes,

> Hence, *this binomial polygon and the normal curve of frequency have a very close relationship to each other, of a geometrical nature, which is quite independent of the magnitude of n.* In short their slopes are given by an identical relation.... No stress seems hitherto to have been laid upon the fact that the normal curve of errors besides being the limit of a symmetrical point-binomial has also this intimate geometrical relationship with it.

It is well known that Pearson applied the same method to the hypergeometric distribution and thus found a differential equation, leading to his four-parameter system of frequency curves.

Bernoulli's and Pearson's proofs are not completely satisfactory from a mathematical point of view; a rigorous proof was first given by Jensen and Rootzén (1986).

From 1755 on, a probabilistic error and estimation theory was developed by Simpson, Lambert, Laplace, and Lagrange. Bernoulli (1778) contributed with a paper on the properties of error distributions and, choosing a semicircle with known diameter as distribution, he estimated the position of the center by the method of maximum likelihood.

As pointed out by Sheynin (1972), Bernoulli (1780) made a further contribution to the theory of errors by using the symmetric binomial and the normal approximation for a description of the random errors of time measurements by a pendulum clock. He assumes that there are n oscillations of the pendulum in a given time period, that the time of an oscillation may be either too long or too short by the same amount, and that these two outcomes are equally probable. Denoting the random error of a single oscillation by α, the total error for x positive and $n - x$ negative elementary errors becomes

$$e(x) = x\alpha + (n - x)(-\alpha) = 2\alpha(x - \tfrac{1}{2}n),$$

which occurs with probability $b(x, n, \tfrac{1}{2})$. Using the true time for one oscillation as time unit, Bernoulli concludes that the observed time will be approximately normally distributed about n with a probable error of $2\alpha(0.4715\sqrt{\tfrac{1}{2}n})$. As an example Bernoulli considers a pendulum with a period of oscillation of $1''$ so that the number of oscillations per day becomes $n = 86{,}400$. Setting $\alpha = 0.01''$, he finds the probable error of the time measurement per day to be $2''$.

He does not compare his theoretical distribution with observations, as he did for the sex ratio.

This is an early example, perhaps the first, of the normal distribution used as an error distribution based on the hypothesis that the observational error is the sum of a large number of elementary errors. The hypothesis of elementary errors was generalized by Laplace and led to the central limit theorem. A history of the central limit theorem has been given by Adams (1974).

Todhunter (p. 223) mentions another contribution by Daniel Bernoulli. In 1734 he attempted to show that the small mutual inclinations of the planetary orbits cannot be attributed to chance. Using the plane of the ecliptic as reference and assuming that all inclinations are equally likely he found the probability, p say, of an inclination smaller than the largest of the five observed. Since p^5 is very small, he concluded that the mutual inclinations are the results of original design. This is an early example of a test of significance, in the same spirit as Arbuthnott's test (1712). For details we refer to Todhunter.

CHAPTER 25

The Insurance Mathematics of de Moivre and Simpson, 1725–1756

25.1 INTRODUCTION

The industrial revolution, the increasing overseas trade, the growth of the British Empire, the accumulation of capital for investment and speculation, the establishment of the Bank of England and of joint-stock companies at the end of the 17th century, in short, the increasing capitalist structure of the British economy also led to the foundation of private insurance companies, at first for marine and fire insurance and a little later for life insurance in various forms.

Life assurances were short-term assurances, for the time of a single voyage or for a single year, say. Because of poor evaluation of the risks, life assurance contained a great deal of gambling and possibility of fraud. The wave of speculation in the first decades of the 18th century also led to speculation on the lives of other persons, for example, kings or other public persons; however, this combination of insurance and gambling was stopped by the Life Assurance Act of 1774, which forbade the insurance of lives in which the insured had no interest.

Several life assurance companies were started at the beginning of the 18th century, but only one, the Amicable Society for a Perpetual Assurance Office in 1705, survived. The Amicable was a primitive group life assurance. The number of members of a group was limited to 2000; on entrance, members should be between 12 and 45 years of age and of good health; the yearly premium was the same for all members independent of age; and the benefit to be paid at death was to be about 20 times the yearly premium. Compared with the prevailing individual life assurances, it had the advantage of being built on a group and being "perpetual."

It was not until 1762, when the Equitable Society was established, that an assurance company became based on "such persons as shall be qualified and be willing to become mutually contributors for equitable assurances on lives and survivorships upon premiums proportionate to the chance of death attending the age of life to be assured, and to the time such assurance is to continue," as stated in the preamble of the deed.

As noted in Chapter 9, state-guaranteed annuities were in general use by governments for raising money. To finance the War of the Spanish Succession, 1701–1713, the British government sold annuities which came to constitute the larger part of the national debt after the war. By an agreement between the government and the South Sea Company in 1720, the company took over the national debt against some trade concessions and payments from the government. Shortly afterwards the company got the greater number of the annuitants to exchange their annuities for company stock during a period with rapidly increasing prices due to speculation. When the slump came later in the year, numerous annuitants and other speculants were ruined.

Economic contracts that depended on the lifetimes of the parties involved were important parts of everyday life in Great Britain. Besides life annuities sold by the government, there were pensions granted by the government, the Church, municipalities, parishes, and private persons; reversions specified by wills and marriage settlements; rules for entailed estates, copyholds, advowsons, and many other forms of succession. Such contracts were exceedingly difficult to evaluate, and the need for a more detailed mathematical analysis of these problems than that provided by Halley was obvious.

This challenge was taken up by de Moivre who published the first textbook on life insurance mathematics under the title *Annuities upon Lives: or, The Valuation of Annuities upon any Number of Lives; as also of Reversions. To which is added, An Appendix concerning the Expectations of Life, and Probabilities of Survivorship*, 1725.

As described in §9.3, Halley had given the formula for the value of an annuity, both for a single life and for several lives, and he had provided a table of the value of single-life annuities for every fifth year of age at an interest rate of 6%. He characterizes this work as being "a most laborious Calculation" and writes that "I took the pains to compute the following table, being the short Result of a not ordinary number of Arithmetical Operations." He gave up the similar work for two lives because of the immense number of arithmetical operations involved and left the problem with the following remark: "I have sought, if it were possible, to find a Theorem that might be more concise than the Rules there laid down, but in vain."

De Moivre took over where Halley stopped. He had a genius for developing mathematical approximations, so he boldly proposed to approximate Halley's life table by a (piecewise) linear function and proved that the value of an annuity under this hypothesis will be a linear function of an annuity certain. Hence it is not necessary to tabulate the value of annuities, since everybody can calculate such values for any rate of interest by means of existing tables of annuities certain. Furthermore, he showed how the value of a joint-life annuity could be expressed approximately by means of the values of the corresponding single-life annuities so that joint-life annuities could also be easily handled. At one stroke he thus changed the whole outlook; what had seemed nearly impossible now became easy. Besides these approximations he gave a systematic exposition of formulae for the value of reversionary annuities, annuities on successive lives, and survivorship annuities. By these results he laid the foundation of modern life insurance mathematics.

De Moivre remarks that one of the consequences that flows from his theory will be to obtain "that Equity which ought to preside in Contracts."

For de Moivre life insurance mathematics is applied probability theory. His treatise begins as follows: "In estimating the Values of *Annuities upon Lives*, Regard must be had to the Interest which Money bears, and to the Probability of the Lives continuing a longer or shorter Time. The Rate of Interest is generally regulated by Law, but the greater or less Probability of the Duration of Life must be deduced from Observation."

Without discussion he accepts Halley's life table and a yearly rate of interest of 5% as the foundation for his analysis. He gives a probabilistic and economic analysis of a great number of contracts from everyday life for the purpose of deriving a formula for the valuation of each type of contract and to find approximate formulae to make the numerical work feasible.

The main contents of de Moivre's *Annuities upon Lives* (1725) may be summarized as follows:

The contents are divided into 38 problems, each with a verbal solution and a numerical example followed by a mathematical demonstration. He uses the same popular style as that in the *Doctrine of Chances*.

Much of the material from the 1725 edition is included in the 1738 edition of the *Doctrine* (pp. 211–231, 251–255). He adds some new material on successive lives, on the value of annuities for children, on temporary life assurances, and a table of "The present Value of an Annuity of One Pound, to continue as long as a Life of a given Age is in being, Interest being estimated at five per cent."

Between 1725 and 1750 several books and tables on annuities were published, but they are all inferior to de Moivre's text, except for Thomas Simpson's *The Doctrine of Annuities and Reversions, Deduced from General and Evident Principles: With Useful Tables, Shewing the Values of Single and Joint Lives, etc. at different Rates of Interest*, 1742.

In the preface Simpson criticizes authors who "without troubling themselves or their readers about observations, etc. have taken upon them to prescribe methods of their own, that have neither foundation in experience nor in reason." He continues, "Yet I would not be thought to condemn any hypothesis grounded upon reason and matters of fact, because such are oftentimes made use of to very great advantage, of which Mr. *de Moivre's* excellent book on this subject is an instance."

Simpson's book is to a large extent built on de Moivre's; he uses the same framework and nearly the same problems. However, his exposition is shorter and clearer, many of his proofs are more general, and he makes three important new contributions: (1) a life table based on the London bills of mortality; (2) tables of values of single- and joint-life annuities for nominees of the same age based on this life table; and (3) rules for calculating joint-life annuities for different ages from the tabulated joint-life annuities. Simpson shows that de Moivre's formula for the value of joint-life annuities is not sufficiently accurate, and he provides the first satisfactory solution to the problem of calculating values of annuities for two and three lives.

In 1743 de Moivre published the second edition of *Annuities on Lives, Plainer, Fuller and more Correct than the former*. It contains a few more problems than discussed in 1725 and 1738 and some improved solutions. The exposition is reorganized so that the demonstrations are transferred to an Appendix. The 5% table of annuities is supplemented by tables for 4% and 6%. In the preface he defends his linear hypothesis, and about the value of joint-life annuities he states, "The greatest Difficulty that occurred to me in this Speculation, was to invent practical Rules that might easily be applied to the Valuation of several Lives; which however, was happily overcome, the Rules being so easy, that by the help of them, more can be perform'd in a

Quarter of an Hour, than by any Method before extant, in a Quarter of a Year."

He ends the preface with a denunciation of Simpson:

> After the pains I have taken to perfect this Second Edition, it may happen, that a certain Person, whom I need not name, out of *Compassion to the Public*, will publish a Second Edition of his Book on the same subject, which he will afford at a *very moderate Price*, not regarding whether he mutilates my Propositions, obscures what is clear, makes a Shew of new Rules, and works by mine; in short confounds in his usual way, every thing with a croud of useless Symbols; if this be the Case, I must forgive the indigent Author, and his disappointed Bookseller.

This is clearly an overreaction, which can be understood only if one remembers Simpson's plagiarism of de Moivre's *Doctrine of Chances* two years before, see §22.4. In both of his books, Simpson refers to de Moivre in the preface, but he does not refer to him in the text, even though he has used nearly all of de Moivre's results. At the time the rules for crediting other authors were not as strict as they are today, but Simpson's omission of references to de Moivre is nevertheless very conspicuous compared with the behavior of Montmort, Nicholas Bernoulli, and de Moivre himself. Of course, de Moivre ought to have mentioned Simpson's new results in his book and modified his own exposition accordingly, but he did not do so, neither in the second nor in the following editions.

Simpson reacted by publishing a 16-page "Appendix, containing Some Remarks on Mr. Demoivre's Book on the same Subject, with Answers to some Personal and Malignant Misrepresentations, in the Preface there-off"..."to clear myself from a charge so highly injurious, and do justice to the foregoing work." He defends himself effectively against de Moivre's unreasonable charges and launches a counterattack by giving a detailed discussion of several errors in both editions of de Moivre's *Annuities*. He ends the *Appendix* with the remark, "Lastly, I appeal to all mankind, whether in his treatment of me, he has not discovered an air of self-sufficiency, ill-nature, and inveteracy, unbecoming a gentleman."

De Moivre did not answer directly but removed the accusation against Simpson from the preface of the following editions.

In a letter to W. Jones, a vice president of the Royal Society, communicated to the Royal Society in 1744 and published in the *Phil. Trans.* (1746), de Moivre finds an approximation to the value of a complete life annuity and gives a better explanation of the use of his piecewise linear mortality hypothesis for calculating the value of annuities than in the first edition.

The final edition of his treatise is to be found at the end of the 1756 edition of the *Doctrine of Chances* under the title *A Treatise of Annuities on Lives* with the running head "The Doctrine of Chances applied to the Valuation of Annuities." This edition contains the text from the second edition with the addition of a Chapter IX: "Serving to render the Solutions in this Treatise more general, and more correct," a reprint of the letter to W. Jones, an extension of the tables of annuities to interest rates of 3 and $3\frac{1}{2}\%$, and some new empirical life tables constructed by other authors. Between the second and the last editions, the third and the fourth editions were published in 1750 and 1752, respectively, being essentially reprints of the second edition.

Most authors commenting on de Moivre's work have naturally used the 1756 edition. However, to understand this edition it is important to consider the first edition and the additions given in the *Doctrine of Chances* (1738).

In 1752 Simpson published *The Valuation of Annuities for single and joint Lives, with a Set of new Tables, far more extensive than any extant*, being Part VI of *Select Exercises for Young Proficients in the Mathematicks*, which is a popular exposition of his previous book with many new examples and with three important additions: (1) a table of the value of annuities on two lives of different ages; (2) a rule by which the value of an annuity on three lives may be expressed by annuities on two lives; and (3) rules for calculating survivorship insurances in which he corrects one of de Moivre's mistakes.

Finally, we note the great investigations of annuitant mortality made in the 1740s by Nicolaas Struyck (1687–1769) and Willem Kersseboom (1691–1771) in Holland and by Antoine Deparcieux (1703–1768) in France. The resulting life tables have been discussed by many authors; we refer to Braun (1925), Westergaard (1936), and K. Pearson (1978). The life tables constructed by Halley, Simpson, Kersseboom, and Deparcieux, along with some commentary by de Moivre, may be found in the *Doctrine of Chances* (1756, pp. 345–348).

De Moivre advocated using a mathematical law of mortality, whereas Simpson used an observed life table. In the first instance, Simpson won; the following generation of actuaries used Simpson's approach, although they did not use his tables but based their calculations on better life tables constructed from the experiences of life insurance offices. At the same time, however, many attempts were made to find a more satisfactory law than that proposed by de Moivre, but it was not until 1825 that Benjamin Gompertz (1779–1865) succeeded in formulating a law that won universal acceptance; it was improved by William Maitland Makeham (1826–1891) in 1860 and has since been an important tool for graduating mortality observations and calculating the value of life insurances. A survey of laws of mortality has been given by J. du Saar (1917).

A polemic history of insurance mathematics in Britain in the 18th century may be found in the preface to Baily (1813) who also discusses the results of de Moivre and Simpson in his text.

The present chapter is a continuation of Chapters 8 and 9, and we shall use the notation employed there. De Moivre and other actuaries in the 18th century did not introduce specific symbols for probabilities of life, expectations, annuities, and so on; they used arbitrary letters, such as M, P, Q for annuities and a, b, c for the number of survivors in the life table; in particular, they did not use subscripts to denote the age of annuitants, they merely mentioned the age in the text. We shall chiefly use the notation from the *Institute of Actuaries' Text Book* by G. King (1902), which also contains many of the results of de Moivre and Simpson, with modern proofs.

25.2   THE LIFE OF THOMAS SIMPSON

Thomas Simpson (1710–1761) was the son of a weaver who wanted Thomas to take up the same profession and therefore did not much care for his son's literacy. Conflicting interests led the 14-year-old Thomas to leave home and settle in a neighboring village at the lodging house of Mrs. Swinfield, the widow of a tailor. There he continued to educate himself and worked as a weaver. From a peddler and fortune teller who occasionally lodged at the same house, Simpson borrowed a book on arithmetic and one on astrology; he soon mastered both and became known as an able fortune teller. He married Mrs. Swinfield, his senior by about 30 years with two children his own age. An unfortunate case of fortune-telling caused Thomas and his family to move to Derby, where he continued his trade as a weaver, became instructor at an evening school, and continued his study of mathematics. At the age of 25 he moved to London, still working as a weaver and teaching mathematics in his spare time. He became a member of one of the many mathematical clubs and won a reputation as an able teacher and a writer of textbooks.

In 1743 he left London to become Second Master of Mathematics at the Royal Military Academy, Woolwich, which gave him regular teaching hours, a regular income, and more time for studying and writing.

He contributed many mathematical problems and solutions to the *Ladies' Diary*, a popular annual publication with a section on elementary mathematics; in 1754 he became editor and compiler of this magazine, continuing for six years.

His teaching, writing, and editorial work resulted in a large correspondence about mathematical and actuarial problems.

He became a successful writer of textbooks and of some books of essays resulting from his research. His astounding output consists of eleven books published between 1737 and 1757. The subject matter and the year of publication of each are as follows:

on the theory of fluxions, 1737;

on the laws of chance, 1740;

speculative and mixed mathematics, 1740;

annuities and reversions, 1742;

on a variety of physical and analytical subjects, 1743;

algebra, 1745;

geometry, 1747;

trigonometry, 1748;

on fluxions, 1750;

select exercises in mathematics, 1752;

miscellaneous tracts on mechanics, physical astronomy, and speculative mathematics, 1757.

Several of these books are of an elementary nature and were designed for courses taught at the military academy. His success as a textbook writer is evidenced by the fact that several of his books had numerous editions and were also published in America, France, and Germany.

Simpson's writings aroused much controversy and many accusations of plagiarism. Besides the controversy with de Moivre, he had a controversy with the First Master at the Military Academy over their books on geometry, and another controversy with the previous editor of the *Ladies' Diary*.

As stated, Simpson made original and important contributions to actuarial science, which we shall detail in the following sections. He also made an important contribution to statistical error theory. It had become customary among astronomers to use the mean of several observations to estimate the true value, but an adequate theoretical background for this procedure was missing. The beginnings of such a theory were provided by Simpson (1755, 1757), who derived the error distribution of the mean under some simple assumptions on the distribution of the individual errors, using de Moivre's technique of generating functions, see Seal (1949) and Stigler (1986).

F. M. Clarke (1929) has written a biography of Simpson on which the sketch above is based.

25.3   DE MOIVRE'S LINEAR AND PIECEWISE LINEAR APPROXIMATION TO HALLEY'S LIFE TABLE

Except for Graunt's table, the only existing empirical life table at the time was Halley's table for the population of Breslau, see Table 9.3.2. Hence, de Moivre had no other choice but to use Halley's table if he wanted to base his analysis on real observations rather than on hypothesis, as the distinction was then formulated. He was of course aware that the mortality rate in London was different from that in Breslau, but he had no means of finding the London mortality rate because the Bills of Mortality still did not contain information on the age at death. He does not discuss the construction of Halley's table, and he uses the table as if the numbers tabulated were l_x and not L_x, see §9.3.

He begins with a discussion of the difference $d_x = l_x - l_{x+1}$ and notes that Halley's life table is piecewise linear from $x = 12$ to 77, with intervals of varying length. In 1725 he gives a discussion in the text only; in 1756, p. 346, he puts an asterisk in the table each time the value of d_x changes. As shown in Fig. 25.3.1, there are nine intervals with constant differences varying from 6 to 11.

De Moivre's great discovery was that the value of an annuity may be expressed as a linear function of an annuity certain if the life table is linear, and that an analogous theorem holds for a temporary annuity if the life table is linear for the age interval in question. He could thus avoid the calculation of the many products and sums which had troubled Halley by using his formula for each linear section of the life table. It will be seen that this idea is the same as that used by de Witt, whose work, it is presumed, was unknown to de Moivre.

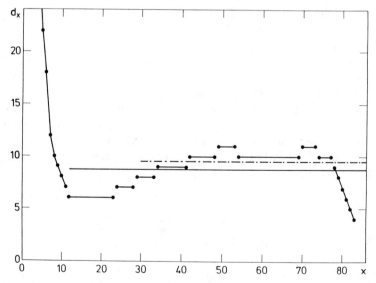

Fig. 25.3.1. A graph of the differences d_x in Halley's life table. The average differences for $x \geqslant 12$ and $x \geqslant 30$ are indicated.

De Moivre writes (1725, p. 11), "Let us therefore consider, 1° what would be the result of an *Hypothesis* that makes the Probabilities of Life to decrease in *Arithmetical Progression*: 2° how far the Calculations deduced from it agree with the Tables: 3° what Corrections are necessary to be made to it where it varies from the Tables." He realized that his hypothesis could not be used below the age of 12.

De Moivre's linear hypothesis relates to the probabilities of life; he assumes that

$$_tp_x = \frac{l_{x+t}}{l_x} = 1 - \frac{t}{\omega - x}, \qquad 12 \leqslant x < \omega, \quad 0 \leqslant t \leqslant \omega - x, \tag{1}$$

and that $_tp_x = 0$ for $x \geqslant \omega$, $t \geqslant 0$. He calls $\omega - x$ for the complement of life. He chooses $\omega = 86$ for the following reasons (1725, p. VIII):

> my Meaning, in fixing that Term, imports no more than this, *viz.* that the Time of contracting for *Annuities* being commonly very remote from the 86th Year of the Life purchased, it is not likely that any Consideration will then be given for the Chance of receiving the Rent of that year, which will produce the very same Conclusions in *Theory*, as if the Extremity of that Year were never attainable.

To distinguish between the numbers in the life table and de Moivre's approximation, let us denote the former by λ_x and the latter by l_x. The linear hypothesis may then be written as

$$l_{x+t} = \lambda_x \left(1 - \frac{t}{\omega - x}\right), \qquad 12 \leqslant x < \omega, \quad 0 \leqslant t \leqslant \omega - x, \tag{2}$$

and $l_{x+t} = 0$ for $x + t \geqslant \omega$. It is clear from Fig. 25.3.1 that the goodness of fit varies considerably with x, being rather poor for $x = 12$ and good for $x = 30$, see also Fig. 25.4.1.

Most authors write about the linear hypothesis as if de Moivre assumed that $l_x = \omega - x$ for $0 \leqslant x \leqslant \omega$, but that is not so.

De Moivre also worked with a finer hypothesis based on the assumption that the life table may be approximated by a piecewise linear function. For the interval $(x, x + s)$ this approximation becomes

$$l_{x+t} = \left(1 - \frac{t}{s}\right)\lambda_x + \frac{t}{s}\lambda_{x+s}, \qquad 0 \leqslant t \leqslant s. \tag{3}$$

He notes that (3) may be considered as the first section of length s of (2) if

ω is chosen as

$$\omega = x + \frac{s\lambda_x}{\lambda_x - \lambda_{x+s}}, \tag{4}$$

so that the piecewise linear function may be composed of sections of linear functions of the type (2) with varying values of ω.

De Moivre gives a sketch of a piecewise linear life table which is omitted from the following editions, but a verbal description of this "polygon" is included in the last edition (1756, p. 325).

It will be seen from Fig. 25.3.1 and Fig. 25.4.1 that a piecewise linear function for 10-year intervals beginning at the age of 12 will give a good approximation, except for the last interval.

25.4 SIMPSON'S LIFE TABLE FOR THE POPULATION OF LONDON

At the initiative of John Smart, the parish clerks in London began in 1728 to record the age of the deceased, and Smart used the observations for the period 1728–1737 to construct a life table.

Simpson observes that this table does not represent true mortality in London because Smart has not taken migrations into account. He therefore modifies Smart's table for ages below 25, but unfortunately his description of the method used is so vague that it is impossible to see what he actually did. He writes (1742, pp. 2–3),

> In doing this, I have supposed the number of persons coming to live in town, after 25 years of age, to be inconsiderable, with respect to the whole number of inhabitants; and therefore the probabilities of life, for all ages above 25 years, the same as this author has made them; but then have increased the numbers of living, corresponding to all ages below 25; so that they may, as near as possible, be in the same proportion one to another, as they would be, were they to be deduced from observations on the mortality of those persons *only*, that was born within the bills. Which was done, by comparing together the number of christenings and burials, and observing, by help of Dr. *Halley*'s table, the proportion which there is between the degrees of mortality at *London* and *Breslau*, in the other parts of life, where the ages are greater than 25.

Smart had $l_0 = 1000$ and $l_{25} = 426$. Simpson did not change Smart's life table for $x \geqslant 25$ but increased l_0 to 1280. Taking $d_0 = 410$, so that $l_1 = 870$, he interpolated between 870 and 426, taking Halley's table into account. The crucial point is how he found l_0 and d_0; he gives no information on this step.

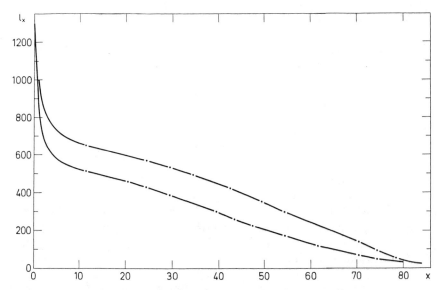

Fig. 25.4.1. Halley's and Simpson's life tables. Halley's table begins with $l_1 = 1000$, Simpson's with $l_0 = 1280$. The curves are linear between dots.

Westergaard (1901, pp. 43–45; 1936, p. 67) has suggested how Simpson may have reached his result.

Simpson does not give a detailed discussion of his table but only states that mortality in London is higher than in Breslau.

We have compared Halley's and Simpson's life tables in Fig. 25.4.1. There are several interesting features. First, Simpson's l_x is smaller than Halley's, and the two curves are nearly parallel for ages between 10 and 60. Second, like Halley's table, Simpson's is approximately linear above 10 years of age. Third, Simpson estimates mortality in the first year to 32%.

Deparcieux criticized Simpson for using too high a mortality for children, to which Simpson (1752, p. 311) replies, "He [Deparcieux] does not seem to be appriz'd, that there is not, perhaps, a City in the World so fatal to the Infant State as the City of *London* (owing, too much, to the Intemperance of Parents and profuse, irregular manner of Living)."

Simpson's table soon became obsolete for insurance purposes because it gives the mortality of the whole population, not of annuitants only.

25.5   SINGLE-LIFE ANNUITIES

As noted by Halley, the fundamental quantity for calculating the value of an annuity for a life (x) is the value of the amount 1 payable in t years if (x)

is then alive. Setting $v = (1 + i)^{-1}$, we have

$$_tE_x = v^t{}_tp_x, \tag{1}$$

$$a_x = \sum_{t=1}^{\infty} v^t{}_tp_x. \tag{2}$$

The sum is finite because $_tp_x = 0$ for $x + t \geqslant \omega$. Halley's principle for calculating the value of a life contingency thus consists in finding the present expected value of each year's benefit and adding these amounts over the period in question. This principle was adopted by de Moivre and Simpson, so the problems they faced were to find the probability of the contingency and to evaluate the sum, either mathematically or numerically.

Halley calculated $_tE_x$ by logarithms and obtained a_x by summation. De Moivre gives a much simpler method for tabulating a_x by means of the backward recursion formula

$$a_x = vp_x(1 + a_{x+1}). \tag{3}$$

To prove this formula de Moivre notes that (x) will get the amount 1 in one year if he is then alive, and that he will also be entitled to all the succeeding rents which have the value a_{x+1}. Discounting the two amounts with respect to both interest and mortality, (3) follows. He gives a numerical example for $x = 29$.

Beginning with $a_{\omega-1} = 0$, a complete table of a_x may be calculated, the amount of work being about the same for the whole table as for calculating a_1 by Halley's method. De Moivre does not tabulate a_x; he writes (1725, p. 8),

> By help of the Method hitherto explained, those that will take the Pains may, if they think fit, compose *Tables* of the several Values of those *Annuities* for any Age proposed, and for any Rate of Interest that shall be fixed upon. *But* 'till that be done, it will be convient to try whether there be not some easy practical Method whereby the Values of those *Annuities* may be determined very near as accurately as if they had been deduced Year by Year from the *Tables of Observations.*

The backward recursion formula is not included in the later editions of de Moivre's treatise, perhaps because they contain tables of a_x. This omission has led to the misunderstanding that the formula is due to Simpson (1742, pp. 18–20). Simpson's contribution is to point out that the formula also holds for joint lives and to "take the Pains" to use the formula for calculating tables of values of annuities based on his own life table.

We note that the backward recursion is a generalization of Nicholas Bernoulli's formula for the calculation of the expectation of life, see §8.2.

By introducing a mathematical hypothesis about $_tp_x$, de Moivre obtains a simple formula for a_x. To justify this approximation he states "that any small Difference that may arise from the two Methods of Calculation is not to be regarded in a Subject of this Nature," thus alluding to the uncertainty inherent in Halley's table and its application to annuitants in London.

He begins with the following problem (1725, p. 11): "Supposing the Probabilities of Life to decrease in *Arithmetic Progression* to find the Value of an Annuity upon such a Life."

For $x < \omega$, the solution is

$$a_x = \frac{1}{i}\left(1 - \frac{1+i}{n} a_{\overline{n}|}\right), \qquad n = \omega - x. \tag{4}$$

Proof. Like de Moivre we shall set $n = \omega - x$ so that $_tp_x = 1 - t/n$. Hence,

$$a_x = \sum_{t=1}^{n} v^t\left(1 - \frac{t}{n}\right) = a_{\overline{n}|} - \frac{1}{n}(Ia)_{\overline{n}|},$$

where $a_{\overline{n}|}$ denotes the value of an annuity certain

$$a_{\overline{n}|} = \sum_{t=1}^{n} v^t = \frac{1 - v^n}{i}, \tag{5}$$

and $(Ia)_{\overline{n}|}$ denotes the value of an increasing annuity certain

$$(Ia)_{\overline{n}|} = \sum_{t=1}^{n} v^t t = \left(n + 1 + \frac{1}{i}\right)a_{\overline{n}|} - \frac{n}{i}. \tag{6}$$

To prove the last result de Moivre refers to a theorem in the *Doctrine of Chances* (1718) on the summation of recurring series but does not carry out the proof. Following his hint we note that

$$b_t = v^t t, \qquad t = 0, 1, \ldots,$$

is a recurring series satisfying the relation

$$b_t = 2vb_{t-1} - v^2 b_{t-2}, \qquad t = 2, 3, \ldots.$$

It follows that $S = b_1 + \cdots + b_n$ satisfies the equation

$$S - b_1 = 2v(S - b_n) - v^2(S - b_{n-1} - b_n),$$

which leads to

$$S(1 - v)^2 = v(1 - v^n) - v(1 - v)nv^n.$$

Eliminating v^n by means of (5), we get (6), which completes the proof.

In subsequent editions de Moivre states that the derivation of (4) requires "something more than an ordinary skill in the Doctrine of Series," so that he will show the reader that the formula is correct by inserting the series for $a_{\overline{n}|}$ and $1/i$ in terms of v^t. This is a more tedious procedure than his original proof.

As explained in §9.2, de Witt's formula for the value of an annuity is

$$a_x = \frac{1}{l_x} \sum_{t=1}^{\omega - x - 1} a_{\overline{t}|} d_{x+t}.$$

Under de Moivre's hypothesis we have $l_x = n$ and $d_{x+t} = 1$, so that

$$a_x = \frac{1}{n} \sum_{t=1}^{n-1} a_{\overline{t}|} = \frac{1}{ni} \sum_{t=1}^{n-1} (1 - v^t),$$

which immediately leads to (4). However, de Moivre and Simpson did not consider this simple proof because they strictly followed Halley's principle.

For later use we note that

$$a_{\omega - s} = \sum_{t=1}^{s} v^t \left(1 - \frac{t}{s}\right) = \frac{1}{i}\left(1 - \frac{1+i}{s} a_{\overline{s}|}\right) = a(s), \tag{7}$$

say, since the value of this annuity depends on s only.

De Moivre next solves the following problem (1725, p. 20): "To find the Value of an Annuity for a limited Interval of Time, during which the Decrements of Life may be considered as equal."

Denoting the value of a temporary annuity to (x) for s years as $a_{x\overline{s}|}$, the solution becomes

$$a_{x\overline{s}|} = a(s) + {}_sp_x[a_{\overline{s}|} - a(s)].$$ (8)

Proof. De Moivre's proof is very simple. From (3.3) we have

$$_tp_x = \left(1 - \frac{t}{s}\right) + {}_sp_x\left[1 - \left(1 - \frac{t}{s}\right)\right], \qquad 0 \le t \le s, \quad {}_sp_x = \frac{\lambda_{x+s}}{\lambda_x},$$

which inserted into

$$a_{x\overline{s}|} = \sum_{t=1}^{s} v^t {}_tp_x,$$

immediately proves (8).

To link the temporary annuities together de Moivre uses the backward recursion (1725, pp. 25–26)

$$a_x = a_{x\overline{s}|} + v^s {}_sp_x a_{x+s}.$$ (9)

His proof of this formula is analogous to the proof of (3), which is obtained for $s = 1$.

Young (1908) suggests that both de Moivre and Simpson failed to observe "the implication [of (3)] of the necessary start with the oldest age attained in the observations"; instead he ascribes this observation to Euler. Young's conclusion seems unwarranted in view of the facts that de Moivre used (9) for $s = 10$ to calculate the value of annuities, beginning with the age of 86, and that Simpson calculated all his tables by means of (3). Simpson's published life table ends with $l_{80} = 29$, but he has of course (like Halley) just continued the table to some higher age for which $l_x = 0$; his table of annuities ends at the age of 75.

To evaluate the accuracy of the approximation de Moivre calculates some values of a_x by (4) and compares them with the values given by Halley. As shown in the table below the percentage error is at most 4.5 for ages between 10 and 70. We do not know how he found $a_1 = 12.57$; it has not been calculated from (4), and in the tables published later he gave the value as 10.80.

He also uses the backward recursion (9) for 10-year intervals to calculate a_x for $x = 86, 76, \ldots, 6$ and $i = 0.05$, but since he does not calculate the corresponding values from Halley's table, he cannot find the errors. We have therefore used (9) for $i = 0.06$ to calculate the values corresponding to those given by Halley, starting from $a_{70} = 5.50$. The table below shows that the percentage error is at most 0.5 for ages between 10 and 60.

Comparison of Exact and Approximate Values[a] of a_x

x	HALLEY	EQUATION 4		EQUATION 9	
		DE MOIVRE	PERCENTAGE ERROR	DE MOIVRE	PERCENTAGE ERROR
1	10.28	(12.57)	—	—	—
10	13.44	12.84	−4.5	13.45	+0.1
20	12.78	12.30	−3.8	12.75	−0.2
30	11.72	11.61	−0.9	11.70	−0.2
40	10.57	10.70	+1.2	10.55	−0.2
50	9.21	9.49	+3.0	9.23	+0.2
60	7.60	7.83	+3.0	7.64	+0.5
70	5.32	5.50	+3.4	—	—

[a]The exact values are calculated from Halley's table; the approximate values are calculated from de Moivre's linear and piecewise linear hypotheses using (4) and (9), respectively.

De Moivre states that the piecewise linear hypothesis will lead to a good approximation for any life table and any rate of interest. However, he is content with the approximation obtained from the linear hypothesis for $x \geqslant 10$, and calculates all his tables of a_x for various rates of interest under this assumption.

For $x < 10$, special methods have to be used because the linear hypothesis does not hold. De Moivre shows how to find $a_{1\overline{11}|}$ from Halley's table and then to find a_1 by means of (9) and a_{12}.

De Moivre's first table of a_x is to be found in the *Doctrine of Chances* (1738), where he also makes the following remark about the value of a_x for $x < 10$:

But before we proceed any farther, it will be convenient to observe, that in Children the Probability of Life increases instead of decreasing; and that therefore those that have a mind to use the Rule given in our LXXIII[d] Problem for finding an Annuity upon the Lives of Children of 1, 2, 3, 4, 5, 6, 7, 8, 9 years of Age, ought to calculate them as if they were of the respective ages of 39, 31, 24, 20, 17, 14, 12, 11, 10.

His rule of conversion is thus very simple: a_1 should be set equal to a_{39}, a_2 equal to a_{31}, and so on, but his motivation is difficult to understand. He refers to a problem in which he assumes that the life table is geometrically decreasing, which means that $_t p_x = p^t$ for a suitably chosen value of p. It follows that

$$a_{\overline{xs}} = \sum_{t=1}^{s} (vp)^t = \frac{1 - (vp)^s}{j}, \qquad j = \frac{i + (1 - p)}{p},$$

$$a_x = a_{x\overline{10-x}} + (vp)^{10-x} a_{10}, \qquad x = 1, \ldots, 9. \tag{10}$$

To determine p from Halley's table we have to solve the equation

$$p^{10-x} = \frac{\lambda_{10}}{\lambda_x}.$$

For $i = 0.05$ we find $a_1 = 11.99$, in agreement with de Moivre who gives the value 11.96 in the table. However, for $x = 2, 3, \ldots, 7$, we get values about 2% larger than those given by de Moivre. It seems that de Moivre has imposed the restriction that a_x should have its maximum for $x = 10$ and that he has smoothed the values between a_1 and a_{10} such that the conversion table gets reasonably decreasing differences with respect to x.

It is clear that the rule of conversion depends on the rate of interest, but nevertheless de Moivre uses the same rule of conversion in all his tables. Since the remark on the conversion rule quoted above is to be found in the 1738 edition only, readers of all the following editions may have wondered how the first nine values in the tables have been found.

A comparison of the values of annuities from the tables of de Moivre and Simpson is given in the following table:

Table of a_x for $i = 0.05$

				x			
	10	20	30	40	50	60	70
De Moivre	14.60	13.89	12.99	11.83	10.35	8.39	5.77
Simpson	14.3	13.0	11.6	10.3	9.2	7.9	6.2
Ratio	0.98	0.94	0.89	0.87	0.89	0.94	1.07

Source: de Moivre (1738, p. 255) and Simpson (1742, pp. 38–39).

It will be seen that Simpson's life table gives values that are about 12% lower than Halley's for ages between 30 and 50.

Simpson (1752, pp. 274–276) had the knack of finding simple approximations. As an example we shall quote his formula for the value of an annuity based on Halley's table,

$$a_x \cong \frac{85 - x}{0.8i(85 - x) + 2},$$

which "from the Age of Eight, To Eighty, will, for the general Part, come within less than 1/8 of an Year's Purchase of the Truth." It is remarkable that Simpson nearly always makes such statements about the accuracy of the many approximations he proposes. He gives a similar formula for values based on his own life table. Clearly, these formulae have no theoretical foundation; they are presumably found by trial and error.

In the second edition of the *Annuities* (1743), de Moivre omitted the backward recursion formulae (3) and (9) and the investigation of the accuracy of (4). From (9) it is clear that he knew the formula for the value of a temporary annuity,

$$a_{x\overline{s}|} = a_x - v^s {}_sp_x a_{x+s}, \tag{11}$$

but nevertheless he forgot the factor ${}_sp_x$ when solving Problem 23 (1743). Of course, Simpson (1743, p. 140) pointed out the error and gave the correct formula. Instead of quoting (11) in later editions, de Moivre derived a formula directly from the linear hypothesis. Assuming that l_{x+t} is linear for $0 \leqslant t \leqslant \omega - x$, we have ${}_sp_x = 1 - s/n$, so that (8) becomes

$$a_{x\overline{s}|} = \frac{1}{n}\left[\frac{s}{i} + \left(n - s - 1 - \frac{1}{i}\right)a_{\overline{s}|}\right], \qquad n = \omega - x, \tag{12}$$

as given by de Moivre (1756, p. 292). This is a generalization of (4), which is obtained for $s = n$.

De Moivre also gave an incorrect formula for the value of a temporary reversionary annuity; Simpson pointed out the error and gave the correct formula

$$a_{x|\overline{s}|} = a_{\overline{s}|} - a_{x\overline{s}|} \tag{13}$$

(see also de Moivre, 1756, p. 293).

At the time, a distinction was made between an annuity "secured by money" and "secured by land," the former being an ordinary annuity and the latter a complete annuity, i.e., an ordinary annuity supplemented by a proportionate part of the yearly payment for the time between the last payment and the moment of death. In his letter to W. Jones in 1744, de Moivre derives the value of a complete annuity.

Suppose that (x) dies at age $x + t$ and set $t = k + \epsilon$, where $k = [t]$. The value of the payments will then be

$$\frac{1 - v^k}{i} + \epsilon v^{k+\epsilon} = \frac{1 - v^t}{i} + \frac{1}{2}\epsilon(1 - \epsilon)iv^t - \cdots \cong \frac{1 - v^t}{i}.$$

De Moivre begins his proof by stating that it is well known that the value of an annuity certain payable during the time t is $(1 - v^t)/i$. However, as shown above, this is only an approximation, except for the case where t is an integer. He proceeds to find the expected value of this annuity certain under the linear hypothesis, which means that the probability of dying between t and $t + dt$ equals $dt/n, n = \omega - x$. Hence,

$$E\left\{\frac{1 - v^t}{i}\right\} = \frac{1}{ni}\int_0^n (1 - v^t)dt = \frac{1}{i} - \frac{a_{\overline{n}|}}{n\delta}, \qquad n = \omega - x, \quad \delta = \ln(1 + i). \quad (14)$$

Since t is continuous, de Moivre naturally uses integration to find the value of the annuity; he had previously used this technique to find the expectation of life. About the integral of v^t, he states that "I do not know, whether the same method has been made use of by others." His method is to make the transformation $y = v^{-t}$ so that

$$\int v^t \, dt = \frac{1}{\delta}\int \frac{dy}{y^2} = -\frac{1}{y\delta} = -\frac{v^t}{\delta}.$$

It is also remarkable that here he uses de Witt's principle instead of Halley's to find the value of the annuity.

De Moivre does not consider the value of a continuous annuity, which is the expected value of a continuous annuity certain

$$\bar{a}_x = \frac{1}{n}\int_0^n \bar{a}_{\overline{t}|} \, dt = \frac{1}{n\delta}\int_0^n (1 - v^t) \, dt = \frac{1}{\delta} - \frac{\bar{a}_{\overline{n}|}}{n\delta}, \qquad n = \omega - x,$$

a formula analogous to (4). This result follows immediately from (14) by multiplication by i/δ. A proof of this result is due to Simpson (1752, pp. 323–324).

De Moivre and Simpson thus succeeded in developing a rather complete theory of single-life annuities supplemented with adequate tables.

25.6   JOINT-LIFE ANNUITIES

De Moivre presumably began his work on joint-life annuities by deriving formulae for the values of annuities on two and three lives under the linear hypothesis. For three lives the formula is rather unwieldy, and for that reason de Moivre looked for another approach. Surprisingly, he begins the discussion in his book by considering a fictitious life with a geometrically decreasing probability of life and proves the following result (1725, p. 33):

Suppose that (x) and (y) are independent and have geometrically decreasing probabilities of life. Then, the value of an annuity granted for the time of their joint continuance equals

$$a_{xy} = \frac{a_x a_y (1 + i)}{(a_x + 1)(a_y + 1) - a_x a_y (1 + i)}. \tag{1}$$

Proof. Suppose that ${}_t p_x = p_1^t$ and ${}_t p_y = p_2^t$ so that ${}_t p_{xy} = (p_1 p_2)^t$. We then have

$$a_x = \sum_{t=1}^{\infty} (v p_1)^t = \frac{v p_1}{1 - v p_1} = \frac{p_1}{1 + i - p_1}.$$

Solving for p_1, de Moivre finds

$$p_1 = \frac{a_x(1 + i)}{a_x + 1},$$

similar expressions being valid for a_y and a_{xy}. Eliminating the p's from

$$a_{xy} = \frac{p_1 p_2}{1 + i - p_1 p_2},$$

(1) follows. He proves an analogous formula for three lives.

This elegant therorem enables him to calculate the value of a joint-life annuity from the values of the single-life annuities under the assumption stated. However, he knows the assumption to be false; in relation to Halley's life table and the linear hypothesis, the formula is only an approximation that has to be justified by further investigation.

In the following sections de Moivre derives formulae for compound survivorship annuities in terms of single- and joint-life annuities. His proofs depend on the compound probability theorem which he had proved in the *Doctrine of Chances* (1718), see §19.4 and §19.6. However, he does not refer to the *Doctrine* but derives the necessary formulae directly in the *Annuities upon Lives*.

He begins with the following problem (1725, p. 36): "The Values of two *single Lives* being given, to find the Value of an *Annuity* upon the *longest* of them; that is to continue so long as either of them is in being."

The solution is

$$a_{\overline{xy}} = a_x + a_y - a_{xy}. \tag{2}$$

Proof. The probability of at least one of the lives (x) and (y) being alive after t years is

$$1 - (1 - {}_tp_x)(1 - {}_tp_y) = {}_tp_x + {}_tp_y - {}_tp_{xy}.$$

multiplication by v^t and summation proves (2).

De Moivre generalizes (2) to any number of lives.

Of course, (2) is exact and valid for any life table; however, to use it in practice de Moivre calculated a_{xy} by means of his approximation (1). Instead of investigating the accuracy of (1) directly, de Moivre investigates the accuracy of (2) by comparing with the results obtained from the linear hypothesis. He states his objective clearly (1725, pp. 46–47):

> To the End that the *Readers* may be freed from any Scruple they may entertain, that the converting *real Lives* into *fictitious* ones, and then combining them together, from thence to deduce the Values of *Annuities* upon the *real Lives*, may perhaps not be altogether to be trusted to; I shall here annex a Calculation of an *Annuity* upon two *Lives* whose Decrements are in *Arithmetic Progression*; which consequently very nearly agree with the Values as deduced from the *Tables*; and then compare the Result with the Value of an *Annuity* upon the two *Lives* considered as *fictitious*.

To carry out this excellent program he first gives a formula without proof,

$$a_{\overline{xy}} = \frac{1}{i} - \frac{1+i}{ni}(a_{\overline{n}|} + a_{\overline{m}|}) + \frac{1+i}{nmi^2}[2m - (2+i)a_{\overline{m}|}], \tag{3}$$

for $n = \omega - x, m = \omega - y, m \leqslant n$.

Since

$$a_{xy} = a_x + a_y - a_{\overline{xy}},$$

it is easy to derive the corresponding formula for a_{xy} by means of (5.4) and (3). We thus get the following result:

Suppose that (x) and (y) are independent and have arithmetically decreasing probabilities of life with complements of life equal to $n = \omega - x$ and $m = \omega - y, m \leqslant n$. Then,

$$a_{xy} = \frac{1}{i} - \frac{1+i}{ni}\left[\left(n - m - 1 - \frac{2}{i}\right)\frac{a_{\overline{m}|}}{m} + \frac{2}{i}\right]. \tag{4}$$

Presumably, de Moivre knew this theorem in 1725 even if he did not state it explicitly. It is given, with an indication of a proof, by Simpson (1742, p. 16), who also gives the formula for three lives.

Proof of (4). From the assumptions it follows that

$$a_{xy} = \frac{1}{nm}\sum_{t=1}^{m} v^{t}(n-t)(m-t).$$

Simpson states that this series may be summed and will be found equal to (4). We shall indicate a proof analogous to the proof of (5.4).
    Let us set

$$b_t = v^t(n-t)(m-t), \qquad t = 0, 1, \ldots, m,$$

so that

$$S = \sum_{t=1}^{m} b_t = nma_{xy}.$$

It is easy to see that b_t satisfies the recursion

$$b_t = 3vb_{t-1} - 3v^2 b_{t-2} + v^3 b_{t-3}, \qquad t = 3, \ldots, m,$$

which by summation from 3 to m gives

$$S - b_1 - b_2 = 3v(S - b_1) - 3v^2(S - b_{m-1}) + v^3(S + b_0 - b_{m-2} - b_{m-1}),$$

so that

$$S(1-v)^3 = v[nm(1-v)^2 - (n+m-1)(1-v)(1-v^m) \\ - 2m(1-v)v^m + 2v(1-v^m)].$$

Inserting $v = (1+i)^{-1}$ and $v^m = 1 - ia_{\overline{m}|}$, we get (4).

To investigate the accuracy of (2), de Moivre gives one numerical example in which he finds a relative error of 0.1%, which would have been 4.6% if he had carried out his calculations correctly. One may wonder why he gives one example only instead of carrying out a systematic investigation like the one he carried out for (5.4). Furthermore, why does he discuss the accuracy of the formula for $a_{\overline{xy}}$ instead of a_{xy}, which is the fundamental quantity? Is that because the relative error of $a_{\overline{xy}}$ is smaller than that of a_{xy}?

As shown in the following table, Simpson carries out the numerical investigation which de Moivre ought to have made.

Simpson's Table[a] of the Values of Joint-Life Annuities Compared with the Approximations (1) and (4) for $i = 0.04$

	VALUE OF a_{xx} FROM		
a_x	LIFE TABLE	(1)	(4)
7	4.7	3.9	4.9
9	6.1	5.3	6.4
11	7.6	6.9	7.9
13	9.5	8.7	9.7
15	11.5	10.6	11.6

[a] The table above gives only every second value from Simpson's table.

Source: Simpson (1742, p. 61).

It will be seen that both formulae give systematic errors as applied to Simpson's life table. Formula (4) gives too large a value, the absolute error being between 0.1 and 0.3. Formula (1) gives too small a value, the error being between 0.7 and 0.9. In addition, a comparison of the third and fourth columns shows that de Moivre's formula (1) based on the geometrically decreasing life table gives a systematically smaller value of about 1.0 than (4) based on the arithmetically decreasing life table.

Simpson does not use any of the approximations. By means of backward recursion he tabulates a_{xx}, $a_{\overline{xx}}$, a_{xxx}, and $a_{\overline{xxx}}$ for $x = 6(1)75$ and $i = 0.03$, 0.04, 0.05 and gives rules for finding the values when the ages differ. To find a_{xy}, he gives the formula

$$a_{xy} \cong a_{yy} + \frac{\frac{1}{2} a_{yy}(a_{xx} - a_{yy})}{a_{xx}}, \qquad x < y,$$

which he states without proof; he has presumably derived the formula by trial and error. If both x and y are between 25 and 50, he states that $a_{xy} \cong a_{ww}$, where $w = \frac{1}{2}(x + y)$. For three lives the general rule is considerably more complicated.

Simpson stresses that his results are based on real observations and that his method of tabulating the value of annuities may be used for any life table.

Simpson's criticism of de Moivre's rule for calculating the value of joint-life annuities must have been a hard blow to de Moivre, who at the time was preparing the second edition of his book. He must have realized that he had let himself been carried astray by the mathematical and numerical simplicity of his approximation. However, instead of admitting his error by rewriting a section of his book, he tried to conceal it. He removes formula (3) for $a_{\overline{xy}}$ and the corresponding discussion of the accuracy of the approximation from the second edition, so that the reader is left with (1) without any indication of its accuracy. Furthermore, he supplements (1) with the slightly simpler formula

$$a_{xy} \cong \frac{a_x a_y}{a_x + a_y - i a_x a_y},\tag{5}$$

stating without proof that (5) has been derived from (1) and "that whether one or the other is used, the Conclusions will very little differ" (1743, p. 91). He adds that (1) is "better adapted to Annuities paid in Money," whereas (5) is "better fitted to Annuities paid by a Grant of Lands," by which he implies that (5) gives slightly larger values than (1). Dividing (5) by (1) we get

$$\frac{1+j}{1+i}, \qquad j = (a_x + a_y - i a_x a_y)^{-1},$$

so that (5) gives the larger value if $j > i$, which is easily shown to be true.

Simpson (1743, p. 138) comments sarcastically on de Moivre's discussion of the small difference between (1) and (5), noting that "must it not seem a little strange that he should here make such an extraordinary *shew* of exactness, and appear so solicitous about a small difference..., when at the same time, both his methods differ from the truth by more than five times as much?."

In his 1752 publication Simpson made further progress in determining the value of joint-life annuities. First, he tabulates a_{xy} for $x = 10(5)75$ and $y = x, x + 5, \ldots, 75$, so that a_{xy} may be found by linear interpolation for any x and y larger than 10. Next, he considers the following problem (1752, p. 279): "To find the Value of an Annuity for three joint Lives, A, B, and C. *Solution.* Let A be the youngest, and C the oldest, of the three proposed Lives: Take the Value of the two *joint Lives* B and C (by Tab. VIII) and find the Age of a single Life D, of the same Value (by Tab. V) then find the Value of the *joint Lives* A and D; which will be the Answer."

In our notation Simpson's rule for finding a_{xyz}, $x < y < z$, becomes,

$$\text{Find } w \text{ such that} \quad a_w = a_{yz} \quad \text{and set} \quad a_{xyz} \cong a_{xw}. \quad (6)$$

Simpson states that "The Reasonableness of the Method of proceeding is evident from the Nature of the Subject, without calling in the Assistance of any Kind of Computation: And in a Number of Examples, respecting Lives of different Ages, I scarce ever found the Error to exceed 1/8 of an Years-Purchase."

Obviously, de Moivre could have made tables of the value of joint-life annuities based on the linear hypothesis but never did, perhaps because of the labor involved and because he wanted to find a simple mathematical solution that could replace his unsatisfactory formula from 1725. It was not until 1756 that the resulting formula was published. In the new Chapter IX we first find the following concession:

> To preserve somewhat of Elegance and Uniformity in my Solutions, as well as to avoid an inconvenient multiplicity of *Canons* and *Symbols,* I did transfer the Decrement of Life from an *Arithmetical* to a *Geometrical* Series: which however, in many Questions concerning *Combined* Lives, creates an error too considerable to be neglected. This hath not escaped the Observation of my Friends, no more than it had my own: but the same Persons might have observed likewise, that such Errors may, when it is thought necessary, be corrected by my own Rules; particularly upon this obvious principle, That, *if money is supposed to bear no Interest, the Values of Lives will coincide with what I call their Expectations.*

He continues by stating a "General Rule for the Valuation of joint Lives," namely,

$$a_{xy} \cong a_w, \quad w = 86 - 2\bar{e}_{x+1:y+1} = y + 1 + \frac{(m-1)^2}{3(n-1)}, \quad (7)$$

$$\bar{a}_{xy} \cong \bar{a}_w, \quad w = 86 - 2\bar{e}_{xy} = y + \frac{m^2}{3n}, \quad (8)$$

where

$$x < y, \quad n = 86 - x, \quad m = 86 - y,$$

and $\bar{e}$ denotes the expectation of life, see §25.9. De Moivre formulates the rule for any number of lives.

Following the hint quoted above we shall assume that w as a function of (x, y) does not depend strongly on i so that we may replace the equation between the a's with the corresponding one between the e's. Since $\bar{e}_w = (86 - w)/2$, (8) follows immediately. Noting that

$$e_w \cong \bar{e}_w - \tfrac{1}{2} = \bar{e}_{w+1}, \quad \text{and} \quad e_{xy} \cong \bar{e}_{xy} - \tfrac{1}{2} = \bar{e}_{x+1:y+1},$$

and setting $e_w = e_{xy}$, we get

$$w + 1 = 86 - 2(\bar{e}_{xy} - \tfrac{1}{2}) = 86 - 2\bar{e}_{x+1:y+1},$$

which proves (7).

De Moivre does not take the trouble to show that his new rule (7) leads to nearly the same result as (4). He gives two numerical examples in which his old rule (5) gives 9% and 15% smaller values, respectively, than his new rule.

With Simpson in mind, but without mentioning him, de Moivre ends his discussion by recommending mathematical rules instead of arithmetical results: "For we do not here aim at an Accuracy beyond what the determination of our main *Data*, the Probabilities of human Life, and the conformity of our Hypothesis to nature, can bear; nor do we give our Conclusions for perfectly exact, as is required in such as are *purely* arithmetical, but only as very near Approximations; upon which business may be transacted, without considerable Loss to any party concerned."

As mentioned above, de Moivre gives a rule for calculating the value of an annuity on the longest of any number of lives. This rule obviously corresponds to (19.4.9) for $k = 1$ for finding the probability of the occurrence of at least one event among n independent events. Without proof, Simpson (1742, pp. 25–26) states the formula for the value of an annuity continuing as long as at least m of n lives survive.

Essentially through Simpson's works (1742, 1752), a theory for annuities on two and three lives had been developed, and tables for their evaluation had been provided.

25.7 REVERSIONARY ANNUITIES

A reversionary annuity to (x) after (y), which will be denoted by $a_{y|x}$, is an annuity to (x) after the death of (y) if (x) is then alive. Since

$$(1 - {}_tp_y){}_tp_x = {}_tp_x - {}_tp_{xy}, \tag{1}$$

it follows that

$$a_{y|x} = a_x - a_{xy}, \tag{2}$$

which may also be seen directly.

It is clear that (2) may be generalized by letting x and y represent combinations of lives, so that the reversionary annuity becomes an annuity payable for the remainder of the status represented by x, after the failure of the status represented by y.

In 1725 de Moivre gives the formulae for several reversionary annuities without presenting the proofs because the results are easy to see directly from the definitions. In 1738 he adds some formulae involving up to four lives and indicates the proofs by calculating probabilities like (1). He also introduces a symbol for annuities on joint lives. Let X, Y, and Z represent the values of single-life annuities, $\overline{XY}$, $\overline{XZ}$, and $\overline{YZ}$ the values of the corresponding joint-life annuities, and so on. He then gives a list of formulae for different kinds of annuities expressed in terms of joint-life annuities. For example, his expression for a reversionary annuity to (u) after the longest of the three lives (x), (y), and (z) becomes

$$U - \overline{XU} - \overline{YU} - \overline{ZU} + \overline{XYU} + \overline{XZU} + \overline{YZU} - \overline{XYZU},$$

which clearly corresponds to the relation between the probabilities involved.

Simpson (1742) proves four general theorems on reversionary annuities. Considering two groups of three lives each he derives the formulae for the annuities characterized by the symbols $\overline{xyz}|abc$, $xyz|abc$, $\overline{xyz}|abc$, and $xyz|\overline{abc}$. From these formulae all the results of de Moivre and many other results are easily found. Simpson indicates the formulae for any number of lives and gives many examples of applications.

25.8 LIFE ASSURANCES, REVERSIONS, AND SUCCESSIVE LIVES

To simplify the exposition let us first introduce some formulae for assurances and reversions in modern notation. Let A_x denote the present value of the sum 1 payable at the end of the assurance year in which (x) dies, i.e., the value of a life assurance. We then have

$$A_x = \frac{1}{l_x} \sum_{t=0}^{\infty} v^{t+1} d_{x+t}, \tag{1}$$

since of l_x insured persons, d_{x+t} die during the tth year. Inserting

$$d_{x+t}/l_x = {}_tp_x - {}_{t+1}p_x,$$

it follows that

$$A_x = v(1 + a_x) - a_x = \frac{1 - ia_x}{1 + i}.$$

Hence, because of the identity

$$1 - ia_x = (1 + i)A_x, \tag{2}$$

a theory about the reversion $1 - ia_x$ is at the same time a theory about the life assurance A_x. The factor $1 + i$ is due to the fact that $1 - ia_x$ represents a reversion for ever with a yearly payment of i after the death of (x) and that an evaluation of this perpetuity at the end of the year in which (x) dies gives $1 + i$. A reversion for ever after (x) with a yearly payment of 1 has the present value $1/i - a_x$, which is the limiting value of the temporary reversionary annuity (5.13) for $s \to \infty$.

The first example of a life assurance given by de Moivre is to be found in the following problem (1725, p. 59): "If there be three equal Lives, and A or his Heirs are to have the Sum S paid upon the Vacancy of any of those Lives, what is the Expectation of A worth in present Money."

The problem is obviously to find the value of the assurance on the three joint lives. De Moivre solves the problem under the assumption that the life table is geometrically decreasing. Let $p_x = p$ for all x. The probability that the three lives all exist after t years and that they do not all exist after $t + 1$ years is $(p^3)^t(1 - p^3)$, and de Moivre therefore gives the solution as

$$A_{xxx} = \sum_{t=0}^{\infty} v^{t+1} p^{3t}(1 - p^3) = \frac{1 - p^3}{1 + i - p^3}.$$

Using the result

$$p = \frac{(1 + i)a_x}{a_x + 1}$$

from the proof of (6.1), de Moivre expresses A_{xxx} in terms of a_x. He also solves two somewhat more complicated problems under the same assumption.

If he had solved the problem for one life only he would have got the result

$$A_x = \frac{1 - p}{1 + i - p} = \frac{1 - ia_x}{1 + i},$$

in agreement with (2).

The next example may be found in the introduction to the section "Of annuities on Lives" in the *Doctrine of Chances* (1738, pp. 212–213), where he writes, "But let us suppose, that instead of an Annuity upon a Life whose Age is given, there should be the Expectation of a Sum (which we may call 1) payable once for all whenever it happens that the Life ceases within a limited time." He derives the formula for a temporary life assurance

$$A^{1}_{x\overline{s}|} = \frac{1}{l_x} \sum_{t=0}^{s-1} v^{t+1} d_{x+t}, \tag{3}$$

and taking the values of l_x and d_x from Halley's table, he finds the value of the temporary life assurance for a life aged 10 and a duration of 11 years to 0.0797. He adds that it would have been possible to extend this analysis if there had been a "Table of Observations concerning a Man's marrying and getting an Heir-Male between 16 and 21."

Immediately after this discussion of the life assurance he formulates the linear mortality hypothesis and derives the value of a_x. He does not, however, remark that the analogous result for the assurance equals

$$A^{1}_{x\overline{s}|} = \frac{a_{\overline{s}|}}{n}, \qquad n = \omega - x, \tag{4}$$

which follows from (3) for $d_{x+t}/l_x = 1/n$.

De Moivre's two examples of life assurances have usually been overlooked because they are omitted from the following editions of the *Annuities*.

After this promising beginning it is strange that neither de Moivre nor Simpson developed a theory of life assurances corresponding to the one they had given for annuities. Perhaps they did not feel the need for such a theory because they usually expressed the benefits in terms of annuities.

Reversions for ever are important for the evaluation of contracts on successive lives, such as copyholds, leases, and advowsons.

De Moivre begins by solving a problem on successive lives (1725, p. 57): "If A enjoys an Annuity for his Life, and at his Decease has the Nomination of a Successor who is likewise to enjoy the Annuity for his Life; to find the Value of the two successive Lives."

Without demonstration he gives a slightly incorrect solution; the correct solution is given in 1738 (p. 225) as

$$a_x + a_y - i a_x a_y. \tag{5}$$

Of course, this solution presupposes that (y) takes over the annuity at the end of the assurance year in which (x) dies.

De Moivre mentions that the present value of a reversion for ever after (x) equals $1/i - a_x$ and uses this quantity as a "discounting factor." He may therefore have found (5) simply as

$$a_x + (1 - ia_x)a_y.$$

However, in the following editions of the *Annuities* he gives a proof based on his concept of an equivalent annuity certain (1738, p. 231). He sets

$$a_x = \frac{1 - v^k}{i}, \tag{6}$$

and solving for v^k, he gets the discounting factor

$$v^k = 1 - ia_x.$$

His characterization of (6) as an equivalent annuity certain is not quite correct because he implies that k is an integer. He interprets k as an "average duration" of the life annuity. If the average durations for (x) and (y) are k and m, respectively, so that the total average duration of the successive lives becomes $k + m$, then the value of the corresponding annuity becomes

$$\frac{1 - v^{k+m}}{i} = a_x + a_y - ia_xa_y,$$

which is de Moivre's proof (1743, pp. 104–105) of (5).

This proof is rather artificial compared with the simple and natural proof given by Simpson (1742, pp. 87–89), which runs as follows:

$$a_x + \sum_{t=0}^{\infty} v^t({}_tp_x - {}_{t+1}p_x)a_y = a_x + a_y[1 + a_x - (1 + i)a_x]$$

$$= a_x + a_y(1 - ia_x) = \frac{1 - (1 - ia_x)(1 - ia_y)}{i}.$$

The both generalize to any number of lives.

As pointed out by Baily (1813), they did not distinguish clearly between the present value of a sum payable at the end of the year in which any number of lives become extinct and the present value of an annuity to commence at the same time.

Simpson (1742, p. 76) generalizes de Moivre's problem about A_{xxx} to a problem about A_{xyz} but gives the solution as $1 - ia_{xyz}$. As mentioned above, de Moivre omits this problem and its solution from the 1743 edition. Instead, he assumes that "upon the Failing of any one of them, that Life shall be immediately replaced, and I then receive a Sum S agreed upon, and that to perpetuity for me and my Heirs." Like Simpson he gives the value of the first sum as $S(1 - ia_{xxx})$ and the value of the whole contract as

$$\frac{S(1 - ia_{xxx})}{ia_{xxx}};$$

his proof is straightforward using the equivalent annuity certain. He should instead have used A_{xxx}, and he would then have reached the correct solution $SA_{xxx}/(1 - A_{xxx})$.

Both de Moivre and Simpson use the value of the reversion $1 - ia_x$ in the same manner as today we use A_x to find the present value of a copyhold and of a perpetual advowson.

25.9 SURVIVORSHIP PROBABILITIES AND EXPECTATIONS OF LIFE

Modern texts on life contingencies all begin with a discussion of the life table, survivorship probabilities, and expectations of life, whereas the books by de Moivre and Simpson end with these problems. The explanation of this peculiar order is given by de Moivre (1725, p. VII):

It may perhaps seem somewhat strange that the Speculation concerning *Survivorship* should be postponed to the Rules given for Settling the Values of *Reversions*; when those Values seem to suppose that the Chance of *Survivorship* is already known. But it will not be difficult to account for the Steps taken in this Matter, for although the Notion of *Survivorship* may successfully be employed in determining the Values of *Reversions*, yet it will be found that the common Cases belonging to that Subject, are so easily derived from a proper Combination of *single* and of *joint Lives*, that any other Consideration would have been entirely superfluous. Nevertheless, the Probability of *Survivorship* may be useful in resolving some extraordinary Cases relating to *Reversions*.

De Moivre gives one example of such cases.

The comprehensive discussion of survivorship probabilities given in the first edition (1725, pp. 74–106) is considerably reduced in the later editions, where de Moivre omits nearly all the proofs; on the other hand, he adds some general formulae.

In the previous parts of the *Annuities*, de Moivre expresses all the results as sums; here, however, he considers the life table as continuous and derives the results by integration.

For a life (x) with a complement of life $n = \omega - x$, he illustrates his reasoning by means of a straight line with time, t, say, moving from the origin to n. According to his hypothesis, the probability of life $_tp_x = (n - t)/n$ corresponds to the ratio of two segments of the line. He supposes that the complement of life is "divided into an infinite Number of equal Parts, representing Moments." The probability that (x) is alive at time t and dies between t and $t + dt$ is then

$$_tp_x\mu_{x+t}\,dt = \frac{n-t}{n}\frac{dt}{n-t} = \frac{dt}{n}, \qquad 0 \leqslant t \leqslant n,$$

and 0 otherwise. The symbol $\mu_{x+t}\,dt$, not used by de Moivre, denotes the conditional probability that (x) dies between $x + t$ and $x + t + dt$, given that he is alive at age $x + t$. We shall use this symbol to save many verbal explanations.

In the following we shall use three lives (x), (y), and (z), ordered according to increasing age, so that the complements of life are $n \geqslant m \geqslant k$, say.

To find the probability that (x) survives (y), de Moivre first finds the "Probability of the first *Life*'s continuing during the Time AB $[0, t]$ or beyond it, and of the Second's failing just at the End of that Time," which is

$$_tp_x\,_tp_y\mu_{y+t}\,dt = \frac{n-t}{n}\frac{dt}{m}, \qquad 0 \leqslant t \leqslant m,$$

and 0 otherwise. He considers the quantity $(n - t)/nm$ as the ordinate of a curve and finds its area so that the probability in question becomes

$$\Pr\{T_x > T_y\} = \frac{1}{nm}\int_0^m (n - t)\,dt = 1 - \frac{m}{2n}. \tag{1}$$

This proof demonstrates his method and is included in all editions. In later editions (1756, p. 324) he only adds, "By the same method of arguing, we may proceed to the finding the Probability of any one of any Number of given Lives surviving all the rest, and thereby verify what we have said in Prob. XVIII and XIX." Because of his linear hypothesis the probability density of any status of r lives at time t will be a polynomial in t of at most the rth degree, and the survivorship probability is then found by integration.

As an example we shall give his proof from the first edition of the probability that (x) survives both (y) and (z). The probability is composed of three parts

$$\int_0^k \frac{n-t}{n}\frac{t}{m}\frac{dt}{k} + \int_0^k \frac{n-t}{n}\frac{t}{k}\frac{dt}{m} + \int_k^m \frac{n-t}{n}\frac{dt}{m} = 1 - \frac{m}{2n} - \frac{k^2}{6nm}, \tag{2}$$

the first integral giving the probability that (x) survives (z) and that (z) survives (y) within the interval $(0, k)$, the second integral giving the probability that (x) survives (y) and that (y) survives (z) within the interval $(0, k)$, and the third integral giving the probability that (x) survives (y) within the interval (k, m).

In 1738 he gives the general rule for finding the probability that any one of a given number of lives survives the rest and states the formulae for five lives. He also gives the six probabilities that three lives will die in any given order. For example, the probability that the order will be $\{(z), (y), (x)\}$ is found as the probability that (y) survives (z) minus the probability that (y) survives both (z) and (x), which by means of (1) and (2) gives

$$\Pr\{T_x > T_y > T_z\} = 1 - \frac{k}{2m} - \frac{m}{2n} + \frac{k^2}{6nm}. \tag{3}$$

De Moivre remarks that the formulae for the expectations of life may be obtained from those for annuities setting $i = 0$. However, he does not discuss these formulae but turns to the derivation of the expectation of life for a continuous life table.

The expectation for (x) of living through the "moment" dt is $_t p_x\, dt$. Hence, the expectation of life becomes

$$\bar{e}_x = \int_0^n \frac{n-t}{n}\, dt = \frac{1}{2}n, \tag{4}$$

which may also be seen directly. The expectation for two joint lives becomes

$$\bar{e}_{xy} = \int_0^m \frac{n-t}{n}\frac{m-t}{m}\, dt = \frac{m}{2} - \frac{m^2}{6n}.$$

He generalizes this result to any number of lives.

The expectation of life for the longest of two lives is

$$\bar{e}_{\overline{xy}} = \bar{e}_x + \bar{e}_y - \bar{e}_{xy},$$

and so on analogously to the formulae for annuities.

To find the expectation of life in general,

$$\bar{e}_x = \frac{1}{l_x} \int_0^{\omega - x} l_{x+t}\, dt,$$

de Moivre uses the piecewise linear approximation, which means that he approximates the integral by using the trapezoidal rule.

Similarly, he approximates the probability that (y) survives (x)

$$\Pr\{T_y > T_x\} = \frac{1}{l_x l_y} \int_0^{\omega - y} l_{y+t}(-dl_{x+t}),$$

by a sum of products of the form

$$\frac{\frac{1}{2}(l_{y+t} + l_{y+t+h})(l_{x+t} - l_{x+t+h})}{l_x l_y}.$$

The reader should compare this with the methods used by Huygens and Nicholas Bernoulli, see §§8.1–8.2.

In 1743 de Moivre adds one more problem: To find the expected time for the rth death among s lives of the same age.

Without any indication of proof he gives the result as $nr/(s + 1)$, where n is the complement of life. This is an extension of the result due to Nicholas Bernoulli (1709), which is obtained for $r = s$, see §8.2. Perhaps de Moivre has used the same method of proof as Bernoulli. He may have found the probability of the rth life dying between t and $t + dt$, which equals

$$s\binom{s-1}{r-1}\left(\frac{t}{n}\right)^{r-1}\left(1 - \frac{t}{n}\right)^{s-r}\frac{dt}{n},$$

and found the expectation of t by integration by parts. De Moivre ends his comments to this problem very effectfully: "This Speculation might be carried to any Number of unequal Lives: but my Design not being to perplex the Reader with too great Difficulties, I shall forbear at present to prosecute the thing any farther." He does not return to this problem; we shall leave its solution to the reader.

Simpson follows de Moivre closely in his discussion of survivorship probabilities; for the evaluation of the integrals involved he uses the Newton–Cotes formulae for numerical integration instead of the trapezoidal rule used by de Moivre. Instead of discussing expectations of life, he solves the following related problem (1742, pp. 124–128): "To determine from a table of observations, and the bills of mortality of any place, the number of souls contained in that place."

For a given interval of time he considers the number of persons entering and leaving the population as defined by the life table, i.e., he defines the stationary population corresponding to the life table. He finds the size of the population as the yearly number of births times the expectation of life,

$$\bar{e}_0 = \frac{1}{l_0} \int_0^\infty l_x \, dx.$$

For London he finds $\bar{e}_0 = 25{,}500/1280$, "whence it appears that the number of the living, at any one time, born within the bills of mortality of this city, is to the number of births happening yearly within the same bills (taken at a medium) as 20 to 1, very near." He adds that the yearly number of burials exceeds the yearly number of births because of migrations.

Young (1908) is of the opinion that Simpson is "apparently entitled to the merit of the first systematic application of the Continuous principle to life-contingency *questions*." He reaches this conclusion because he does not consider de Moivre's geometrical approach in 1725 as an application of the method of fluxions. However, de Moivre's use of "moments" implies the method of fluxions. Simpson took over de Moivre's reasoning with the only modification of introducing the dot notation for the differential element. Perhaps the credit should go to Nicholas Bernoulli (1709) who used the calculus to find the expected time for the last death among a given number of lives, see §8.2. It follows from the *Doctrine of Chances* (1718, p. XIV), but not from the later editions, that de Moivre knew Bernoulli's thesis; he does not, however, refer to Bernoulli in the *Annuities*.

25.10 SURVIVORSHIP INSURANCES

After having derived survivorship probabilities and expectations of life one would have expected de Moivre to continue with the corresponding annuities and assurances according to Halley's principle, i.e., multiplying the probabilities by the discounting factor v^t and evaluating the integrals. However, he did not do so but gave a simpler formula based on an incorrect argument.

Consider an assurance $\bar{A}_x$ payable at the moment of death of (x) and let it be split into two parts according to whether (x) dies before or after (y) so that

$$\bar{A}_x = \bar{A}^1_{xy} + \bar{A}^2_{xy} \tag{1}$$

with an obvious notation. Expressed in terms of integrals we have

$$\int_0^\infty v^t\, {}_tp_x\mu_{x+t}\, dt = \int_0^\infty v^t\, {}_tp_y\, {}_tp_x\mu_{x+t}\, dt + \int_0^\infty v^t(1 - {}_tp_y)\, {}_tp_x\mu_{x+t}\, dt. \qquad (2)$$

The corresponding survivorship probabilities are

$$1 = \bar{Q}^1_{xy} + \bar{Q}^2_{xy} = \int_0^\infty {}_tp_y\, {}_tp_x\mu_{x+t}\, dt + \int_0^\infty (1 - {}_tp_y)\, {}_tp_x\mu_{x+t}\, dt, \qquad (3)$$

so that

$$\bar{A}_x = \bar{Q}^1_{xy}\bar{A}_x + \bar{Q}^2_{xy}\bar{A}_x. \qquad (4)$$

De Moivre's mistake consists in setting $\bar{A}^1_{xy}$ equal to $\bar{Q}^1_{xy}\bar{A}_x$. His reasoning has been shown in the following table:

CONDITION	BENEFIT	PROBABILITY
(x) dies before (y)	0	$\bar{Q}^1_{xy} = m/2n$
(x) dies after (y)	$\bar{A}_x$	$1 - \bar{Q}^1_{xy} = 1 - m/2n$
Expectation:	$\bar{A}_x(1 - m/2n)$	

The probability is calculated under the linear mortality hypothesis, see (9.1). The table above corresponds to the simplest example given by de Moivre (1756, Problem 17) with the modification that he as usual replaces $\bar{A}_x$ by $1 - ia_x$.

In general, de Moivre's principle consists in multiplying the unconditional assurance by the probability of the condition being fulfilled. He gives several such examples in the second and later editions; in the first edition he gives one example only. His principle was adopted by Simpson (1742). Perhaps they considered this solution as an approximation only but they do not say so.

In 1752, however, Simpson observes that de Moivre's principle is wrong. He writes (1752, pp. 322–323),

'Tis true the Manner of proceeding by first finding the Probability of Survivorship (which Method is used in my former Work, and which a celebrated Author on this Subject has largely insisted on in three successive Editions) may be applied to good Advantage, when the given Ages are nearly equal: But then, it is certain, that this is not a genuine Way of going to Work; and that the Conclusions hence derived are, at the best, but near Approximations. The Rate of Interest that Money bears must be compounded with the Probability of Survivorship, and the Expectation on each particular Year must be determined, in order to have a true Solution.

After having stated the right method for finding survivorship insurances, Simpson faced the difficult problem of evaluating the sums and integrals involved. He naturally wanted to express the survivorship insurances in terms of his tabulated joint-life annuities. We shall demonstrate his way of reasoning by two examples.

Consider first the assurance A^1_{xy} with the benefit being paid at the end of the year so that

$$A^1_{xy} = \frac{1}{l_x l_y} \sum_{t=0}^{\infty} v^{t+1}(d_{x+t} l_{y+t+1} + \tfrac{1}{2} d_{x+t} d_{y+t}), \tag{5}$$

where the last term in the bracket presupposes that the deaths within each year are uniformly distributed. Expressing the d's in terms of the l's and setting $_t p_x = {}_{t+1} p_{x-1}/p_{x-1}$ we get

$$A^1_{xy} = \frac{1}{2} \left\{ \frac{1 - ia_{xy}}{1+i} + \frac{a_{x-1:y}}{p_{x-1}} - \frac{a_{x:y-1}}{p_{y-1}} \right\}, \tag{6}$$

which gives A^1_{xy} in terms of three joint-life annuities. This result is due to Baily (1813, pp. 183–185); it is the formula that Simpson was looking for but did not find.

Simpson does not give (5) explicitly, but it is implied in his derivation of A^1_{xy} under the linear hypothesis. Setting $l_{x+t} = n - t$ and $l_{y+t} = m - t$, $m \leq n$, he finds (1752, pp. 316–318)

$$A^1_{xy} = \sum_{t=0}^{m} v^{t+1} \left(\frac{m-t-1}{nm} + \frac{1}{2nm} \right) = \frac{1}{n} \sum_{t=0}^{m} v^{t+1} \frac{2m - 2t - 1}{2m}. \tag{7}$$

He could of course have expressed the sum as a linear function of an annuity certain but he does not do so because this result cannot be generalized; he only uses that A^1_{xy} varies inversely as $n = 2\bar{e}_x$. Noting that

$$A^1_{xy} + A^1_{xy} = A_{xy},$$

he gets

$$A^1_{yy} = \tfrac{1}{2} A_{yy} \cong \tfrac{1}{2}(1 - ia_{yy}),$$

from which he obtains the approximation

$$A_{xy}^1 \cong \tfrac{1}{2}(1 - ia_{yy})\frac{\bar{e}_y}{\bar{e}_x}, \qquad x \leqslant y, \tag{8}$$

which is easily applicable, since he has tabulated a_{xx} and $\bar{e}_x$ based on his life table. The last part of his argument is of course unsatisfactory, and he does not give an evaluation of the accuracy of the formula. It will be seen that he does not reach Baily's formula because he specializes to the linear hypothesis.

As another example consider an insurance in which (z) gets an annuity after (y), if (y) survives (x). Simpson (1752, pp. 318–321) first finds the probability that (x) dies before (y) in the sth year as

$$\frac{(l_x - l_{x+s})d_{y+s} + \tfrac{1}{2}d_{x+s}d_{y+s}}{l_x l_y} = \frac{s}{n}\frac{1}{m} + \frac{1}{2nm} = \frac{2s+1}{2nm},$$

which by summation gives the probability that (x) dies before (y) before the end of the tth year as $t^2/2nm$, $t < m$. The value in question then becomes

$$\frac{1}{2nm}\sum_{t=1}^{k} v^t\, {}_t p_z t^2 = \frac{1}{2nmk}\sum_{t=1}^{k} v^t(k-t)t^2, \qquad k < m < n.$$

Since

$$(1 - {}_t p_x)(1 - {}_t p_y) = \frac{t^2}{nm},$$

Simpson writes the result as

$$\tfrac{1}{2}(a_z - a_{z:\overline{xy}}) = \tfrac{1}{2}a_{\overline{xy}|z},$$

which may be calculated by means of his rule for reducing an annuity on three lives to an annuity on two lives. This result, which is exact for a linear life table, is then applied to his own life table.

Simpson's results for survivorship insurances are thus valid only for a linear life table. Nevertheless, his contributions represent an essential step forward, and his clearly formulated method of proof became of great importance for the following generation of actuaries.

25.11 THE SCOTTISH MINISTERS' WIDOWS' FUND OF 1744

Various schemes of provision for ministers' widows were established about the beginning of the 18th century, some based on a single premium, others on yearly premiums to be distributed as benefits in the same year. Between 1741 and 1744 two Scottish ministers, Alexander Webster and Robert Wallace, with some assistance from Colin Maclaurin worked out a plan for the Scottish Ministers' Widows' Fund of 1744, which has functioned ever since. Based on yearly premiums from all Scottish ministers, annuities were paid to widows and children. Since the income during the first years of operation greatly exceeded the expenses, a fund was built up so that, when the stationary state arrived, premiums and interest of the fund would balance the benefits and administrative expenses. The actuarial calculations were based on statistics on the average number of ministers, widows and orphans for the period 1722–1741, on Halley's life table, and on an interest rate of 4% as shown in a memorandum probably written by Webster (1748). The construction of this fund has served as model for many later pension funds; its history has been written by Dunlop (1971) and Dow (1975).

25.12 PROBLEMS

1. Assuming that the linear mortality hypothesis holds, evaluate the accuracy of de Moivre's approximations (6.1), (6.5), and (6.7) by comparison with (6.4). Extend this analysis to three lives.

2. From the identity $a_w = a_{xy}$ it follows that $_t p_w = {_t}p_{xy}$ for all t. Show that this equation is satisfied if and only if Gompertz's law of mortality holds, i.e., if $\log {_t}p_x = c^x(c^t - 1)\log g$, so that w may be found from the equation $c^w = c^x + c^y$, see King (1902, Chapter 12).

3. Prove that the value of a complete annuity approximately equals $a_x + \frac{1}{2}A_x(1 + i)^{1/2}$, and compare this result with de Moivre's approximation (5.14).

4. Let (x_i), $i = 1, 2, \ldots$, be a given number of lives ordered according to increasing age, and let $n_i = \omega - x_i$ be the complements of life. Show that the probability of (x_1) surviving all the other equals

$$1 - \frac{n_2}{2n_1} - \frac{n_3^2}{6n_1 n_2} - \frac{n_4^3}{12n_1 n_2 n_3} - \frac{n_5^4}{20n_1 n_2 n_3 n_4} - \cdots,$$

see de Moivre (1738, p. 223).

5. Under the linear mortality hypothesis prove that

$$A_{xy} = \frac{1}{nm} \left\{ (n + m - 1)a_{\overline{n}|} - \frac{2}{i}(a_{\overline{n}|} - nv^n) \right\}, \qquad m = \omega - y \leqslant n = \omega - x.$$

Compare with (6.4).

6. Evaluate $\bar{A}_x$ and $\bar{A}^1_{xy}$ for $_tp_x = (n - t)/n$ and $_tp_y = (m - t)/m$, $m \leqslant n$, and discuss the accuracy of de Moivre's approximation to $\bar{A}^1_{xy}$. Analyze the same problem for $_tp_x = p^t_1$ and $_tp_y = p^t_2$.

7. Evaluate A^1_{xy}, $x \leqslant y$, under the linear hypothesis and discuss the accuracy of Simpson's approximation (10.8).

8. "If A, B, C agree amongst themselves to buy an Annuity to be by them equally divided whilst they live together, then to be divided equally between the two next Survivors, then to belong entirely to the last Survivor, for his life, to find what each of them ought to contribute towards the Purchase." Problem posed by de Moivre (1725, p. 52; 1738, p. 222).

   Answer: A's contribution should be $A - \frac{1}{2}\overline{AB} - \frac{1}{2}\overline{AC} + \frac{1}{3}\overline{ABC}$.

9. "A and B enjoy an annuity, to which a third person C, after the decease of A, is to have the sole right of possession for life, provided B be then extinct; otherwise it is to be equally divided between him and B, during their joint lives, and then to belong entirely to C, for life, if he be the last survivor: To find the value of the right of C in that annuity." Problem posed by Simpson (1742, p. 70).

   Answer: $C - \overline{AC} - \frac{1}{2}\overline{BC} + \frac{1}{2}\overline{ABC}$.

10. "D, whilst in Health, makes a Will, whereby he bequeaths 500£ to E, and 300£ to F, with this Condition, that if either of them dies before him, the whole is to go to the Survivor of the two; what are the Values of the Expectations of E and F, estimated from the time that the Will was writ?" Problem posed by de Moivre (1743, p. 50).

    Show that the expectation of E according to the method of de Moivre equals 331.6£. Find the correct value under the linear mortality hypothesis, expressing the integrals in terms of continuous annuities certain. Show that this gives 327.1£ for E's expectation.

References

Acsádi Gy., and J. Nemeskéri, (1970). *History of human life span and mortality.* Akadémiai Kiado, Budapest.

Adams, W. J. (1974). *The life and times of the central limit theorem.* Kaedmon, New York.

Aiton, E. J. (1969). Kepler's second law of planetary motion. *Isis,* **60,** 75–90.

Algemeene Maatschappij van Levensverzekering en Lijfrente (1898). *Mémoires pour servir à l'histoire des assurances sur la vie et des rentes viagères aux Pays-Bas.* Amsterdam.

Algemeene Maatschappij van Levensverzekering en Lijfrente (1900). Notice succincte sur la marche de la science actuarielle dans les Pays-Bas (Hollande), depuis ses début, jusqu'à la fin du XIXe siècle. *Troisième Congrès Intern. d'Actuaires,* Paris, 1900, pp. 885–927.

Anonymous (1700). A calculation of the credibility of human testimony. *Phil. Trans.,* **21,** 359–365. Comm. 1699.

Arbuthnott, J. (1692). *Of the Laws of Chance, or, A Method of Calculation of the Hazards of Game.* London. 4th ed., revised by John Ham, 1738.

Arbuthnott, J. (1712). An argument for Divine Providence, taken from the constant regularity observ'd in the births of both sexes. *Phil. Trans.,* **27,** 186–190. Comm. 1710. Reprinted in Kendall and Plackett (1977).

Archibald, R. C. (1926). A rare pamphlet of Moivre and some of his discoveries. *Isis,* **8,** 671–683.

Armitage, A. (1966). *Edmond Halley.* Nelson, London.

Arnauld, A., and P. Nicole, (1662). *La logique, ou l'art de penser.* Paris. Many later editions. Reprinted by Flammarion, Paris, 1970. Translated as *The art of thinking* by J. Dickoff and P. James, 1964, Bobbs-Merrill, Indianapolis.

Baily, F. (1813). *The doctrine of Life-Annuities and Assurances.* Richardson, London.

Bartholomew, D. J. (1984). *God of chance.* SCM Press, London.

Bartholomew, D. J. (1988). Probability, statistics and theology (with discussion). *J. Roy. Statist. Soc. Ser. A,* **151,** 137–178.

549

Barton, D. E. (1958). The matching distribution: Poisson limiting forms and derived methods of approximation. *J. Roy. Statist. Soc. Ser. B*, **20**, 73–92.

Bayes, T. (1764). An essay towards solving a problem in the doctrine of chances. *Phil. Trans.*, 1763, **53**, 370–418. Reprinted in facsimile in *Two papers by Bayes*, ed. by W. E. Deming, 1940, The Graduate School, Dept. of Agriculture, Washington. Reprinted in *Biometrika*, 1958, **45**, 293–315, and in Pearson and Kendall (1970).

Bellhouse, D. R. (1988). Probability in the sixteenth and seventeenth centuries: An analysis of Puritan Casuistry. *Intern. Statist. Rev.*, **56**, 63–74.

Bernoulli, D. *Die Werke von Daniel Bernoulli*, ed. by D. Speiser, Vol. 2, Analysis, Wahrscheinlichkeitsrechung, 1982, bearbeitet und kommentiert von L. P. Bouckaert und B. L. van der Waerden. Birkhäuser, Basel.

Bernoulli, D. (1734). Quelle est la cause physique de l'inclination des plan des orbites des planètes. *Recueil des Pièces qui on Remporté le Prix de l'Académie Royale des Sciences*, **3**, 95–122.

Bernoulli, D. (1738). Specimen theoriae novae de mensura sortis. *Commentarii Acad. Sci. Imp. Petrop.*, 1730–1731, **5**, 175–192. Reprinted in *Werke*, **2**, 1982, 223–234. Translated into English as "Exposition of a new theory on the measurement of risk," *Econometrica*, 1954, **22**, 23–36.

Bernoulli, D. (1770–1771). Mensura sortis ad fortuitam successionem rerum naturaliter contingentium applicata. *Novi Comm. Acad. Sci. Imp. Petrop.*, 1769, **14**, 26–45; 1770, **15**, 3–28. Reprinted in *Werke*, **2**, 1982, 325–360.

Bernoulli, D. (1778). Diiudicatio maxime probabilis plurium observationem discrepantium atque verisimillima inductio inde formanda. *Acta Acad. Sci. Imp. Petrop.*, 1777, **1**, 3–23. Reprinted in *Werke*, **2**, 1982, 361–375. Translated into English by C. G. Allen as "The most probable choice between several discrepant observations and the formation therefrom of the most likely induction," *Biometrika*, 1961, **48**, 1–18; reprinted in Pearson and Kendall (1970).

Bernoulli, D. (1780). Specimen philosophicum de compensationibus horologicis, et veriori mensura temporis. *Acta Acad. Sci. Imp. Petrop.*, 1777, **2**, 109–128. Reprinted in *Werke*, **2**, 1982, 376–390.

Bernoulli, J. *Die Werke von Jakob Bernoulli*, Vol. 3, Wahrscheinlichkeitsrechnung, ed. by B. L. van der Waerden, 1975. Birkhäuser, Basel.

Bernoulli, J. (1684–1690). Meditationes (Aus den Meditationes von Jakob Bernoulli). In *Die Werke von Jakob Bernoulli*, Vol. 3, 21–89, Birkhäuser, Basel, 1975.

Bernoulli, J. (1685). Problème proposé par M. Bernoulli. *Journal des Sçavans*, 1685, p. 314. Reprinted in *Werke*, **3**, 91.

Bernoulli, J. (1690). Quaestiones nonnullae de usuris, cum solutione problematis de sorte alearum. *Acta Eruditorum*, 1690, pp. 219–223. Reprinted in *Werke*, **3**, 91–93.

Bernoulli, J. (1713). *Ars Conjectandi*. Thurnisius, Basilea. Reprinted in *Editions Culture et Civilisation*, Bruxelles, 1968, and in *Die Werke von Jakob Bernoulli*, Vol. 3, Birkhäuser, Basel, 1975. German translation by Haussner (1899). Part 1 translated into French by Vastel (1801) and into Italian by Dupont and Roero (1984). English translation of Part 2 by Maseres (1795) and of Part 4 by Bing Sung (1966). Russian translation of Part 4 by J. V. Uspensky (1913), reprinted in 1986.

Bernoulli, J. (1986). *On the law of large numbers. Ars conjectandi, Part 4*, translated from Latin into Russian by J. V. Uspensky, ed. by Yu. V. Prohorov. Nauka, Moscow.

Bernoulli, John (1710). Letter to Montmort. See Montmort (1713, pp. 283–298).

Bernoulli, N. (1709). *De Usu Artis Conjectandi in Jure*. Basel. Reprinted in *Die Werke von Jakob Bernoulli*, Vol. 3, 287–326, 1975, Birkhäuser, Basel. Translated into English, with notes, by T. Drucker, 1976 (unpublished).

Bernoulli, N. (1710–1713). Letters to Montmort. See Montmort (1713).

Bernoulli, N. (1717). Solutio generalis Problematis XV propositi à D. de Moivre in tractatu *de Mensura Sortis* inserto Actis Philosophicis Anglicanis No. 329 pro numero quocunque Collusorum. *Phil. Trans.*, **29**, 133–144. Comm. 1714.

Bernoulli, N. (1774). Letters to 'sGravesande. Written in 1712, published in *Oeuvres de 'sGravesande*, 1774, Vol. 2, 221–236.

Berry, A. (1898). *A short history of astronomy*. Murray, London. Reprinted by Dover, New York, 1961.

Biermann, K.-R. (1957). Eine Aufgabe aus den Anfängen der Wahrscheinlichkeitsrechnung. *Centaurus*, **5**, 142–150.

Biggs, N. L. (1979). The roots of combinatorics. *Hist. Math.*, **6**, 109–136.

Birnbaum, A., and C. Eisenhart. (1967). Tercentennials of Arbuthnot and de Moivre. *Amer. Statist.*, **21**, 22–29.

Böckh, R. (1893). Halley als Statistiker. *Bull. Intern. Statist. Inst.*, **7**, 1–24.

Bortkewitsch, L. von (1898). *Das Gesetz der kleinen Zahlen*. Teubner, Leipzig.

Bos, H. J. M. (1972). Christiaan Huygens. In *Dictionary of scientific biography*, ed. by C. C. Gillispie, Vol. 6, 597–613.

Bouckaert, L. P. (1982). Einleitung zur Analysis. Séries récurrentes. *Die Werke von Daniel Bernoulli*, Vol. 2, 21–29. Birkhäuser, Basel.

Brahe, T. (1573). *De nova stella*. Copenhagen.

Brahe, T. (1602). *Astronomiae instauratae progymnasmata*, ed. by J. Kepler. Prague.

Brakel, J. van (1976). Some remarks on the prehistory of the concept of statistical probability. *Arch. Hist. Ex. Sci.*, **16**, 119–136.

Braun, H. (1925). *Geschichte der Lebensversicherung und der Lebensversicherungstechnik*. Koch, Nürnberg.

Briggs, H. (1624). *Arithmetica Logarithmica*. Jones, London.

Briggs, H. (1633). *Trigonometria Britannica*, ed. by H. Gellibrand. Rammasenius, Gaudae.

Browne, W. (1714). *Christiani Hugenii Libellus de Ratiociniis in Ludo Aleae. Or, the Value of all Chances in Games of Fortune; Cards, Dice, Wagers, Lotteries etc. Mathematically Demonstrated*. London.

Byrne, E. F. (1968). *Probability and opinion*. Nijhoff, The Hague.

Cantor, M. (1880–1908). *Vorlesungen über Geschichte der Mathematik*. 4 vols. Teubner, Leipzig. Reprinted by Johnson Reprint Corporation, New York.

Caramuel, J. (1670). *Mathesis Biceps*. Campania.

Cardano, G. *Opera Omnia Hieronymi Cardani*, cura Caroli Sponii, Lyon, 1663. 10 vols. Reprinted by Johnson Reprint Corporation, New York, 1967.

Cardano, G. (1539). *Practica Arithmetice et Mensurandi singularis*. Milan. Reprinted in *Opera Omnia*, Vol. 4, 1663.

Cardano, G. (c. 1564). *Liber de Ludo Aleae*. First printed in *Opera Omnia*, Vol. 1, 1663. Translated into English by S. H. Gould in Ore (1953), reprinted in *The book on games of chance*, 1961, Holt, Rinehart and Winston, New York.

Cardano, G. (1570). *Opus Novum de Proportionibus Numerorum*. Basel.

Cardano, G. (1575). *De Vita Propria Liber*. First published in 1643, Paris, reprinted in *Opera Omnia*, Vol. 1, 1663. Engl. transl. by J. Stoner as *The book of my life*, 1931, Dent, London. Reprinted by Dover, New York, 1962.

Catalan, E. (1837). Solution d'un problème de probabilité, relatif au jeu de rencontre. *J. Math. Pure et Appl.*, **2**, 469–482.

Chateleux, P. J. L. de, and J. P. van Rooijen (1937). *Le rapport de Johan de Witt sur le calcul des rentes viagerès*. Traduction française avec commentaire et historique. Nijhoff, La Haye. Also in *Verzekerings-Archief*, 1937, **18**, 41–85.

Cheyne, G. (1703). *Fluxionum Methodus Inversa*. London.

Chiaramonti, S. (1628). *De tribus novis stellis quae annis 1572, 1600, 1604 comparuere*. Cesena.

Chiaramonti, S. (1633). *Difesa al suo Antiticone e libro delle tre nuove stelle dall' opposozione dell' autore de' due massimi sistemi*. Firenze.

Chiaramonti, S. (1643). *Antiphilolaos*. Cesena.

Chrystal, G. (1900). *Algebra. An elementary text-book*. 2nd ed. Reprinted by Chelsea, New York, 1964.

Clarke, F. M. (1929). *Thomas Simpson and his times*. New York.

Clausius, R. (1849). Ueber die Natur derjenigen Bestandtheile der Erdatmosphäre, durch welche die Lichtreflexion in derselben bewirkt wird. *Ann. Phys. Chem.*, **76**, 161–188.

Cohen, I. B. (1980). *The Newtonian revolution*. Cambridge Univ. Press.

Copernicus, N. (1543). *De Revolutionibus Orbium Coelestium*. Reprinted in facsimile by Johnson Reprint Corporation, New York, 1965. Translated into English in *Occasional notes of the Royal Astronomical Society*, **2**, No. 10, London, 1947.

Cotes, R. (1722). Aestimatio Errorum in Mixta Mathesi, per Variationes Partium Trianguli Plani et Sphaerici. In *Opera Miscellanea*, 1722, Cambridge.

Coumet, E. (1965). Le problème des partis avant Pascal. *Arch. Intern. d'Histoire des Sciences*, **18**, 245–272.

Coumet, E. (1970). La théorie du hasard est-elle née par hasard? *Annales: Économies, Sociétés, Civilisations*, **25**, 574–598.

Cournot, A. A. (1843). *Exposition de la théorie des chances et des probabilités*. Hachette. Paris.

Craig. J. (1699). *Theologiae Christianae Principia Mathematica*. London. Reprinted with commentary by Daniel Titius in Leipzig in 1755. Reprinted in part, with

English translation, as "Craig's rules of historical evidence" in *History and Theory: Studies in the Philosophy of History*, Supp. Vol. 4, pp. 1–31, 1964.

Cramér, H. (1928). On the composition of elementary errors. *Skand. Aktuarietidskr.*, **11**, 13–74; 141–180.

Cramér, H. (1937). *Random variables and probability distributions.* Cambridge Univ. Press. 2nd ed. 1962. 3rd. ed. 1970.

Cramér, H. (1946). *Mathematical methods of statistics.* Princeton Univ. Press.

Cramér, H. (1972). On the history of certain expansions used in mathematical statistics. *Biometrika*, **59**, 205–207. Reprinted in Kendall and Plackett (1977).

Cullen, M. J. (1975). *The Statistical Movement in Early Victorian Britain.* Harvester Press, Sussex, and Barnes & Noble, New York.

Czuber, E. (1891). *Theorie der Beobachtungsfehler.* Teubner, Leipzig.

Czuber, E. (1899). *Die Entwicklung der Wahrscheinlichkeitstheorie und ihrer Anwendungen.* Jahresbericht der Deutschen Mathematiker-Vereinigung. Vol. 7. Teubner, Leipzig.

Daston, L. J. (1980). Probabilistic expectation and rationality in classical probability theory. *Hist. Math.*, **7**, 234–260.

Daston, L. J. (1987). The domestication of risk: Mathematical probability and insurance 1650–1830. In *The probabilistic revolution*, ed. by L. Krüger et al., Vol. 1, 237–260.

Daston, L. (1988). *Classical Probability in the Enlightenment.* Princeton Univ. Press, New Jersey.

David, F. N. (1955). Dicing and gaming (a note on the history of probability). *Biometrika*, **42**, 1–15. Reprinted in Pearson and Kendall (1970).

David, F. N. (1962). *Games, gods and gambling.* Griffin, London.

David, F. N., and D. E. Barton. (1962). *Combinatorial chance* Griffin, London.

Daw, R. H., and E. S. Pearson. (1972). Abraham De Moivre's 1733 derivation of the normal curve: a bibliographical note. *Biometrika*, **59**, 677–680. Reprinted in Kendall and Plackett (1977).

Derham, W. (1713). *Physico-Theology: A Demonstration of the Being and Attributes of God from his Works of Creation.* London.

Descartes, R. (1637). *La géométrie*, Leiden. Translated into Latin with commentaries by F. van Schooten, 1649, Amsterdam; second enlarged edition in two volumes, 1659–1661. The original is reprinted by Hermann, Paris, 1886. Reprinted with English translation by Dover, New York, 1954.

Dijksterhuis, E. J. (1953). Christiaan Huygens. *Centaurus*, **2**, 265–282.

Dow, J. B. (1975). Early actuarial work in eighteenth-century Scotland (with discussion). *Trans. Fac. Actuaries, Edinburgh*, **33**, 193–229.

Drake, S. (1978). *Galileo at work.* Univ. Chicago Press.

Drake, S. (1985). Galileo's accuracy in measuring horizontal projections. *Annali dell'Istituto e Museo di Storia della Scienza di Firenze*, **10**, 1–14.

Dresher, M. (1961). *Games of strategy.* Prentice-Hall, Englewood Cliffs, New Jersey.

Dreyer, J. L. E. (1890). *Tycho Brahe. A picture of scientific life and work in the sixteenth century.* Black, Edinburgh.

Drucker, T. (1976). English translation, with notes, of *Nicholaus Bernoulli's De Usu Artis Conjectandi in Jure.* Unpublished.

Dunlop, A. I. (1971). Provision for ministers' widows in Scotland—eighteenth century. *Rec. Scottish Church Hist. Soc.,* **17**, 233–248.

Dupont, P. (1975–1976). Concetti probabilistici in Roberval, Pascal e Fermat. *Rend. Seminario Matematico dell'Università di Torino,* **34**, 235–245.

Dupont, P. (1979). I fondamenti del calcolo delle probabilità in Blaise Pascal. *Acad. Sci. Torino,* **113**, 243–253.

Dupont, P. (1982). *Abraham de Moivre, De Mensura Sortis.* Testo originale latino con traduzione, interpretazione, complementi e commenti. Parte prima: I primi 19 problemi. *Quaderni di Matematica,* **38**, 1982, Università di Torino.

Dupont, P., and C. S. Roero. (1984). Il Trattato "De Ratiociniis in Ludo Aleae" di Christiaan Huygens con le "Annotationes" di Jakob Bernoulli ("Ars Conjectandi", Parte I) Presentati in Traduzione Italiana, con Commento Storico-Critico e Risoluzioni Moderne. *Memorie della Acad. Scienze Torino, Serie V,* Vol. 8.

Dutka, J. (1953). Spinoza and the theory of probability. *Scripta Math.,***19**, 24–33.

Dutka, J. (1988). On the St. Petersburg Paradox. *Arch. Hist. Ex. Sci.,* **39**, 13–39.

Edgeworth, F. Y. (1905). The law of error. *Trans. Cambr. Phil. Soc.,* **20**, 35–65; 113–141.

Edwards, A. W. F. (1982a). Sums of powers of integers. *Mathl. Gaz.,* **66**, 22–28.

Edwards, A. W. F. (1982b). Pascal and the problem of points. *Intern. Statist. Rev.,* **50**, 259–266. Reprinted in Edwards (1987).

Edwards, A. W. F. (1983). Pascal's problem: The "Gambler's Ruin." *Intern. Statist. Rev.,* **51**, 73–79. Reprinted in Edwards (1987).

Edwards, A. W. F. (1986). Is the reference to Hartley (1749) to Bayesian inference? *Amer. Statist.,* **40**, 109–110.

Edwards, A. W. F. (1987). *Pascal's arithmetical triangle.* Griffin, London.

Eggenberger, J. (1894). Beiträge zur Darstellung des Bernoullischen Theorems, der Gammafunktion und des Laplaceschen Integrals. *Mitth. Naturforsch. Ges. Bern,* **50**, 110–182. Comm. 1893.

Eisenhart, C. (1974). The background and evolution of the method of least squares. Unpublished. Distributed to participants of the ISI meeting, 1963, revised 1974.

Ellis, R. L. (1844). On the solution of equations in finite differences. *Camb. Math. J.,* **4**, 182–190. Reprinted in *The mathematical and other writings of Robert Leslie Ellis,* Deighton, Bell and Co., Cambridge, 1863.

Eneström, G. (1879). Differenskalkylens Historia. *Upsala Universitets Årsskrift 1879. Matematik och Naturvetenskap.* **1**, 1–69.

Eneström, G. (1896). Ett bidrag till mortalitetstabellernes historia före Halley. (A contribution to the history of mortality-tables before Halley.). *Öfversigt Kongl. Vetenskaps-Akademiens Förhandlingar,* Stockholm.

Eneström, G. (1898). Sur la méthode de Johan de Witt (1671) pour le calcul de rentes viagères. *Archief Verzekerings-Wetenschap*, **3**, 62–68.

Euler, L. *Opera omnia. Commentationes Algebraicae ad Theoriam Combinationum et Probabilitatum Pertinentes*, Ser. 1, Vol. 7. Ed. by L. G. du Pasquier. Teubner, Leipzig, 1923.

Euler, L. (1753). Calcul de la probabilité dans le jeu de rencontre. *Mém. Acad. Sci. Berlin*, **7**, 255–270. Comm. 1751. Reprinted in *Opera Omnia*, Ser. 1, Vol. 7, 11–25, 1923.

Euler, L. (1785). Solutio quarundam quaestionum difficiliorum in calcolo probabilium. *Opuscula Analytica*, **2**, 331–346. Reprinted in *Opera Omnia*, Ser. 1, Vol. 7, 408–424, 1923.

Euler, L. (1811). Solutio quaestionis curiosae ex doctrina combinationum. *Mém. Acad. Sci. St.-Petersbourg*, **3**, (1809–1810), 57–64. Comm. 1779. Reprinted in *Opera Omnia*, Ser. 1, Vol. 7, 435–440, 1923.

Feigenbaum, L. (1985). Brook Taylor and the Method of Increments. *Arch. Hist. Ex. Sci.*, **34**, 1–140.

Feller, W. (1970). *An introduction to probability theory and its applications.* 3rd ed., revised printing. Wiley, New York.

Fermat, P. de. *Oeuvres*, 4 vols. ed. by P. Tannery and C. Henry, 1891–1922. Gauthier-Villars, Paris. The 1654 correspondence with Pascal is in Vol. 2, 1894, 288–314, and the 1656 correspondence with Carcavi and Huygens on 320–331.

Fermat, P. de (1679). *Varia opera mathematica.* Toulouse. Contains the letters from Pascal to Fermat.

Fieller, E. C. (1931). The duration of play. *Biometrika*, **22**, 377–404.

Fisher, R. A. (1925). *Statistical methods for research workers.* Oliver and Boyd, Edinburgh. Numerous later editions.

Fisher, R. A. (1934). Randomisation, and an old enigma of card play. *Math. Gazette*, **18**, 294–297. Reprinted in *Collected papers of R. A. Fisher*, ed. by J. H. Bennett, Univ. of Adelaide, 1973, Vol. 3, 162–165.

Fleckenstein, J. O. (1949). *Johann und Jakob Bernoulli.* Beiheft Nr. 6 zur *Elemente der Mathematik*, Birkhäuser, Basel. Reprinted 1977.

Fleckenstein, J. O. (1973). Nikolaus I Bernoulli. In *Dictionary of scientific biography*, ed. by C. C. Gillispie, Vol. 2, 56–57.

Fontenelle, B. de (1721). Éloge de M. de Montmort. *Hist. Acad. Roy. Sci.*, 1719, 83–93.

Forestani, L. (1603). *Practica d'arithmetica e geometria.* Siena.

Fraser, D. C. (1927). Newton and interpolation. In *Isaac Newton, 1642–1727.* A memorial volume ed. by W. J. Greenstreet. Bell, London.

Fraser, D. C. (1928). *Newton's interpolation formulas.* Layton, London.

Fréchet, M. (1940). *Les probabilités associées a un système d'événements compatibles et dépendants. I. Événements en nombre fini fixe.* Actualités Scientifiques et Industrielles, No. 859. Hermann, Paris.

Fréchet, M. (1943). *Les probabilités associées a un système d'événements compatibles et dépendants. II. Cas particuliers et applications.* Actualités Scientifiques et Industrielles, No. 942. Hermann, Paris.

Galilei, G. (c. 1620). *Sopra le Scoperte dei Dadi.* First printed in *Opera Omnia,* 1718. English translation in David (1962).

Galilei, G. (1632). *Dialogo sopra i due massimi sistemi del mondo, Tolemaico, e Copernicano.* (Dialogue concerning the Two Chief World Systems—Ptolemaic and Copernican.) Translated into German by E. Strauss, 1891, Teubner, Leipzig. Translated into English by Stillman Drake, 1953, 2nd ed. 1967, Univ. California Press, Berkeley and Los Angeles.

Gani, J. (1982). Newton on "a Question touching ye different Odds upon certain given Chances upon Dice." *Math. Scientist,* **7**, 61–66.

Garber, D., and S. Zabell. (1979). On the emergence of probability. *Arch. Hist. Ex. Sci.,* **21**, 33–53.

Gerson, Levi ben (1321). *The Work of the Computer.*

Gillies, D. A. (1987). Was Bayes a Bayesian? *Hist. Math.,* **14**, 325–346.

Gillispie, C. C. (1970–1980). *Dictionary of scientific biography,* Vols. I–XIV. Scribner, New York.

Glass, D. V. (1950). Graunt's life table. *J. Inst. Actuaries,* **76**, 60–64.

Glass, D. V. (1964). John Graunt and his "Natural and Political Observations." *Proc. Roy. Soc. London, Ser. B,* **159**, 2–37.

Goldstine, H. H. (1977). *A History of numerical analysis from the 16th through the 19th century.* Springer, New York.

Gouraud, C. (1848). *Histoire du Calcul des Probabilités depuis ses Origines jusqu'a nos Jours.* Durand, Paris.

Graetzer, J. (1883). *Edmund Halley und Caspar Neumann: Ein Beitrag zur Geschichte der Bevölkerungs-statistik.* Schottländer, Breslau.

Graunt, J. (1662). *Natural and Political Observations made upon the Bills of Mortality.* Martyn, London. 2nd ed. 1662; 3rd. ed. 1665; 4th ed. 1665; 5th ed. 1676. First edition reprinted in W. Willcox (ed.): *Natural and political observations,* Baltimore, 1939; in *J. Inst. Actuaries,* 1964, **90**, 1–61, with a Foreword by B. Benjamin; and in *Pioneers of demography. The earliest classics: John Graunt and Gregory King,* Gregg Intern. Publ., 1973. Fifth edition reprinted in C. H. Hull (ed.): *The economic writings of Sir William Petty,* 1899, Cambridge Univ. Press; reprinted by Kelly, Fairfield, New Jersey, 1986.

'sGravesande, G. J. (1774). *Oeuvres Philosophiques et Mathématiques.* Ed. by J. N. S. Allamand, Michel Rey, Amsterdam.

'sGravesande, G. J. (1774). Démonstration mathématique de la direction de la providence divine. Written in 1712. See *Oeuvres,* 1774, Vol. 2, 221–236.

Greenwood, M. (1940). A statistical Mare's nest? *J. Roy. Statist. Soc.,* **103**, 246–248.

Greenwood, M. (1941–1943). Medical statistics from Graunt to Farr. *Biometrika,* **32**, 101–127; **32**, 203–225; **33**, 1–24. Reprinted in Pearson and Kendall (1970).

Haaften, M. van (1925). Johan de Witt en de Levensverzekering. *De Levensverzekering,* **2**, 171–185.

Hacking, I. (1965). *Logic of statistical inference.* Cambridge Univ. Press.

Hacking, I. (1971). Jacques Bernoulli's art of conjecturing. *Brit. J. Phil. Sci.*, **22**, 209–229.

Hacking, I. (1975). *The Emergence of Probability.* Cambridge Univ. Press.

Hacking, I. (1980). From the emergence of probability to the erosion of determinism. In *Probabilistic thinking, thermodynamics and the interaction of the history and philosophy of science*, ed. by J. Hintikka et al., Vol. 2, 105–123. Reidel, Dordrecht.

Haight, F. A. (1967). *Handbook of the Poisson distribution.* Wiley, New York.

Hald, A. (1984). Nicholas Bernoulli's theorem. *Intern. Statist. Rev.*, **52**, 93–99.

Hald, A. (1984). Commentary on "De Mensura Sortis." *Intern. statist. Rev.*, **52**, 229–236.

Hald, A. (1986). Galileo's statistical analysis of astronomical observations. *Intern. Statist. Rev.*, **54**, 211–220.

Hald, A. (1987). On the early history of life insurance mathematics. *Scand. Actuarial J.*, 1987, 4–18.

Hald, A. (1988). On de Moivre's solutions of the problem of duration of play 1708–1718. *Arch. Hist. Ex. Sci.*, **38**, 109–134.

Hald, A., and S. Johansen. (1983). On de Moivre's recursion formulae for the duration of play. *Intern. Statist. Rev.*, **51**, 239–253.

Hall, A. R. (1983). *The revolution in science 1500–1750.* Longman, London.

Halley, E. (1694). An Estimate of the Degrees of the Mortality of Mankind, drawn from curious Tables of the Births and Funerals at the City of Breslaw; with an Attempt to ascertain the Price of Annuities upon Lives. *Phil. Trans.*, **17**, 596–610. Comm. 1693. Some further Considerations on the Breslaw Bills of Mortality. *Phil. Trans.*, **17**, 654–656. Reprinted in *J. Inst. Actuaries*, 1874, **18**, 251–265, and in facsimile in 1985, **112**, 278–301 with an introduction by G. Heywood.

Hasover, A. M. (1967). Random mechanisms in Talmudic literature. *Biometrika*, **54**, 316–321. Reprinted in Pearson and Kendall (1970).

Haussner, R. (1899). *Wahrscheinlichkeitsrechnung (Ars Conjectandi) von Jakob Bernoulli.* Anmerkungen von R. Haussner. Ostwald's Klassiker, Nr. 107–108. Engelmann, Leipzig.

Hendriks, F. (1851–1853). *Contributions of the history of insurance and of the theory of life contingencies.* Layton, London. Also in *The Assurance Magazine*, **2**, (1852), 121–150, 222–258; **3** (1853), 93–120.

Henny, J. (1975). Niklaus und Johann Bernoullis Forschungen auf dem Gebiet der Wahrscheinlichkeitsrechnung in ihrem Briefwechsel mit Pierre Rémond de Montmort. In *Die Werke von Jakob Bernoulli*, **3**, 457–507. Birkhäuser, Basel. Dissertation, 1973.

Heyde, C. C., and E. Seneta. (1977). *I. J. Bienaymé: Statistical Theory Anticipated.* Springer, Berlin.

Holgate, P. (1984). The influence of Huygens' work in dynamics on his contribution to probability. *Intern. Statist. Rev.*, **52**, 137–140.

Hooper, G. (1700). See Anonymous (1700).

L'Hospital, G. F. A. de (1696). *Analyse des Infiniment Petits.* Paris.

Hudde, J. (1665). Letter to Huygens on the fifth problem. *Oeuvres de Huygens,* **5,** 470–471.

Hull, C. H. (ed.) (1899). *The Economic Writings of Sir William Petty together with the Observations upon the Bills of Mortality more probably by Captain John Graunt.* 2 vols. Cambridge Univ. Press. Reprinted by Kelley, Fairfield, New Jersey, 1986.

Huygens, C. *Oeuvres Complètes.* 22 volumes. Société Hollandaise des Sciences. Nijhoff, La Haye. 1888–1950.

Huygens, C. (1657). *De Ratiociniis in Ludo Aleae,* printed in *Exercitationum Mathematicarum* by F. van Schooten, Elsevirii, Leiden. Reprinted in *Oeuvres,* Vol.. 14 (1920) and in Bernoulli (1713). The Dutch version *Van Rekeningh in Spelen van Geluck,* written in 1656, was first published in 1660. Reprinted in *Oeuvres,* Vol. 14 (1920), together with a French translation. English translation by Arbuthnott (1692); German translation by Haussner (1899); Italian translation by Dupont and Roero (1984).

Irwin, J. O. (1955). A unified derivation of some well-known frequency distributions of interest in biometry and statistics. *J. Roy. Statist. Soc. Ser. A,* **118,** 389–404.

Iversen, L. (1910). *Dødeligheden blandt Forsørgede.* (The Mortality of Annuitants.) Lehmann and Stage, København.

Jensen, E. L., and H. Rootzén. (1986). A note on de Moivre's limit theorems: Easy proofs. *Statist. Probability Letters,* **4,** 231–232.

John, V. (1884). *Geschichte der Statistik.* Enke, Stuttgart. Reprinted 1968 by Sändig Reprint Verlag, Vaduz, Liechtenstein.

Johnson, N. L. (1957). A note on the mean deviation of the binomial distribution. *Biometrika,* **44,** 532–533.

Jordan, K. (1972). *Chapters on the classical calculus of probability.* Akadémiai Kiadó, Budapest. Hungarian edition in 1956.

Jordan, M. C. (1867). De quelques formules de probabilité. *C. R. Acad. Sci. Paris,* **65,** 993–994.

Jorland, G. (1987). The Saint Petersburg paradox 1713–1937. In *The probabilistic revolution,* ed. by Krüger et al., Vol. 1, 157–190.

Keiding, N. (1987). The method of expected number of deaths, 1786–1886–1986. *Intern. Statist. Rev.,* **55,** 1–20.

Kendall, M. G. (1956). The beginnings of a probability calculus. *Biometrika,* **43,** 1–14. Reprinted in Pearson and Kendall (1970).

Kendall, M. G. (1957). A note on playing cards. *Biometrika,* **44,** 260–262. Reprinted in Pearson and Kendall (1970).

Kendall, M. G. (1960). Where shall the history of statistics begin? *Biometrika,* **47,** 447–449. Reprinted in Pearson and Kendall (1970).

Kendall, M. G. (1963). Isaac Todhunter's History of the Mathematical Theory of Probability. *Biometrika,* **50,** 204–205. Reprinted in Pearson and Kendall (1970).

Kendall, M. G. (1968). Thomas Young on coincidences. *Biometrika,* **55,** 249–250. Reprinted in Pearson and Kendall (1970).

Kendall, M. G., and A. G. Doig. (1968). *Bibliography of statistical literature Pre-1940*. Oliver and Boyd, Edinburgh. Reprinted by Arno Press, New York.

Kendall, M., and R. L. Plackett, eds. (1977). *Studies in the history of statistics and probability*, Vol. 2. Griffin, London.

Kepler, J. *Gesammelte Werke*. Ed. by M. Caspar, 1937 ff. Beck, München.

Kepler, J. (1609). *Astronomia Nova*. Ges. Werke, **3**, 1937.

Kepler, J. (1619). *Harmonice Mundi*. Ges. Werke, **6**, 1940.

Kiefer, J. (1961). On large deviations of the empiric D. F. of vector chance variables and a law of the iterated logarithm. *Pacific J. Math.*, **11**, 649–660.

King, G. (1902). *Institute of Actuaries' text book of the principles of interest, life annuities, and assurances, and their practical application, Part II, life contingencies*. Layton, London. 2nd. ed.

Kline, M. (1972). *Mathematical thought from ancient to modern times*. Oxford Univ. Press, New York.

Knapp, G. F. (1874). *Theorie des Bevölkerungs-Wechsels*. Vieweg, Braunschweig.

Knobloch, E. (1974). The mathematical studies of G. W. Leibniz on combinatorics. *Hist. Math.*, **1**, 409–430.

Kohli, K. (1975a). Zur Publikationsgeschichte der Ars Conjectandi. In *Die Werke von Jakob Bernoulli*, Vol. 3, 391–401. Birkhäuser, Basel.

Kohli, K. (1975b). Spieldauer: Von Jakob Bernoullis Lösung der fünften Aufgabe von Huygens bis zu den Arbeiten von de Moivre. In *Die Werke von Jakob Bernoulli*, Vol. 3, 403–455. Birkhäuser, Basel. Dissertation, 1967.

Kohli, K. (1975c). Aus dem Briefwechsel zwischen Leibniz und Jakob Bernoulli. In *Die Werke von Jakob Bernoulli*, Vol. 3, 509–513. Birkhäuser, Basel.

Kohli, K. (1975d). Kommentar zur Dissertation von Niklaus Bernoulli: De Usu Artis Conjectandi in Jure. In *Die Werke von Jakob Bernoulli*, Vol. 3, 541–556. Birkhäuser, Basel.

Kohli, K., und B. L. van der Waerden. (1975). Bewertung von Leibrenten. In *Die Werke von Jakob Bernoulli*, Vol. 3, 515–539. Birkhäuser, Basel.

Korteweg, D. J. (1920). Aperçy de la Genèse de l'Ouvrage "De Ratiociniis in Ludo Aleae" et des Recherches subséquentes de Huygens sur les Questions de Probabilité. *Oeuvres de Huygens*, Vol. 14, 3–48.

Kreager, P. (1988). New light on Graunt. *Population Studies*, **42**, 129–140.

Krüger, L. et al. (eds.) (1987). *The probabilistic revolution*. Vol. 1: Ideas in History, Vol. 2: Ideas in the Sciences. MIT Press, Cambridge, Massachusetts.

Kuhn, T. S. (1957). *The Copernican revolution*. Harvard Univ. Press., Cambridge, Massachusetts.

Lacroix, S. F. (1819). *Traité du Calcul Différentiel et du Calcul Intégral*, Vol. 3. 2nd ed. Courcier, Paris.

Lagrange, J. L. *Oeuvres*, Vols. 1–14. 1867–1892. Gauthier-Villars, Paris.

Lagrange, J. L. (1759). Sur l'intégration d'une équation différentielle a différences finies, qui contient la théorie des suites récurrentes. In *Miscellanea Taurinensia*, Vol. 1, 33–42. Reprinted in *Oeuvres*, Vol. 1, 1867, 23–36.

Lagrange, J. L. (1772). Sur une nouvelle espèce de calcul relatif a la différentiation et a l'intégration des quantités variables. *Nouv. Mém. Acad. Roy. Sci. Berlin*, 1772. Reprinted in *Oeuvres*, Vol. 3, 1869, 441–476.

Lagrange, J. L. (1777). Recherches sur les suites récurrentes dont les termes varient de plusieurs manières différentes, ou sur l'intégration des équations linéaires aux différences finies et partielles; et sur l'usage de ces équations dans la théorie des hasards. *Nouv. Mém. Acad. Roy. Sci. Berlin*, **6**, 1775, 183–272. Comm. 1776. Reprinted in *Oeuvres*, Vol. 4, 1869, 151–251.

Lambert, J. H. (1773). Examen d'une espèce de superstition ramenée au calcul des probabilités. *Nouveau Mém. Acad. Roy. Sci. et Belle-Lettres de Berlin*, 1771, 411–420.

Laplace, P. S. de. *Oeuvres*, Vols. 1–14. 1843–1912. Vol. 1–4, *Imprimerie Royale, Paris*; Vols. 5–14, Gauthier-Villars, Paris.

Laplace, P. S. de (1771). Recherches sur le calcul intégral aux différences infiniment petites, & aux différences finies. *Miscellanea Taurinensia*, 1766–1769, **4**, 273–345.

Laplace, P. S. de (1774a). Mémoire sur les suites récurro-récurrentes et sur leurs usages dans la théorie des hasards. *Mém. Acad. Roy. Sci. Paris*, **6**, 353–371. Reprinted in *Oeuvres*, Vol. 8, 1891, 5–24.

Laplace, P. S. de (1774b). Mémoire sur la probabilité des causes par les événements. *Mém. Acad. Roy. Sci. Paris*, **6**, 621–656. Reprinted in *Oeuvres*, Vol. 8, 1891, 27–65. Translated into English with an introduction by S. M. Stigler in *Statistical Sci.*, 1986, **1**, 359–378.

Laplace, P. S. de (1776). Recherches sur l'intégration des équations différentielles aux différences finies et sur leur usage dans la théorie des hasards. *Mém. Acad. Roy. Sci. Paris*, **7**. Comm. 1773. Reprinted in *Oeuvres*, Vol. 8, 1891, 69–197.

Laplace, P. S. de (1782). Mémoire sur les suites. *Mém. Acad. Roy. Sci. Paris*, 1779. Reprinted in *Oeuvres*, Vol. 10, 1894, 1–89.

Laplace, P. S. de (1786). Mémoire sur les approximations des formules qui sont fonctions de très grand nombres. *Mém. Acad. Roy. Sci. Paris*, 1783. Reprinted in *Oeuvres*, Vol. 10, 1894, 295–338.

Laplace, P. S. de (1812). *Théorie Analytique des Probabilités*. Paris. 2nd. ed. 1814; 3rd. ed. 1820. Reprinted in *Oeuvres*, Vol. 7, 1886.

Laplace, P. S. (1814). *Essai Philosophique sur les Probabilités*. Paris. 6th ed. translated 1902 by F. W. Truscott and F. L. Emory as *A philosophical essay on probabilities*. Reprinted 1951 by Dover, New York.

Leibniz, G. W. (1666). *De Arte Combinatoria*. Reprinted in Leibnizens mathematische Schriften, Vol. 5, ed. by C. I. Gerhardt, Halle, 1858.

Lexis, W. (1875). *Einleitung in die Theorie der Bevölkerungsstatistik*. Strassburg.

Locke, J, (1690). *An Essay Concerning Human Understanding*. Routledge, London.

Lucas, H.-C. and M. J. Petry (1982). *Ergänzungsband zum Sämtliche Werke [von Spinoza] in sieben Bänden*, Philosophische Bibliothek, Vol. 350. Felix Meiner, Hamburg.

MacMahon, P. A. (1915–1916). *Combinatory Analysis*, Vols. I and II. Cambridge Univ. Press. Reprinted by Chelsea, New York, 1960.

Mahoney, M. S. (1973). *The Mathematical Career of Pierre de Fermat* (1601–1665). Princeton Univ. Press, New Jersey.

Maistrov, L. E. (1974). *Probability Theory. A Historical Sketch*. Academic Press, New York.

Markoff, A. A. (1912), *Wahrscheinlichkeitsrechnung*. Teubner, Leipzig.

Markov, A. A. ed. (1913). *J. Bernoulli: On the law of large numbers*. Translated by J. V. Uspensky and with a preface by A. A. Markov, Moscow. Reprinted in 1986 with new notes and commentaries under the editorship of Yu. V. Prohorov. Nauka, Moscow.

Markov, A. A. (1924). *The Calculus of Probabilites*. 4th ed. Moscow. In Russian.

Maseres, F. (1795). *Mr. James Bernoulli's Doctrine of Permutations and Combinations, and some other Useful Mathematical Tracts*. White, London.

Maty, M. (1755). Mémoire sur la Vie et sur les Ecrits de Mr. Abraham de Moivre. *J. Brittanique*, 1755. Reprinted by Scheurleer, La Haye, 1760.

McClintock, B. (1984). On the measurement of chance, or, on the probability of events in games depending upon fortuitous chance. Translation of *De Mensura Sortis* by A. de Moivre (1712). *Intern. Statist. Rev.*, **52**, 237–262.

Meitzen, A. (1886). *Geschichte, Theorie und Technik der Statistik*. 2nd. ed. 1903. Cotta'sche Buchhandlung, Berlin.

Mesnard, J. (1970). Introduction to Pascal's *Traité du Triangle Arithmétique et Traité Connexes*. In *Blaise Pascal; Oeuvres Complètes*, Vol. 2, 1166–1175, ed. by J. Mesnard. Desclée de Brouwer, Bruges.

Moivre, A. de (1698). A Method of Raising an infinite Multinomial to any given Power, or Extracting any given Root of the same. *Phil. Trans.*, **19**, 619–625. Comm. 1697.

Moivre, A. de (1699). A Method of extracting the Root of an Infinite Equation. *Phil. Trans.*, **20**, 190–193. Comm. 1698.

Moivre, A. de (1704). *Animadversiones in D. Georgii Cheynaei Tractatum de Fluxionum Methodo Inversa*. London

Moivre, A. de (1712). De Mensura Sortis, seu, de Probabilitate Eventuum in Ludis a Casu Fortuito Pendentibus. *Phil. Trans.*, **27**, 213–264. Comm. 1711. Reprinted by Johnson Reprint Corporation, New York. Translated into English by B. McClintock in *Intern. Statist. Rev.*, 1984, **52**, 237–262. Problems 1–19 translated into Italian by P. Dupont (with commentaries) in *Quaderni di Matematica*, **38**, 1982, Univ. di Torino.

Moivre, A. de (1717). Solutio generalis altera praecedentis Problematis, ope Combinationum et Serierum infinitarum. *Phil. Trans.*, **29**, 145–158. Comm. 1714.

Moivre, A. de (1718). *The Doctrine of Chances: or, A Method of Calculating the Probability of Events in Play*. Pearson, London.

Moivre, A. de (1724a). De Fractionibus Algebraicis Radicalitate immunibus ad Fractiones Simpliciores reducendis, deque summandis Terminis quarundam Serierum æquali Intervallo a se distantibus. *Phil. Trans.*, **32**, 162–178. Comm. 1722.

Moivre, A. de (1724b). De Sectione Anguli. *Phil. Trans.*, **32**, 228–230. Comm. 1722.

Moivre, A. de (1725). *Annuities upon Lives: or, The Valuation of Annuities upon any Number of Lives; as also, of Reversions. To which is added, An Appendix concerning the Expectations of Life, and Probabilites of Survivorship.* Fayram, Motte and Pearson, London.

Moivre, A. de (1730). *Miscellanea Analytica de Seriebus et Quadraturis.* Tonson & Watts, London. *Miscellaneis Analyticis Supplementum,* 22 pp.

Moivre, A. de (1733). *Approximatio ad Summam Terminorum Binomii* $(a + b)^n$ *in Seriem expansi.* Printed for private circulation.

Moivre, A. de (1738). *The Doctrine of Chances.* The second edition, fuller, clearer, and more correct than the first. Woodfall, London. Reprinted by Cass, London, 1967.

Moivre, A. de (1743). *Annuities on Lives: Second Edition, plainer, fuller, and more correct than the former. With Several Tables, exhibiting at one View, the Values of Lives, for several Rates of Interest.* Woodfall, London. 3rd ed., 1750; 4th ed., 1752.

Moivre, A. de (1746). A letter from Mr. Abraham De Moivre, F. R. S. to William Jones, Esquire, F. R. S. concerning the easiest method for calculating the value of annuities upon lives, from tables of observations. *Phil. Trans.*, **43**, 65–78. Comm. 1744.

Moivre, A. de (1756). *The Doctrine of Chances.* The third edition, fuller, clearer, and more correct than the former. Millar, London. Reprinted by Chelsea, New York, 1967.

Moivre, A. de (1756). *A Treatise of Annuities on Lives.* In *The Doctrine of Chances,* pp. 261–328. Translated into Italian with a commentary by Gaeta and Fontana, 1776, Milan. Translated into German with a commentary by Czuber, 1906, Vienna.

Montmort, P. R. de (1708). *Essay d'Analyse sur les Jeux de Hazard.* Quillau, Paris. Published anonymously.

Montmort, P. R. de (1713). *Essay d'Analyse sur les Jeux de Hazard.* Seconde Edition. Revûe et augmentée de plusieurs Lettre. Quillau, Paris. Reprinted 1714. Published anonymously. Reprinted by Chelsea, New York, 1980.

Montmort, P. R. de (1720). De Seriebus infinitis Tractatus. *Phil. Trans.*, **30**, 633–675. Comm. 1717.

Mora-Charles, M. S. de (1986). Leibniz et le problème des partis. Quelques papiers inédits. *Hist. Math.*, **13**, 352–369.

Morgan, A. de (1837). *Treatise on the theory of probabilities.* (Extracted from the *Encyclopædia Metropolitana.*) Clowes, London.

Napier, J. (1614). *Mirifici Logarithmorum Canonis Descriptio.* Hart, Edinburgh. Translated into English by E. Wright (1616) as *A description of the admirable Table of Logarithmes.* Okes, London.

Napier, J. (1619). *Mirifici Logarithmorum Canonis Constructio.* Hart, Edinburgh. Translated into English with notes by W. R. MacDonald (1889) as *The construction of the wonderful Canon of Logarithms.* Blackwood, Edinburgh.

Napier Tercentenary Memorial Volume, ed. by C. G. Knott, (1915). Longman, London.

Naux, C. (1966, 1971). *Histoire des logarithmes de Neper a Euler*. Vol. 1 (1966); Vol. 2 (1971). Blanchard, Paris.

Netto, E. (1901). *Lehrbuch der Combinatorik*. Teubner, Leipzig. 2nd. ed. (1927). Reprinted by Chelsea, New York, 1964.

Neumann, J. von, and O. Morgenstern. (1944). *Theory of games and economic behavior*. 2nd ed. (1947). Princeton Univ. Press, New Jersey.

Newton, I. *The mathematical papers of Isaac Newton*, ed. by D. T. Whiteside. 8 Vols. Cambridge Univ. Press (1967–1981).

Newton, I. (1687). *Philosophiae Naturalis Principia Mathematica*. London. Translated into English by A. Motte (1729). Revised translation with notes by F. Cajori (1934), Univ. California Press, Berkeley.

Newton, I. (1704). *Opticks*. London. Reprinted by Dover, New York, 1979.

Newton, I. (1711). *De Analysi per Aequationes Numero Terminorum Infinitas*. London. Circulated as manuscript in 1669.

Newton, I. (1711). *Methodus Differentialis*. Newton's papers on interpolation are published in *The mathematical papers of Isaac Newton*, ed. by D. T. Whiteside, Vol. 4, 14–73; Vol. 8, 236–257, Cambridge Univ. Press (1967, 1981).

Newton, I. (1728). *The Chronology of Ancient Kingdoms Amended*. London.

Newton, I. (1733). *Observations upon the Prophesies*. London.

Newton, I. (1736). *Tractatus de Methodus Serierum et Fluxionum*. London. Written in 1671.

Nieuwentyt, B. (1715). *Het regt gebruik der Wereldbeschouwingen*. Amsterdam. English translation as *The Religious Philosopher: Or, the Right Use of Contemplating the Works of the Creator*, London.

Oettinger, L. (1837). *Die Lehre von den Combinationen*. Gross, Freiburg.

Oettinger, L. (1852). *Die Wahrscheinlichkeits-Rechnung*. Reimer, Berlin

Ore, O. (1953). *Cardano: The gambling scholar*. Princeton Univ. Press, New Jersey. Reprinted by Dover, New York, 1965.

Ore, O. (1960). Pascal and the invention of probability theory. *Amer. Math. Monthly*, **67**, 409–419.

Pacioli, L. (1494). *Summa de arithmetica, geometria, proportioni et proportionalita*. Venezia.

Parzen, E. (1960). *Modern probability theory and its applications*. Wiley, New York.

Pascal, B. *Oeuvres complètes*. Many editions from 1779 on, the latest by J. Mesnard, Vol. 1 (1964), Vol. 2 (1970), Desclée de Brouwer, Bruges. Pascal's work on probability is in Vol. 2.

Pascal, B. (1654). Correspondence with Fermat. Reprinted in *Oeuvres complètes* (1779) and in many later editions. For English translations, see Smith (1929) and David (1962).

Pascal, B. (1665). *Traité du triangle arithmétique, avec quelques autres petits traités sur la même matière*. Desprez, Paris. Reprinted in *Oeuvres complètes*. English translation of the first part in Smith (1929), pp. 67–79.

Pasquier, L. G. du (1910). Die Entwicklung der Tontinen bis auf die Gegenwart. *Z. schweiz. Statistik*, **46**, 484–513.

Pearson, E. S., and M. Kendall, (eds.) (1970). *Studies in the history of statistics and probability*, Vol. 1. Griffin, London.

Pearson, E. S., and R. H. Daw. (1972). Abraham de Moivre's 1733 derivation of the normal curve: a bibliographical note. *Biometrika*, **59**, 677–680. Reprinted in Kendall and Plackett (1977).

Pearson, K. (1895). Contributions to the mathematical theory of evolution. II. Skew variation in homogeneous material. *Phil. Trans.*, A **186**, 343–414. Reprinted in *Karl Pearson's early statistical papers*, pp. 41–112. Cambridge Univ. Press. (1948).

Pearson, K. (1924). Historical note on the origin of the normal curve of errors. *Biometrika*, **16**, 402–404.

Pearson, K. (1925). James Bernoulli's theorem. *Biometrika*, **17**, 201–210.

Pearson, K. (1926). Abraham de Moivre. *Nature*, **117**, 551–552.

Pearson K. (1928). Biometry and chronology. *Biometrika*, **20A**, 241–262, 424.

Pearson, K. (1978). *The history of statistics in the 17th and 18th centuries*, ed. by E. S. Pearson. Lectures by Karl Pearson given at University College London during the academic sessions 1921–1933. Griffin, London.

Petty, W. (1662). *A Treatise of Taxes and Contributions*. Brooke, London. Reprinted in *The Economic Writings of Sir William Petty*; see C. H. Hull, (1899).

Petty, W. (1683–1687). *Essays in Political Arithmetick*. London. Reprinted in *The Economic Writings of Sir William Petty*; see C. H. Hull (1899).

Petty, W. (1687). *Five Essays in Political Arithmetick*. Mortlock, London. Reprinted in *The Economic Writings of Sir William Petty*; see C. H. Hull (1899).

Petty, W. (1690). *Political Arithmetick*. Clavel, London. Reprinted in *The Economic Writings of Sir William Petty*; see C. H. Hull (1899).

Peverone, G. F. (1558). *Due brevi e facili trattati, il primo d'arithmetica, l'altro di geometria*. Lyon.

Plackett, R. L. (1958). The principle of the arithmetic mean. *Biometrika*, **45**, 130–135. Reprinted in Pearson and Kendall (1970).

Plackett, R. L. (1986). The old statistical account. *J. Roy. Statist. Soc.*, A **149**, 247–251.

Plackett, R. L. (1988). Data analysis before 1750. *Intern. Statist. Rev.*, **56**, 181–195.

Poisson, S. D. (1837). *Recherches sur la Probabilité des Jugements en Matière Criminelle et en Matière Civile, précédées des Règles Générales du Calcul des Probabilités*. Bachelier, Paris.

Popper, K. R. (1959). The logic of scientific discovery. Hutchinson, London.

Prohorov, Yu. V. (ed.) (1986). *J. Bernoulli: On the law of large numbers*. Nauka, Moscow. In Russian.

Prohorov. Yu. V. (1986). The law of large numbers and the evaluation of the probability of large deviations. In *J. Bernoulli: On the law of large numbers*, pp. 116–150. Nauka, Moscow. In Russian.

Ptolemy, C. (c. 140). *The Almagest*.

Rabinovitch, N. L. (1969). Probability in the Talmud. *Biometrika*, **56**, 437–441. Reprinted in Kendall and Plackett (1977).

Rabinovitch, N. L. (1970a). Combinations and probability in rabbinic literature. *Biometrika*, **57**, 203–205. Reprinted in Kendall and Plackett (1977).

Rabinovitch, N. L. (1970b). Rabbi Levi Ben Gershon and the origins of mathematical induction. *Arch. Hist. Ex. Sci.*, **6**, 237–248.

Rabinovitch, N. L. (1973). *Probability and statistical inference in ancient and medieval Jewish literature*. Univ. Toronto Press, Toronto.

Riccioli, J. B. (1651). *Almagestum Novum*. Bologne.

Riddell, R. C. (1980). Parameter disposition in pre-Newtonian planetary theories. *Arch. Hist. Ex. Sci.*, **23**, 87–157.

Riordan, J. (1958). *An introduction to combinatorial analysis*. Wiley, New York.

Ronan, C. A. (1972). Edmond Halley, in *Dictionary of scientific biography*, ed. by C. C. Gillespie, Vol. 6, pp. 67–72.

Saar, J. du (1917). *Over Sterfteformules en Lijfrenten*. Noordhoff, Groningen.

Sambursky, S. (1956). On the possible and probable in Ancient Greece. *Osiris*, **12**, 35–48. Reprinted in Kendall and Plackett (1977).

Saurin, J. (1706). Eloge de M. Bernoulli, cy-devant Professeur de Mathématique à Bâle. *J. des Sçavans*, 1706, 81–89. Reprinted in part in Kohli (1975a).

Schneider, I. (1968). Der Mathematiker Abraham de Moivre. (1667–1754). *Arch. Hist. Ex. Sci.*, **5**, 177–317.

Schneider, I. (1972). *Die Entwicklung des Wahrscheinlichkeitsbegriffs in der Mathematik von Pascal bis Laplace*. Habilitationsschrift. München.

Schneider, I. (1974). Clausius' erste Anwendung der Wahrscheinlichkeitsrechnung im Rahmen der atmosphärischen Lichtstreuung. *Arch. Hist. Ex. Sci.*, **14**, 143–158.

Schneider, I. (1975). Abraham de Moivre. *Enzyklopädie "Die Grossen der Weltgeschichte,"* Vol. 4, 334–347. Kindler, Zürich.

Schneider, I. (1976). The introduction of probability into mathematics. *Hist. Math.*, **3**, 135–140.

Schneider, I. (1980a). Christiaan Huygens's contribution to the development of a calculus of probabilities. *Janus*, **67**, 269–279.

Schneider, I. (1980b). Why do we find the origin of a calculus of probabilities in the seventeenth century? In *Probabilistic thinking, thermodynamics and the interaction of the history and philosophy of science*, ed. by J. Hintikka et al., Vol. 2, 3–24. Reidel, Dordrecht.

Schneider, I. (1981). Leibniz on the probable. In *Mathematical perspectives*, pp. 201–219. Academic Press, New York.

Schneider, I. (1983a). Jakob Bernoulli und Johannes Faulhaber über arithmetische Reihen höherer Ordnung. In *Jahrbuch der Technischen Universität München 1982*, pp. 132–140.

Schneider, I. (1983b). Potenzsummenformeln im 17. Jahrhundert. *Hist. Math.*, **10**, 286–296.

Schneider, I. (1984). The role of Leibniz and Jakob Bernoulli for the development of probability theory. *LLULL*, **7**, 69–89.

Schneider, I. (1985). Luca Pacioli und das Teilungsproblem: Hintergrund und Lösungsversuche. In *Mathemata*, ed. by M. Folkerts und U. Lindgren, Vol. 12, 237–246. Steiner, Wiesbaden.

Schooten, F. van (1657). *Exercitationum Mathematicarum*. Elsevirii, Leiden.

Schooten, F. van (1660). *Mathematische Oeffeningen*. G. van Goedesbergh, Amsterdam.

Seal, H. L. (1949). The historical development of the use of generating functions in probability theory. *Bull. de l'Assoc. des Actuaires suisses*, **49**, 209–228. Reprinted in Kendall and Plackett (1977).

Seal, H. L. (1980). Early uses of Graunt's life table. *J. Inst. Actuaries*, **107**, 507–511.

Seneta, E. (1979). Pascal and probability. In *Interactive statistics*, ed. by D. McNeil, pp. 225–233. North-Holland, Amsterdam.

Seneta, E. (1983). Modern probabilistic concepts in the work of E. Abbe and A. de Moivre. *Math. Scientist*, **8**, 75–80.

Shafer, G. (1976). Review of Ian Hacking's The Emergence of Probability. *J. Amer. Statist. Assoc.*, **71**, 519–521.

Shafer, G. (1978). Non-additive probabilities in the work of Bernoulli and Lambert. *Arch. Hist. Ex. Sci.*, **19**, 309–370.

Shapiro, B. J. (1983). *Probability and certainty in seventeenth-century England*. Princeton Univ. Press, New Jersey.

Sheynin, O. B. (1970). Daniel Bernoulli on the normal law. *Biometrika*, **57**, 199–202. Reprinted in Kendall and Plackett (1977).

Sheynin, O. B. (1971). Newton and the classical theory of probability. *Arch. Hist. Ex. Sci.*, **7**, 217–243.

Sheynin, O. B. (1972). D. Bernoulli's work on probability. *RETE Strukturgeschichte der Naturwissenschaften*, **1**, 273–300. Reprinted in Kendall and Plackett (1977).

Sheynin, O. B. (1973). Mathematical treatment of astronomical observations (a historical essay). *Arch. Hist. Ex. Sci.*, **11**, 97–126.

Sheynin, O. B. (1974). On the prehistory of the theory of probability. *Arch. Hist. Ex. Sci.*, **12**, 97–141.

Sheynin, O. B. (1975). J. Kepler as a statistician. *Bull. Inst. Internat. Statist.*, **46**(2), 341–354.

Sheynin, O. B. (1977). Early history of the theory of probability. *Arch. Hist. Ex. Sci.*, **17**, 201–259.

Sheynin, O. B. (1986). J. Bernoulli and the beginnings of probability theory. In *J. Bernoulli: On the law of large numbers*; see Prohorov (1986), pp. 83–115.

Shoesmith, E. (1985). Nicholas Bernoulli and the argument for divine providence. *Intern. Statist. Rev.*, **53**, 255–259.

Shoesmith, E. (1987). The continental controversy over Arbuthnot's argument for divine providence. *Hist. Math.*, **14**, 133–146.

Simpson, T. (1740). *The Nature and Laws of Chance. The Whole after a new, general, and conspicuous Manner, and illustrated with a great Variety of Examples.* Cave, London. Reprinted 1792.

Simpson, T. (1742). *The Doctrine of Annuities and Reversions, Deduced from General and Evident Principles: With useful Tables, shewing the Values of Single and Joint Lives, etc. at different Rates of Interest.* Nourse, London. 2nd ed., 1775.

Simpson, T. (1743). *Appendix* [to The Doctrine of Annuities and Reversions, 1742] containing Some Remarks on Mr. DeMoivre's Book on the same Subject, with Answers to some Personal and Malignant Misrepresentations, in the Preface thereof. Nourse, London.

Simpson, T. (1752). *Select Exercises for Young Proficients in the Mathematicks. Part VI. The Valuation of Annuities for single and joint Lives, with a Set of new Tables, far more extensive than any extant.* Nourse, London. Part VI reprinted in 1791 as *A Supplement to the Doctrine of Annuities and Reversions.* Wingrave, London.

Simpson, T. (1756). A letter to the Right Honourable George Earl of Macclesfield, President of the Royal Society, on the advantage of taking the mean of a number of observations, in practical astronomy. *Phil. Trans., **49**,* 82–93. Comm. 1755.

Simpson, T. (1757). *Miscellaneous Tracts on Some Curious, and Very Interesting Subjects in Mechanics, Physical-Astronomy, and Speculative Mathematics.* Nourse, London.

Smith, D. E. (1929). *A source book in mathematics.* McGraw-Hill, New York. Reprinted by Dover, New York, 1959.

Spiess, O. (1975). Zur Vorgeschichte des Petersburger Problems. In *Die Werke von Jakob Bernoulli,* Vol. 3, 557–567.

Spinoza, B. de (1687). *Reeckening van Kanssen.* Van Dyck, 'sGravenhage. French translation in Huygens' *Oeuvres,* Vol. 14, 29–31. English translation in Dutka (1953). German translation in Lucas and Petry (1982).

Stigler, S. M. (1977). Eight centuries of sampling inspection: The trial of the Pyx. *J. Amer. Statist. Assoc., **72**,* 493–500.

Stigler, S. M. (1978). Laplace's early work: Chronology and citations. *Isis,* **69,** 234–254.

Stigler, S. M. (1980). Stigler's law of eponomy. *Trans. New York Acad. Sci.,* 2nd ser., **39,** 147–157.

Stigler, S. M. (1982). Poisson on the Poisson distribution. *Statist. Probability Letters,* **1,** 33–35.

Stigler, S. M. (1983). Who discovered Bayes's theorem? *Amer. Statist., **37**,* 290–296.

Stigler, S. M. (1986). John Craig and the probability of history: From the death of Christ to the birth of Laplace. *J. Amer. Statist. Assoc., **81**,* 879–887.

Stigler, S. M. (1986). *The history of statistics: The measurement of uncertainty before 1900.* Harvard Univ. Press, Cambridge, Massachusetts.

Stigler, S. M. (1988). The dark ages of probability in England: The seventeenth century work of Richard Cumberland and Thomas Strode. *Intern. Statist. Rev.,* **56,** 75–88.

Stirling, J. (1730). *Methodus Differentialis.* London.

Strode, T. (1678). *A Short Treatise of the Combinations, Elections, Permutations and*

Composition of Quantities. Illustrated by Several Examples, with a New Speculation of the Diffferences of the Powers of Numbers. Godbid, London.

Struyck, N. *Les Oeuvres de Nicolas Struyck* (1687–1769), qui se rapportent au calcul des chances, à la statistique générale, à la statistique des décès et aux rentes viagères, tirées des Oeuvres Complètes. Traduites par J. A. Vollgraff. La Société Génerale Néerlandaise d'Assurances sur la Vie et de Rentes Viagères. Amsterdam, 1912.

Struyck, N. (1716). *Uytreekening der Kansen in het speelen, door de Arithmetica en Algebra, beneevens eene Verhandeling van Looterijen en Interest.* Amsterdam. Reprinted in *Oeuvres* (1912), pp. 1–164.

Struyck, N. (1740). *Calcul des Rentes viagères.* Reprinted in *Oeuvres* (1912), pp. 194–210.

Sung, Bing (1966). *Translations from James Bernoulli* (with a preface by A. P. Dempster). Department of Statistics, Harvard University, Cambridge, Massachusetts.

Süssmilch, J. P. (1741). *Die Göttliche Ordnung in den Veränderungen des menschlichen Geschlechts, aus der Geburt, Tod, und Fortpflanzung desselben erwiesen.* Berlin. 2nd. ed. (1761); 3rd. ed. (1765); 4th ed. (1775–1776) by Baumann, with a supplementary volume.

Sutherland, I. (1963). John Graunt: A tercentenary tribute. *J. Roy. Statist. Soc. Ser. A*, **126**, 537–556.

Takács, L. (1967). On the method of inclusion and exclusion. *J. Amer. Statist. Assoc.*, **62**, 102–113.

Takács, L. (1969). On the classical ruin problems. *J. Amer. Statist. Assoc.*, **64**, 889–906.

Takács, L. (1980). The problem of coincidences. *Arch. Hist. Ex. Sci.*, **21**, 229–244.

Takács, L. (1981). On the "Problème des Ménages." *Discrete Math.*, **36**, 289–297.

Tartaglia, N. (1556). *General trattato di numeri e misure.* Venezia.

Taton, R. (1974). Blaise Pascal. In *Dictionary of scientific biography*, ed. by C. C. Gillispie, Vol. 10, 330–342.

Taylor, B. (1715). *Methodus Incrementorum Directa et Inversa.* London.

Thatcher, A. R. (1957). A note on the early solutions of the problem of the duration of play. *Biometrika*, **44**, 515–518. Reprinted in Pearson and Kendall (1970).

Todhunter, I. (1865). *A History of the Mathematical Theory of Probability from the Time of Pascal to that of Laplace.* Macmillan, London. Reprinted by Chelsea, New York, 1949.

Trembley, J. (1804). Observations sur le calcul d'un jeu de hasard. *Mém. Acad. Roy. Berlin*, 86–102. Comm. 1802.

Trenerry, C. F. (1926). *The origin and early history of insurance.* King. London.

Uspensky, J. V. (1913). Translation of *Ars Conjectandi*, Part 4, into Russian. See A. A. Markov (1913).

Uspensky, J. V. (1937). *Introduction to mathematical probability.* McGraw-Hill, New York.

Vastel, L. G. F. (1801). *L'Art de Conjecturer, Traduit du Latin de Jacques Bernoulli; Avec des Observations, Éclaircissemens et Additions. Première Parti.* Caen.

Vernon, P. E. (1936). The matching method applied to investigations of personality. *Psych. Bull.*, **33**, 149–177.

Vollgraff, J. A. (1950). Biographie de Chr. Huygens. In *Oeuvres de Huygens*, Vol. 22, 383–778.

Waerden, B. L. van der (ed.) (1975). *Die Werke von Jakob Bernoulli*, Vol. 3. Birkhäuser, Basel.

Waerden, B. L. van der (1975a). Historische Einleitung. In *Die Werke von Jakob Bernoulli*, Vol. 3, 2–18. Birkhäuser, Basel.

Waerden, B. L. van der (1975b). Kommentar zu den Meditationes und der Ars Conjectandi. In *Die Werke von Jakob Bernoulli*, Vol. 3, 353–383, Birkhäuser, Basel.

Waerden, B. L. van der (1975c). Die gedruckten Vorarbeiten zur Ars Conjectandi und die Datierung der Meditationes. In *Die Werke von Jakob Bernoulli*, Vol. 3, 385–389, Birkhäuser, Basel.

Walker, H. M. (1934). Abraham de Moivre. *Scripta Mathematica*, **2**, 316–333. Reprinted in A. de Moivre, *The doctrine of chances*, Chelsea, New York, 1967.

Wallis, J. (1655). *Arithmetica Infinitorum.* Oxford.

Webster, A. (1748). *Calculations with the Principles and Data on which they are instituted shewing the Rise and Progress of the Fund.* Lumisden, Edinburgh.

Wesley, W. G. (1978). The accuracy of Tycho Brahe's instruments. *J. Hist. Astronomy*, **9**, 42–53.

Westergaard, H. (1901). *Die Lehre von der Mortalität und Morbilität.* Second completely revised edition. Fischer, Jena.

Westergaard, H. (1932). *Contributions to the history of statistics.* King, London.

Westfall, R. S. (1958). *Science and religion in seventeenth-century England.* Yale Univ. Press., New Haven. Reprinted as an Ann Arbor Paperback, 1973.

Westfall, R. S. (1980). *Never at rest. A biography of Isaac Newton.* Cambridge Univ. Press.

Whitworth, W. A. (1878). *Choice and Chance.* 3rd ed., Cambridge Univ. Press; 5th ed. (1901), Deighton Bell, Cambridge. Reprinted by Hafner, New York, 1959.

Willcox, W. F. (1937). The founder of statistics. *Intern. Statist. Rev.*, **5**, 321–328.

Wilson, C. (1968). Kepler's derivation of the elliptical path. *Isis*, **59**, 4–25.

Witt, J. de (1671). *Waerdye van Lyf-Renten Naer proportie van Los-Renten.* s'Graven-Hage. Reprinted in *Die Werke von Jakob Bernoulli*, Vol. 3, 327–350, 1975. English translation by F. Hendriks (1852–1853). French translation by Chateleux and Rooijen (1937).

Witt, J. de (1671). *Letters to Jan Hudde.* In *Brieven van Johan de Witt*, Vol. 4, 1913, Fruin, Amsterdam. English translation by F. Hendriks (1852–1853). French translation in *Mémoires* published by Algemeene Maatschappij van Levensverzekering (1898).

Wollenschläger, K. (1933). Der mathematische Briefwechsel zwischen Johann I Bernoulli und Abraham de Moivre. *Verh. Naturforschenden Ges. Basel,* **43,** 151–317.

Young, T. (1819). Remarks on the probabilities of error in physical observations, and on the density of the earth, considered, especially with regard to the reduction of experiments on the pendulum. In a letter to Capt. Henry Kater, F.R.S. *Phil. Trans.,* 1819, 70–95.

Young, T. E. (1908). Historical notes relating to the discovery of the formula, $a_x = vp_x(1 + a_{x+1})$; and to the introduction of the calculus in the solution of actuarial problems. *J. Inst. Actuaries,* **42,** 188–205.

Youshkevitch, A. P. (1986a). Biography of J. Bernoulli. In *J. Bernoulli: On the law of large numbers,* ed. by Yu. V. Prohorov, pp. 157–161. Nauka, Moscow. In Russian.

Youshkevitch, A. P. (1986b). Nicholas Bernoulli and the publication of James Bernoulli's *Ars Conjectandi.* In Russian. English translation in *Theory Probab. Appl.,* **31,** 286–303, 1987.

Zabell, S. L. (1988). The probabilistic analysis of testimony. *J. Statistical Planning and Inference,* **20,** 327–354.

Index

References of secondary importance are collected under the sub-entry *mentioned*, placed last among the sub-entries.